D1590222

Introductory FINITE ELEMENT METHOD

Mechanical Engineering Series

Frank Kreith - Series Editor

Published Titles

Energy Audit of Building Systems: An Engineering Approach
Moncef Krarti
Entropy Generation Minimization
Adrian Bejan
Finite Element Method Using MATLAB, 2nd Edition
Young W. Kwon & Hyochoong Bang
Fundamentals of Environmental Discharge Modeling
Lorin R. Davis
Introductory Finite Element Method
Chandrakant S. Desai & Tribikram Kundu
Intelligent Transportation Systems: New Principles and Architectures
Sumit Ghosh & Tony Lee
Mathematical & Physical Modeling of Materials Processing Operations
Olusegun Johnson Ileghus, Manabu Iguchi & Walter E. Wahnsiedler
Mechanics of Composite Materials
Autar K. Kaw
Mechanics of Fatigue
Vladimir V. Bolotin
Mechanism Design: Enumeration of Kinematic Structures According to Function
Lung-Wen Tsai
Nonlinear Analysis of Structures
M. Sathyamoorthy
Practical Inverse Analysis in Engineering
David M. Trujillo & Henry R. Busby
Principles of Solid Mechanics
Rowland Richards, Jr.
Thermodynamics for Engineers
Kau-Fui Wong
Viscoelastic Solids
Roderic S. Lakes

Forthcoming Titles

Distributed Generation: The Power Paradigm for the New Millennium
Anne-Marie Borbely & Jan F. Kreider
Engineering Experimentation
Euan Somerscales
Fluid Power Circuits and Control: Fundamentals and Applications
John S. Cundiff
Heat Transfer in Single and Multiphase Systems
Greg F. Naterer
Mechanics of Solids & Shells
Gerald Wempner & Demosthenes Talaslidis

Introductory FINITE ELEMENT METHOD

Chandrakant S. Desai
Tribikram Kundu

CRC Press
Boca Raton London New York Washington, D.C.

Library of Congress Cataloging-in-Publication Data

Desai, C. S. (Chandrakant S.), 1936-
Introductory finite element method / Chandrakant S. Desai, Tribikram Kundu.
p. cm.
Includes bibliographical references and index.
ISBN 0-8493-0243-9 (alk. paper)
1. Finite element method. I. Kundu, T. (Tribikram). II. Title.

TA347.F5 .D48 2001
620'.001'1535--dc21 2001017466

Direct all inquiries to CRC Press LLC, 2000 N.W. Corporate Blvd., Boca Raton, Florida 33431.

Visit the CRC Press Web site at www.crcpress.com

International Standard Book Number 0-8493-0243-9
Library of Congress Card Number 2001017466
Printed in the United States of America 2 3 4 5 6 7 8 9 0
Printed on acid-free paper

Dedication

To

Our wives: Patricia Desai
and Nupur Kundu

Children: Maya and Sanjay; Ina and Auni

and Parents: Sankalchand Desai and Kamala Desai;
Makhan Lal Kundu and Sandhya Rani Kundu

Preface

The finite element method has gained tremendous attention and popularity. The method is now taught at most universities and colleges, is researched extensively, and is used by the practicing engineers, industry, and government agencies. The teaching of the method has been concentrated at the postgraduate level. In view of the growth and wide use of the method, however, it becomes highly desirable and necessary to teach it at the undergraduate level.

There are a number of books and publications available on the finite element method. It appears that almost all of them are suitable for the advanced students and require a number of prerequisites such as theories of constitutive or stress-strain laws, mechanics, and variational calculus. Some of the introductory treatments have presented the method as an extension of matrix methods of structural analysis. This viewpoint may no longer be necessary, since the finite element method has reached a significant level of maturity and generality. It has acquired a sound theoretical basis, and in itself has been established as a general procedure relevant to engineering and mathematical physics. These developments permit its teaching and use as a general technique from which applications to topics such as mechanics, structures, geomechanics, hydraulics, and environmental engineering arise as special cases. It is therefore essential that the method be treated as a general procedure and taught as such.

This book is intended mainly for the undergraduate and beginning graduate students. Its approach is sufficiently elementary so that it can be introduced with the background of essentially undergraduate subjects. At the same time, the treatment is broad enough so that the reader or the teacher interested in various topics such as stress-deformation analysis, fluid and heat flow, potential flow, time-dependent problems, diffusion, torsion, and wave propagation can use and teach from it. The book brings out the intrinsic nature of the method that permits confluence of various disciplines and provides a distinct and rather novel approach for teaching the finite element method at an elementary level. The book can be used for any student with no prior exposure to the finite element method. The prerequisites for understanding the material will be undergraduate mathematics, strength of materials, and undergraduate courses in structures, hydraulics, geotechnical engineering, and matrix algebra. Introductory knowledge of computer programming is desirable but not necessary. The text is written in such a way that no prior knowledge of variational principles is necessary. Over a period of the last 30 years or so, the authors have taught, based on these prerequisites, an undergraduate course and a course for user groups composed of beginners. This experience has shown that undergraduates or beginners equipped with these prerequisites, available to them in the undergraduate curricula at most academic institutions, can understand and use the material presented in this book.

The first chapter presents a rather philosophical discussion of the finite element method and often defines various terms on the basis of eastern and western concepts from antiquity. The second chapter gives a description of the eight basic steps and fundamental principles of variational calculus. Chapters 3 to 5 cover one-dimensional problems in stress-deformation analysis and steady and time-dependent flow of heat and fluids. The fundamental generality of the method is illustrated by showing the common characteristics of the formulation for these topics and by indicating the fact that their governing equations are essentially similar. The generality is further established by including computer codes in Chapter 6 that can solve different types of problems.

Understanding and using the finite element method are closely linked with the use of the computer. It is the belief of the authors that strictly theoretical teaching of the method may not give the student an idea of the details and the ranges of applicability of the technique. Consequently this text endeavors to introduce the student, gradually and simultaneously with the theoretical teaching, to the use and understanding of computer codes. The codes presented in Chapter 6 are thoroughly documented and detailed so that they can be used and understood without difficulty. Details of these codes, designed for the beginner, are given in Appendix 3. It is recommended that these or other available codes be used by the student while learning various topics in this book.

Chapter 7 introduces the idea of higher-order approximation for the problem of beam bending and beam-column. One-dimensional problems in mass transport (diffusion-convection) and wave propagation are covered in Chapters 8 and 9, respectively. These problems illustrate, by following the general procedure, formulations for different categories of time-dependent problems.

Chapter 10 presents the basic finite element formulation for two- and three-dimensional problems. Then in Chapters 11 to 14 different types of two-dimensional problems are presented. The chapters on Torsion (Chapter 11) and Other Field Problems (Chapter 12) have been chosen because they involve only one degree-of-freedom at a point. Chapters 13 and 14 cover two-dimensional stress-deformation problems involving two and higher degrees-of-freedom at a point.

The text presents the finite element method by using simple problems. It must be understood, however, that it is for the sake of easy introduction that we have used relatively simple problems. The main thrust of the method, on the other hand, is for solving complex problems that cannot be easily solved by the conventional procedures.

For a thorough understanding of the finite element method, it is essential that the students perform hand calculations. With this in mind, most chapters include a number of problems to be solved by hand calculations. They also include problems for home assignments and self-study.

The formulations have been presented by using both the variational and residual procedures. In the former, the potential, complementary, hybrid, and mixed procedures have been discussed. In the residual procedures, main attention has been given to Galerkin's method. A number of other residual methods are also becoming popular. They are described, therefore, in Appendix 1, which gives descriptions, solutions, and comparisons for a problem by using a number of methods: closed-form, Galerkin, collocation, subdomain, least squares, Ritz, finite difference, and finite element.

Formulations by the finite element method usually result in algebraic simultaneous equations. Detailed description of these methods is beyond the scope of this book. Included in Appendix 2, however, are brief introductions to the commonly used direct and iterative procedures for the solution of algebraic simultaneous equations.

Appendix 3 presents details of a number of computer codes relevant to various topics in the text.

The book can be used for one or two undergraduate courses. The second course may overlap with or be an introductory graduate course. Although a number of topics have been covered in the book, a semester or quarter course could include a selected number of topics. For instance, a quarter course can cover Chapters 1 to 6, and then one or two topics from the remaining chapters. For a class interested in mechanics and stress-deformation analyses, the topics can be Beam Bending and Beam-Column (Chapter 7), and Two-Dimensional Stress Deformation (Chapters 10 and 13). If time is available (in the case of a semester course), Chapter 9 on One-Dimensional Stress Wave Propagation, Chapter 11 on Torsion, and/or Chapter 14 on Multicomponent Systems can be added. A class oriented toward field problems and hydraulics can choose one or more of Chapters 8, 9, 11, and 12 in addition to Chapters 1 to 6 and 10.

We would like to express special appreciation to Jose Franscisco Perez Avila for his assistance in the manuscript preparation and Shashank Pradhan for implementing some of the computer codes and preparation of user's manuals.

We realize that it is not easy to write at an elementary level for the finite element method with so many auxiliary disciplines. The judgment of this book is better left to the readers.

Many natural systems can be considered continuous or interconnected, and their behavior is influenced by a large number of parameters. In order to understand such a system, we must understand all the parameters. Since this is not possible we make approximations, by selecting only the significant of them and neglecting the others. Such a procedure allows understanding of the entire system by comprehending its components taken one at a time. These approximations or models obviously involve errors, and we strive continuously to improve the models and reduce the errors.

Chandrakant S. Desai
Tribikram Kundu
Tucson, Arizona

Authors

Chandrakant S. Desai is a Regents' Professor and Director of the Material Modeling and Computational Mechanics Center, Department of Civil Engineering and Engineering Mechanics, University of Arizona, Tucson. He was a Professor in the Department of Civil Engineering, Virginia Polytechnic Institute and State University, Blacksburg, from 1974 to 1981, and a Research Civil Engineer at the U.S. Army Engineer Waterways Experiment Station, Vicksburg, MS from 1968 to 1974.

Dr. Desai has made original and significant contributions in basic and applied research in material modeling and testing, and computational methods for a wide range of problems in civil engineering, mechanics, mechanical engineering, and electronic packaging. He has authored/edited 20 books and 18 book chapters, and has been the author/coauthor of over 270 technical papers. He was the founder and General Editor of the *International Journal for Numerical and Analytical Methods in Geomechanics* from 1977 to 2000, and he has served as a member of the editorial boards of 12 journals.

Dr. Desai has also been a chair/member of a number of committees of various national and international societies. He is the President of the International Association for Computer Methods and Advances in Geomechanics. Dr. Desai has also received a number of recognitions: Meritorious Civilian Service Award by the U.S. Corps of Engineers, Alexander von Humboldt Stiftung Prize by the German Government, Outstanding Contributions Medal in Mechanics by the International Association for Computer Methods and Advances in Geomechanics, Distinguished Contributions Medal by the Czech Academy of Sciences, Clock Award by ASME (Electrical and Electronic Packaging Division), Five Star Faculty Teaching Finalist Award, and the El Paso Natural Gas Foundation Faculty Achievement Award at the University of Arizona, Tucson.

Professor T. Kundu received his bachelor degree in mechanical engineering from the Indian Institute of Technology, Kharagpur in 1979. His M.S. and Ph.D. were in the field of mechanics from the Mechanical and Aerospace Engineering Department of the University of California, Los Angeles in 1980 and 1983, respectively. He joined the University of Arizona as an assistant professor in 1983 and was promoted to full professor in 1994.

Dr. Kundu has made significant and original contributions in both basic and applied research in computational mechanics and nondestructive evaluation (NDE) of materials by ultrasonic and acoustic microscopy techniques. He is editor or coeditor of 8 books, coauthor of a textbook, author of a book chapter, and author/coauthor of over 130 technical papers; half of those have been published in refereed journals. He is a Fellow of ASME (American Society of Mechanical Engineers) and ASCE (American Society of Civil Engineers). He has received a number of awards, including the President's Gold Medal from IIT, the UCLA Alumni Award, the Humboldt Fellowship from Germany, and the Best Paper Award from the International Society for Optical Engineering (SPIE). He has extensive research collaborations with international and U.S. scientists. He has spent 21 months as an Alexander von Humboldt Scholar in the Department of Biology, J. W. Goethe University, Frankfurt, Germany. He has also spent several months as a visiting professor at a number of other institutes — Department of Mechanics, Chalmers University of Technology, Gothenberg, Sweden; Acoustic Microscopy Center, Semienov Institute of Chemical Physics, Russian Academy of Science, Moscow; Department of Civil Engineering, EPFL (Swiss Federal Institute of Technology in Lausanne), Switzerland; Department of Mechanical Engineering, University of Technology of Compiegne, France; Materials Laboratory, University of Bordeaux, France; LESiR Laboratory, Ecole Normale Superior (ENS), Cachan, France; Aarhus University Medical School, Aarhus, Denmark; Wright-Patterson Material Laboratory, Dayton, OH.

Contents

1

Introduction

Basic Concept

In its current form the finite element (FE) method was formalized by civil engineers. The method was proposed and formulated previously in different manifestations by mathematicians and physicists.

The basic concept underlying the finite element method is not new. The principle of discretization is used in most forms of human endeavor. Perhaps the necessity of discretizing, or dividing a thing into smaller manageable things, arises from a fundamental limitation of human beings in that they cannot see or perceive things surrounding them in the universe in their entirety or totality. Even to see things immediately surrounding us, we must make several turns to obtain a jointed mental picture of our surroundings. In other words, we discretize the space around us into small segments, and the final assemblage that we visualize is one that simulates the real continuous surroundings. Usually such jointed views contain an element of error.

In perhaps the first act toward a rational process of discretization, man divided the matter of the universe into five interconnected basic essences (*Panchmahabhuta*), namely, sky or vacuum, air, water, earth, and fire, and added to them perhaps the most important of all, time, by singing

> Time created beings, sky, earth,
> Time burns the sun and time will bring
> What is to come. Time is the master
> of everything.[1*]

We conceived the universe to be composed of an innumerable (perhaps finite) number of solar systems, each system composed of its own stars, planets, and galaxies. In our solar system we divided the planet earth into interconnected continents and oceans. The plate of earth we live on is composed of interconnected finite plates.

---

* The superscript number indicates references at the end of the chapter.

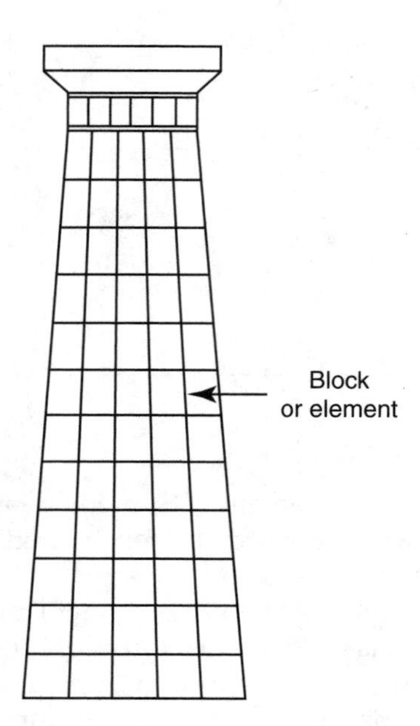

FIGURE 1.1
Building column composed of blocks or elements.

When man started counting, the numeral system evolved. To compute the circumference or area of a circle, early thinkers drew polygons of progressively increasing and decreasing size inside and outside the circle, respectively, and found the value of π to a high degree of accuracy. In (civil) engineering we started buildings made of blocks or elements (Figure 1.1).

When engineers surveyed tracts of land, the tract was divided into smaller tracts, and each small tract was surveyed individually (Figure 1.2). The connecting of the individual surveys provided an approximate survey of the entire tract. Depending on the accuracy of the survey performed, a closing error would be involved. In aerial photography a survey of the total area is obtained by matching or patching together a number of photographs.

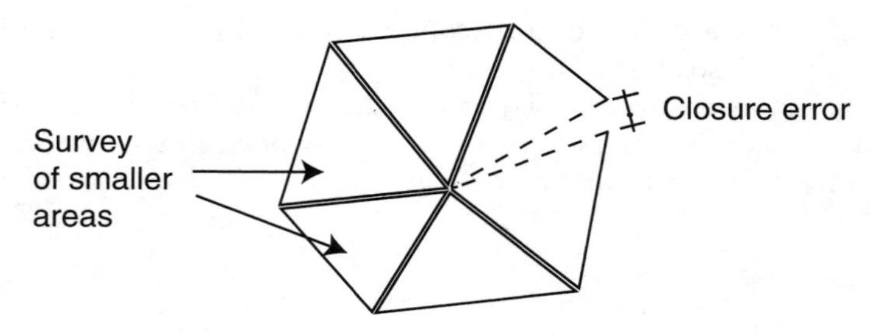

FIGURE 1.2
Discretization in surveying and closure error in survey of subdivided plot.

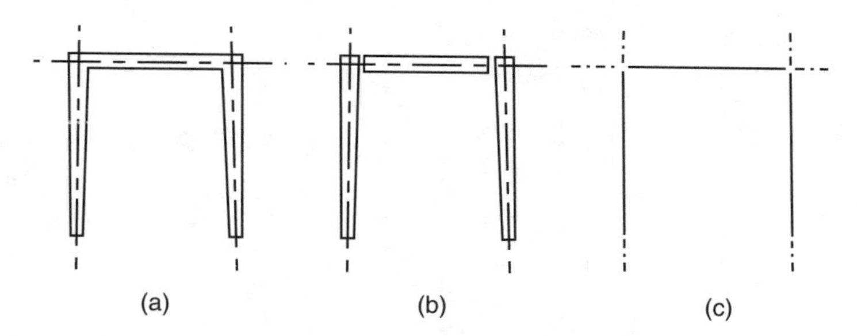

FIGURE 1.3
Discretization of engineering structure: (a) Actual structure, (b) Discretized structure, and (c) Idealized one-dimensional model.

For stress analysis of modern framed structures in civil engineering classically, methods such as slope deflection and moment distribution were used. The structure was divided into component elements, each component was examined separately, and (stiffness) properties were established (Figure 1.3). The parts were assembled so that the laws of equilibrium and physical condition of continuity at the junctions were enforced.

Although a system or a thing could be discretized in smaller systems, components, or finite elements, we must realize that the original system itself is indeed a whole. Our final aim is to combine the understandings of individual components and obtain an understanding of the wholeness or continuous nature of the system. In a general sense, as the modern scientific thinking recognizes, which the Eastern philosophical and metaphysical concepts had recognized in the past, all systems or things are but parts of the ultimate continuity in the universe!

The foregoing abstract and engineering examples make us aware of the many activities of man that are based on discretization.

Process of Discretization

Discretization implies approximation of the real and the continuous. We use a number of terms to process the scheme of discretization such as subdivision, continuity, compatibility, convergence, upper and lower bounds, stationary potential, minimum residual, and error. As we shall see later, although these terms have specific meanings in engineering applications, their conception has deep roots in man's thinking. In the following we discuss some of these terms; a number of figures and aspects have been adopted from Russell.[2]

Subdivision

Zeno argued that space is finite and infinitely divisible and that for things to exist they must have magnitudes. Figure 1.4(a) shows the concept of finite

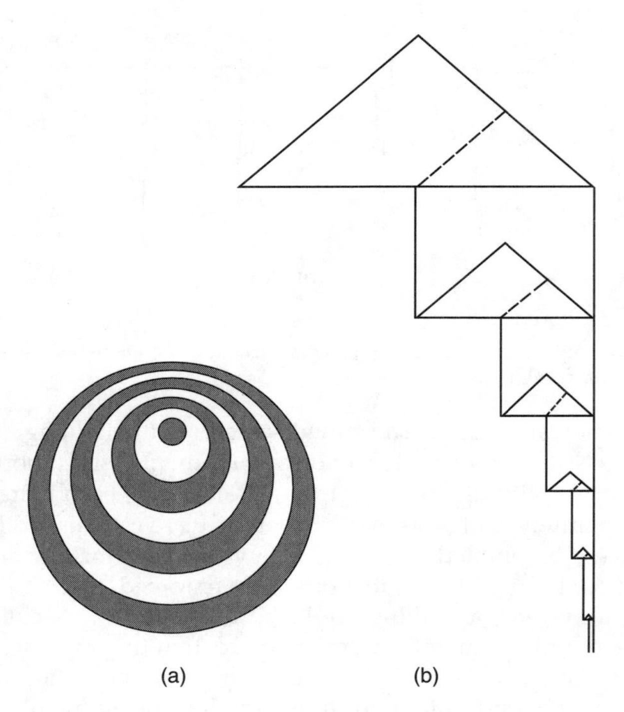

FIGURE 1.4
Finiteness and divisibility: (a) Infinite space and (b) Infinitely divisible triangle. (From Ref. 2, by permission of Aldus Books Limited, London.)

space. If the earth were contained in space, what contained the space in turn?[2] Figure 1.4(b) illustrates this idea for the divisibility of a triangle into a number of component triangles.

Continuity

Aristotle said that a continuous quantity is made up of divisible elements. For instance, other points exist between any two points in a line, and other moments exist between two moments in a period of time. Therefore, space and time are continuous and infinitely divisible,[2] and things are consecutive, contiguous, and continuous (Figure 1.5).

These ideas of finiteness, divisibility, and continuity allow us to divide continuous things into smaller components, units, or elements.

Convergence

For evaluating the approximate value of π, or the area of a circle, we can draw polygons within (Figure 1.6(a)) and around (Figure 1.6(b)) the circle. As we make a polygon, say, the outside one, smaller and smaller, with a

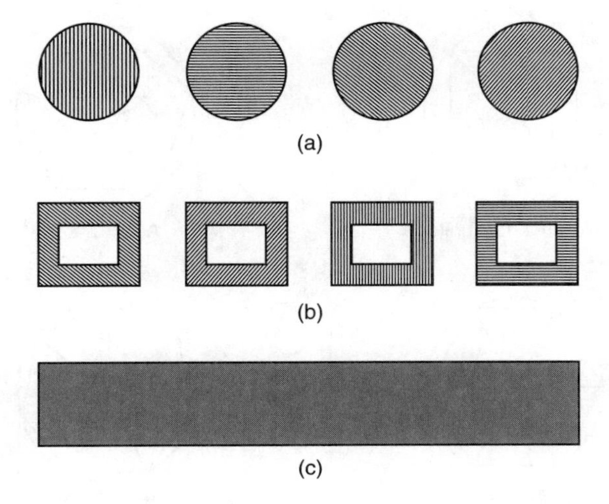

FIGURE 1.5
Concepts of continuity: (a) Consecutive, (b) Contiguous, and (c) Continuous. From Ref. 2, by permission of Aldus Books Limited, London.

greater number of sides, we approach the circumference or the area of the circle. This process of successively moving toward the exact or correct solution can be termed convergence.

The idea is analogous to what Eudoxus and Archimedes called the method of exhaustion. This concept was used to find areas bounded by curves; the available space was filled with simpler figures whose areas could be easily calculated. Archimedes employed the method of exhaustion for the parabola (Figure 1.7); here by inscribing an infinite sequence of smaller and smaller triangles, one can find the exact numerical formula for the parabola. Indeed, an active practitioner of the finite element method soon discovers that the pursuit of convergence of a numerical procedure is indeed fraught with exhaustion!

In the case of the circle (Figure 1.6), convergence implies that as the inside or outside polygon is assigned an increasingly greater number of sides, we approach or converge to the area of the circle. Figure 1.6(c) shows the plots of successive improvement in the values of the area of the circle from the two procedures: polygons of greater sides drawn inside and outside. We can see that as the number of sides of the polygons is increased, the approximate areas converge or approach or tend toward the exact area.

Physical models: The student can prepare pictorial or physical (cardboard or plastic) models to illustrate convergence from the example of the area of the circle.

Bounds

Depending on the course of action that we take from within or from outside, we approach the exact solution of the area of the circle. However, the value

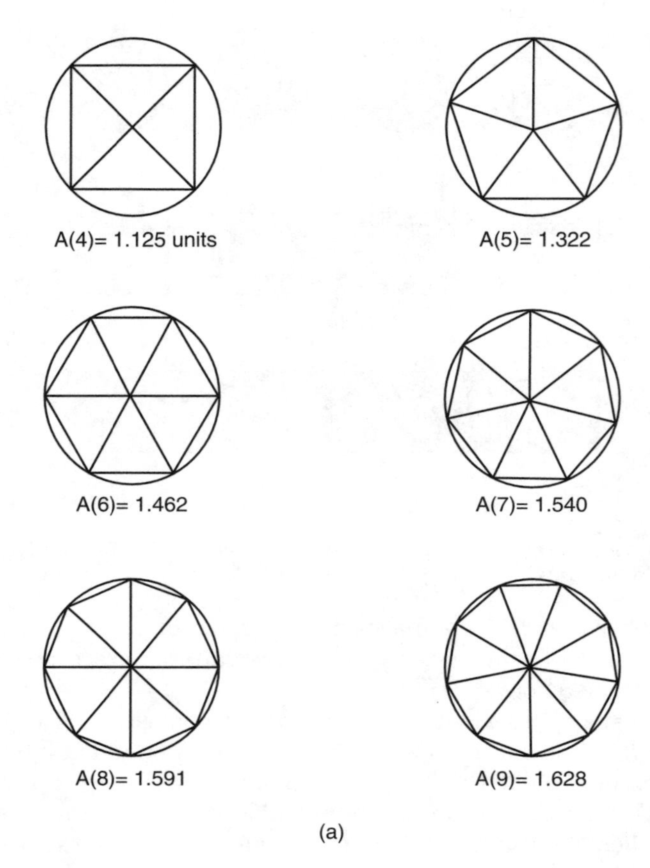

FIGURE 1.6(a)
Convergence and bounds for approximate area of circle: (a) Polygons inside circle.

from each method will be different. Figure 1.6(c) shows that convergence from within the circle gives a value lower than the exact, while that from outside gives a value *higher.* The former yields the lower bound and the latter, the upper bound. Figure 1.8 depicts the process of convergence to upper and lower bound solutions.

Error

It should be apparent that discretization involves approximation. Consequently, what we obtain is not the exact solution but an approximation to that solution. The amount by which we differ can be termed the error. For example, the areas (or perimeters) of the polygons inscribed in the circle (Figure 1.6) are always less than the area (or perimeter) of the circle, and the areas (or perimeters) of the circumscribed polygons are always greater than the area (or perimeter) of the circle. The difference between the approximation and the exact perimeter is the error, which becomes smaller and smaller

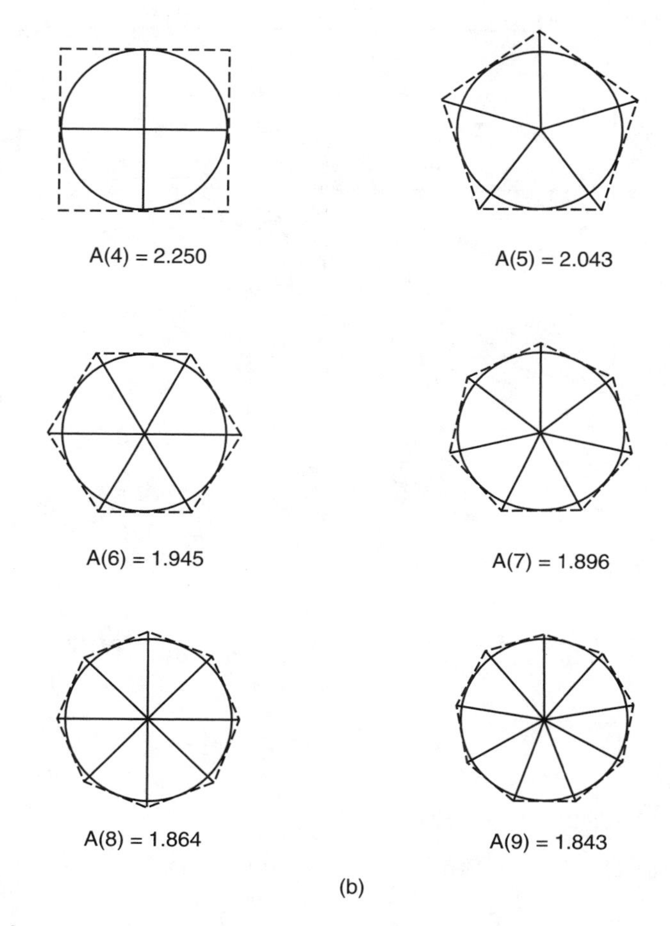

FIGURE 1.6(b)
Convergence and bounds for approximate area of circle: (b) Polygons outside circle.

as the number of sides of the polygons increases. We can express error in the area as

$$A^* - A = \varepsilon, \tag{1.1}$$

where A^* = the exact area, A = approximate area, and ε = error.

Principles and Laws

To describe the behavior of things or systems around us, we need to establish laws based on principles. A law can be a statement or can be expressed by

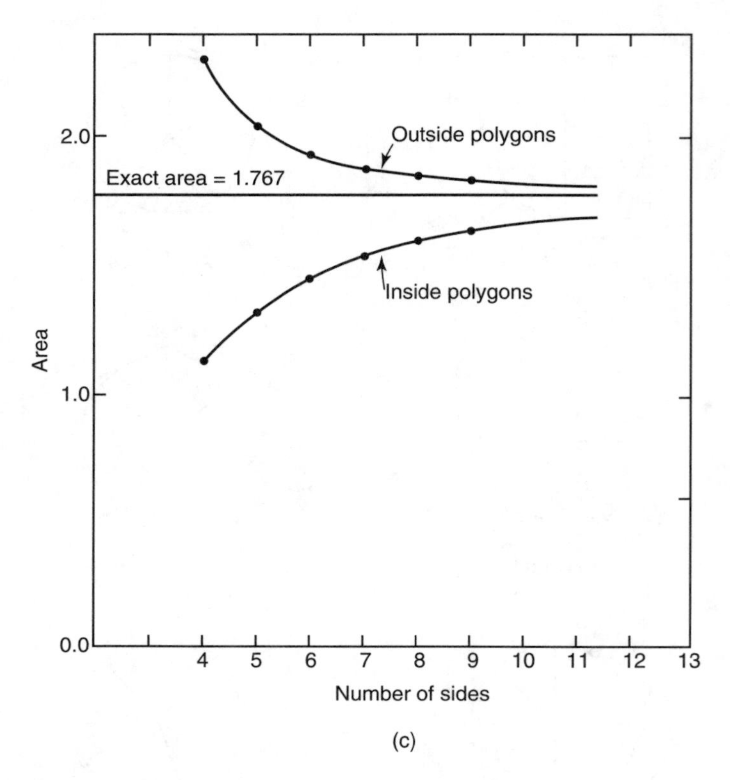

FIGURE 1.6(c)
Convergence for approximate area of circle (also see Figures 1.6a and 1.6b).

a mathematical formula. Principles have often been proposed by intuition, hypothesized, and then proved. Newton's second law states

$$F = ma \tag{1.2a}$$

or

$$F = m\frac{d^2u}{dt^2}, \tag{1.2b}$$

where F = force, m = mass, and a = acceleration or second derivative of displacement u with respect to time t. The principle is that at a given time the body is in dynamic equilibrium and a measure of energy contained in the body assumes a stationary value.

A simple statically loaded linear elastic column (Figure 1.9(a)) follows the principle that at equilibrium, under given load and boundary constraints, the potential energy of the system assumes a stationary (minimum) value and that the equation governing displacement is

$$v = \frac{PL}{AE}, \tag{1.3}$$

where v = displacement in the vertical y direction, P = applied load, L = length of the column, A = cross-sectional area, and E = modulus of elasticity.

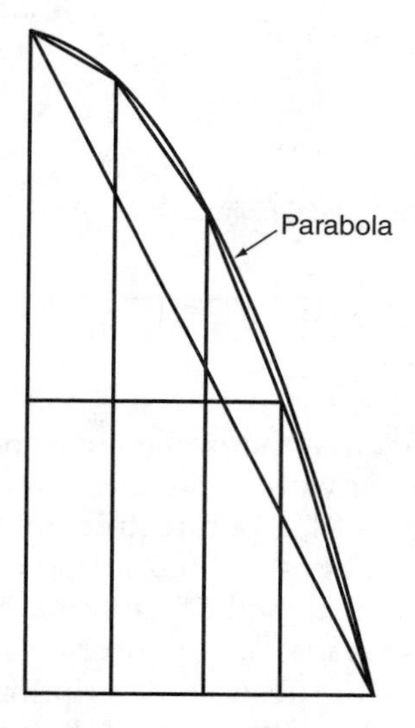

FIGURE 1.7
Concept of convergence or exhaustion. (From Ref. 2, by permission of Aldus Books Limited, London.)

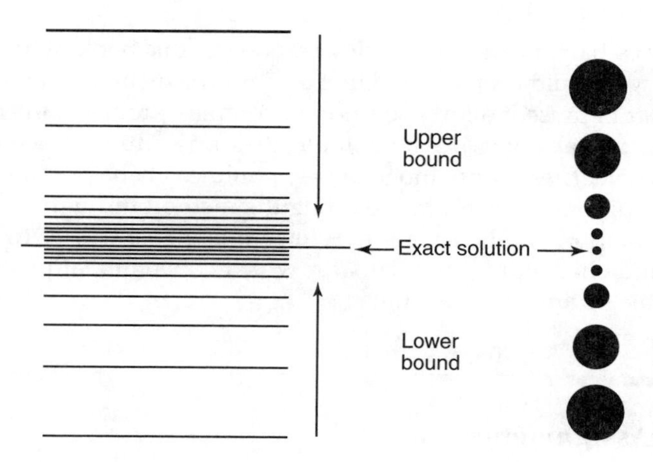

FIGURE 1.8
Concept of bounds.

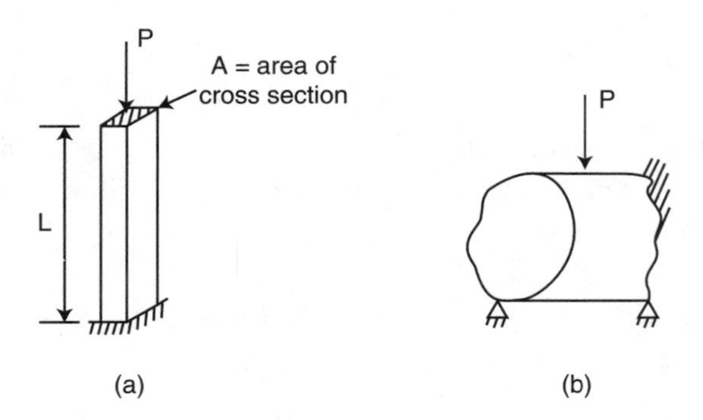

FIGURE 1.9
Structures subjected to loads (causes): (a) Column and (b) Body or structure.

Cause and Effect

The essence of all investigations is the examination and understanding of causes and their effects. The effect of work is tiredness and that of too much work is fatigue or stress. The effect of load on a structure (Figure 1.9(b)) is to cause deformations, strains, and stresses; too much load causes fatigue and failure.

When studying finite element methods, our main concern is the cause and effect of forcing functions (loads) on engineering systems.

The foregoing offers a rather abstract description of ideas underlying the process of discretization, inherent in almost all human endeavors. Comprehending these ideas significantly helps us to understand and extend the finite element concept to engineering; that is the goal of this text.

Important Comment

Although we have presented the descriptions in this book by using simple problems, we should keep in mind that the finite element method is powerful and popular because it allows solution of complex problems in engineering and mathematical physics. The complexities arise due to factors such as irregular geometries, nonhomogeneities, nonlinear behavior, and arbitrary loading conditions. Hence, after learning the method through simple examples, computations, and derivations, our ultimate goal will be to apply it to complex and challenging problems for which conventional procedures are not available or are very difficult.

Review Assignments

In the beginning stages of the study, it may prove very useful to assign the student homework that requires review of some of the basic laws, principles, and equations. This will facilitate understanding the method and also reduce the necessity of reviews by the teacher. The following are two suggested home assignments that cover topics from many undergraduate curricula.

Home Assignment 1

1. Define:
   (a) Stress at a point
   (b) Strain
   (c) Hooke's law
2. Define:
   (a) Principal stresses and strains
   (b) Invariants of stresses and strains
3. (a) Define potential energy as a sum of strain energy and potential of applied loads.
   (b) Give examples for analysis in (civil)engineering in which you have used the concept of potential energy.
4. Define:
   (a) Darcy's law and coefficient of permeability
   (b) Coefficient of thermal conductivity and coefficient of thermal expansion
5. Derive the Laplace equation for steady-state flow.
6. Derive the fourth-order differential equation governing beam bending,

$$\frac{d^2}{dx^2}\left(EI\frac{d^2w}{dx^2}\right) = p - k_s w,$$

where w = displacement, p = applied load, and $k_s w$ = support reaction, k_s = spring constant, E = modulus of elasticity, I = moment of inertia, and x = coordinate along beam axis. See Figure 1.10.

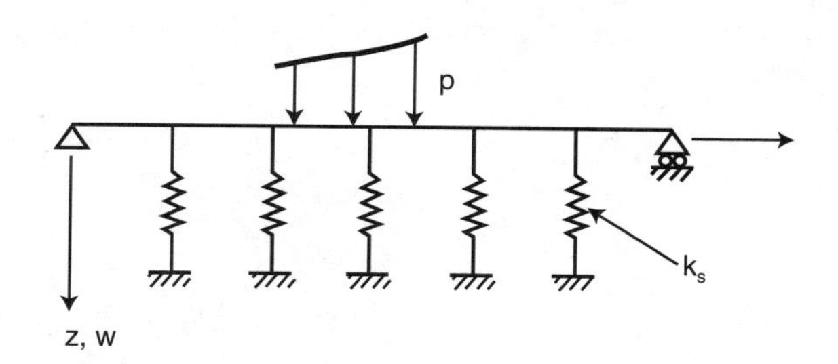

FIGURE 1.10

Home Assignment 2

1. Define:
   (a) Determinant
   (b) Row, column, and rectangular matrices
   (c) Matrix addition and subtraction
   (d) Matrix multiplication
   (e) Inverse of a matrix
   (f) Transpose of a matrix
   (g) Symmetric matrix
   (h) Sparsely populated and banded matrices
2. Define:
   (a) A set of algebraic simultaneous equations
   (b) Describe Gaussian elimination with respect to the following equations,

$$2x_1 + 3x_2 = 14,$$

$$4x_1 + 5x_2 = 10,$$

   and find the value of x_1 and x_2.
3. Define:
   (a) Total derivative
   (b) Partial derivative for one variable and two variables and the chain rule of differentiation

References

1. *The Upnishads — Praise of Time*, 1964. See available translations of *The Upnishads* from Sanskrit to English, e.g., *The Upnishads* by Swami Nikhilananda, Harper Torchbooks, New York.
2. Russell, B., 1959. *Wisdom of the West*, Crescent Books, Inc., Rathbone Books Ltd., London.

2

Steps in the Finite Element Method

Introduction

Formulation and application of the finite element method are considered to consist of eight basic steps. These steps are stated in this chapter in a general sense. The main aim of this general description is to prepare for complete and detailed consideration of each of these steps in this and subsequent chapters. At this stage, the reader may find the general description of the basic steps in this chapter somewhat overwhelming. However, when these steps are followed in detail with simple illustrations in the subsequent chapters, the ideas and concepts will become clear.

Mathematical foundations of the variational formulation and the residual formulation (Galerkin's method) are given in more detail after a brief description of the eight steps in the finite element method. A good comprehension of these procedures is necessary for a thorough understanding of the derivation of the element equations.

General Idea

Engineers are interested in evaluating effects such as deformations, stresses, temperature, fluid pressure, and fluid velocities caused by forces such as applied loads or pressures and thermal and fluid fluxes. The nature of distribution of the effects (deformations) in a body depends on the characteristics of the force system and of the body itself. Our aim is to find this distribution of the effects. For convenience, we shall often use displacements or deformations u (Figure 2.1) in place of effects. Subsequently, when other problems such as heat and fluid flow are discussed they will involve distribution of temperature and fluid heads and their gradients.

We assume that it is difficult to find the distribution of u by using conventional methods and decide to use the finite element method, which is based on the concept of discretization, as explained in Chapter 1. We divide the

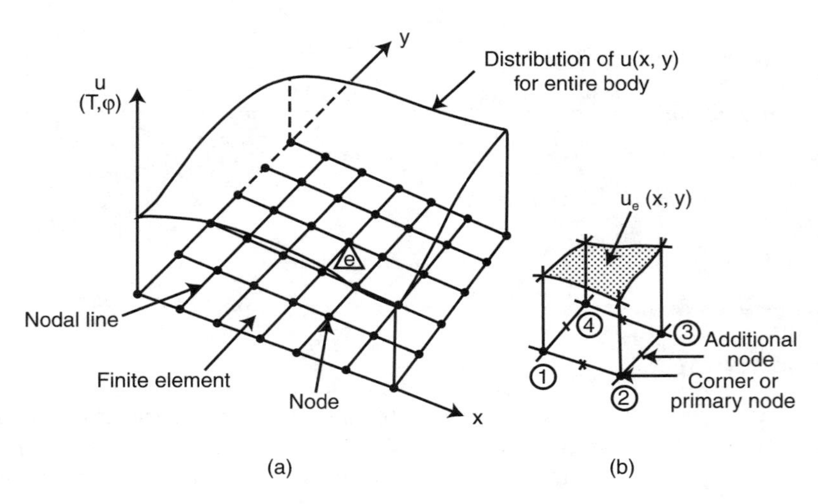

FIGURE 2.1
Distribution of displacement u, temperature T, or fluid head φ. (a) Discretization of two-dimensional body and (b) Distribution of u_e over a generic element e.

body into a number of smaller regions (Figure 2.1(a)) called finite elements.[1,2] A consequence of such subdivision is that the distribution of displacement is also discretized into corresponding subzones (Figure 2.1(b)). The subdivided elements are now easier to examine as compared to the entire body and distribution of u over it.

For stress-deformation analysis of a body in equilibrium under external loading, the examination of the elements involves derivation of the stiffness–load relationship. To derive such a relationship, we make use of the laws and principles governing the behavior of the body. Since our primary concern is to find the distribution of u, we contrive to express the laws and principles in terms of u. We do this by making an advance choice of the pattern, shape, or outline of the distribution of u over an element. In choosing a shape, we follow certain rules dictated by the laws and principles. For example, one law says that the loaded body, to be reliable and functional, cannot experience breaks anywhere in its regime. In other words, the body must remain continuous. Let us now describe in detail various steps involved in the foregoing qualitative statements.

Step 1. Discretize and Select Element Configuration

This step involves subdividing the body into a suitable number of small bodies, called finite elements. The intersections of the sides of the elements are called nodes or nodal points, and the interfaces between the elements are called nodal lines and nodal planes. Often we may need to introduce additional node points along the nodal lines and planes (Figure 2.1(b)).

An immediate question that arises is how small should the elements chosen be? In other words, how many elements would approximate the continuous

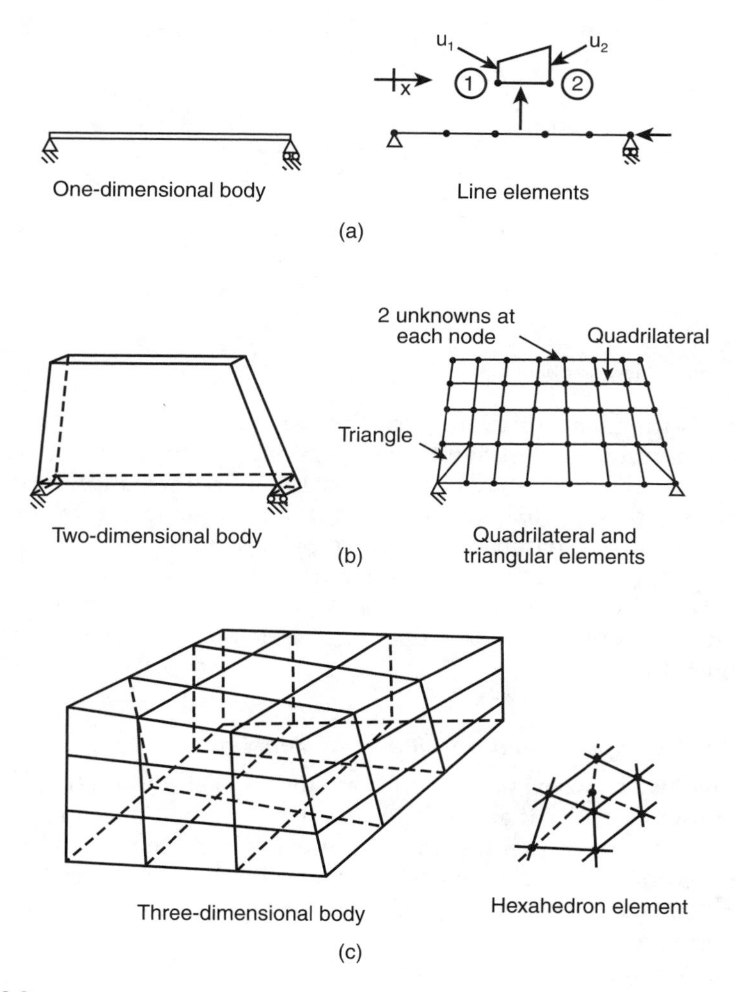

FIGURE 2.2
Different types of elements: (a) One-dimensional element, (b) Two-dimensional elements, and (c) Three-dimensional elements.

medium as closely as possible? This depends on a number of factors, which we shall discuss.

What type of element should be used? This will depend on the characteristics of the continuum and the idealization that we may choose to use. For instance, if a structure or a body is idealized as a one-dimensional line, the element we use is a line element (Figure 2.2(a)). For two-dimensional bodies, we use triangles and quadrilaterals (Figure 2.2(b)); for three-dimensional idealization, a hexahedron with different specializations (Figure 2.2(c)) can be used.

Although we could subdivide the body into regular-shaped elements in the interior (Figure 2.3), we may have to make special provisions if the

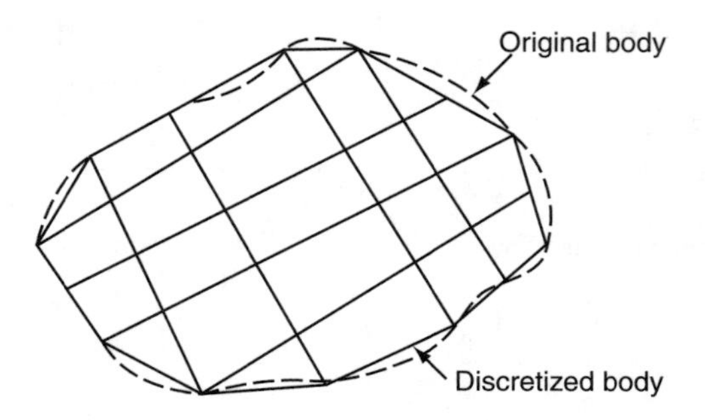

FIGURE 2.3
Discretization for irregular boundary.

boundary is irregular. For many cases, the irregular boundary can be approximated by a number of straight lines (Figure 2.3). On the other hand, for many other problems, it may be necessary to use mathematical functions of sufficient order to approximate the boundary. For example, if the boundary shape is similar to a parabolic curve, we can use a second-order quadratic function to approximate that boundary. The concept of isoparametric elements that we shall discuss later makes use of this idea. It may be noted that inclusion of irregular boundaries in a finite element formulation poses no great difficulty.

Step 2. Select Approximation Models or Functions

In this step, we choose a pattern or shape for the distribution (Figure 2.1) of the unknown quantity that can be a displacement and/or stress for stress-deformation problems, temperature in heat flow problems, fluid pressures and/or velocity for fluid flow problems, and both temperature (fluid pressure) and displacement for coupled problems involving effects of both flow and deformation.

The nodal points of the element provide strategic points for writing mathematical functions to describe the shape of the distribution of the unknown quantity over the domain of the element. A number of mathematical functions such as polynomials and trigonometric series can be used for this purpose, especially polynomials because of the ease and simplification they provide in the finite element formulation. If we denote u as the unknown, the polynomial interpolation function can be expressed as

$$u = N_1 u_1 + N_2 u_2 + N_3 u_3 + \cdots + N_m u_m. \tag{2.1}$$

Here $u_1, u_2, u_3, \ldots, u_m$ are the values of the unknowns at the nodal points and $N_1, N_2, \ldots, N_m$ are the interpolation functions; in subsequent chapters we shall give details of these functions. For example, in the case of the line element with two end nodes (Figure 2.2(a)) we can have u_1 and u_2 as unknowns or degrees of freedom and for the triangle (Figure 2.2(b)) we can have $u_1, u_2, \ldots, u_6$

as unknowns or degrees of freedom if we are dealing with a plane deformation problem where there are two displacements at each node.

A degree of freedom can be defined as an independent (unknown) displacement that can occur at a point. For instance, for the problem of one-dimensional deformation in a column (Figure 2.2(a)), there is only one way in which a point is free to move, that is, in the uniaxial direction. Then a point has one degree of freedom. For a two-dimensional problem (Figure 2.2(b)), if deformations can occur only in the plane of the body (and bending effects are ignored), a point is free to move only in two independent coordinate directions; thus a point has two degrees of freedom. In Chapter 7, when bending is considered, it will be necessary to consider rotations or slopes as independent degrees of freedom.

We note here that after all the steps of the finite element method are accomplished, we shall find the solution as the values of the unknowns u at all the nodes, that is $u_1, u_2, \ldots, u_m$. To initiate action toward obtaining the solution, however, we have assumed *a priori* or in advance a shape or pattern that we hope will satisfy the conditions, laws, and principles of the problem at hand.

The reader should realize that the solution obtained will be in terms of the unknowns only at the nodal points. This is one of the outcomes of the discretization process. Figure 2.4 shows that the final solution is a combination of solutions in each element patched together at the common boundaries. This is further illustrated by sketching a cross section along *A–A*. It can be seen that the computed solution is not necessarily the same as the

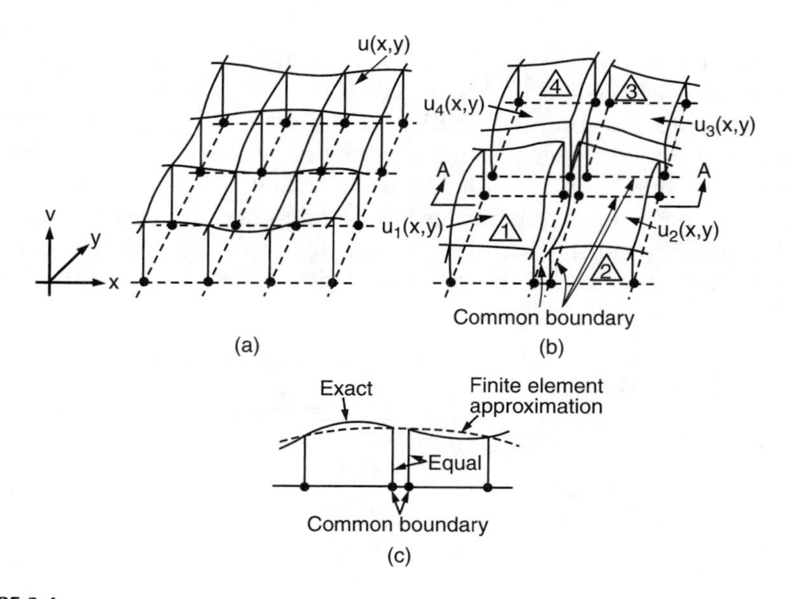

FIGURE 2.4
Approximate solution as patchwork of solutions over elements: (a) Assemblage, (b) Neighboring elements, and (c) Section along A–A.

exact continuous solution shown by the solid curve. The statement in Ch'nter 1 that discretization yields approximate solutions can be visualized from this schematic representation (Figure 2.4). Obviously, we would like the discretization to be such that the computed solution is as close as possible to the exact solution; that is, the error is a minimum.

Step 3. Define Strain (Gradient)-Displacement (Unknown) and Stress–Strain (Constitutive) Relationships

To proceed to the next step, which uses a principle (say, the principle of minimum potential energy) for deriving equations for the element, we must define appropriate quantities that appear in the principle. For stress-deformation problems one such quantity is the strain (or gradient) of displacement. For instance, in the case of deformation occurring only in one direction y (Figure 2.5(a)), the strain ε_y assumed to be small is given by

$$\varepsilon_y = \frac{dv}{dy}, \tag{2.2}$$

where v is the deformation in the y direction. For the case of fluid flow in one direction, such a relation is the gradient g_x of fluid head (Figure 2.5(b)):

$$g_x = \frac{d\varphi}{dx}. \tag{2.3}$$

Here φ is the fluid head or potential and g_x is the gradient of φ, that is, rate of change of φ with respect to the x coordinate.

In addition to the strain or gradient, we must also define an additional quantity, the stress or velocity; usually, this is done by expressing its relationship with the strain. Such a relation is called a stress–strain law. In a generalized sense, it is a constitutive law and describes the response or effect (displacement, strain) in a system due to applied cause (force). The stress–strain law is one of the most vital parts of finite element analysis.

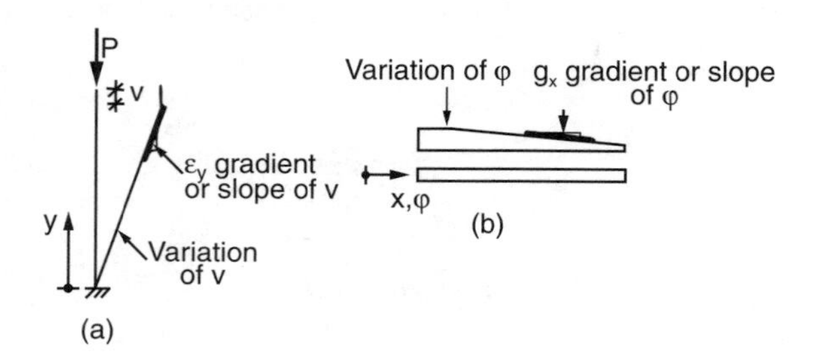

FIGURE 2.5
Problems idealized as one-dimensional: (a) One-dimensional stress-deformation and (b) One-dimensional flow.

Unless it is defined to reflect precisely the behavior of the material or the system, the results from the analysis can be of very little significance. As an elementary illustration, consider Hooke's law, which defines the relationship of stress to strain in a solid body:

$$\sigma_y = E_y \varepsilon_y, \tag{2.4a}$$

where σ_y = stress in the vertical direction and E_y = Young's modulus of elasticity. If we substitute ε_y from Equation 2.2 into Equation 2.4a, we have the expression for stress in terms of displacements as

$$\sigma_y = E_y \frac{dv}{dy}. \tag{2.4b}$$

One of the other simple linear constitutive laws is Darcy's law for fluid flow through porous media:

$$v_x = -k_x g_x, \tag{2.4c}$$

where k_x = coefficient of permeability, v_x = velocity, and g_x = gradient. In electrical engineering the corresponding law is Ohm's law.

Step 4. Derive Element Equations

By invoking available laws and principles, we obtain equations governing the behavior of the element. The equations here are obtained in general terms and hence can be used for all elements in the discretized body.

A number of alternatives are possible for the derivation of element equations. The two most commonly used are the energy methods and the residual methods.

Use of the energy procedures requires knowledge of variational calculus. At this stage of our study of the finite element method, we shall postpone detailed consideration of variational calculus and in a somewhat less rigorous manner introduce the ideas simply through the use of differential calculus.

Energy Methods

These procedures are based on the idea of finding consistent states of bodies or structures associated with stationary values of a scalar quantity assumed by the loaded bodies. In engineering, usually this quantity is a measure of energy or work. The process of finding stationary values of energy requires use of the mathematical disciplines called variational calculus involving use of variational principles. In this introductory book, detailed treatment of variational calculus is considered to be not warranted. However, a description of the energy approach used in the book is given below, and an introductory description of variational calculus is given at the end of this chapter.

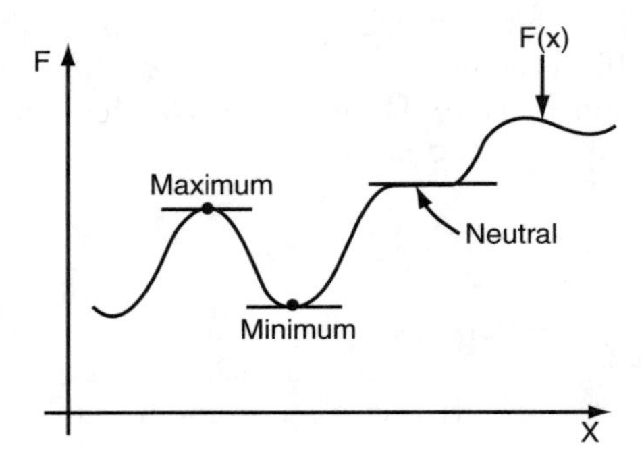

FIGURE 2.6
Stationary values of a function.

Within the realm of energy methods, there are a number of methods and variational principles, e.g., the principle of stationary potential and complementary energies, Reissner's mixed principle, and hybrid formulations, which are commonly used in finite element applications.[3-6]

Stationary Value

In simple words, the term stationary can imply a maximum, minimum, or saddle point of a function *F(x)* (Figure 2.6). Under certain conditions, the function may simply assume a minimum or a maximum value. To find the point of a stationary value, we equate the derivative of F to zero:

$$\frac{dF}{dx} = 0. \tag{2.5}$$

Potential Energy

In the case of stress-deformation analysis, the function F is often represented by one of the energy functions stated previously. For instance, we can define F to be the potential energy in a body under load. If the body, say, a simple column under the given support conditions (Figure 2.5(a)), is linear and elastic and if it is in equilibrium, it can be shown that the column will assume minimum potential energy. To comply with the commonly used notation we denote potential energy by the symbol Π_p, where the subscript denotes potential energy.

The potential energy is defined as the sum of the internal strain energy U and the potential of the external loads W_p; the latter term denotes the capacity of load P to perform work through a deformation v of the column. Therefore,

$$\Pi_p = U + W_p. \tag{2.6a}$$

When we apply the principle of minimum potential energy, we essentially take the derivative (or variation) of Π_p and equate it to zero. We assume that the load remains constant while taking the derivative; then

$$\delta\Pi_p = \delta U - \delta W_p = 0. \tag{2.6b}$$

The symbol δ denotes variation of the potential energy Π_p. As indicated subsequently in Equation 2.9, we can interpret the variation or change as composed of a series of partial differentiation of Π_p. Here we use the relation between the variation in potential of external loads and in work done by the loads as

$$\delta W = -\delta W_p. \tag{2.6c}$$

Note that the negative sign in Equations 2.6b and 2.6c arises because the potential of the external loads in Equation 2.6a is lost through the work done by the external load.

The fact that for linear, elastic bodies in equilibrium the value of Π_p is a minimum can be verified by showing that the second derivative or variation of Π_p is greater than zero; that is,

$$\delta^2\Pi_p = \delta^2 U - \delta^2 W_p > 0. \tag{2.7}$$

Proof of Equation 2.7 can be found in advance treatments on energy methods and is not included in this text. The symbol δ is a compact symbol used to denote variation or a series of partial differentiations. For our purpose, we shall interpret it simply as a symbol that denotes derivatives of Π_p with respect to the independent coordinates or unknowns in terms of which it is expressed. For example, if

$$\Pi_p = \Pi_p(u_1, u_2, \ldots, u_n), \tag{2.8}$$

where $u_1, u_2, \ldots, u_n$ are the total number of unknowns (at the nodes), then $\delta\,\Pi_p = 0$ implies

$$\begin{aligned} \frac{\partial \Pi_p}{\partial u_1} &= 0, \\ \frac{\partial \Pi_p}{\partial u_2} &= 0, \\ &\vdots \\ \frac{\partial \Pi_p}{\partial u_n} &= 0. \end{aligned} \tag{2.9}$$

Here n = total number of unknowns.

In subsequent chapters we shall illustrate the use of the principle of stationary energy and other energy principles for finite element formulations of various problems.

Method of Weighted Residuals

One of the two major alternatives for formulating the finite element method is the method of weighted residuals (MWR). A number of schemes are employed under the MWR, among which are collocation, subdomain, least squares, and Galerkin's methods.[3,7] For many problems with certain mathematical characteristics (discussed in later chapters), Galerkin's method yields results identical to those from variational procedures and is closely related to them. Galerkin's method has been the most commonly used residual method for finite element applications.

The MWR is based on minimization of the residual left after an approximate or trial solution is substituted into the differential equations governing a problem. As a simple illustration, let us consider the following differential equation:

$$\frac{\partial^2 u^*}{\partial x^2} - \frac{\partial u^*}{\partial t} = f(x), \tag{2.10a}$$

where u^* is the unknown, x is the coordinate, t is the time, and $f(x)$ is the forcing function. In mathematical notation, Equation 2.10a is written as

$$Lu^* = f, \tag{2.10b}$$

where

$$L \equiv \frac{\partial^2}{\partial x^2} - \frac{\partial}{\partial t}$$

is the differential operator.

We are seeking an approximate solution to Equation 2.10 and denote an approximate or trial function u for u^* as

$$\begin{aligned} u &= \varphi_0 + \sum_{i=1}^{n} \alpha_i \varphi_i \\ &= \varphi_0 + \alpha_1 \varphi_1 + \alpha_2 \varphi_2 + \cdots + \alpha_n \varphi_n. \end{aligned} \tag{2.11}$$

Here $\varphi_1, \varphi_2, \ldots, \varphi_n$ are known functions chosen in such a way as to satisfy the homogeneous boundary conditions; φ_0 is chosen to satisfy the essential, geometric, or forced boundary conditions; and α_i are parameters or constants to be determined. Categories of boundary conditions are explained subsequently and discussed in Chapter 3. Often, for convenience, u in Equation 2.11 is written as

$$u = \sum_{i=1}^{n} \alpha_i \varphi_i \tag{2.12}$$

in which $\alpha_1 = 1$ and $\varphi_1 = \varphi_0$. If the approximate solution u is substituted into Equation 2.10, we are left with a residual

$$R(x) = Lu - f, \tag{2.13}$$

which is zero if $u = u^*$.

In the method of weighted residuals, the aim is to find an approximate solution u for u^* such that the residual $R(x)$ in Equation 2.13 is made as small as possible or is minimized. In other words the error between the approximate solution u^* and the exact solution u is minimized. A number of schemes are possible to achieve the aim of minimization of $R(x)$; details of some of the major schemes — collocation, subdomain, least squares, and Galerkin methods — are given in Appendix 1. Moreover, in the subsequent chapters, Galerkin's method is used to derive finite element equations for a number of simple problems. For the sake of completeness, only brief and general statements for these methods are given below.

Mathematically the idea of minimization can be expressed as

$$\int_D R(x) W_i(x) dx = 0, \quad i = 1, 2, \ldots, n, \tag{2.14}$$

where D denotes the domain of a structure or body under consideration. For the one-dimensional column problem, the domain is simply the linear extent of the column.

In Equation 2.14 the W_i, denote weighting functions. Various residual schemes such as collocation, subdomain, and Galerkin use different weighting functions. For instance, in the case of the collocation method, $W_i = 1$. The expression in Equation 2.14 implies that the weighted value of $R(x)$ over the domain of a structure vanishes. Figure 2.7 shows a schematic representation of Equation 2.14. The shaded areas in Figure 2.7(b) denote error between the approximate and the exact solution $u^* - u$ over the domain D. The residual $R(x)$ over D (Figure 2.7(a)) is related to the error $u^* - u$, and the sum or the integral of $R(x)$ over D is minimized.

As a simple illustration, let us consider the following second-order differential equation that governs the problems of one-dimensional stress deformation in a column and flow in Chapters 3 and 4, respectively:

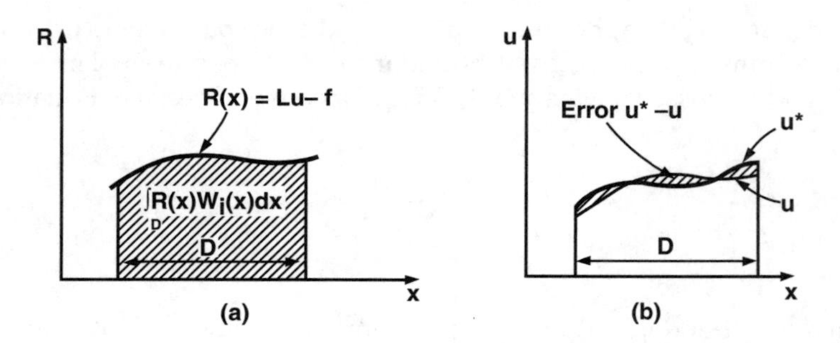

Figure 2.7
Schematic representation of residual: (a) Integration of R over D and (b) Error $u^* - u$ over D.

$$c\frac{d^2v^*}{dy^2} = f, \tag{2.15a}$$

where v^* is the unknown (deformation), y is the coordinate axis, c denotes a material property, and f is the forcing function. For a column, $c = EA$, f is the applied load, E is the modulus of elasticity, and A is the cross-sectional area. Assume that $EA = 1$ and $f = 10$ units of load; then Equation 2.15a specializes to

$$\frac{d^2v^*}{dy^2} = 10. \tag{2.15b}$$

An approximate solution for v^* can be written as a special case of Equation 2.11 as

$$v = \alpha_1 + \alpha_2 y + \alpha_3 y^2 = \sum \alpha_i \varphi_i, \tag{2.16}$$

where $\varphi_1 = 1$, $\varphi_2 = y$, and $\varphi_3 = y^2$. This function should be chosen so as to satisfy the boundary conditions of the problem. Then the residual $R(y)$ is given by

$$R(y) = \frac{d^2v}{dy^2} - 10. \tag{2.17a}$$

According to the MWR, Equation 2.14 leads to

$$\int_0^L R(y)W_i(y)dy = 0, \quad i = 1, 2, 3, \tag{2.17b}$$

or

$$\int_0^L \left(\frac{d^2 v}{dy^2} - 10 \right) W_i(y) dy = 0, \tag{2.17c}$$

or

$$\int_0^L \left[\left(\frac{d^2 \Sigma \alpha_i \varphi_i}{dy^2} \right) - 10 \right] W_i(y) dy = 0. \tag{2.17d}$$

Here L is the length of the column which represents the domain D. Now, we can substitute for the second derivative in Equation 2.17 by differentiating v in Equation 2.16 twice. The final results will yield three simultaneous equations in α_1, α_2, and α_3 as

$$\begin{aligned}
&\int_0^L \left[\frac{d^2 (\Sigma \alpha_i \varphi_i)}{dy^2} - 10 \right] W_1(y) dy = 0, \\
&\int_0^L \left[\frac{d^2 (\Sigma \alpha_i \varphi_i)}{dy^2} - 10 \right] W_2(y) dy = 0, \\
&\int_0^L \left[\frac{d^2 (\Sigma \alpha_i \varphi_i)}{dy^2} - 10 \right] W_3(y) dy = 0,
\end{aligned} \tag{2.18}$$

which are solved for α_1, α_2, and α_3. When these values of α_1, α_2, and α_3 are substituted into Equation 2.16, we obtain the approximate solution for v^*.

At this stage, it is not necessary to go into the details of the steps required to proceed from Equation 2.17 to Equation 2.18; they are given in Appendix 1 and in the subsequent chapters.

Element Equations

Use of either of the two foregoing methods will lead to equations describing the behavior of an element, which are commonly expressed as

$$[\mathbf{k}]\{\mathbf{q}\} = \{\mathbf{Q}\} \tag{2.19}$$

where $[\mathbf{k}]$ = element property matrix, $\{\mathbf{q}\}$ = vector of unknowns at the element nodes, and $\{\mathbf{Q}\}$ = vector of element nodal forcing parameters. Equation 2.19 is expressed in a general sense; for the specific problem of stress analysis, $[\mathbf{k}]$ = stiffness matrix, $\{\mathbf{q}\}$ = vector of nodal displacements, and $\{\mathbf{Q}\}$ = vector of nodal forces. Details of the matrices in Equation 2.19 will be developed and described fully in subsequent chapters.

Step 5. Assemble Element Equations to Obtain Global or Assemblage Equations and Introduce Boundary Conditions

Our final aim is to obtain equations for the entire body that define approximately the behavior of the entire body or structure. In fact, as will be discussed in various chapters, use of the variational or residual procedure is relevant to the entire body; it is for simplicity that we view the procedure in Step 4 as having been applied to a single element.

Once the element equations, Equation 2.19, are established for a generic element, we are ready to generate equations recursively for other elements by using Equation 2.19 again and again. Then we add them together to find global equations. This assembling process is based on the law of compatibility or continuity (Chapter 1). It requires that the body remain continuous; that is, the neighboring points should remain in the neighborhood of each other after the load is applied (Figure 2.4). In other words, the displacements of two adjacent or consecutive points must have identical values (Figure 2.8(a)). Depending on the type and nature of the problem, we may need to enforce the continuity conditions more severely. For instance, for deformations occurring in a plane, it may be sufficient to enforce continuity of the displacements only. On the other hand, for bending problems, the physical properties of the deformed body under the load requires that in addition to

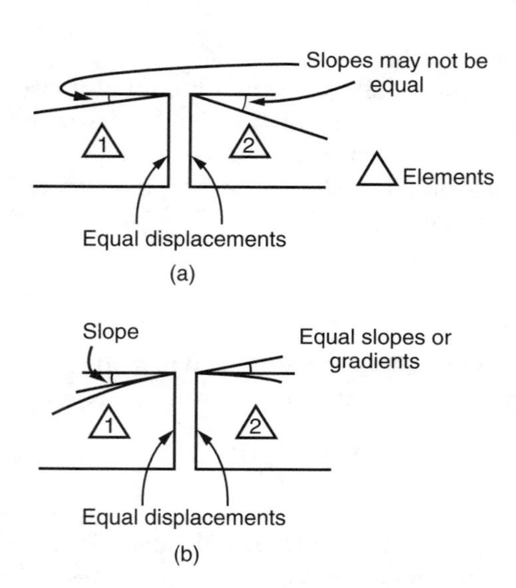

FIGURE 2.8
Interelement compatibility: (a) Compatibility for plane problems and (b) Compatibility for bending-type problems.

the continuity of displacements we ensure that the slopes or the first derivative of displacements are also continuous or compatible at adjacent nodes (Figure 2.8(b)). Often it may become necessary to satisfy compatibility of the curvatures or the second derivative also.

Finally, we obtain the assemblage equations, which are expressed in matrix notation as

$$[\mathbf{K}]\{\mathbf{r}\} = \{\mathbf{R}\}, \tag{2.20}$$

where **[K]** = assemblage property matrix, **{r}** = assemblage vector of nodal unknowns, and **{R}** = assemblage vector of nodal forcing parameters.

For stress-deformation problems, these quantities are the assemblage stiffness matrix, nodal displacement vector, and nodal load vector, respectively.

Boundary Conditions

Until now we have considered only the properties of a body or structure (Figure 2.9(a)). Equation 2.20 tells us about the capabilities of the body to withstand applied forces. It is just like saying that one is an engineer. How he will perform his engineering duties will depend on the surroundings and the problems he faces; these aspects can be called *constraints*. In the case of

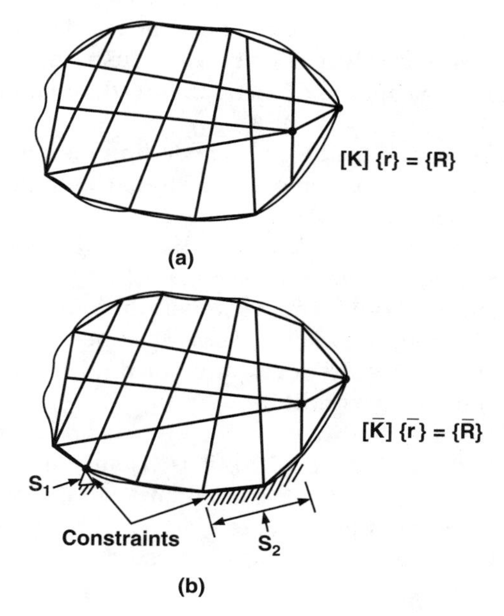

FIGURE 2.9
Boundary conditions or constraints: (a) Body without constraints and (b) Body with constraints.

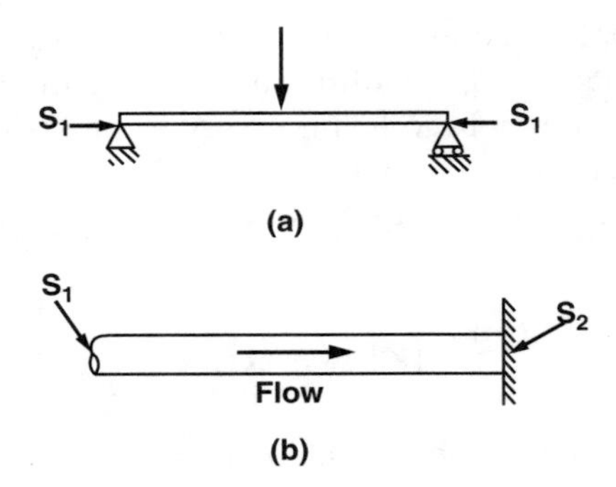

FIGURE 2.10
Examples of boundary conditions: (a) Beam with boundary conditions and (b) Pipe flow with boundary conditions.

engineering bodies, the surroundings or the constraints are the boundary conditions. Only when we introduce these conditions can we decide how the body will perform.

Boundary conditions, then, are the physical constraints or supports (Figure 2.9(b)) that must exist so that the structure or body can stand in space uniquely. These conditions are commonly specified in terms of known values of the unknowns on a part of the surface or boundary S_1 and/or gradients or derivatives of the unknowns on S_2. Figure 2.10(a) depicts a beam. In the case of the simply supported beam, the boundary S_1 is the two end points where the displacements are given. This type of constraint expressed in terms of displacements is often called the essential, forced, or geometric boundary conditions.

At the simple supports of the beam, the bending moment is zero; that is, the second derivative of displacement vanishes. This type of constraint is often called a natural boundary condition.

Figure 2.10(b) shows a cylinder through which fluid or temperature flows. On the boundary S_1 temperature or fluid head is known; this is the essential boundary condition. The right end is impervious to water, or insulated against heat; then the boundary condition is specified as fluid or heat flux, which is proportional to the first derivative of fluid head or temperature. This is the natural boundary condition.

To reflect the boundary conditions in the finite element approximation of the body represented by Equation 2.20, it is usually necessary to modify these equations only for the geometric boundary conditions. Further details and procedure for such modification are given in subsequent chapters. The final modified assemblage equations are expressed by inserting overbars as

$$[\overline{\mathbf{K}}]\{\bar{\mathbf{r}}\} = \{\overline{\mathbf{R}}\} \tag{2.21}$$

Step 6. Solve for the Primary Unknowns

Equation 2.21 is a set of linear (or nonlinear) simultaneous algebraic equations, which can be written in standard familiar form as

$$\begin{aligned} K_{11}r_1 + K_{12}r_2 + \cdots + K_{1n}r_n &= R_1, \\ K_{21}r_1 + K_{22}r_2 + \cdots + K_{2n}r_n &= R_2, \\ &\vdots \\ K_{n1}r_1 + K_{n2}r_2 + \cdots + K_{nn}r_n &= R_n. \end{aligned} \tag{2.22}$$

These equations can be solved by using the well-known Gaussian elimination or iterative methods. Detailed coverage of these methods is beyond the scope of this book; however, later when individual problems are considered, some of these methods will be illustrated. Moreover, brief descriptions of these solution methods are given in Appendix 2. At the end of this step, we have solved for the unknowns (displacements) $r_1, r_2, \ldots, r_n$. These are called primary unknowns because they appear as the first quantities sought in the basic Equation 2.22. The designation of the word primary will change depending on the unknown quantity that appears in Equation 2.22. For instance, if the problem is formulated by using stresses as unknowns, the stresses will be called the primary quantities. For the flow problem the primary quantity can be the fluid or velocity head or potential.

Step 7. Solve for Derived or Secondary Quantities

Very often additional or secondary quantities must be computed from the primary quantities. In the case of stress-deformation problems such quantities can be strains, stresses, moments, and shear forces; for the now problem they can be velocities and discharges. It is relatively straightforward to find the secondary quantities once the primary quantities are known, since we can make use of the relations between the strain and displacement and stress and strain that are defined in Step 3.

Step 8. Interpretation of Results

The final and the important aim is to reduce the results from the use of the finite element procedure to a form that can be readily used for analysis and design. The results are usually obtained in the form of printed output from the computer. We then select critical sections of the body and plot the values of displacement and stresses along them, or we can tabulate the results. It is often very convenient and less time consuming to use (available) routines and ask the computer to plot or tabulate the results.

Introduction to Variational Calculus

There are a number of reasons for including the consideration of the subject of calculus of variations in the study of the finite element method. Two main reasons are[8]

1. "The field equations (ordinary or partial differential equations) and the associated boundary conditions of many problems can be derived from variational principles. In formulating an approximate theory, the shortest and clearest derivation is usually obtained through variational calculus."
2. "The 'direct' method of solution of variational problems is one of the most powerful tools for obtaining numerical results in practical problems of engineering importance."

In this section variational calculus is described in a simplified manner. This knowledge is necessary to obtain the appropriate functional from which element equations are derived. This functional can be obtained from the governing equation and natural boundary conditions of the problem of interest following the principles of variational calculus. Interested readers are referred to other sources[7,9-12] for detailed discussions of variational calculus.

Definitions of Functions and Functionals

Figure 2.11 shows a function f of x. Mathematically it is denoted as $f(x)$. The figure simply shows how the function value f changes as x varies. Here x is

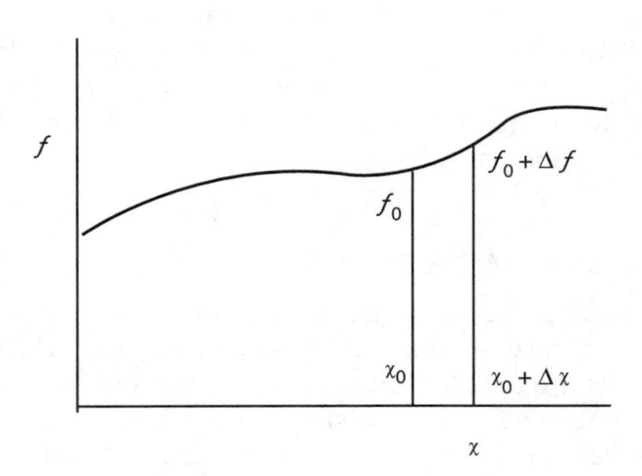

FIGURE 2.11
The function value varies by Δf as the independent variable x changes by Δx.

called the independent variable and f is called the function value. Any function f can be dependent on one independent variable x as shown in Figure 2.11, or can be a function of more than one independent variables x_1, x_2, x_3, … x_n. It is then denoted as $f = f(x_1, x_2, x_3, \ldots x_n)$.

A functional is a function of functions that gives a specific value for a given function. It is denoted as $F = F(f)$, where f is a function as defined above. The value of the functional $F(f)$ depends on f. The integral of f between two limits is a simple example of a functional. Note that if F is defined in the following manner

$$F = \int_0^5 f dx \tag{2.23}$$

then the value of F depends on the function expression f. If $f = x^2$ then $F = 41.67$, and if $f = x^3$ then $F = 156.25$. Note the similarities between a function and a functional — both have an independent variable. For a function, such as $f = x^2$, the independent variable is a parameter x that changes from, say, 2 to 3, then the function value f changes from 4 to 9. Similarly, the functional F, defined in Equation 2.23, has an independent variable f which is a function, and as f is changed from x^2 to x^3 the functional value F varies from 41.67 to 156.25.

A functional can be a function of a number of functions as shown below:

$$F = \int_0^5 f^2 dx + \int_1^3 g dx + h(5) - h(0) \tag{2.24}$$

In Equation 2.24 the functional F is defined as a function of three different functions f, g, and h. The above functional can be denoted as $F = F(f,g,h)$. In Equation 2.24, the functions f and g are integrated between two limits and h are evaluated at two points 5 and 0.

Variations of Functions

Let the value of the function $f(x)$ at $x = x_0$ be given by f_0 and at $x = x_0 + \Delta x$ it is $f_0 + \Delta f$ as shown in Figure 2.11.

Then $f_0 + \Delta f$ can be expressed in terms of f_0, Δx and derivatives of f at x_0 from the Taylor's series expansion as given below:

$$\begin{aligned} f_0 + \Delta f &= f(x_0 + \Delta x) \\ &= f(x_0) + \Delta x \cdot f'(x_0) + \frac{1}{2}(\Delta x)^2 f''(x_0) + \frac{1}{6}(\Delta x)^3 f'''(x_0) + \ldots. \end{aligned} \tag{2.25}$$

where f', f'', f''' of the above equation denote the first, second, and third derivatives, respectively, of the function f. Note that the Taylor's series expansion is an infinite expansion; however, the terms containing the higher order derivatives can be ignored for small Δx since the higher order derivatives are multiplied by $(\Delta x)^n$ where n is the order of the derivative.

The above equation can be rewritten in the following form,

$$f_0 + \Delta f = f_0 + \delta f + \frac{1}{2}\delta^2 f + \frac{1}{6}\delta^3 f + \ldots \tag{2.26}$$

where

$\delta f = f'(x_0)\Delta x$ = first variation of f,
$\delta^2 f = f''(x_0)(\Delta x)^2$ = second variation of f,
$\delta^3 f = f'''(x_0)(\Delta x)^3$ = third variation of f.

Similarly for a function of two variables, one obtains from the Taylor's series expansion,

$$f_0 + \Delta f = f(x_0 + \Delta x, y_0 + \Delta y) = f_0 + \delta f + \frac{1}{2}\delta^2 f + \ldots \tag{2.27}$$

where

$\delta f = f_x(x_0, y_0)\Delta x + f_y(x_0, y_0)\Delta y$ = first variation of f,
$\delta^2 f = f_{xx}(x_0, y_0)(\Delta x)^2 + 2 f_{xy}(x_0, y_0)(\Delta x \Delta y) + f_{yy}(x_0, y_0)(\Delta y)^2$ = second variation of f,

and so on.

In the above expressions f_x and f_y denote partial derivatives of the function with respect to x and y respectively; f_{xx}, f_{xy}, and f_{yy} represent second partial derivatives. Again, if Δx and Δy are small then higher order variations can be ignored and the total variation Δf can be approximately equated to the first variation (δf) only.

A similar idea can be applied to the variation of a functional. If one considers a functional $F = F(f, g, h)$ then for small variations of the functions f, g, and h one can write

$$\Delta F \approx \delta F = F_f \cdot \Delta f + F_g \cdot \Delta g + F_h \cdot \Delta h \tag{2.28}$$

where ΔF is the total variation of the functional, δF is the first variation, F_f, F_g, and F_h are partial derivatives of the functional expression with respect to the functions f, g, and h, respectively, and Δf, Δg, and Δh are variations of the

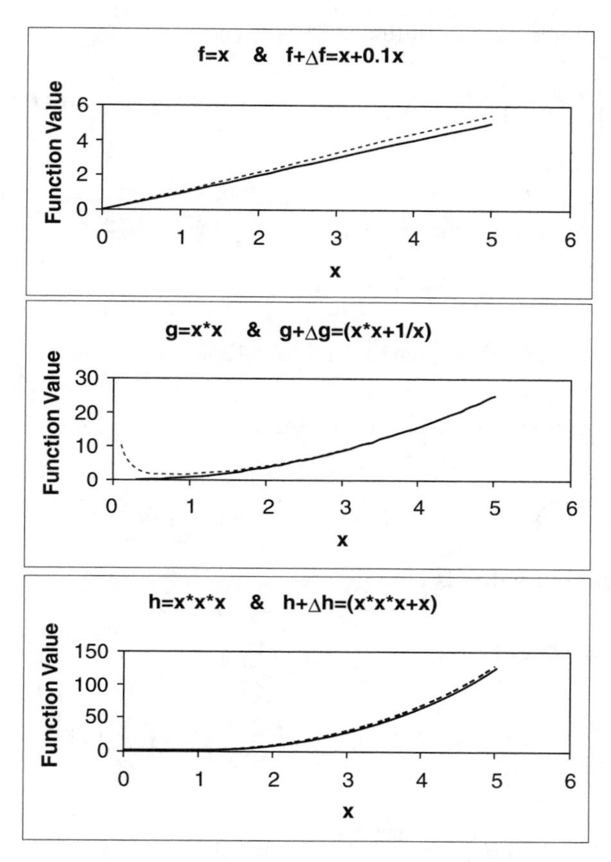

FIGURE 2.12
Functions f, g, h (continuous lines) and $f + \Delta f$, $g + \Delta g$, $h + \Delta h$ (dotted lines).

functions f, g, and h, respectively. When computing δF using Equation 2.28 for the functional defined in Equation 2.24 the integral operators of Equation 2.24 should operate on $F_f . \Delta f$ and $F_g \cdot \Delta g$. The function $F_h \cdot \Delta h$ should be computed at $x = 5$ and $x = 0$. Note that Δf, Δg, and Δh are independent arbitrary functions. This will be clear from the following example.

Example 2.1

Consider the functional defined in Equation 2.24 and let f, g, and h be x, x^2, and x^3, respectively. Obtain the value of the functional. Then assign the variations $0.1x$, $1/x$ and x to the functions f, g, and h, respectively. Calculate the total variation ΔF and the first variation δF and compare these two values. Functions f, g, h, and $f + \Delta f$, $g + \Delta g$, $h + \Delta h$ are shown in Figure 2.12 by continuous and dotted lines, respectively.

Solution: The functional, Equation 2.24 is given by

$$F(f,g,h)=\int_0^5 f^2dx+\int_1^3 gdx+h(5)-h(0),$$ where $f = x$, $g = x^2$ and $h = x^3$ as stated in the problem above.

Clearly, $F(x,x^2,x^3)=\int_0^5 x^2dx+\int_1^3 x^2dx+x^3\Big|_{x=5}-x^3\Big|_{x=0}=\dfrac{x^3}{3}\Big|_0^5+\dfrac{x^3}{3}\Big|_1^3+125-0,$

or

$$F(x,\,x^2,\,x^3)=41.67+8.67+125-0=175.34.$$

Now if we give small variations to the functions f, g, and h as shown in Figure 2.12

$$f+\Delta f=x+0.1x,\; g+\Delta g=x^2+1/x,\; h+\Delta h=x^3+x$$

then the functional value is changed as

$$F(x,\,x^2,\,x^3)+\Delta F=F(x+0.1x,\,x^2+1/x,\,x^3+x)$$

$$=\int_0^5 (x+0.1x)^2dx+\int_1^3\left(x^2+\frac{1}{x}\right)dx+(x^3+x)\Big|_{x=5}-(x^3+x)\Big|_{x=0}$$

$$=1.1\frac{x^3}{3}\Big|_0^5+\frac{x^3}{3}\Big|_1^3+\ln x\Big|_1^3+130-0$$

$$=50.42+9.77+130=190.19$$

Hence, the total variation ΔF of the functional = 190.19 – 175.34 = 14.84.

Now, let us calculate the first variation δF. From its definition, Equation 2.28

$$\delta F=\int_0^5 2f\cdot\Delta f\cdot dx+\int_1^3 \Delta g\cdot dx+\Delta h(5)-\Delta h(0)$$

or

$$\delta F=\int_0^5 2x(0.1x)dx+\int_1^3\frac{1}{x}dx+5-0=8.33+1.1+5=14.43.$$

Comparing the total variation ΔF with the first variation δF one can state that for this example more than 97% of the total variation has been captured by the first variation term.

Example 2.2

Let the function f be a function of two independent variables x and y. Functions g and h of Equation 2.24 are defined as partial derivatives of f with respect to x and y, respectively. Then derive the expression of the first variation of this functional.

Solution: Substitute g and h of Equation 2.24 by f_x and f_y, respectively. Then we obtain

$$F(f, f_x, f_y) = \int_0^5 f^2 dx + \int_1^3 f_x dx + f_y(5) - f_y(0).$$

The first variation of this functional is given by

$$\delta F = \frac{\partial F}{\partial f}\Delta f + \frac{\partial F}{\partial f_x}\Delta f_x + \frac{\partial F}{\partial f_y}\Delta f_y = \int_0^5 2f \cdot \Delta f \cdot dx + \int_1^3 \Delta f_x \cdot dx + \Delta f_y(5) - \Delta f_y(0).$$

Example 2.3

Obtain the first variation δF and the total variation ΔF of the functional

$$F(f, f_x, f_y) = \int_1^8 \int_1^8 \frac{1}{f} dxdy + \int_0^3 (f_x)^2 dy + \int_0^2 f_y dx + \frac{1}{f(1, 0.5)}$$

where $f = xy$, and the variations of f, f_x, and f_y are $\Delta f = 0.1xy$, $\Delta f_x = 0.2y$, and $\Delta f_y = 0.3x^2$.

Solution: $F(f, f_x, f_y) = \int_1^8 \int_1^8 \frac{1}{xy} dxdy + \int_0^3 (y)^2 dy + \int_0^2 x dx + \frac{1}{f(1, 0.5)}$,

or

$$F = \ln x\Big|_1^8 \cdot \ln y\Big|_1^8 + \frac{y^3}{3}\Big|_0^3 + \frac{x^2}{2}\Big|_0^2 + \frac{1}{0.5} = 4.32 + 9 + 2 + 2 = 17.3$$

Now,

$$F(f + \Delta f, f_x + \Delta f_x, f_y + \Delta f_y) =$$

$$\int_1^8 \int_1^8 \frac{1}{(xy + 0.1xy)} dxdy + \int_0^3 (y + 0.2y)^2 dy + \int_0^2 (x + 0.3x^2) dx + \frac{1}{f(1, 0.5) + \Delta f(1, 0.5)},$$

or

$$F+\Delta F=\frac{1}{1.1}\left(\ln x\big|_1^8\cdot\ln y\big|_1^8\right)+1.44\frac{y^3}{3}\bigg|_0^3+\frac{x^2}{2}\bigg|_0^2+0.3\frac{x^3}{3}\bigg|_0^2+\frac{1}{0.5+0.1(0.5)}$$

$$=3.93+12.96+2+0.8+1.82=21.51.$$

Hence, the total variation $\Delta F = 21.51 - 17.32 = 4.19$.

Now let us calculate the first variation δF,

$$\delta F=\frac{\partial F}{\partial f}\Delta f+\frac{\partial F}{\partial f_x}\Delta f_x+\frac{\partial F}{\partial f_y}\Delta f_y=\int_1^8\int_1^8-\frac{\Delta f}{f^2}dxdy+\int_0^3 2f_x\Delta f_x dy+\int_0^2\Delta f_y dx-\frac{\Delta f(1,0.5)}{\left(f(1,0.5)\right)^2}$$

$$=\int_0^8\int_0^8-\frac{0.1}{xy}dxdy+\int_0^3(2y)(0.2y)dy+\int_0^2 0.3x^2dx-\frac{0.1(0.5)}{0.25}$$

$$=-0.1(4.32)+0.4(9)+0.8-0.2=-0.43+3.6+0.8-0.2=3.77$$

Note that in this example 90% of the total variation ΔF (= 4.19) has been captured by the first variation term δF (= 3.77).

Stationary Values of Functions and Functionals

When a function reaches an extremum (maximum, minimum, or saddle point) then its derivative at that point vanishes. For a function of one independent variable, this is obvious from its plot, Figure 2.6.

Extremum points are also known as stationary points because the function value is stationary at these points. Since the slope $\frac{df}{dx} = 0$ at stationary points the first variation of the function $\delta f = \frac{df}{dx}\cdot\Delta x$ must be zero at those points.

Now let us consider a function of two variables, $f = f(x,y)$. Partial derivatives of this function with respect to x and y must vanish at stationary points. As a result, its first variation $\delta f = \frac{\partial f}{\partial x}\cdot\Delta x + \frac{\partial f}{\partial y}\cdot\Delta y$ must also vanish at stationary points. In the same manner if a functional attains a stationary value then its first variation must be equal to zero.

Example 2.4

What condition must be satisfied for the functional

$$F(f,f_x)=\frac{1}{2}\int_{x_1}^{x_2}Ak(f_x)^2dx-\int_{x_1}^{x_2}qfdx+v_2f_2-v_1f_1$$ to reach a stationary value?

Where v_1 and v_2 are two constants, f_1 and f_2 are values of the function f at $x = x_1$ and x_2, respectively, and A, k, q are functions of x.

Solution: The first variation of the functional must be equal to zero for the functional to reach a stationary value. First variation of this functional when equated to zero gives

$$\delta F = \frac{1}{2}\int_{x_1}^{x_2} 2Ak(f_x)\Delta f_x dx - \int_{x_1}^{x_2} q\Delta f dx + v_2 \cdot \Delta f\big|_{x_2} - v_1 \Delta f\big|_{x_1}$$

$$= \int_{x_1}^{x_2} Ak(f_x)\Delta f_x dx - \int_{x_1}^{x_2} q\Delta f dx + v_2 \cdot \Delta f_2 - v_1 \Delta f_1 = 0$$

where Δf_1 and Δf_2 are values of Δf at $x = x_1$ and x_2, respectively.

Hence, the condition for the above functional to reach a stationary value is

$$\int_{x_1}^{x_2} Ak(f_x)\Delta f_x dx - \int_{x_1}^{x_2} q\Delta f dx + v_2 \cdot \Delta f_2 - v_1 \Delta f_1 = 0$$

Δf_x to Δf Conversion

In the above example, the first integral contains the function Δf_x. We can rewrite that integral in terms of Δf by carrying out the integration by parts as shown below:

$$\int_{x_1}^{x_2} Ak(f_x)\Delta f_x dx = Ak(f_x)\Delta f\big|_{x_1}^{x_2} - \int_{x_1}^{x_2} (Akf_x)_x (\Delta f) dx \tag{2.29}$$

where $(Akf_x)_x$ is the partial derivative of the function Akf_x with respect to x. Note that if A and k are independent of x then $(Akf_x)_x = Akf_{xx}$ where f_{xx} is the double derivative of the function f with respect to x.

Example 2.5

Simplify the solution of Example 2.4 when A and k are independent of x such that the final expression does not contain any variation of f_x.

Solution: Using Equation 2.29 the final equation of Example 2.4

$$\int_{x_1}^{x_2} Ak(f_x)\Delta f_x dx - \int_{x_1}^{x_2} q\Delta f dx + v_2 \cdot \Delta f_2 - v_1 \Delta f_1 = 0$$

can be rewritten in the following form,

$$Ak(f_x)\delta f\Big|_{x_1}^{x_2} - \int_{x_1}^{x_2} (Akf_x)_x(\Delta f)dx - \int_{x_1}^{x_2} q(\Delta f)dx + v_2 \cdot \Delta f_2 - v_1 \cdot \Delta f_1 = 0$$

or

$$(Akf_x + v_2)\Delta f\Big|_{x=x_2} - (Akf_x + v_1)\Delta f\Big|_{x=x_1} - \int_{x_1}^{x_2} Akf_{xx}(\Delta f)dx - \int_{x_1}^{x_2} q(\Delta f)dx = 0$$

or

$$(Akf_x + v_2)\Delta f\Big|_{x=x_2} - (Akf_x + v_1)\Delta f\Big|_{x=x_1} - \int_{x_1}^{x_2} (Akf_{xx} + q)\Delta f dx = 0.$$

For the above equation to be satisfied for arbitrary values of Δf, the following conditions must be satisfied:

1. $Akf_{xx} + q = 0$ between x_1 and x_2
2. $Akf_x + v_2 = 0$ at $x = x_2$
3. $Akf_x + v_1 = 0$ at $x = x_1$

Note that if $Akf_{xx} + q$ is not equal to zero in the entire domain between x_1 and x_2 then Δf can be chosen such that the integral value becomes nonzero. It should be also noted here that if Δf is arbitrary between x_1 and x_2 but zero at the two boundary points x_1 and x_2 then the second and third conditions listed above are not required.

More on the Stationary Value of a Functional — Physical Interpretation

Among different possible expressions of the function f, if the functional $F(f, f_x) = \frac{1}{2}\int_{x_1}^{x_2} Ak(f_x)^2 dx - \int_{x_1}^{x_2} qf dx + v_2 f_2 - v_1 f_1$ has the minimum value for $f = g$ then which equations must be satisfied by g?

From Examples 2.4 and 2.5 one can say that g must satisfy the conditions

1. $Akg_{xx} + q = 0$ between x_1 and x_2
2. $Akg_x + v_2 = 0$ at $x = x_2$
3. $Akg_x + v_1 = 0$ at $x = x_1$

since the functional reaches a stationary value with $f = g$. In other words, if a problem is governed by the equation $Akf_{xx} + q = 0$, and boundary conditions $Akf_x + v_1 = 0$ at $x = x_1$ and $Akf_x + v_2 = 0$ at $x = x_2$, then that problem can be solved by minimizing the functional

$$F(f, f_x) = \frac{1}{2}\int_{x_1}^{x_2} Ak(f_x)^2 dx - \int_{x_1}^{x_2} qf dx + v_2 f_2 - v_1 f_1 .$$

The function g that makes this functional stationary satisfies the governing equation and both the boundary conditions; hence, it must be the solution of the problem.

This gives us an alternative method of solving a problem. In this method it is not necessary to solve a boundary value problem in the traditional way of first solving the differential equations and then evaluating the unknown constants from the boundary conditions. In the new method we need to minimize a functional; this minimization guarantees the satisfaction of governing equations and some boundary conditions.

Natural and Forced Boundary Conditions

Some boundary conditions are automatically satisfied from the minimization of the functional and are called natural boundary conditions. These boundary conditions contain derivatives of the unknown function f. Some boundary conditions cannot be satisfied by simply minimizing the functional and must be specified at the boundary. These boundary conditions do not contain derivatives of the primary unknown and are called forced boundary conditions since these are not naturally obtained from minimizing the functional. Forced boundary conditions are also known as geometric, Dirichlet type, or first kind boundary conditions. Natural boundary conditions are also known as gradient, Neumann type, or second kind boundary conditions. If on a boundary both types of boundary conditions are specified then those boundary conditions are called mixed or third kind boundary conditions.

Example 2.6

The governing equation for one-dimensional fluid flow problem is given by

$$Ak\frac{\partial^2 h}{\partial x^2} + f(x) = 0,$$

where h is the fluid head, A is the cross-sectional area, k is the permeability in the flow direction, and $f(x)$ is the fluid influx per unit length. Fluid heads at the end points of the one-dimensional domain ($x = x_1$ and x_2) are specified as h_1 and h_2, respectively. What functional should be minimized for satisfying this governing equation?

Solution: In this example, the governing equation has been specified and we want to obtain the functional. Hence, this situation is opposite to Examples 2.4 and 2.5 where the functional was given and we derived the governing equations. Also note that the boundary conditions specified in this example are forced boundary conditions, so we do not expect to automatically satisfy them by minimizing the functional.

Comparing the final expression of δF in Example 2.5, and the governing equation of this problem, we can obtain the following expression for δF:

$$\delta F = Akh_x \Delta h\Big|_{x=x_2} - Akh_x \Delta h\Big|_{x=x_1} - \int_{x_1}^{x_2} \left(Akh_{xx} + f\right)\Delta h dx$$

In the final expression of Example 2.5, f has been replaced by h, q has been replaced by f, and terms containing v_1 and v_2 have been dropped to obtain the above expression. If δF is equated to zero for arbitrary value of Δh between x_1 and x_2, then $Akh_{xx} + f = 0$ must be satisfied between x_1 and x_2.

Since h is specified at $x = x_1$ and x_2, Δh must be zero at the two boundaries. Hence, the first two terms of the right-hand side of the above equation should vanish, and if the governing equation is satisfied then the third term also should vanish. Now our task is to obtain the functional expression F from its first variation δF. Note that δF can be rewritten as

$$\delta F = Akh_x \Delta h\Big|_{x_1}^{x_2} - \int_{x_1}^{x_2} Akh_{xx} \Delta h dx - \int_{x_1}^{x_2} f \Delta h dx = \int_{x_1}^{x_2} Akh_x \Delta h_x dx - \int_{x_1}^{x_2} f \Delta h dx$$

or

$$F\left(h, h_x\right) = \frac{1}{2}\int_{x_1}^{x_2} Akh_x^2 dx - \int_{x_1}^{x_2} fh dx.$$

Hence, by minimizing the functional, $F(h, h_x) = \frac{1}{2}\int_{x_1}^{x_2} Akh_x^2 dx - \int_{x_1}^{x_2} fh dx$, one can guarantee the satisfaction of the given governing equation.

Example 2.7

For the one-dimensional fluid flow problem, in Example 2.6, if the forced boundary conditions at $x = x_1$ and $x = x_2$ are replaced by natural boundary conditions, $Akh_x = -v_1$ at $x = x_1$, and $Akh_x = -v_2$ at $x = x_2$ then what should be the corresponding functional expression?

Solution: Comparing the final expression of δF in Example 2.5, and the governing equation and natural boundary conditions of this problem we can derive the following expression for δF.

$$\delta F = \left(Akh_x + v_2\right)\Delta h\Big|_{x=x_2} - \left(Akh_x + v_1\right)\Delta h\Big|_{x=x_1} - \int_{x_1}^{x_2} \left(Akh_{xx} + f\right)\Delta h dx.$$

If δF is equated to zero for arbitrary values of Δh at the boundary points, $x = x_1$ and x_2 as well as for x between x_1 and x_2, then $Akh_x = -v_1$ at x_1, $Akh_x = -v_2$ at x_2 and $Akh_{xx} + f = 0$ between x_1 and x_2.

Now our task is to obtain the functional expression F from its first variation δF. Note that δF can be rewritten as

$$\delta F = v_2 \Delta h\Big|_{x_2} - v_1 \Delta h\Big|_{x_1} + Akh_x \Delta h\Big|_{x_1}^{x_2} - \int_{x_1}^{x_2} Akh_{xx} \Delta h dx - \int_{x_1}^{x_2} f \Delta h dx$$

$$= \int_{x_1}^{x_2} Akh_x \Delta h_x dx - \int_{x_1}^{x_2} f \Delta h dx + v_2 \Delta h\Big|_{x_2} - v_1 \Delta h\Big|_{x_1}$$

or

$$F(h, h_x) = \frac{1}{2}\int_{x_1}^{x_2} Akh_x^2 dx - \int_{x_1}^{x_2} fh dx + v_2 h(x_2) - v_1 h(x_1).$$

The only difference between the functional expressions of Examples 2.6 and 2.7 is that the last two terms of the functional expression of Example 2.7 are missing in Example 2.6.

Two-Dimensional Problems

For two-dimensional fluid flow, heat flow, or stress-deformation problems the functional expression contains area integrals and line integrals. The area integrals are associated with the governing equation in the two-dimensional domain (an area) of the problem and the line integrals are associated with the boundary (line) of the problem.

Example 2.8

Evaluate the first variation of the following functional

$$F(f, f_x, f_y) = \frac{1}{2}\iint_A \left[f_x^2 + f_y^2\right] dA - \iint_A Q \cdot f \cdot dA - \oint_S q \cdot f \cdot dS$$

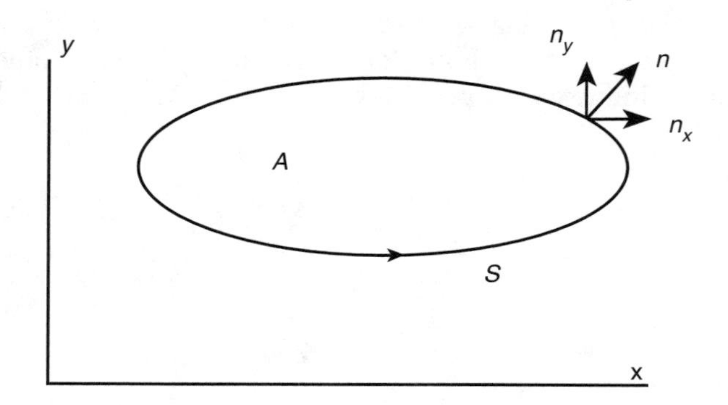

FIGURE 2.13
Area A and boundary S of a two-dimensional domain. Outward normal n has two components, n_x and n_y.

where f is a function of x and y, and f_x and f_y are partial derivatives of f with respect to x and y respectively. Q and q are two functions of x and y. Area and boundary of the two dimensional domain are defined by A and S, respectively, see Figure 2.13.

Solution:

$$\delta F = \frac{\partial F}{\partial f_x}\Delta f_x + \frac{\partial F}{\partial f_y}\Delta f_y + \frac{\partial F}{\partial f}\Delta f$$

$$= \frac{1}{2}\iint_A \left[2f_x\Delta f_x + 2f_y\Delta f_y\right]dA - \iint_A Q\cdot\Delta f\cdot dA - \oint_S q\cdot\Delta f\cdot dS.$$

One can simplify the above expression further in the following manner

$$\left\{ \begin{aligned} f_x\cdot\Delta f_x &= \left(f_x\cdot\Delta f_x + f_{xx}\cdot\Delta f\right) - f_{xx}\cdot\Delta f \\ &= \frac{\partial\left(f_x\cdot\Delta f\right)}{\partial x} - f_{xx}\cdot\Delta f = \left(f_x\cdot\Delta f\right)_x - f_{xx}\cdot\Delta f. \end{aligned} \right\}$$

In the above expressions subscript x indicates partial derivatives with respect to x. Then

$$\delta F = \iint_A \left[\left(f_x\Delta f\right)_x - f_{xx}\Delta f + \left(f_y\Delta f\right)_y - f_{yy}\Delta f\right]dA - \iint_A Q\cdot\Delta f\cdot dA - \oint_S q\cdot\Delta f\cdot dS$$

$$= \iint_A \left[\left(f_x\Delta f\right)_x + \left(f_y\Delta f\right)_y\right]dA - \iint_A \left(f_{xx} + f_{yy} + Q\right)\cdot\Delta f\cdot dA - \oint_S q\cdot\Delta f\cdot dS.$$

Now let us apply the Green's theorem to the first integral of the above expression. Note that the Green's theorem in two dimensions is given by

$$\iint_A \frac{\partial f}{\partial x} dA = \oint_S f \cdot n_x \cdot dS$$

$$\iint_A \frac{\partial f}{\partial y} dA = \oint_S f \cdot n_y \cdot dS$$

where n_x and n_y are x and y components of the unit normal vector n at the boundary, Figure 2.13.

Applying Green's theorem,

$$\begin{aligned}\delta F &= \oint_S \left(f_x n_x + f_y n_y\right)\Delta f \cdot dS - \iint_A \left(f_{xx} + f_{yy} + Q\right) \cdot \Delta f \cdot dA - \oint_S q \cdot \Delta f \cdot dS \\ &= \oint_S \left(f_x n_x + f_y n_y - q\right)\Delta f \cdot dS - \iint_A \left(f_{xx} + f_{yy} + Q\right) \cdot \Delta f \cdot dA\end{aligned}$$

Example 2.9

What governing equation and boundary conditions are satisfied by minimizing the following functional?

$$F\left(f, f_x, f_y\right) = \frac{1}{2}\iint_A \left[f_x^2 + f_y^2\right] dA - \iint_A Q \cdot f \cdot dA - \oint_S q \cdot f \cdot dS.$$

Solution: Minimizing the above functional or equating its first variation (see Example 2.8) to zero gives rise to the following equation.

$$\delta F = \oint_S \left(f_x n_x + f_y n_y - q\right)\Delta f \cdot dS - \iint_A \left(f_{xx} + f_{yy} + Q\right) \cdot \Delta f \cdot dA = 0.$$

If the above equation has to be satisfied for arbitrary Δf then the expressions $(f_x n_x + f_y n_y - q)$ and $(f_{xx} + f_{yy} + Q)$ of the boundary and area integrals must individually vanish. Hence, in the two-dimensional domain A, $f_{xx} + f_{yy} + Q = 0$; this is the governing equation of the problem. Along the boundary S, $f_x n_x + f_y n_y - q = 0$; this is the natural boundary condition.

Example 2.10

Obtain the governing equation and natural boundary conditions by minimizing the following functional:

$$F(T,T_x,T_y)=\frac{1}{2}\iint_A\left[T_x^2+T_xT_y+QT\right]dA+\frac{1}{3}\oint_{S_1}T^2dS+\oint_{S_2}TT_0dS$$

where Q and T_0 are known functions. S_1 and S_2 are the two parts of the boundary S, or $S = S_1 \cup S_2$ (in other words, S is the summation of S_1 and S_2, or S_1 union S_2).

Solution:

$$\delta F=\frac{1}{2}\iint_A\left[2T_x\cdot\Delta T_x+T_x\cdot\Delta T_y+\Delta T_x\cdot T_Y+2Q\cdot\Delta T\right]dA+\frac{1}{3}\oint_{S_1}2T\cdot\Delta T\cdot dS+\oint_{S_2}\Delta T\cdot T_0dS$$

$$=\frac{1}{2}\iint_A\left[2(T_x\cdot\Delta T)_x-2T_{xx}\Delta T+(T_x\cdot\Delta T)_y-T_{xy}\cdot\Delta T+(T_y\cdot\Delta T)_x-T_{xy}\cdot\Delta T+2Q\cdot\Delta T\right]dA$$

$$+\frac{1}{3}\oint_{S_1}2T\cdot\Delta T\cdot dS+\oint_{S_2}T_0\cdot\Delta T\cdot dS$$

$$=\frac{1}{2}\iint_A\left[2(T_x\cdot\Delta T)_x+(T_x\cdot\Delta T)_y+(T_y\cdot\Delta T)_x\right]dA-\frac{1}{2}\iint\left[2T_{xx}+2T_{xy}-2Q\right]\cdot\Delta T\cdot dA$$

$$+\frac{1}{3}\oint_{S_1}2T\cdot\Delta T\cdot dS+\oint_{S_2}T_0\cdot\Delta T\cdot dS.$$

Applying the two-dimensional Green's theorem to the first integral we obtain

$$\delta F=\frac{1}{2}\oint_S\left[2T_x\cdot n_x+T_x\cdot n_y+T_y\cdot n_x\right]\cdot\Delta T\cdot dS$$

$$+\frac{1}{3}\oint_{S_1}2T\cdot\Delta T\cdot dS+\oint_{S_2}T_0\cdot\Delta T\cdot dS-\iint_A\left[T_{xx}+T_{xy}-Q\right]\cdot\Delta T\cdot dA$$

$$=\oint_S\left[T_x\cdot n_x+\frac{1}{2}T_x\cdot n_y+\frac{1}{2}T_y\cdot n_x\right]\cdot\Delta T\cdot dS$$

$$+\frac{1}{3}\oint_{S_1}2T\cdot\Delta T\cdot dS+\oint_{S_2}T_0\cdot\Delta T\cdot dS-\iint_A\left[T_{xx}+T_{xy}-Q\right]\cdot\Delta T\cdot dA$$

$$=\oint_{S_1}\left[T_x\cdot n_x+\frac{1}{2}T_x\cdot n_y+\frac{1}{2}T_y\cdot n_x+\frac{2T}{3}\right]\cdot\Delta T\cdot dS$$

$$+\oint_{S_2}\left[T_x\cdot n_x+\frac{1}{2}T_x\cdot n_y+\frac{1}{2}T_y\cdot n_x+T_0\right]\cdot\Delta T\cdot dS$$

$$-\iint_A\left[T_{xx}+T_{xy}-Q\right]\cdot\Delta T\cdot dA$$

If this first variation is equated to zero for arbitrary ΔT then the following conditions must be satisfied,

In domain A (governing equation): $T_{xx} + T_{yy} - Q = 0$

Along boundary S_1 (natural boundary condition):

$$T_x \cdot n_x + \frac{1}{2}\left(T_x \cdot n_y + T_y \cdot n_x\right) + \frac{2T}{3} = 0$$

Along boundary S_2 (natural boundary condition):

$$T_x \cdot n_x + \frac{1}{2}\left(T_x \cdot n_y + T_y \cdot n_x\right) + T_o = 0$$

Example 2.11

Find a functional $F(f, f_x, f_y)$ such that minimizing that functional will guarantee the satisfaction of the following governing equation and boundary conditions.

Governing Equation:

In domain A, $f_{xx} + f_{yy} - Q = 0$

Boundary Conditions:

Along boundary S_1, $f_x \cdot n_x + f_y \cdot n_y + q = 0$

Along the rest of the boundary (S_2), $f = f_0$.

Solution: Note that in this problem natural and forced boundary conditions have been specified along boundaries S_1 and S_2, respectively. The natural boundary conditions should be obtained from $\delta F = 0$ equation; however, the forced boundary condition will only be obtained in the form $\Delta f = 0$ along S_2. This problem is, to some extent, an inverse problem of Example 2.9.

We can satisfy the governing equation and boundary conditions by equating the following δF to zero.

$$\delta F = \oint_{S_1}\left(f_x n_x + f_y n_y + q\right)\Delta f \cdot dS + \oint_{S_2}\left(f_x n_x + f_y n_y\right)\Delta f \cdot dS - \iint_A \left(f_{xx} + f_{yy} - Q\right)\cdot \Delta f \cdot dA,$$

or

$$\delta F = \oint_{S_1} q\Delta f \cdot dS + \oint_{S}\left(f_x n_x + f_y n_y\right)\Delta f \cdot dS - \iint_A \left(f_{xx} + f_{yy} - Q\right)\cdot \Delta f \cdot dA,$$

or

$$\delta F = \oint_{S_1} q\Delta f \cdot dS + \iint_A \left[(f_x \cdot \Delta f)_x + (f_y \cdot \Delta f)_y \right] dA - \iint_A (f_{xx} + f_{yy} - Q) \cdot \Delta f \cdot dA,$$

or

$$\delta F = \oint_{S_1} q\Delta f \cdot dS + \iint_A \left[f_{xx} \cdot \Delta f + f_x \cdot \Delta f_x + f_{yy} \Delta f + f_y \cdot \Delta f_y \right] dA - \iint_A (f_{xx} + f_{yy} - Q) \cdot \Delta f \cdot dA$$

$$= \oint_{S_1} q\Delta f \cdot dS + \iint_A \left[f_x \cdot \Delta f_x + f_y \cdot \Delta f_y + Q \cdot \Delta f \right] dA;$$

hence,

$$F(f, f_x, f_y) = \oint_{S_1} q \cdot f \cdot dS + \frac{1}{2} \iint_A \left[f_x^2 + f_y^2 + 2Q \cdot f \right] dA$$

Example 2.12

Obtain the appropriate functional $F(T, T_x)$ such that minimizing that functional will guarantee the satisfaction of the following governing equation and boundary conditions.

Governing equation, $T_{xx} - T_t + q = 0$, for x between x_1 and x_2.

At boundary $x = x_1$, $T_x = Q_1$ and at boundary $x = x_2$, $T_x = Q_2$.

In the above equations subscripts x and t indicate partial derivatives with respect to x and t respectively. Treat T and T_x as two independent functions, T_t and q as some known functions, and Q_1, Q_2 are two constants.

Solution: To satisfy the given governing equation and natural boundary conditions we need to equate the following δF to zero for arbitrary ΔT.

$$\delta F = -\int_{x_1}^{x_2} [T_{xx} - T_t + q] \Delta T \cdot dx + (T_x - Q_2) \Delta T\Big|_{x_2} - (T_x - Q_1) \Delta T\Big|_{x_1}.$$

One can rewrite the above δF in the following form,

$$\delta F = T_x \cdot \Delta T\big|_{x_2} - T_x \cdot \Delta T\big|_{x_1} - \int_{x_1}^{x_2} T_{xx} \cdot \Delta T \cdot dx + \int_{x_1}^{x_2} (T_t - q)\Delta T \cdot dx - Q_2 \cdot \Delta T\big|_{x_2} + Q_1 \cdot \Delta T\big|_{x_1}$$

$$= T_x \cdot \Delta T\big|_{x_1}^{x_2} - \int_{x_1}^{x_2} T_{xx} \cdot \Delta T \cdot dx + \int_{x_1}^{x_2} (T_t - q)\Delta T \cdot dx - Q_2 \cdot \Delta T\big|_{x_2} + Q_1 \cdot \Delta T\big|_{x_1}$$

$$= \int_{x_1}^{x_2} T_x \cdot \Delta T_x \cdot dx + \int_{x_1}^{x_2} (T_t - q)\Delta T \cdot dx - Q_2 \cdot \Delta T\big|_{x_2} + Q_1 \cdot \Delta T\big|_{x_1}.$$

Hence,

$$F(T, T_x) = \frac{1}{2}\int_{x_1}^{x_2} T_x^2 \cdot dx + \int_{x_1}^{x_2} (T_t - q)T \cdot dx - Q_2 \cdot T\big|_{x_2} + Q_1 \cdot T\big|_{x_1}$$

or

$$F(T, T_x) = \int_{x_1}^{x_2} \left(\frac{T_x^2}{2} + T_t T - qT \right) dx - Q_2 \cdot T\big|_{x_2} + Q_1 \cdot T\big|_{x_1}.$$

Summary

Eight steps of finite element analysis and the basic knowledge of variational calculus are presented here. Different terms, such as primary unknown, secondary unknown, element equation, vector of nodal unknowns, vector of nodal forcing parameters, functional, its total and first variations, have been explained. It is shown how governing equations and natural boundary conditions can be obtained by minimizing a functional or, in other words, equating the first variation of the functional to zero. In the example problems several governing equations and natural boundary conditions are obtained from different functionals. A number of functionals are also obtained in these examples from the governing equations and natural boundary conditions for both one-dimensional and two-dimensional problems.

Problems

2.1. Consider the functional $F = \int_0^5 f^2 dx + \int_1^3 g dx + h(5) - h(0)$ and let f, g, and h be $x^{1/2}$, x, and x^2, respectively. Obtain the value of the functional. Then give the variations $0.1x$, $1/x$, and x to the functions f, g, and h, respectively. Calculate the total variation ΔF and the first variation δF and compare these two values.

2.2. Obtain the first variation δF and the total variation ΔF of the functional

$$F(f, f_x, f_y) = \int_1^8 \int_1^8 \frac{1}{f} dx dy + \int_0^3 (f_x)^2 dy + \int_0^2 f_y dx + \frac{1}{f(1,0.5)}$$

where $f = xy$, and the variations of f, f_x, and f_y are $\Delta f = 0.2xy$, $\Delta f_x = 0.1y^2$, and $\Delta f_y = 0.2x$.

2.3. For problems 2.1 and 2.2, calculate the second variation term $\delta^2 F$ (next higher order term in the Taylor's series expansion), then investigate how closely the total variation (ΔF) matches with the summation of first and second variations ($\delta F + \delta^2 F$).

2.4. Obtain the governing equation and natural boundary conditions by minimizing the following functional

$$F(v, v_y) = \frac{AE}{2} \int_{y_1}^{y_2} (v_y)^2 dy - A \int_{y_1}^{y_2} Y \cdot v \cdot dy - \int_{y_1}^{y_2} T \cdot v \cdot dy - P_1 v\big|_{y_1} - P_2 v\big|_{y_2}$$

where v_y is the derivative of v with respect to y. P_1 and P_2 are two constants; Y and T are known functions of y.

2.5. Obtain the governing equation and natural boundary conditions by minimizing the following functional

$$F(T, T_x) = A \int_{x_1}^{x_2} \left[\frac{1}{2} \alpha (T_x)^2 + T \cdot T_t + q \cdot T \right] dx$$

where T_x and T_t are derivatives of T with respect to x and t respectively. α is a constant and q is a known function of x.

2.6. Obtain the governing equation that corresponds to the following functional

$$\Omega(w, w_{xx}) = \int_{x_1}^{x_2} \left[\frac{1}{2} F \cdot (w_{xx})^2 - p \cdot w \right] dx$$

where w_{xx} is the second derivative of w with respect to x. F and p are two known functions of x.

2.7. Find a functional $F(v, v_t)$ such that minimizing this functional gives rise to the following differential equation and boundary conditions.

$$\begin{aligned} &m \cdot v_{tt} + kv = 0 && t_1 \le t \le t_2 \\ &v = 0 && \text{for } t = t_1 \\ &v_t = v_0 && \text{for } t = t_2 \end{aligned}$$

where v_t and v_{tt} are single and double derivatives of v with respect to t.

2.8. Obtain the governing equation and natural boundary conditions by minimizing the following functional

$$F(\psi_x, \psi_y) = \frac{hG}{2} \iint_A \left[(\psi_x - y)^2 + (\psi_y + x)^2 \right] \theta^2 dA$$

where ψ_x and ψ_y are derivatives of ψ with respect to x and y respectively. h, G are constants and θ is a known function of x and y.

2.9. Obtain the governing equation and natural boundary conditions by minimizing the following functional

$$F(\psi, \psi_x, \psi_y) = \iint_A \frac{k}{2} \left[(\psi_x)^2 + (\psi_y)^2 \right] dA - \iint_A Q \cdot \psi \cdot dA - \oint_S q \cdot \psi \cdot dS$$

where ψ_x and ψ_y are derivatives of ψ with respect to x and y respectively. k is a constant, and Q and q are two known functions of x and y. S is the boundary of domain A.

2.10. Obtain a functional $F(T, T_x, T_y)$ such that minimizing that functional should give rise to the following differential equations and boundary conditions.

$$\begin{aligned} &k_x T_{xx} + k_y T_{yy} + Q = 0 && \text{in two-dimensiona domain } A \\ &T = T_1 && \text{on boundary } S_1 \\ &k_x T_x n_x + k_y T_y n_y + \alpha(T - T_0) + q = 0 && \text{on boundary } S_2 \end{aligned}$$

where x and y subscripts of T indicate derivatives of T with respect to x and y, respectively, and x and y subscripts of k and n indicate x and y components of k and n, respectively. α, Q, q, T_1 and T_0 are different constants. S_1 and S_2 are two parts of the boundary S that surrounds the domain A. n is the outward normal vector on boundary S, and n_x and n_y are two components of n.

2.11. Obtain the governing equation and natural boundary conditions corresponding to the following functional.

$$\Omega = \iint_A \left[\frac{1}{2}(\phi_x)^2 + \frac{1}{2}(\phi_y)^2 + \phi^3 + \phi^2 + c\phi\right] dA + \frac{1}{2}\int_S (\phi_0 - \phi)^2 \, dS$$

where, c and ϕ_0 are known functions. A and S are enclosed area and boundary, respectively, of the two-dimensional problem domain.

2.12. Consider the functional

$$\Omega(\phi, \phi_x) = \int_0^1 (\phi_x^2 - 4\phi) \, dx$$

where ϕ is a function of x, ϕ_x is the derivative of ϕ with respect to x.

(a) Obtain the value of this functional for $\phi = x^2$.

(b) Evaluate $\Omega(\phi + \delta\phi, \phi_x + \delta\phi_x)$ for $\delta\phi = 0.05x$, $\delta\phi_x = 0.1x$ and $\phi = x^2$.

(c) Evaluate $\delta\Omega$ for above ϕ, $\delta\phi$, and $\delta\phi_x$, where $\delta\Omega$ is the first variation of Ω.

2.13. Construct the governing equation and natural boundary conditions associated with the following functional,

$$\Omega = \frac{1}{3}\iint_A \left[\phi_x^3 + \phi_y^3 - 3Q\phi\right] dA - \int_S q\phi \, dS$$

where ϕ is a function of x and y, ϕ_x and ϕ_y are derivatives of ϕ with respect to x and y, respectively, and q is a known function of x and y.

References

1. Turner, M. J., Clough, R. W., Martin, H. C., and Topp, L. C., 1956. "Stiffness and deflection analysis of complex structures," *J. Aero. Sci.*, 23, 9, 805-823.
2. Clough, R. W., 1960. "The finite element method in plane stress analysis," in *Proc. 2nd Conf. on Electronic Computation*, ASCE, Pittsburgh.

3. Crandall, S. H., 1956. *Engineering Analysis,* McGraw-Hill, New York.
4. Argyris, J. H., 1960. *Energy Theorems and Structural Analysis,* Butterworth's, London.
5. Desai, C. S. and Abel, J. F., 1972. *Introduction to the Finite Element Method,* Van Nostrand Reinhold, New York.
6. Pian, T. H. H. and Tong, P., 1972. "Finite element methods in continuum mechanics," in *Advances in Applied Mechanics,* 12, Academic Press, New York.
7. Finlayson, B. A., 1972. *The Method of Weighted Residuals and Variational Principles,* Academic Press, New York.
8. Fung, Y.C., *Foundations of Solid Mechanics,* Prentice-Hall, Englewood Cliffs, NJ.
9. Bliss, G. A., 1944. *Calculus of Variations,* The Open Court Publishing Company, LaSalle, IL.
10. Gelfand, I. M. and Fomin, S. V., 1963. *Calculus of Variations,* Prentice-Hall, Englewood Cliffs, NJ.
11. Lanczos, C., *The Variational Principles of Mechanics*, 1949. University Press, Toronto.
12. Pars, L., 1962. *An Introduction to the Calculus of Variations,* John Wiley & Sons, New York.

3

One-Dimensional Stress Deformation

Introduction

From here through Chapter 9 we shall consider engineering problems idealized as one-dimensional. The main motive for treating these simple problems is to introduce the reader to the details of various steps so that basic concepts can be understood and assimilated thoroughly without undue complex and lengthy derivations. An advantage to this approach is that hand calculations can be performed; for two- and three-dimensional problems this can become increasingly difficult.

Although simple problems are treated (Chapters 3–9), we introduce many concepts and terms that are general and relevant to advanced theory and applications. These concepts are explained and defined in easy terms, often with intuitive and physical explanations. It may be mentioned that one-dimensional idealizations permit not only simple derivations but often provide satisfactory solutions for many practical problems.

As our first problem we consider the case of a column, strut, or bar of uniform cross section subjected to purely axial loading (Figure 3.1(a)). Under these conditions we can assume that the deformations will occur only in one, vertical direction. Consequently, we can further assume that the column can be replaced by a line with the axial stiffness EA lumped at the centerline (Figure 3.1(b)). Now we consider derivations of the finite element method in the step-by-step procedure as described in Chapter 2.

Step 1. Discretization and Choice of Element Configuration

Before we proceed further it is necessary to describe the coordinates or geometry of the column by using a convenient coordinate system. In the one-dimensional approximation, it is necessary to use only one coordinate along the vertical direction. We call this the y-axis. Because this coordinate system is defined to describe the entire column (or structure), it can be called the global coordinate system.

We now discretize the column into an arbitrary number of smaller units that are called finite elements (Figure 3.l(c)). The intersections of elements are called nodes or nodal points.

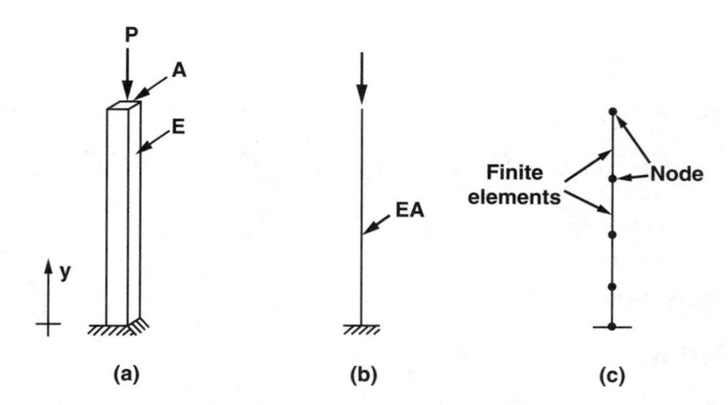

FIGURE 3.1
Axially loaded column: (a) Actual column, (b) One-dimensional idealization, and (c) Discretization.

At this stage, before Step 2, it is useful to introduce the concept of a local or element coordinate system. There are a number of advantages to using a local system for deriving element equations (Steps 2-4); its use, particularly for multidimensional problems, makes the required derivatives and integrations extremely simple to handle. Indeed, it is possible to obtain all derivations by using the global system, but use of the local system (as we shall see subsequently) facilitates the derivatives and is economical.

Explanation of Global and Local Coordinates

For a simple explanation, let us consider an example, Figure 3.2(a); here we need to define or survey a plot of land and relate it to a standard or global point or benchmark *A*. Surveying the plot could entail locating each point in it and establishing its distance from point *A*. Let us assume that point *A* is far away from the plot and that it is difficult to establish a direct relation with *A*. A local point *B* is available, and its distance from *A* is known, however. Point *B* is accessible from all points in the plot. We can first define the distances of each point in the plot from the point *B*; then, knowing the distance $\mathbf{x}_{ab}$ from point *A* to point *B*, it is possible to define the distances of the points in the plot with respect to point *A*. For example, if the distance of any point *P* from *B* is $\mathbf{x}_b$, then its distance from point *A* is

$$\mathbf{x}_a = \mathbf{x}_b + \mathbf{x}_{ab}. \tag{3.1a}$$

Here we can call the measurements with respect to point *B* local coordinates, and those from point *A*, global coordinates. We can see that the measurements with respect to point *B*, which is in the vicinity, make the definition

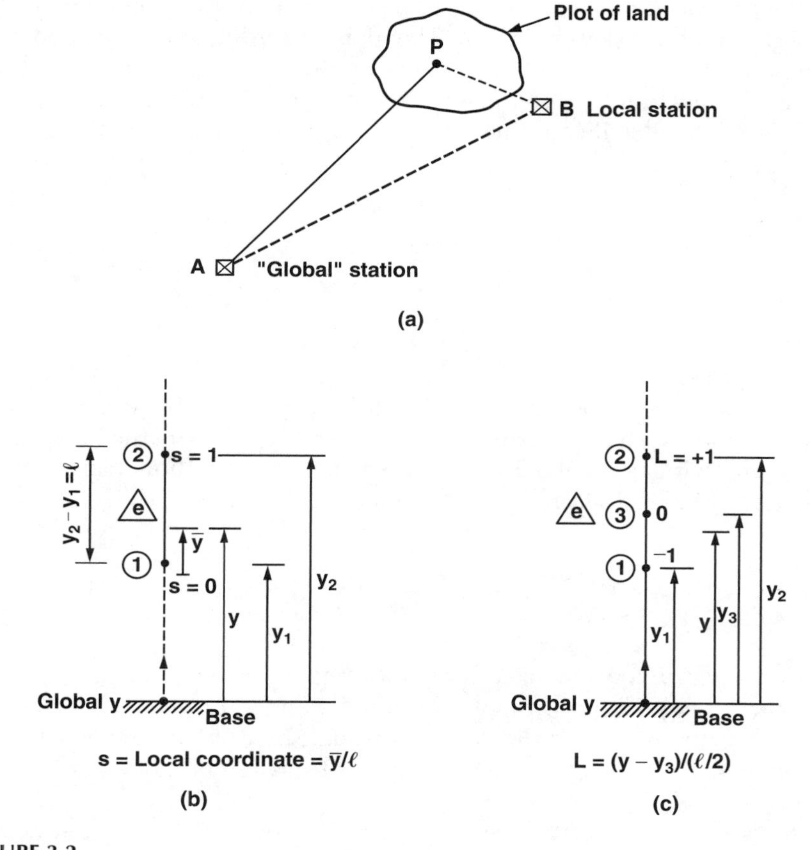

FIGURE 3.2
Global and local coordinates: (a) Concept of global and local coordinate systems, (b) Local coordinate measured from node point 1, and (c) Local coordinate measured from midnode 3.

of the plot much simpler than those with respect to point *A*, which is more difficult to reach. There can be a number of possible points like *B* with respect to which we can define the local coordinate system; such choices will depend on the nature of the problem and the convenience and ease of measurements.

The basic idea of the use of local coordinate systems in the finite element method is very similar to the foregoing concept.

Local and Global Coordinate System for the One-Dimensional Problem

As mentioned above, there are a number of ways in which we can define local coordinate systems for the one-dimensional problem (Figure 3.1). We consider here two local coordinate systems, Figures 3.2(b) and (c).

In the first case, we measure the local coordinate from the node point 1 of a generic element e (Figure 3.2(b)). The global coordinate is measured from the base of the column. Note that node 1 is analogous to point B and that the base is analogous to point A in Figure 3.2(a). We call the local coordinate $\bar{y}$; then the global coordinate of any point in the element is

$$y = \bar{y} + y_1; \tag{3.1b}$$

hence

$$\bar{y} = y - y_1. \tag{3.1c}$$

Often it is more convenient to express the local coordinate as a nondimensional number; such a procedure can considerably facilitate the integrations and differentiations involved in the subsequent computation. Here we nondimensionalize by dividing $\bar{y}$ by the length of the element; thus

$$s = \frac{y - y_1}{y_2 - y_1} = \frac{\bar{y}}{l}, \tag{3.1d}$$

where s = nondimensionalized local coordinate, l = length of the element, and y_1 and y_2 = global coordinates of nodes 1 and 2, respectively. Note that because we nondimensionalized the coordinate by dividing by the length of the element, $l = y_2 - y_1$, the value of s varies from zero at node 1 to unity at node 2.

In the second alternative, we can attach the origin of the local system at an intermediate point in the element, say, at the midpoint (Figure 3.2(c)). Here the local coordinate is written as

$$L = \frac{y - y_3}{l/2}. \tag{3.2}$$

The values of L range from –1 at point 1 to 0 at point 3 to 1 at point 2.

An important property of these local coordinates is that they are in nondimensionalized form, and their values are expressed as numbers and often lie between zero and unity. It is this property that imparts simplicity to the subsequent derivations.

Step 2. Select Approximation Model or Function for the Unknown (Displacement)

One of the main ideas in the finite element method that the reader should grasp at this stage is the *a priori* or in advance selection of mathematical functions to represent the deformed shape of an element under loading. This implies that since it is difficult to find a closed form or exact solution, we guess a solution shape or distribution of displacement by using an appropriate mathematical function. In choosing this function, we must follow the laws, principles, and constraints or boundary conditions inherent in the problem.

The most common functions used are polynomials. In the initial stages of the finite element method, the polynomials used were expressed in terms of generalized coordinates; however, now most finite element work is done by using interpolation functions, which can often be considered as transformed generalized coordinate functions.

Generalized Coordinates

The simplest polynomial that we can use is the one that gives linear variation of displacements within the element (Figure 3.3(b)),

$$v = \alpha_1 + \alpha_2 y, \tag{3.3a}$$

or in matrix notation,

$$\{\mathbf{v}\} = [1 \quad y]\begin{Bmatrix}\alpha_1 \\ \alpha_2\end{Bmatrix} \tag{3.3b}$$

or

$$\{\mathbf{v}\} = [\phi]\{\alpha\}, \tag{3.3c}$$

where α_1 and α_2 = generalized coordinates, y = the coordinate of any point in the element, and v = displacement at any point in the element. The α's contain the displacements at nodes v_1 and v_2 and the coordinates of nodes y_1 and y_2. To show this, we first evaluate v for points 1 and 2 by substituting for y,

$$v_1 = \alpha_1 + \alpha_2 y_1, \tag{3.4a}$$

$$v_2 = \alpha_1 + \alpha_2 y_2, \tag{3.4b}$$

or in matrix notation,

$$\begin{Bmatrix}v_1 \\ v_2\end{Bmatrix} = \begin{bmatrix}1 & y_1 \\ 1 & y_2\end{bmatrix}\begin{Bmatrix}\alpha_1 \\ \alpha_2\end{Bmatrix} \tag{3.4c}$$

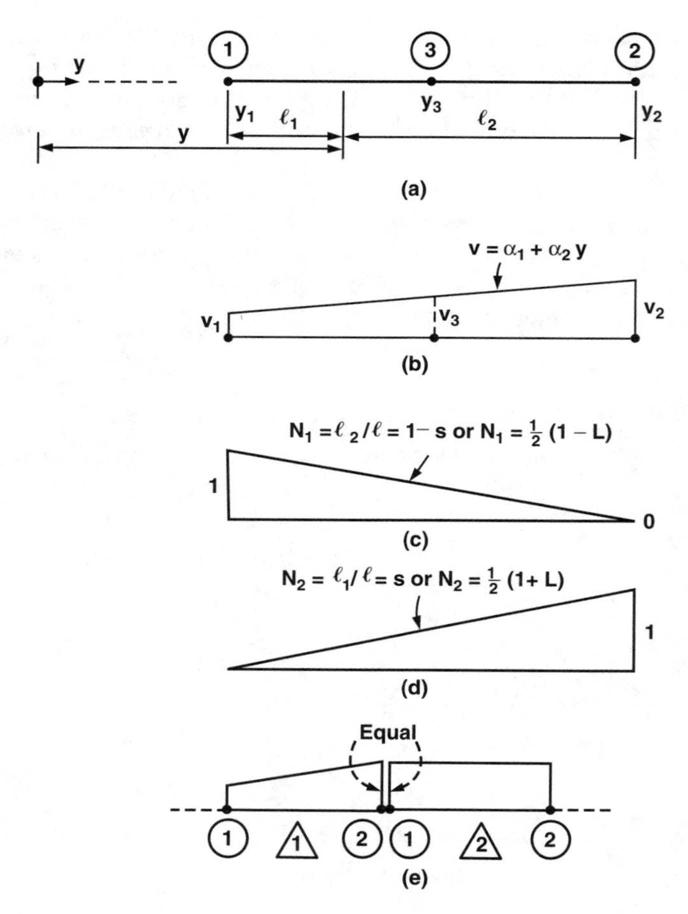

FIGURE 3.3
Linear interpolation functions and interelement compatibility.

or

$$\{\mathbf{q}\} = [\mathbf{A}]\{\alpha\}. \tag{3.4d}$$

Here $\{\mathbf{q}\}^T = [v_1\ v_2]$ is the vector of nodal displacements. Second, we solve for $\{\alpha\}$ as

$$\{\alpha\} = [\mathbf{A}]^{-1}\{\mathbf{q}\}, \tag{3.5a}$$

where

$$[\mathbf{A}]^{-1} = \frac{1}{|J|}\begin{bmatrix} y_2 & -y_1 \\ -1 & 1 \end{bmatrix} \tag{3.5b}$$

$$= \frac{1}{y_2 - y_1} \begin{bmatrix} y_2 & -y_1 \\ -1 & 1 \end{bmatrix}. \tag{3.5c}$$

Here $|J|$ = Jacobian determinant $= y_2 - y_1 = l$ equals the length of the element. Thus

$$\underset{(2\times 1)}{\begin{Bmatrix} \alpha_1 \\ \alpha_2 \end{Bmatrix}} = \frac{1}{l} \underset{(2\times 2)}{\begin{bmatrix} y_2 & -y_1 \\ -1 & 1 \end{bmatrix}} \underset{(2\times 1)}{\begin{Bmatrix} v_1 \\ v_2 \end{Bmatrix}} \tag{3.5d}$$

Therefore

$$\alpha_1 = \frac{y_2 v_1 - y_1 v_2}{l}, \qquad \alpha_2 = \frac{-v_1 + v_2}{l}, \tag{3.5e}$$

which shows that α's are functions of and made up of $y_1, y_2, v_1,$ and v_2. Note that α's are related to but are not explicit functions of nodal displacements. This is one of the reasons we call them generalized coordinates.

Now we can substitute $\{\alpha\}$ into Equation 3.3 to express v in terms of the nodal displacements:

$$v = \frac{y_2 v_1 - y_1 v_2}{l} + \left(\frac{-v_1 + v_2}{l} \right) y \tag{3.6a}$$

$$= \left(\frac{y_2 - y}{l} \right) v_1 - \left(\frac{y_1 - y}{l} \right) v_2 \tag{3.6b}$$

$$= \frac{l_2}{l} v_1 + \frac{l_1}{l} v_2 = \left(1 - \frac{\bar{y}}{l} \right) v_1 + \frac{\bar{y}}{l} v_2 \tag{3.6c}$$

$$= N_1 v_1 + N_2 v_2, \tag{3.6d}$$

where $N_1 = 1 - \bar{y}/l = 1 - s$, $N_2 = \bar{y}/l = s$, and $\bar{y} = l_1$ (Figures 3.2(b) and 3.3). Equation 3.6d leads us to the concept of interpolation function models. In this equation, N_1 and N_2 are called interpolation, shape, or basis functions. The displacement v at any point in the element can now be expressed as

$$v = \begin{bmatrix} N_1 & N_2 \end{bmatrix} \begin{Bmatrix} v_1 \\ v_2 \end{Bmatrix} \tag{3.7a}$$

$$= [\mathbf{N}]\{q\}, \tag{3.7b}$$

where $[\mathbf{N}]$ is called the matrix of interpolation, shape, or basis functions. A property of the interpolation functions is that their sum equals unity. For example, in the above, $N_1 + N_2 = 1$.

Interpolation Functions

Since our aim in the finite element analysis is to find nodal displacements v_1 and v_2, we can see the advantage of using approximation models of the type in Equation 3.7. In contrast to Equation 3.3a, here the displacement v is expressed directly in terms of nodal displacements. Also, the use of interpolation functions makes it quite easy to perform the differentiations and integrations required in the finite element formulations.

An interpolation function bears a value of unity for the degree-of-freedom it pertains to and zero value corresponding to all other degrees-of-freedom. For example, Figures 3.3(c) and (d) show distributions of N_1 and N_2; along the element; the function N_1 pertains to point 1, and N_2 pertains to point 2.

For a given element it is possible to devise and use different types of local coordinates and interpolation functions N_i. Let us consider the second coordinate system (Figure 3.2(c)) described in Step 2 and in Equation 3.2, and express the v as

$$v = \tfrac{1}{2}(1 - L)v_1 + \tfrac{1}{2}(1 + L)v_2 \tag{3.8a}$$

$$= N_1 v_1 + N_2 v_2 \tag{3.8b}$$

$$= \begin{bmatrix} N_1 & N_2 \end{bmatrix} \begin{Bmatrix} v_1 \\ v_2 \end{Bmatrix} \tag{3.8c}$$

$$= [N]\{q\}. \tag{3.8d}$$

Note that N_i $(i = 1, 2)$ in Equation 3.8b are different from N_i $(i = 1, 2)$ in Equation 3.6d; however, they yield the same linear variation within the element. Alternative coordinate systems based on local measurement from other points on the element are possible; for example, see Problem 3.1.

Relation between Local and Global Coordinates

An important point to consider is that a one-to-one correspondence exists between the local coordinate s or L and the global coordinate y of a point in the element. For example, for L in Equation 3.2 we have

$$y = \tfrac{1}{2}(1 - L)y_1 + \tfrac{1}{2}(1 + L)y_2 \tag{3.9a}$$

$$= N_1 y_1 + N_2 y_2 \tag{3.9b}$$

$$= [\mathbf{N}]\{\mathbf{y}_n\}, \tag{3.9c}$$

where $\{\mathbf{y_n}\}^T = [y_1\ y_2]$ is the vector of nodal coordinates.

In fact an explanation of the concept of isoparametric elements, which is the most common procedure now in use, can be given at this stage. A comparison of Equation 3.8 and 3.9 shows that both the displacement v and the coordinate y at a point in the element are expressed by using the same (iso) interpolation functions. An element formulation where we use the same (or similar) functions for describing the deformations in and the coordinates (or geometry) of an element is called the isoparametric element concept.[1] This is rather an elementary example; we shall subsequently look at other and more general isoparametric elements.

Variation of Element Properties

We often tacitly assume that the material properties such as cross-sectional area A and the elastic modulus E are constant within the element. It is not necessary to assume that they are constant. We can introduce required variation, linear or higher order, for these quantities. For instance, they can be expressed as linear functions:

$$E = N_1 E_1 + N_2 E_2 = [\mathbf{N}]\{\mathbf{E}_n\}, \tag{3.10}$$

$$A = N_1 A_1 + N_2 A_2 = [\mathbf{N}]\{\mathbf{A}_n\}, \tag{3.11}$$

where $\{\mathbf{E_n}\}^T = [E_1\ E_2]$ and $\{\mathbf{A_n}\}^T = [A_1\ A_2]$ are the vectors of nodal values of E and A at nodes 1 and 2, respectively.

Requirements for Approximation Functions

As we have stated before, the choice of an approximation function is guided by laws and principles governing a given problem. Thus an approximation function should satisfy certain requirements in order to be acceptable. For general use these requirements are expressed in mathematical language. However, in this introductory treatment, we shall discuss them in rather simple words.

An approximation function should be continuous within an element. The linear function for v (Equations 3.7 and 3.8) is indeed continuous. In other words, it does not yield a discontinuous value of v but rather a smooth variation of v, and the variation does not involve openings, overlaps, or jumps.

The approximation function should provide interelement compatibility up to a degree required by the problem. For instance, for the column problem involving axial deformations, it is necessary to ensure interelement compatibility at least for displacements of adjacent nodes. That is, the approximation function should be such that the nodal displacements between adjacent nodes are the same. This is shown in Figure 3.3(e). Note that for this case, the higher derivatives such as the first derivatives may not be compatible. The displacement at node 2 of element 1 should be equal to the displacement at node 1 of element 2. For the case of the one-dimensional element, the linear Approximation function satisfies this condition automatically.

As indicated in Chapter 2 (Figure 2.8), satisfaction of displacement compatibility by the linear function does not necessarily fulfill compatibility of first derivative of displacement, that is, slope. For axial deformations, however, if we provide for the compatibility up to only the displacement, we can still expect to obtain reliable and convergent solutions. Often, this condition is tied in with the highest order of derivative in the energy function such as the potential energy. For example, in Equation 3.21 below, the highest order of derivative $dv/dy = \varepsilon_y$ is 1; hence, the interelement compatibility should include order of v at least up to 0 (zero), that is, displacement v. In general, the formulation should provide interelement compatibility up to order $n - 1$, where n is the highest order of derivative in the energy function. Approximation functions that satisfy the condition of compatibility can be called conformable.

The other and important requirement is that the approximation function should be complete; fulfillment of this requirement will assure monotonic convergence. Monotonic convergence can be explained in simple terms as a process in which the successive approximate solutions approach the exact solution consistently without changing sign or direction. For instance, in Figure 1.6(c) the approximate areas approach the exact area in such a way that each successive value of the area is smaller or greater than the previous value of area for upper and lower bound solutions, respectively.

Completeness can be defined in a number of ways. One of the ways is to relate it to the characteristics of the chosen approximation function. If the function for displacement approximation allows for rigid body displacements (motions) and constant states of strains (gradients), then the function can be considered to be complete. A rigid body motion represents a displacement mode that the element can experience without development of stresses in it. For instance, consider the general polynomial for v as:

$$v = \alpha_1 + \alpha_2 y \,\Big|\, + \alpha_3 y^2 + \alpha_4 y^3 + \cdots + \alpha_{n+1} y^n \tag{3.3d}$$

In Equations 3.3a and 3.6 we have chosen a linear polynomial by truncating the general polynomial as shown by the vertical dashed line. The linear approximation contains the constant term α_1 which allows for the rigid body displacement mode. In other words, during this mode, the element remains rigid and does not experience any strain or stress, that is, $\alpha_2 y = 0$.

The requirement of constant state of strain (ε_y) for the one-dimensional column deformation is fulfilled by the linear model, Equation 3.3a because of the existence of term $\alpha_2 y$. This condition implies that as the mesh is refined, that is, the elements become smaller and smaller, the strain ε_y in each element approaches a constant value.

In the case of one-dimensional plane deformations in the column, the condition of constant state includes only ε_y — the first derivative or gradient of v. Additional constant strain states may exist in other and more general problems such as beam and plate bending. In such cases, it will be necessary to satisfy the constant strain state requirement for all such generalized strain or gradients of the unknowns involved, e.g., see Chapters 7, 11–14.

In addition to monotonic convergence, we may be interested in the rate of convergence. This aspect is often tied in with the completeness of the polynomial expansion used for the problem. For instance, for the one-dimensional column problem, completeness of the approximation function requires that a linear function, that is, a polynomial of order (n) equal to one, is needed. Completeness of the polynomial expansion requires that all terms including and up to the first order should be included. This is automatically satisfied for the linear model, Equation 3.3a, since all terms up to $n = 1$ are chosen from Equation 3.3d. In the case of beam bending, Chapter 7, we shall see that a cubic approximation function is required to satisfy completeness. Then it is necessary to choose a polynomial expansion such that it includes all terms up to and includes the order 3.

It may happen that an approximation function of a certain order (n) may not include all terms from the polynomial expansion, Equation 3.3d, and still satisfy the requirements of rigid body motion and constant states of strain.

As an example of a complete approximation satisfying rigid body motion and constant state of strain in a two-dimensional problem, but not complete in the sense of the polynomial expansion, see Chapter 12, Equation 12.9.

For two-dimensional problems, the requirement of completeness of polynomial expansion can be explained through the polynomial expansion represented by using Pascal's triangle, see Chapters 10–13.

Here we have given rather an elementary explanation of the requirements for approximation functions. The subject is wide in scope and the reader interested in advanced analysis of finite element method will encounter the subject quite often. For example, the completeness requirement can be further explained by using the so-called patch test developed by Irons;[2] it is discussed in References 3 and 4. Moreover, the approximation model should satisfy the requirements of isotropy or geometric invariance.[5] These topics are considered beyond the scope of this text.

Step 3. Define Strain-Displacement and Stress–Strain Relations

For stress-deformation problems, the actions or causes (Chapter 2) are forces, and the effects or responses become strains, deformations, and stresses. The basic parameter is the strain or rate of change of deformation. The link connecting the action and response is the stress–strain or constitutive law of the material. It is necessary to define relations between strains and displacements and stresses and strains for the derivation of element equations in Step 4. Hence, in this step we consider these two relations. We note at this stage that although we use familiar laws from strength of materials and elasticity for the stress-deformation problem, in later chapters we shall use the relations relevant to specific topics. For instance, in the case of the fluid flow problem (Chapter 4), the relation between gradient and fluid head and Darcy's law will be used.

Returning to the axial deformation of the column element, the strain-displacement relation, assuming small strains, can be expressed as

$$\varepsilon_y = \frac{dv}{dy}, \tag{3.12a}$$

where ε_y = axial strain. Since we have chosen to use the local coordinate L and since our aim is to find dv/dy in the global system, we can use the chain rule of differentiation as

$$\varepsilon_y = \frac{dv}{dy} = \frac{dL}{dy}\frac{dv}{dL}. \tag{3.12b}$$

Now, from Equation 3.2 we have

$$\frac{dL}{dy} = \frac{d}{dy}\left(\frac{y - y_3}{l/2}\right) = \frac{2}{l} \tag{3.12c}$$

and from Equation 3.8

$$\frac{dv}{dL} = \frac{d}{dL}\left[\tfrac{1}{2}(1 - L)v_1 + \tfrac{1}{2}(1 + L)v_2\right] = \tfrac{1}{2}\left[-1 \quad 1\right]\begin{Bmatrix} v_1 \\ v_2 \end{Bmatrix}. \tag{3.12d}$$

Substitution of Equations 3.12b–3.12d into Equation 3.12a leads to

$$\varepsilon_y = \frac{1}{l}\left[-1 \quad 1\right]\begin{Bmatrix} v_1 \\ v_2 \end{Bmatrix} \tag{3.13a}$$

or

$$\underset{(1\times 1)}{\{\varepsilon_y\}} = \underset{(1\times 2)}{[\mathbf{B}]}\ \underset{(2\times 1)}{\{\mathbf{q}\}}, \tag{3.13b}$$

where $[\mathbf{B}] = (1/l) \times [-1 \ 1]$ can be called the strain-displacement transformation matrix. Because this is a one-dimensional problem, the strain vector $\{\varepsilon_y\}$contains only one term, and the matrix $[\mathbf{B}]$ is only a row vector. We shall retain this terminology for multidimensional problems; however, with multidimensional problems these matrices will have higher orders.

The student can easily see that Equation 3.12d indicates constant value of strain within the element; this is because we have chosen linear variation for the displacement. We can then call this element a constant strain-line element.

Stress–Strain Relation

For simplicity, we assume that the material of the column element is linearly elastic (Figure 3.4). This assumption permits use of the well-known Hooke's law,

$$\sigma_y = E\varepsilon_y, \tag{3.14a}$$

or in matrix notation,

$$\underset{(1\times 1)}{\{\sigma_y\}} = \underset{(1\times 1)}{[\mathbf{C}]}\ \underset{(1\times 1)}{\{\varepsilon_y\}}, \tag{3.14b}$$

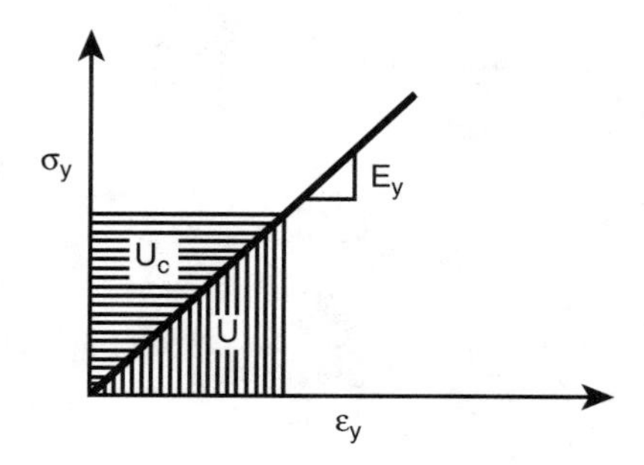

FIGURE 3.4
Linear elastic constitutive or stress–strain (Hooke's) law.

where [**C**] is the stress–strain matrix. Here, for the one-dimensional case, matrices in Equation 3.14b consist of simply one scalar term.

Substitution of Equation 3.13 into Equation 3.14b now allows us to express $\{\sigma_y\}$ in terms of $\{\mathbf{q}\}$ as

$$\{\sigma_y\} = [\mathbf{C}][\mathbf{B}]\{\mathbf{q}\}. \tag{3.15}$$

Step 4. Derive Element Equations

A number of procedures are available for deriving element equations. Among these are the variational and residual methods. Principles based on potential and complementary energies and hybrid and mixed methods are used within the framework of variational methods. As described in Chapter 2 and in Appendix 1, a number of schemes such as Galerkin, collocation, and least squares fall under the category of residual methods. We shall use some of these methods in this chapter and subsequently in other chapters.

Principle of Minimum Potential Energy

In simple words, if a loaded elastic body is in equilibrium under given geometric constraints or boundary conditions, the potential energy of the deformed body assumes a stationary value. In the case of linear elastic bodies in equilibrium, the value is a minimum; since most problems we consider involve this specialization, for convenience we shall use the term minimum.

Figure 3.5 shows a simple axial member represented by a linear spring with spring constant $k(F/L)$. Under a load P, the spring experiences a displacement equal to v.

The potential or the potential energy Π_p of the spring is composed of two parts, strain energy U and potential W_p of the external load (see Chapter 2):

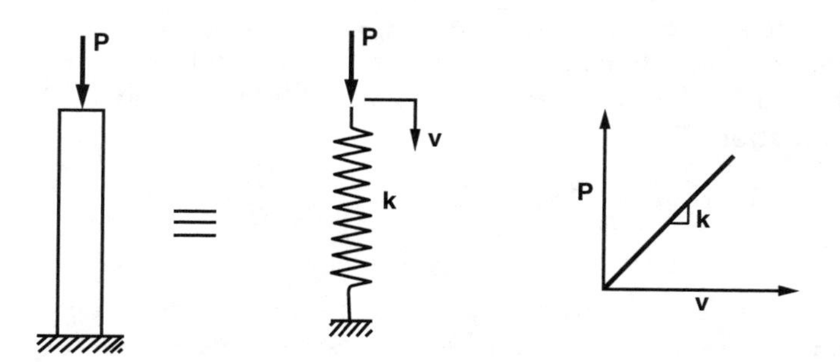

FIGURE 3.5
Idealized linear spring.

$$\Pi_p = U + W_p. \tag{3.16}$$

The strain energy U can be interpreted as the area under the stress–strain curve (Figure 3.4). Mathematically, when we minimize Π_p, we differentiate or take variations of Π_p with respect to the displacement v. While doing this we assume that the force remains constant, and we can relate variation of work done W by the load and the potential of the load as

$$\delta W = -\delta W_p, \tag{3.17}$$

where δ denotes arbitrary change, variation, or perturbation. For our purpose we can consider it to imply a series of partial differentiations. The negative sign in Equation 3.17 occurs because the potential W_p of external loads is lost into work by these loads, W. Then the principle of minimum potential energy is expressed as

$$\delta\Pi_p = \delta U + \delta W_p = \delta U - \delta W = 0. \tag{3.18}$$

There are two ways that we can determine the minimum of Π_p: manual and mathematical. Both involve essentially an examination of the function represented by Π_p until we find the minimum point. For simple understanding, we first consider the manual procedure and write the potential energy for the spring (Figure 3.5), assuming undeformed state of the spring as the datum for potentials, as

$$\begin{aligned}\Pi_p &= \tfrac{1}{2}(kv)v - Pv \\ &= \tfrac{1}{2}kv^2 - Pv,\end{aligned} \tag{3.19a}$$

where kv = force in the spring and $\frac{1}{2}(kv)v$ denotes strain energy as the area under the load-displacement curve (Figure 3.5). Since the load in the spring goes from 0 to kv, we have to use average strain energy. The term Pv denotes the potential of load P; since we have assumed P to be constant, this term does not include $\frac{1}{2}$. We further assume that $P = 10$ units and $k = 10$ units per unit deformation. Then

$$\Pi_p = \tfrac{1}{2}10v^2 - 10v = 5v^2 - 10v. \tag{3.19b}$$

Now we search for the minimum by examining values of the potential Π_p for various values of deformations v. The results are shown in the following table: a positive v is assumed to act in the direction of the applied load:

v	Π_p
–2.000	+40.000
–1.000	+15.000
0.000	0.000
0.125	–1.1719
0.250	–2.1875
0.500	–3.7500
1.000	–5.0000
2.000	0.0000
3.000	15.0000
4.000	40.0000
5.000	75.0000
etc.	

Figure 3.6 shows a plot of Π_p vs. v. It can be seen that Π_p has a minimum value at $v = 1$. Hence, under $P = 10$ units, the spring, when in equilibrium, will deform by 1 unit.

On the other hand, we can perform the procedure of going to and fro on the function Π_p to find its minimum by using mathematics. It is well known that a function assumes a minimum value at a point(s) where its derivative is zero. Applying this principle to Π_p in Equation 3.19a, we obtain

$$\delta\Pi_p = \tfrac{2}{2} kv\delta v - P\delta v = 0$$

or

$$(kv - P)\delta v = 0. \tag{3.20a}$$

Since δv is arbitrary, the term in parentheses must vanish. Therefore,

$$kv - P = 0$$

or

$$kv = P, \tag{3.20b}$$

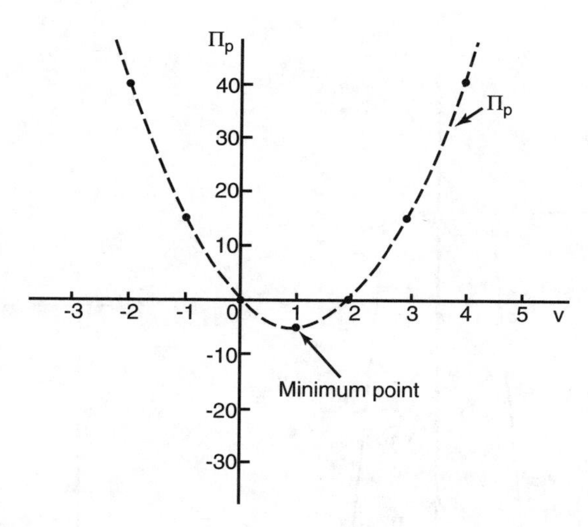

FIGURE 3.6
Variation of potential energy.

which is the equation of equilibrium for the spring. $\delta\Pi_p = 0$ in Equation 3.20 is analogous to equating $d\Pi_p/dv = 0$, which will result in the same equilibrium Equation 3.20b. Substitution of the numerical values gives

$$10v = 10.$$

Therefore, $v = 1.0$ unit, the same answer as before.

We note here that in most problems solved by using the finite element method Π_p is a function of a large number of parameters or nodal displacements. Consequently, it is most economical and direct to use the mathematical methods because the manual procedure is cumbersome and often impossible.

Functional for One-Dimensional Stress Deformation Problem

The appropriate functional (Π_p) for solving the one-dimensional stress–deformation problem can be obtained in two ways: (1) from the total potential energy expression and (2) directly from the governing equation and natural boundary conditions using the variational calculus principles as discussed in Chapter 2.

Total Potential Energy Approach

The element shown in the Figure 3.7 is subjected to three types of loads. These are a surface traction $\bar{T}_y^*$ per unit surface area, a body force $\bar{Y}$ per unit

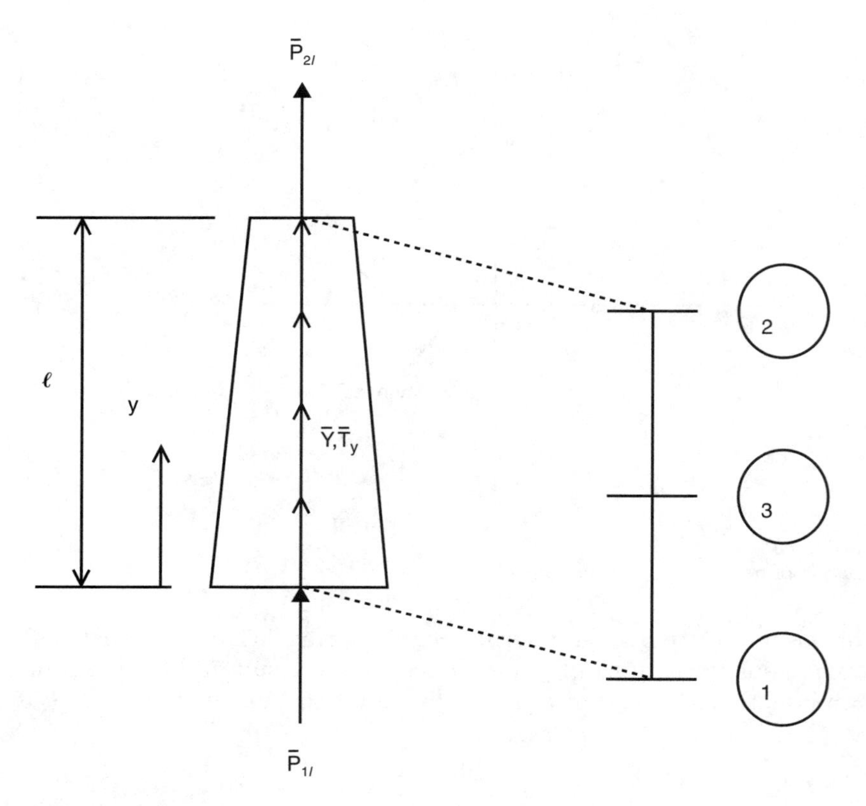

FIGURE 3.7
Generic column element with loads.

volume, and concentrated loads $\bar{P}_{1l}$ and $\bar{P}_{2l}$ at nodes 1 and 2, respectively; here the overbar denotes a prescribed quantity.

From our knowledge of Mechanics of Materials, we can write the total potential energy of this element as[5]

$$\Pi_p = \iiint_V \frac{1}{2}\sigma_y \varepsilon_y dV - \iiint_V \bar{Y} v dV - \iint_S \bar{T}_y^* v dS - \sum_{i=1}^{2} \bar{P}_{il} v_i \tag{3.21}$$

If all variables are functions of y then the above equation can be written in the following form:

$$\Pi_p = \frac{A}{2}\int_{y_1}^{y_2} \sigma_y \varepsilon_y dy - A\int_{y_1}^{y_2} \bar{Y} v dy - C\int_{y_1}^{y_2} \bar{T}_y^* v dy - \sum_{i=1}^{2} \bar{P}_{il} v_i \tag{3.22}$$

where A is the cross-sectional area assumed to be constant and C is the periphery or circumference of the cross section.

If we denote the surface traction per unit length by $\overline{T}_y$ then, $\overline{T}_y = C.\overline{T}_y^*$. Equation 3.22) can be rewritten in the following form

$$\Pi_p = \frac{A}{2}\int_{y_1}^{y_2} \sigma_y \varepsilon_y dy - A\int_{y_1}^{y_2} \overline{Y} v dy - \int_{y_1}^{y_2} \overline{T}_y v dy - \sum_{i=1}^{2} \overline{P}_{il} v_i \tag{3.23}$$

Under stable equilibrium the potential energy is minimum. Hence, one can guarantee the satisfaction of equilibrium equations by minimizing the above expression of the potential energy.

Variational Principle Approach

From our knowledge of the theory of elasticity or advanced mechanics of materials one can write the governing equation or the equilibrium equation of an elemental volume, that is subjected to only normal stress and force in the y-direction:

$$\frac{\partial \sigma_y}{\partial y} + f_y = 0 \tag{3.24}$$

where f_y is the y-direction force per unit volume. For a bar element Equation 3.24 can be written as

$$A\frac{\partial \sigma_y}{\partial y} + Af_y = A\frac{\partial \sigma_y}{\partial y} + F_y = 0 \tag{3.25}$$

where A is the cross-sectional area of the bar and F_y is the force per unit length of the bar. If the bar is subjected to a body force $\overline{Y}$ per unit volume and a surface traction $\overline{T}_y$ per unit length then

$$F_y = A\overline{Y} + \overline{T_y}. \tag{3.26}$$

The equilibrium equation of the bar element can now be written as

$$A\frac{\partial (E\varepsilon_y)}{\partial y} + F_y = AE\frac{\partial \varepsilon_y}{\partial y} + F_y = 0. \tag{3.27}$$

Since, $\varepsilon_y = \frac{\partial v}{\partial y}$ the above equation takes the following form

$$AE\frac{\partial^2 v}{\partial y^2} + F_y = 0. \tag{3.28}$$

Comparing the above equation with the governing equation given in Example 2.6, one can easily see that the appropriate functional for this differential equation should be

$$\Pi\left(v, \frac{\partial v}{\partial y}\right) = \frac{1}{2}\int_{y_1}^{y_2} AE\left(\frac{\partial v}{\partial y}\right)^2 dy - \int_{y_1}^{y_2} F_y v dy. \tag{3.29}$$

Substituting Equation 3.26 into Equation 3.29 we obtain

$$\Pi\left(v, \frac{\partial v}{\partial y}\right) = \frac{1}{2}\int_{y_1}^{y_2} AE\left(\frac{\partial v}{\partial y}\right)^2 dy - \int_{y_1}^{y_2} A\overline{Y} v dy - \int_{y_1}^{y_2} \overline{T}_y v dy. \tag{3.30}$$

Recognizing that the strain is the derivative of displacement (Equation 3.12a) and stress is the strain times the Young's modulus (Equation 3.14a), the above equation can be written as

$$\Pi = \frac{A}{2}\int_{y_1}^{y_2} \sigma_y \varepsilon_y dy - A\int_{y_1}^{y_2} \overline{Y} v dy - \int_{y_1}^{y_2} \overline{T}_y v dy - \sum_{1}^{2} \overline{P}_{il} v_i. \tag{3.31}$$

The last term of Equation 3.31 appears only if the nonzero concentrated force P_{il} exists at the i-th node.

Note that Equations 3.23 and 3.31 are identical. Hence, Π of Equation 3.31 should be the same as Π_p of Equation 3.23.

Equation 3.31 or 3.23 can now be expressed in terms of the local coordinate system by using the transformation of Equation 3.2 as

$$dy = \frac{l}{2} dL. \tag{3.32}$$

Therefore,

$$\Pi_p = \frac{Al}{4}\int_{-1}^{1} \sigma_y \varepsilon_y dL - \frac{Al}{2}\int_{-1}^{1} \overline{Y} v dL - \frac{1}{2}\int_{-1}^{1} \overline{T}_y v dL - \sum_{1}^{M} \overline{P}_{il} v_i. \tag{3.33}$$

Next we substitute for v, ε_y, and σ_y from Equations 3.8, 3.13, and 3.14, respectively, in Equation 3.33 to obtain, in matrix notation,

$$\Pi_p = \frac{Al}{4}\int_{-1}^{1} \underset{(1\times1)}{\{\varepsilon_y\}^T} \underset{(1\times1)}{[\mathbf{C}]} \underset{(1\times1)}{\{\varepsilon_y\}} dL - \frac{Al}{2}\int_{-1}^{+1} \underset{(1\times2)}{[\mathbf{N}]} \underset{(2\times1)}{\{\mathbf{q}\}} \underset{(1\times1)}{\bar{Y}} dL$$
$$- \frac{l}{2}\int_{-1}^{1} \underset{(1\times2)}{[\mathbf{N}]} \underset{(2\times1)}{[\mathbf{q}]} \underset{(1\times1)}{\bar{T}_y} dL - \sum_{1}^{M} \underset{(1\times1)}{\bar{P}_{il}} \underset{(1\times1)}{v_i} \tag{3.34}$$

or

$$\Pi_p = \frac{Al}{4}\int_{-1}^{1} \underset{(1\times2)}{\{\mathbf{q}\}^T} \underset{(2\times1)}{[\mathbf{B}]^T} \underset{(1\times1)}{[\mathbf{C}]} \underset{(1\times2)}{[\mathbf{B}]} \underset{(2\times1)}{\{\mathbf{q}\}} dL$$
$$- \frac{Al}{2}\int_{-1}^{1} \underset{(1\times2)}{[\mathbf{N}]} \underset{(2\times1)}{\{\mathbf{q}\}} \underset{(1\times1)}{\bar{Y}} dL \tag{3.35}$$
$$- \frac{l}{2}\int_{-1}^{1} \underset{(1\times2)}{[\mathbf{N}]} \underset{(2\times1)}{[\mathbf{q}]} \underset{(1\times1)}{\bar{T}_y} dL - \sum_{1}^{M} \underset{(1\times1)}{\bar{P}_{il}} \underset{(1\times1)}{v_i},$$

where $\bar{Y}$ and $\bar{T}_y$ are assumed to be uniform.

Equation 3.35 represents a quadratic function expressed in terms of v_1 and v_2. In matrix notation, transposing in Equation 3.34 is necessary to make the matrix multiplication in $\{\varepsilon_y\}^T$ $[\mathbf{C}]\{\varepsilon_y\}$ consistent so as to yield the scalar (energy) term $\sigma_y\varepsilon_y = E\varepsilon_y^2$ in Equation 3.33. The need for transposing will become clear when we expand the terms in Equation 3.33. The last term denotes summation, $\bar{P}_{1l}\, v_l + \bar{P}_{2l}\, v_2$, if $M = 2$.

Expansion of Terms

We now consider the first three terms in Equation 3.35 one by one and expand them as follows:

First term:

$$U = \frac{Al}{4}\int_{-1}^{1} [v_1 \quad v_2]\frac{1}{l}\begin{Bmatrix} -1 \\ 1 \end{Bmatrix} E\frac{1}{l}[-1 \quad 1]\begin{Bmatrix} v_1 \\ v_2 \end{Bmatrix} dL$$
$$= \frac{AE}{4l}\int_{-1}^{1}\left(v_1^2 - 2v_1v_2 + v_2^2\right) dL. \tag{3.36a}$$

Second term:

$$W_{p1} = \frac{Al\bar{Y}}{2}\int_{-1}^{1}\left[\frac{1}{2}(1-L)v_1 + \frac{1}{2}(1+L)v_2\right]dL. \tag{3.36b}$$

Here $\bar{Y}$ is assumed to be uniform gravity load per unit volume. Similarly, for uniform $\bar{T}_y$ the third term yields

$$W_{p2} = \frac{\bar{T}_y l}{2}\int_{-1}^{1}\left[\frac{1}{2}(1-L)v_1 + \frac{1}{2}(1+L)v_2\right]dL. \tag{3.36c}$$

Finally, the sum of Equations 3.36a, 3.36b, and 3.36c and the last term in Equation 3.35 gives

$$\begin{aligned}\Pi_p &= \frac{AE}{4l}\int_{-1}^{1}\left(v_1^2 - 2v_1v_2 + v_2^2\right)dL \\ &- \frac{Al\bar{Y}}{2}\int_{-1}^{1}\left[\frac{1}{2}(1-L)v_1 + \frac{1}{2}(1+L)v_2\right]dL \\ &- \frac{\bar{T}_y l}{2}\int_{-1}^{1}\left[\frac{1}{2}(1-L)v_1 + \frac{1}{2}(1+L)v_2\right]dL - \sum_{1}^{M}\bar{P}_{il}v_i.\end{aligned} \tag{3.37}$$

Notice that Π_p is a quadratic or second-order function in v_1 and v_2.

Now we invoke the principle of minimum potential energy (see Chapter 2) to find its minimum value by differentiating Π_p with respect to v_1 and v_2 as

$$\begin{aligned}\frac{\partial \Pi_p}{\partial v_1} &= \frac{AE}{4l}\int_{-1}^{1}\left(2v_1 - 2v_2\right)dL - \frac{Al\bar{Y}}{2}\int_{-1}^{1}\frac{1}{2}(1-L)dL \\ &- \frac{\bar{T}_y l}{2}\int_{-1}^{1}\frac{1}{2}(1-L)dL - \bar{P}_{1l} = 0,\end{aligned} \tag{3.38a}$$

$$\begin{aligned}\frac{\partial \Pi_p}{\partial v_2} &= \frac{AE}{4l}\int_{-1}^{1}\left(-2v_1 + 2v_2\right)dL - \frac{Al\bar{Y}}{2}\int_{-1}^{1}\frac{1}{2}(1+L)dL \\ &- \frac{\bar{T}_y l}{2}\int_{-1}^{1}\frac{1}{2}(1+L)dL - \bar{P}_{2l} = 0.\end{aligned} \tag{3.38b}$$

Although we used differential calculus and partial differentiations to find the minimum of Π_p, for advanced applications to multidimensional problems, it is more concise and convenient to use calculus of variations; then both Equation 3.38a and Equation 3.38b are written together as

$$\delta \Pi_p = 0, \tag{3.39}$$

where δ is the variation notation. We shall briefly discuss variational principles at various stages, but the majority of derivations in this book will be obtained by using differential calculus.

Integration

For the one-dimensional problem, the integrations in Equation 3.38 are simple; in fact, the first term in these equations is independent of the coordinate L; hence integrations involve only constant terms. Thus,

$$\frac{AE}{4l}\int_{-1}^{1}(2v_1 - 2v_2)dL = \frac{AE}{4l}(2v_1 - 2v_2)L\Big|_{-1}^{1}$$

$$= \frac{AE}{l}(v_1 - v_2)$$

$$\frac{Al\bar{Y}}{2}\int_{-1}^{1}\frac{1}{2}(1 - L)dL = \frac{Al\bar{Y}}{4}\left(L - \frac{L^2}{2}\right)\Big|_{-1}^{1} = \frac{Al\bar{Y}}{2}.$$

Integrations of other terms in Equation 3.38 lead to

$$\frac{AE}{l}(v_1 - v_2) - \frac{Al\bar{Y}}{2} - \frac{\bar{T}_y l}{2} - \bar{P}_{1l} = 0, \tag{3.40a}$$

$$\frac{AE}{l}(-v_1 + v_2) - \frac{Al\bar{Y}}{2} - \frac{\bar{T}_y l}{2} - \bar{P}_{2l} = 0. \tag{3.40b}$$

In matrix notation,

$$\frac{AE}{l}\begin{bmatrix} 1 & -1 \\ -1 & 1 \end{bmatrix}\begin{Bmatrix} v_1 \\ v_2 \end{Bmatrix} = \frac{Al\bar{Y}}{2}\begin{Bmatrix} 1 \\ 1 \end{Bmatrix} + \frac{\bar{T}_y l}{2}\begin{Bmatrix} 1 \\ 1 \end{Bmatrix} + \begin{Bmatrix} \bar{P}_{1l} \\ \bar{P}_{2l} \end{Bmatrix} \tag{3.41a}$$

or

$$[\mathbf{k}]\{\mathbf{q}\} = \{\mathbf{Q}\}. \tag{3.41b}$$

Here $[\mathbf{k}]$ = stiffness matrix of the element, and with the linear approximation it is identical to the matrix of stiffness influence coefficients in matrix structural analysis; $\{\mathbf{Q}\}$ = element nodal load vector, and it is composed of body force (due to gravity), surface traction or loading, and the joint loads. It is interesting that the finite element derivations with the linear approximation result in lumping of the applied loads (equally) at the two nodes: $Al\bar{Y}$ = weight of the element, where Al = volume of the element, and $\bar{T}_y l$ = total surface load. We note that if higher-order approximations were used, the load vector would not necessarily come out as lumped loading; we shall consider such cases at later stages.

Equation 3.41 provides a general expression that can be used repeatedly to find stiffness relations for all elements in the assemblage.

In the foregoing, we derived element equations by differentiating the expanded function for Π_p. This is possible because there are only two variables, v_1 and v_2. Advanced problems, however, involve a large number of variables, and we commonly write the results directly in terms of matrix equations. For instance, differentiation of Π_p in Equation 3.35

$$\frac{\partial \Pi_p}{\partial v_1} = 0,$$

$$\frac{\partial \Pi_p}{\partial v_2} = 0,$$

leads to

$$\frac{Al}{2}\int_{-1}^{1} \underset{(2\times 1)}{[\mathbf{B}]^T} \underset{(1\times 1)}{[\mathbf{C}]} \underset{(1\times 2)}{[\mathbf{B}]}\, dL\{\mathbf{q}\} = \frac{Al}{2}\int_{-1}^{1} \underset{(2\times 1)}{[\mathbf{N}]^T} \underset{(1\times 1)}{\bar{Y}}\, dL$$

$$+\frac{l}{2}\int_{-1}^{1} \underset{(2\times 1)}{[\mathbf{N}]^T} \underset{(1\times 1)}{\bar{T}_y}\, dL + \underset{\substack{(2\times 1)\\ l=1,2,}}{\{\bar{\mathbf{P}}_{il}\}}$$

or

$$[\mathbf{k}]\{\mathbf{q}\} = \{\mathbf{Q}\}, \tag{3.41b}$$

where

$$[\mathbf{k}] = \frac{Al}{2}\int_{-1}^{1}[\mathbf{B}]^T E[\mathbf{B}]dL$$

and

$$\{\mathbf{Q}\} = \frac{Al}{2}\int_{-1}^{1}[\mathbf{N}]^T \bar{Y} dL + \frac{l}{2}\int_{-1}^{1}[\mathbf{N}]^T \bar{T}_y dL + \{\bar{\mathbf{P}}_{il}\}.$$

The terms in Equation 3.41 are the same as in Equation 3.38, except that here they are arranged in matrix notation. The transpose in $[\mathbf{N}]^T$ arises because after differentiations (Equation 3.38) the force terms yield a load vector that has an order of 2 in the final results. After the process of differentiation on Π_p, the end result is a set of (linear) simultaneous equations (Equation 3.41b).

Often, the stiffness matrix [**k**] is called an operator, which means that when operated on (nodal) displacements, the results are the (nodal) forces.

Comment

The primary task in finite element formulation can be considered to involve derivation of element equations. The subsequent steps essentially involve assembly, patching up or combination of elements, and use of linear algebra for solution of the resulting simultaneous equations.

Step 5. Assemble Element Equations to Obtain Global Equations

Potential Energy Approach (for Assembly) — Although, for simple understanding, we considered the equilibrium of a single element in the foregoing step, it is necessary to emphasize that it is the equilibrium of the entire structure in which we are interested. Consequently, we look at the total potential energy of the assemblage and find its stationary (minimum) value. The procedure of assembling element equations can be interpreted through minimization of total potential energy.

As an example, we consider the column (Figure 3.1) divided into three elements with four nodes (Figure 3.8).

Here we have numbered nodes starting from the top of the column and have measured the global coordinate y as positive downward. This is just for convenience; for instance, since the loads act downward, they are positive, and displacements are positive. It is, however, possible to measure y as positive upward or downward from any convenient point and to number nodes in other consistent fashions.

By using Equation 3.37 we can write the potential energies Π_p for each element in the assemblage and add them together to obtain total potential energy Π_p^t, as

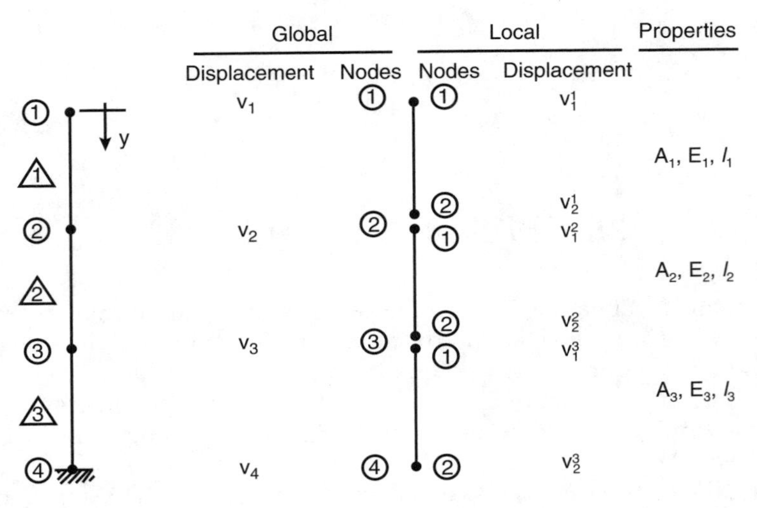

FIGURE 3.8
Discretization of column and numbering.

$$
\begin{aligned}
\Pi_p^t = \sum_{e=1}^{M} \Pi_{pe} = & \frac{A_1E_1}{4l_1}\int_{-1}^{1}\left(v_1^2 - 2v_1v_2 + v_2^2\right)dL \\
& + \frac{A_2E_2}{4l_2}\int_{-1}^{1}\left(v_2^2 - 2v_2v_3 + v_3^2\right)dL \\
& + \frac{A_3E_3}{4l_3}\int_{-1}^{1}\left(v_3^2 - 2v_3v_4 + v_4^2\right)dL \\
& - \frac{A_1l_1\overline{Y}_1}{2}\int_{-1}^{1}\left[\frac{1}{2}(1-L)v_1 + \frac{1}{2}(1+L)v_2\right]dL \\
& - \frac{A_2l_2\overline{Y}_2}{2}\int_{-1}^{1}\left[\frac{1}{2}(1-L)v_2 + \frac{1}{2}(1+L)v_3\right]dL \\
& - \frac{A_3l_3\overline{Y}_3}{2}\int_{-1}^{1}\left[\frac{1}{2}(1-L)v_3 + \frac{1}{2}(1+L)v_4\right]dL \qquad (3.42) \\
& - \frac{\overline{T}_{y1}l_1}{2}\int_{-1}^{1}\left[\frac{1}{2}(1-L)v_1 + \frac{1}{2}(1+L)v_2\right]dL \\
& - \frac{\overline{T}_{y2}l_2}{2}\int_{-1}^{1}\left[\frac{1}{2}(1-L)v_2 + \frac{1}{2}(1+L)v_3\right]dL
\end{aligned}
$$

$$- \frac{\overline{T}_{y3} l_3}{2} \int_{-1}^{1} \left[\frac{1}{2}(1-L)v_3 + \frac{1}{2}(1+L)v_4 \right] dL$$

$$- \overline{P}_{1l}^{1} v_1 - \left(\overline{P}_{2l}^{1} + \overline{P}_{1l}^{2}\right) v_2 - \left(\overline{P}_{2l}^{2} + \overline{P}_{1l}^{3}\right) v_3$$

$$- \overline{P}_{2l}^{3} v_4.$$

Here we assume that A, E, I, $\overline{Y}$, and $\overline{T}_y$ are different for different elements and their subscript denotes an element number, e denotes an element, the superscript on P denotes an element, and M = total number of elements. Note that the local joint or point loads at common nodes are added together and yield the global joint or point loads at these points. For instance, $\overline{P}_{2l}^{1} + \overline{P}_{1l}^{2}$ gives global point load at global node point 2.

For global equilibrium we minimize Π_p^t, with respect to all four nodal displacement unknowns, v_1, v_2, v_3, and v_4. Thus,

$$\frac{\partial \Pi_p^t}{\partial v_1} = \frac{A_1 E_1}{4 l_1} \int_{-1}^{1} (2v_1 - 2v_2) dL - \frac{A_1 l_1 \overline{Y}_1}{2} \int_{-1}^{1} \frac{1}{2}(1-L) dL$$
$$- \frac{\overline{T}_{y1} l_1}{2} \int_{-1}^{1} \left[\frac{1}{2}(1-L) \right] dL - \overline{P}_{1l}^{1} = 0, \tag{3.43}$$

and so on. In combined variational notation we write

$$\delta \Pi_p^t = 0, \tag{3.44a}$$

which denotes

$$\frac{\partial \Pi_p^t}{\partial v_1} = 0,$$
$$\frac{\partial \Pi_p^t}{\partial v_2} = 0,$$
$$\frac{\partial \Pi_p^t}{\partial v_3} = 0, \tag{3.44b}$$
$$\frac{\partial \Pi_p^t}{\partial v_4} = 0.$$

After the required integrations and arranging the four equations in matrix notation, we obtain

$$\begin{bmatrix} \frac{A_1E_1}{l_1} & \frac{-A_1E_1}{l_1} & & 0 \\ \frac{-A_1E_1}{l_1} & \frac{A_1E_1}{l_1} + \frac{A_2E_2}{l_2} & \frac{-A_2E_2}{l_2} & \\ & \frac{-A_2E_2}{l_2} & \frac{A_2E_2}{l_2} + \frac{A_3E_3}{l_3} & \frac{-A_3E_3}{l_3} \\ 0 & & \frac{-A_3E_3}{l_3} & \frac{A_3E_3}{l_3} \end{bmatrix} \begin{Bmatrix} v_1 \\ v_2 \\ v_3 \\ v_4 \end{Bmatrix}$$

$$= \begin{Bmatrix} \frac{A_1l_1\bar{Y}_1}{2} + \frac{\bar{T}_{y1}l_1}{2} + \bar{P}_1^1 \\ \frac{A_1l_1\bar{Y}_1}{2} + \frac{\bar{T}_{y1}l_1}{2} + \frac{A_2l_2\bar{Y}_2}{2} + \frac{\bar{T}_{y2}l_2}{2} + \bar{P}_{2l}^1 + \bar{P}_{1l}^2 \\ \frac{A_2l_2\bar{Y}_2}{2} + \frac{\bar{T}_{y2}l_2}{2} + \frac{A_3l_3\bar{Y}_3}{2} + \frac{\bar{T}_{y3}l_3}{2} + \bar{P}_{2l}^2 + \bar{P}_{1l}^3 \\ \frac{A_3l_3\bar{Y}_3}{2} + \frac{\bar{T}_{y3}l_3}{2} + \bar{P}_{2l}^3 \end{Bmatrix} \tag{3.45a}$$

or

$$[\mathbf{K}]\{\mathbf{r}\} = \{\mathbf{R}\}, \tag{3.45b}$$

where $[\mathbf{K}]$ = assemblage stiffness matrix, $\{\mathbf{r}\}^T = [v_1\, v_2\, v_3\, v_4]$ = assemblage nodal displacement vector, and $\{\mathbf{R}\}$ = assemblage nodal load vector.

Direct Stiffness Method

The foregoing approach can be explained and understood alternatively through the direct stiffness method.[6] We note, however, that the basic idea of assembly evolves essentially as a result of the minimization of total potential energy. A close look at Equation 3.45a shows that the stiffness or influence coefficients corresponding to the common node for elements 1 and 2 are added together; this is indicated by enclosing them in dashed lines. Similarly, loads at the common node are also added together. This interpretation can lead to the familiar concept of obtaining the assemblage matrix by adding individual element matrices of contributing elements through the direct stiffness approach.

Let us express the matrix equations for the three elements by labeling the terms with subscripts as

$$
\begin{array}{ll}
\text{Global} & \rightarrow 1 \quad 2 \\
\downarrow \text{Local} & \rightarrow 1 \quad 2 \;\text{Local Global}
\end{array}
$$

$$
\begin{matrix} 1 & 1 \\ 2 & 2 \end{matrix}\quad \frac{A_1E_1}{l_1}\begin{bmatrix} 1 & -1 \\ -1 & 1 \end{bmatrix}\begin{Bmatrix} v_1^1 \rightarrow v_1 \\ v_2^1 \rightarrow v_2 \end{Bmatrix} = \frac{A_1 l_1 \bar{Y}_1}{2}\begin{Bmatrix} 1 \\ 1 \end{Bmatrix} + \frac{\bar{T}_{y1} l_1}{2}\begin{Bmatrix} 1 \\ 1 \end{Bmatrix} + \begin{Bmatrix} \bar{P}_{1l}^1 \\ \bar{P}_{2l}^1 \end{Bmatrix}, \tag{3.46a}
$$

$$
\begin{array}{ll}
\text{Global} & \rightarrow 2 \quad 3 \\
\downarrow \text{Local} & \rightarrow 1 \quad 2 \;\text{Local Global}
\end{array}
$$

$$
\begin{matrix} 2 & 1 \\ 3 & 2 \end{matrix}\quad \frac{A_2E_2}{l_2}\begin{bmatrix} 1 & -1 \\ -1 & 1 \end{bmatrix}\begin{Bmatrix} v_1^2 \rightarrow v_2 \\ v_2^2 \rightarrow v_3 \end{Bmatrix} = \frac{A_2 l_2 \bar{Y}_2}{2}\begin{Bmatrix} 1 \\ 1 \end{Bmatrix} + \frac{\bar{T}_{y2} l_2}{2}\begin{Bmatrix} 1 \\ 1 \end{Bmatrix} + \begin{Bmatrix} \bar{P}_{1l}^2 \\ \bar{P}_{2l}^2 \end{Bmatrix}, \tag{3.46b}
$$

$$
\begin{array}{ll}
\text{Global} & \rightarrow 3 \quad 4 \\
\downarrow \text{Local} & \rightarrow 1 \quad 2 \;\text{Local Global}
\end{array}
$$

$$
\begin{matrix} 3 & 1 \\ 4 & 2 \end{matrix}\quad \frac{A_3E_3}{l_3}\begin{bmatrix} 1 & -1 \\ -1 & 1 \end{bmatrix}\begin{Bmatrix} v_1^3 \rightarrow v_3 \\ v_2^3 \rightarrow v_4 \end{Bmatrix} = \frac{A_3 l_3 \bar{Y}_3}{2}\begin{Bmatrix} 1 \\ 1 \end{Bmatrix} + \frac{\bar{T}_{y3} l_3}{2}\begin{Bmatrix} 1 \\ 1 \end{Bmatrix} + \begin{Bmatrix} \bar{P}_{1l}^3 \\ \bar{P}_{2l}^3 \end{Bmatrix}. \tag{3.46c}
$$

Here the superscript indicates the element; for instance, v_2^1 is the displacement at node 2 for element 1, and so on (Figure 3.8).

To assemble these three relations, we note that the total global degrees of freedom are 4, and hence the assemblage matrix and load vector are of the order of 4. Consider Equation 3.47 in which we assign

$$
\begin{bmatrix}
\frac{A_1E_1}{l_1} & -\frac{A_1E_1}{l_1} & & 0 \\
-\frac{A_1E_1}{l_1} & \frac{A_1E_1}{l_1} + \frac{A_2E_2}{l_2} & -\frac{A_2E_2}{l_2} & \\
 & -\frac{A_2E_2}{l_2} & \frac{A_2E_2}{l_2} + \frac{A_3E_3}{l_3} & -\frac{A_3E_3}{l_3} \\
0 & & -\frac{A_3E_3}{l_3} & \frac{A_3E_3}{l_3}
\end{bmatrix}
\begin{Bmatrix}
v_1^1 = v_1 \\
v_2^1 = v_1^2 = v_2 \\
v_2^2 = v_1^3 = v_3 \\
v_2^3 = v_4
\end{Bmatrix} \tag{3.47}
$$

$$
= \begin{Bmatrix} \dfrac{A_1 l_1 \overline{Y}_1}{2} + \dfrac{\overline{T}_{y1} l_1}{2} + \overline{P}_{1l}^1 \\ \dfrac{A_1 l_1 \overline{Y}_1}{2} + \dfrac{\overline{T}_{y1} l_1}{2} + \overline{P}_{2l}^1 + \dfrac{A_2 l_2 \overline{Y}_2}{2} + \dfrac{\overline{T}_{y2} l_2}{2} + \overline{P}_{1l}^2 \\ \dfrac{A_2 l_2 \overline{Y}_2}{2} + \dfrac{\overline{T}_{y2} l_2}{2} + \overline{P}_{2l}^2 + \dfrac{A_3 l_3 \overline{Y}_3}{2} + \dfrac{\overline{T}_{y3} l_3}{2} + \overline{P}_{1l}^3 \\ \dfrac{A_3 l_3 \overline{Y}_3}{2} + \dfrac{\overline{T}_{y3} l_3}{2} + \overline{P}_{2l}^3 \end{Bmatrix} \quad (3.47)
$$

blocks of 4×4 for the assemblage stiffness matrix and 4×1 for the assemblage load vector. Now we insert the coefficients of the matrices for the three elements (Equation 3.46) into the proper locations in Equation 3.47; for instance, the coefficient in Equation 3.46b for element 2 corresponding to the local indices (2, 2) is added to the global location (3, 3) and so on. Similarly the nodal loads corresponding to the local index 2 are added to the global location 3 and so on.

We notice that Equation 3.47 is the same as Equation 3.45. We also see that the direct stiffness approach is based essentially on the physical requirement that the displacements at the common nodes between elements are continuous; that is, there exists interelement compatibility of displacements at the common nodes.

Boundary Conditions

Until now we have concentrated only on the properties of the column. Next we consider the physical conditions that support the column in space, because the foregoing stiffness properties are called into action only when the column is supported. This leads us to the concept of boundary conditions or constraints.

As the name implies, a boundary condition denotes a prescribed value of displacement or its gradient(s) on a part of the boundary of the structure or body. A boundary condition tells us how the body (column) is supported in space. After we introduce these conditions, we have a structure that is ready to withstand applied forces; in lieu of these conditions, Steps 1–6 resulting in Equation 3.45 tell us only about the capability of the column to withstand forces and not how they are withstood. In other words, without boundary conditions, the stiffness matrix **[K]** is singular; that is, its determinant vanishes, and there can be an infinite number of possible solutions. Hence, the equations in 3.45 cannot be solved until **[K]** is modified to reflect the boundary conditions.

Types of Boundary Conditions

As discussed in Chapter 2, we encounter three kinds of boundary conditions. They are prescribed (1) displacements (or other relevant unknowns), (2) slopes or gradients of unknowns, or (3) both. They can also be called first or Dirichlet, second or Neumann, and third or mixed boundary conditions, respectively.

For instance, if the column is supported such that the axial displacement at the base is specified, it is the first condition; if the base is fixed such that gradient or slope is specified, it is the second condition; and if the base is fully fixed, it constitutes the mixed condition.

The boundary conditions in terms of the given displacement are often called geometric or forced, while those in terms of gradients are often called natural; the latter can often relate to generalized forces prescribed on the boundary.

Homogeneous or Zero-Valued Boundary Condition

As an illustration, let us assume that the displacement at the column base, that is, at nodal 4 (v_4) (Figure 3.8), is zero; one can specify a nonzero displacement also if the base experiences a given settlement.

The boundary condition can be simulated in the finite element equations by properly modifying the assemblage equations (Equation 3.45). To understand this modification, we shall consider one of the available procedures by writing Equation 3.45 in symbolic form as

$$\begin{bmatrix} k_{11} & k_{12} & 0 & 0 \\ k_{21} & k_{22} & k_{23} & 0 \\ 0 & k_{32} & k_{33} & k_{34} \\ 0 & 0 & k_{43} & k_{44} \end{bmatrix} \begin{Bmatrix} v_1 \\ v_2 \\ v_3 \\ v_4 \end{Bmatrix} = \begin{Bmatrix} R_1 \\ R_2 \\ R_3 \\ R_4 \end{Bmatrix}. \tag{3.48}$$

We delete the fourth row and the fourth column corresponding to v_4, which leads to the modified equations as

$$\begin{bmatrix} k_{11} & k_{12} & 0 \\ k_{21} & k_{22} & k_{23} \\ 0 & k_{32} & k_{33} \end{bmatrix} \begin{Bmatrix} v_1 \\ v_2 \\ v_3 \end{Bmatrix} = \begin{Bmatrix} R_1 \\ R_2 \\ R_3 \end{Bmatrix} \tag{3.49a}$$

or

$$[\overline{\mathbf{K}}]\{\overline{\mathbf{r}}\} = \{\overline{\mathbf{R}}\}. \tag{3.49b}$$

The overbar denotes modified matrices.

The assemblage stiffness matrix (Equation 3.45) is symmetric and banded or sparsely populated. Symmetry implies that $k_{ij} = k_{ji}$, and bandedness implies that nonzero values of k_{ij} occur only on the main diagonal and a few off-diagonals, whereas other coefficients are zero. These properties, which occur in many engineering problems, make solving the equations easier and economical. This is achieved by storing in the computer only the nonzero elements within the banded zone. In Appendix 2 we discuss the solution of systems of such simultaneous equations.

Nonzero Boundary Conditions

If the specified displacement has nonzero value, then the procedure is somewhat different, which can include the homogeneous condition as a special case. For instance, assume that $v_4 = \delta$. Here Equation 3.48 is modified by setting the coefficient k_{44} equal to 1, all other coefficients in row 4 as 0, and $R_4 = \delta$. Therefore

$$\begin{bmatrix} k_{11} & k_{12} & 0 & 0 \\ k_{21} & k_{22} & k_{23} & 0 \\ 0 & k_{32} & k_{33} & k_{34} \\ 0 & 0 & 0 & 1 \end{bmatrix} \begin{Bmatrix} v_1 \\ v_2 \\ v_3 \\ v_4 \end{Bmatrix} = \begin{Bmatrix} R_1 \\ R_2 \\ R_3 \\ \delta \end{Bmatrix}. \tag{3.50a}$$

These equations can now be solved for v_1, v_2, v_3, and v_4. Often, when dealing with large matrices (Appendix 2) when we take advantage of the symmetry of a matrix for reducing storage requirements, it becomes advisable and necessary to restore the symmetry of the matrix in Equation 3.50a that is broken by the modification. It can be done by subtracting from R_1, R_2, and R_3 the quantities $k_{14} \times \delta$, $k_{24} \times \delta$, and $k_{34} \times \delta$, respectively. Thus, the equations reduce to

$$\begin{bmatrix} k_{11} & k_{12} & 0 & 0 \\ k_{21} & k_{22} & k_{23} & 0 \\ 0 & k_{32} & k_{33} & 0 \\ 0 & 0 & 0 & 1 \end{bmatrix} \begin{Bmatrix} v_1 \\ v_2 \\ v_3 \\ v_4 \end{Bmatrix} = \begin{Bmatrix} R_1 - 0 \\ R_2 - 0 \\ R_3 - k_{34} \times \delta \\ \delta \end{Bmatrix}. \tag{3.50b}$$

Here $k_{i4} \times \delta$ $(i = 1, 2, 3)$ is transposed to the right-hand side. For instance, for the third row,

$$0 \times v_1 + k_{32}v_2 + k_{33}v_3 + k_{34} \times \delta = R_3$$

or

$$0 \times v_1 + k_{32}v_2 + k_{33}v_3 + 0 = R_3 - k_{34}\delta. \tag{3.50c}$$

The foregoing boundary conditions expressed in terms of the unknown displacement are the forced or geometric constraints. The natural boundary conditions usually do not need the special consideration required for the geometric boundary condition. The natural boundary condition, if specified as a zero value for a derivative (slope) of the unknown, is satisfied automatically in an integrated sense in the finite element formulation. In other words, the sum of the computed values of the derivative at the boundary vanishes approximately.

Different problems involve different unknowns and different categories of boundary conditions, which we shall discuss.

Step 6. Solve for Primary Unknowns: Nodal Displacements

Equation 3.45 is a set of linear algebraic simultaneous equations. The set for the column problem is linear because the coefficients K_{ij}, which are composed of material properties (E) and geometric properties (l, A), are constants and do not depend on the magnitude or conditions of loading and deformations. For instance, we have assumed linear Hooke's law and small strains and deformations in the formulation of the foregoing equations. We note here that if we assume material behavior to be nonlinear and large strains and deformations, then Equation 3.45 will be nonlinear. Although we shall briefly discuss nonlinear behavior at later stages, detailed discussion of this topic is beyond the scope of this text.

The unknowns in Equation 3.50 are the nodal displacements given by the assemblage vector $\{\mathbf{r}\}^T = [v_1\, v_2\, v_3\, v_4]$ in which v_4 is already known. The equations can be solved by using direct, iterative, or other methods; see Appendix 2. Gaussian elimination and a number of its modifications are one of the common sets of direct methods used for solution of the finite element equations. The direct procedure involves two steps: elimination and back substitution.[5,7]

To illustrate the solution procedure, we now consider two examples.

Example 3.1

We assume the following data for the column problem (Figure 3.8):

Modulus of elasticity	$E = 1000$ kg/cm^2*
Element length	$l = 10$ cm
Cross-sectional area	$A = 1.0$ cm^2
Surface traction	$T_y = 1.0$ kg/cm*
Body weight	$Y = 0.5$ kg/cm^3*
Boundary condition	$v_4 = 0{,}0$

* The metric units are used in this book for some of the examples. They can be converted to SI units, e.g., 1 kg = 9.81 N.

We also assume that the column has uniform cross-section and material properties.

Substitution of these values into Equation 3.41 leads to the following element and assemblage equations:

$$\frac{1000 \times 1}{10}\begin{bmatrix} 1 & -1 \\ -1 & 1 \end{bmatrix}\begin{Bmatrix} v_1 \\ v_2 \end{Bmatrix} = \frac{1 \times 10 \times 0.5}{2}\begin{Bmatrix} 1 \\ 1 \end{Bmatrix} + \frac{10 \times 1}{2}\begin{Bmatrix} 1 \\ 1 \end{Bmatrix}$$

or

$$\begin{bmatrix} 100 & -100 \\ -100 & 100 \end{bmatrix}\begin{Bmatrix} v_1 \\ v_2 \end{Bmatrix} = \begin{Bmatrix} 2.5 \\ 2.5 \end{Bmatrix} + \begin{Bmatrix} 5 \\ 5 \end{Bmatrix}$$

and so on for the other two elements. Assembly of the three elements (Equation 3.45) yields

$$\begin{bmatrix} 100 & -100 & 0 & 0 \\ -100 & 100 & -100 & 0 \\ 0 & -100 & 200 & -100 \\ 0 & 0 & -100 & 100 \end{bmatrix}\begin{Bmatrix} v_1 \\ v_2 \\ v_3 \\ v_4 \end{Bmatrix} = \begin{Bmatrix} 2.5 \\ 5.0 \\ 5.0 \\ 2.5 \end{Bmatrix} + \begin{Bmatrix} 5 \\ 10 \\ 10 \\ 5 \end{Bmatrix}.$$

Introduction of the boundary condition $v_4 = 0$ leads to the modified equations

$$\begin{bmatrix} 100 & -100 & 0 \\ -100 & 200 & -100 \\ 0 & -100 & 200 \end{bmatrix}\begin{Bmatrix} v_1 \\ v_2 \\ v_3 \end{Bmatrix} = \begin{Bmatrix} 2.5 \\ 5.0 \\ 5.0 \end{Bmatrix} + \begin{Bmatrix} 5 \\ 10 \\ 10 \end{Bmatrix} = \begin{Bmatrix} 7.5 \\ 15 \\ 15 \end{Bmatrix}.$$

In the expanded form

$$100v_1 - 100v_2 + 0 = 7.5, \tag{3.51a}$$

$$-100v_1 + 200v_2 - 100v_3 = 15.0, \tag{3.51b}$$

$$0 - 100v_2 + 200v_3 = 15.0. \tag{3.51c}$$

For solving these equations, first we follow the elimination procedure. Equation 3.51a gives

$$100v_1 = 7.5 + 100v_2.$$

Substitution for 100 v_1 in Equation 3.51b yields

$$-7.5 - 100v_2 + 200v_2 - 100v_3 = 15$$

or

$$100v_2 = 22.5 + 100v_3.$$

Now, Equation 3.51c gives

$$-22.5 - 100v_3 + 200v_3 = 15$$

or

$$v_3 = \frac{37.5}{100} \text{ cm.}$$

Back substitution leads to

$$v_2 = \frac{60}{100} \text{ cm}$$

and

$$v_1 = \frac{67.5}{100} \text{ cm.}$$

Example 3.2

Instead of uniform $\bar{Y}$ and $\bar{T}_y$, in this example we apply only a concentrated load equal to 10 kg at the top, that is, at node 1. We can look upon this load as a global joint load P_{1g}. In this case, the final equations are

$$\begin{aligned} 100v_1 - 100v_2 + 0 \quad &= 10, \\ -100v_1 + 200v_2 - 100v_3 &= 0, \\ 0 - 100v_2 + 200v_3 &= 0. \end{aligned} \tag{3.52}$$

Note that the equations on the left-hand side are the same as before; only the right-hand-side load vector is changed. This gives us an idea that for linear problems, where K_{ij} do not change, for a given structure, the elimination process is required only once since we can store the elimination steps in the computer. For different loadings only the back substitution needs to be performed. Thus, we obtain

$$v_1 = \frac{30}{100} \text{ cm} \quad v_4 = 0 \text{ (specified)}$$
$$v_2 = \frac{20}{100} \text{ cm}$$
$$v_3 = \frac{10}{100} \text{ cm}$$

Step 7. Solve for Secondary Unknowns; Strains and Stresses

For the displacement formulation based on potential energy, nodal displacements are the primary unknowns. We call them primary because they are the main unknowns involved in the formulation and in Equation 3.45. The secondary unknowns are those derived from the primary unknowns evaluated in Step 6. For stress analysis problems, these are strains, stresses, moments, shear forces, etc. These designations will change depending on the type of formulation procedure used (see Chapter 11).

Strains and Stresses

By substituting the computed displacements from Step 6 into Equation 3.13, strains in the elements are computed as follows:

Example 3.1 (continued)

For element 1,

$$\varepsilon_y(1) = \frac{1}{10}\begin{bmatrix}-1 & 1\end{bmatrix}\frac{1}{100}\begin{Bmatrix}67.5 \\ 60\end{Bmatrix}$$
$$= \frac{-7.5}{1000}.$$

Similarly, the strains for elements 2 and 3 are

$$\varepsilon_y(2) = \frac{-22.5}{1000}$$

and

$$\varepsilon_y(3) = \frac{-37.5}{1000}.$$

Axial stresses in the three elements can now be derived by using Equation 3.14:

$$\sigma_y(1) = \frac{-7.5}{1000} \times 1000 = -7.5 \text{ kg/cm}^2,$$

$$\sigma_y(2) = \frac{-22.5}{1000} \times 1000 = -22.5,$$

$$\sigma_y(3) = \frac{-37.5}{1000} \times 1000 = -37.5.$$

The negative sign denotes compressive stresses.

Example 3.2 (continued)

The strains and stresses are

$$\varepsilon_y(1) = \frac{-10}{1000},$$

$$\varepsilon_y(2) = \frac{-10}{1000},$$

$$\varepsilon_y(3) = \frac{-10}{1000},$$

and

$$\sigma_y(1) = -10.0 \text{ km/cm}^2,$$

$$\sigma_y(2) = -10.0,$$

$$\sigma_y(3) = -10.0.$$

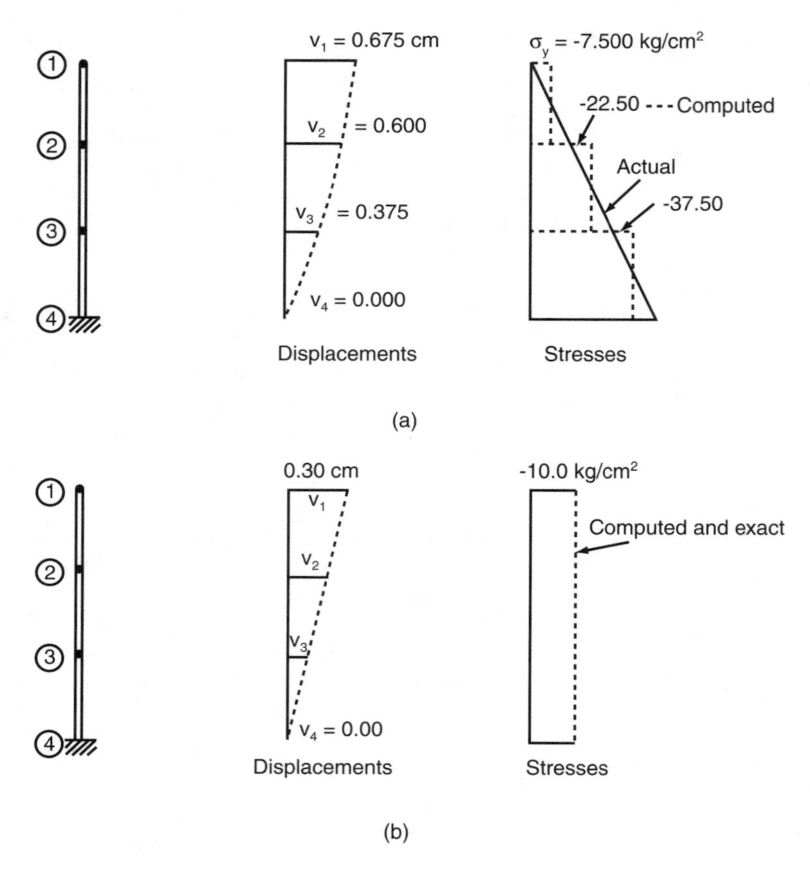

FIGURE 3.9
Distribution of computed displacements and stresses: (a) Results for Example 3.1 and (b) Results for Example 3.2.

Step 8. Interpretation and Display of Results

Figure 3.9 shows plots of displacements and stresses for the two previous examples. For Example 3.1, the distribution of displacements within each element is linear, whereas for the entire column, the displacement distribution need not be linear. The computed stresses are constant within each element, whereas the actual distribution over the entire column can be linear. Existence of the constant stress states is dictated by the assumed displacement model, which is linear.

In Example 3.2, the computed displacements and stresses are the same as the exact value from closed form solutions, because for the exact solution,

$$\sigma_y = -\frac{P}{A} = \frac{-10}{1} = -10 \text{ kg/cm}^2$$

and

$$v_1 = \frac{P(3l)}{AE} = \frac{10 \times 30}{1000 \times 1} = \frac{30}{100}, \text{ etc.}$$

An important observation can be made at this stage. The accuracy of the assumed approximation will depend on the type of loading, geometry, and material properties. In Example 3.1, the loading varies from zero at the top to the highest value at the base for the gravity load $\bar{Y}$ and is uniformly distributed for the surface loading $\bar{T}_y$. The computed stresses are an average of the actual, and the accuracy of the finite element computations is considered satisfactory.

For Example 3.2, the numerical solution is identical with the exact solution, because there is only one concentrated load. A number of other factors can influence the accuracy of numerical predictions and can require additional considerations for improvements to account for the factors.

The accuracy of the predictions can be improved by using two methods: (1) finer mesh and/or (2) higher-order approximation models. The decision as to which approach should be used depends on the characteristics of the problem, and trade-offs exist with respect to accuracy, reliability, and computer cost.

Very often it becomes essential to use higher-order models. If the column (Figure 3.8) has irregular geometry and loading, it may be useful to consider higher-order models. For instance, we can use a quadratic approximation for v as

$$\begin{aligned} v &= \tfrac{1}{2}L(L-1)v_1 + \tfrac{1}{2}L(L+1)v_2 + (1-L^2)v_0 \\ &= [\mathbf{N}]\{\mathbf{q}\}, \end{aligned} \tag{3.53}$$

where

$$[\mathbf{N}] = \left[\tfrac{1}{2}L(L-1) \quad \tfrac{1}{2}L(L+1) \quad (1-L^2)\right]$$

and

$$\{\mathbf{q}\}^T = [v_1 \quad v_2 \quad v_0].$$

The element here has three nodes (Figure 3.10), with the third node 0 in the middle of the element. It is not necessary to have the third node in the middle. In fact, we can choose the node anywhere within the element; then the interpolation functions N_i will be different for such choices.

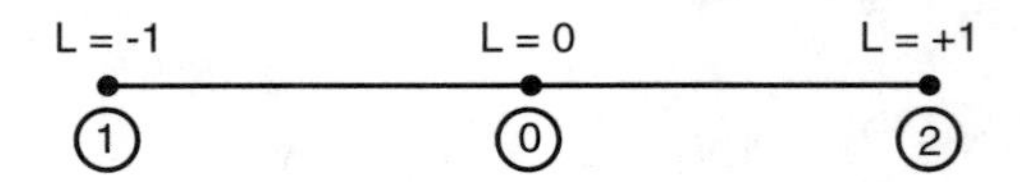

FIGURE 3.10
Element with quadratic or second-order approximation model.

The steps of the finite element formulations can be repeated to derive the required equations for the higher-order element. This is left to the reader as an exercise; see Problem 3.8.

Formulation by Galerkin's Method

In the case of the variational approach, we applied the procedure to a single generic element and used the results recursively for deriving properties of all elements. Then the individual element results were combined to obtain the assemblage equations for the entire discretized body. We also emphasized that in reality the variational procedure was applied to the entire domain, and it was only for convenience that we chose to analyze element by element. To illustrate this concept, the assemblage equations (Equation 3.45) were first derived by considering the total potential energy of the discretized body.

As explained in Chapter 2 and in Appendix 1, Galerkin's residual procedure is also applied so as to minimize the residual over the entire domain. Another important aspect of this procedure is that the approximation function and the functions φ_i (Equation 2.12) are defined over the total domain.

While applying Galerkin's method for finite element analysis, we use the interpolation functions N_i as the weighting functions (Equation 2.14), which are defined over the entire domain. In view of the foregoing characteristics of Galerkin's method and because for the variational procedure we have defined N_i relevant simply for an element, it is necessary to clarify a number of aspects and explain the relevance of N_i over the total domain.

Explanation and Relevance of Interpolation Functions

Figure 3.11(a) shows an approximation v to the exact solution v^* defined for the total domain of a one-dimensional body. The approximate or trial function v for the displacement over the entire domain can be expressed in the sense of Equation 2.12 as

$$v = \sum_{k=1}^{M} \sum_{j=1}^{m} N_j^k v_j^k, \tag{3.54}$$

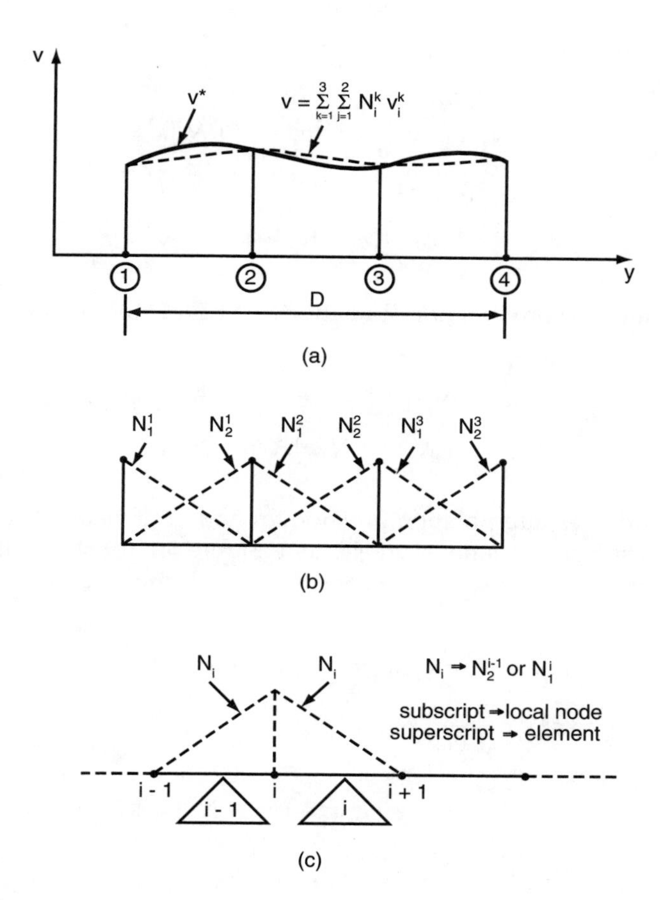

FIGURE 3.11
Approximation and interpolation functions in Galerkin's method: (a) Approximation for v over total domain D, (b) Interpolation functions over D, (c) Interpolation function for node i.

where the superscript k denotes an element, M is the total number of elements, and m is the number of interpolation functions per element.

For the column with three elements (Figure 3.8), Equation 3.54 can be expressed as follows:

Sum for elements, k = 1, 2, 3;

$$v = \sum_{j=1}^{m} N_j^1 v_j^1 + \sum_{j=1}^{m} N_j^2 v_j^2 + \sum_{j=1}^{m} N_j^3 v_j^3. \tag{3.55a}$$

Sum over interpolation functions, j = 1, 2,. …, m:

$$\begin{aligned} v = & N_1^1 v_1^1 + N_2^1 v_2^1 + N_3^1 v_3^1 + \cdots + N_m^1 v_m^1 \\ & + N_1^2 v_1^2 + N_2^2 v_2^2 + N_3^2 v_3^2 + \cdots + N_m^2 v_m^2 \\ & + N_1^3 v_1^3 + N_2^3 v_2^3 + N_3^3 v_3^3 + \cdots + N_m^3 v_m^3 . \end{aligned} \tag{3.55b}$$

For the linear approximation (Equation 3.6) m = 2; therefore

$$\begin{aligned} v = & N_1^1 v_1^1 + N_2^1 v_2^1 + N_1^2 v_1^2 + N_2^2 v_2^2 \\ & + N_1^3 v_1^3 + N_2^3 v_2^3 . \end{aligned} \tag{3.55c}$$

Here N_2^1 denotes interpolation function for node 2 of element 1, and so on, where the superscript denotes an element. Figure 3.11(b) shows the plots of the linear interpolation functions over the three elements. Since $v_1^1 = v_1$,

$$v_2^1 = v_1^2 = v_2,\ v_2^2 = v_1^3 = v_3, \text{ and } v_2^3 = v_4, \tag{3.56}$$

we have

$$\begin{aligned} v & = N_1^1 v_1 + \left(N_2^1 + N_1^2\right) v_2 + \left(N_2^2 + N_1^3\right) v_3 + N_2^3 v_4 \\ & = \sum N_i v_i = N_1 v_1 + N_2 v_2 + N_3 v_3 + N_4 v_4 , \end{aligned}$$

where N_i denotes interpolation functions around node i; for instance, in general $N_2 \Rightarrow N_2^1 + N_1^2$. For the linear case (Figure 3.11(c)),

$$N_i(y) = \begin{cases} \dfrac{y - y_{i-1}}{y_i - y_{i-1}} = N_2^{i-1}, & y_{i-1} \le y \le y_i , \\ \dfrac{y_{i+1} - y}{y_{i+1} - y_i} = N_1^i , & y_i \le y \le y_{i+1} , \end{cases} \tag{3.57a}$$

$$= \begin{cases} s_{i-1}, & s_{i-1} = \dfrac{y - y_{i-1}}{l_{i-1}} , \\ s_i , & s_i = \dfrac{y_{i+1} - y}{l_i} , \end{cases} \tag{3.57b}$$

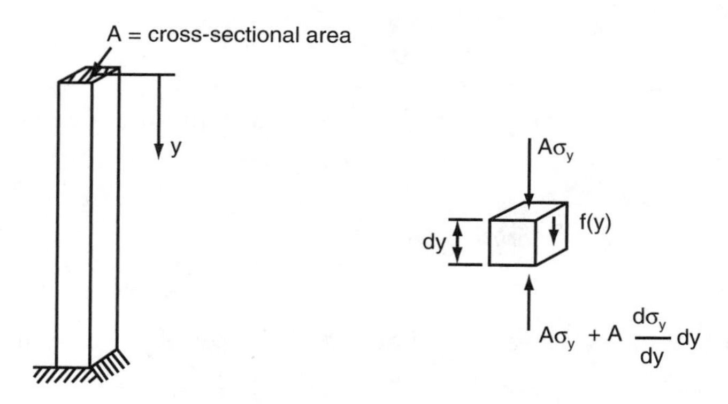

FIGURE 3.12
Equilibrium of column segment.

where s is the local coordinates $0 \leq s \leq 1$, and l_i denotes the length of element *i*. The definition of N_i in Equation 3.57a is valid only for points within the domain D. For each end node one of the values will be relevant. Thus the interpolation function around node *i* has nonzero values only in the two adjoining elements $i - 1$ and *i* and zero values in all other elements.

With the foregoing explanation, we are ready to use Galerkin's method for formulating finite element equations for the column problem. We first derive the governing differential equation for the column (Figure 3.12). By considering the equilibrium of forces for an elemental volume of the column, we have

$$A\sigma_y - A\sigma_y - A\frac{d\sigma_y}{dy}dy + f(y)dy = 0$$

or

$$A\frac{d\sigma_y}{dy} = f(y). \tag{3.58a}$$

Substitution of $d\sigma_y/dy$ from Equation 3.14a after insertion of ε_y from Equation 3.12a gives (after dropping the subscript to *E* for convenience)

$$AE\frac{d^2v^*}{dy^2} = f(y). \tag{3.58b}$$

Here v^* is the exact solution, and we have considered *f*(*y*) as surface traction $\bar{T}(y)$ per unit length per unit surface area. It is possible to consider *f*(*y*) as body force $\bar{Y}(y)$ and joint load P_{il}; the latter can be defined in a function form as

$$P_{il}(y) = P_{il}(y_j) \times \delta(y - y_j), \tag{3.59a}$$

where $P_{il}(y_j)$ = magnitude of joint load at point j and $\delta(y - y_j)$ is the Dirac delta function,

$$\delta(y - y_j)dy = \begin{cases} 1, & y = y_j, \\ 0, & y \neq y_j. \end{cases} \tag{3.59b}$$

The residual for the differential Equation 3.58b and the linear approximation function 3.6 is

$$R(v) = EA\frac{d^2v}{dy^2} - f. \tag{3.60}$$

For convenience we wrote f instead of $f(y)$. Minimization of R with respect to N_i (Equation 3.6) leads to

$$\int_{0=y_1}^{h=y_4} \left(EA\frac{d^2v}{dy^2} - f \right) N_i dy = 0, \tag{3.61}$$

where h is the total length of the column equal to y_4, the coordinate of the last node (Figure 3.8).

Integration by parts for the first term gives

$$\int_{y_1}^{y_4} EA\frac{d^2v}{dy^2} N_i dy = EA\frac{dv}{dy} l_y N_i \bigg|_{y_1}^{y_4} - EA\int_{y_1}^{y_4} \frac{dv}{dy} \cdot \frac{dN_i}{dy} dy, \tag{3.62a}$$

and hence Equation 3.61 is

$$-EA\int_{y_1}^{y_4} \frac{dv}{dy} \cdot \frac{dN_i}{dy} dy = \int_{y_1}^{y_4} f N_i dy - EA\frac{dv}{dy} N_i \bigg|_{y_1}^{y_4}. \tag{3.62b}$$

We inserted l_y, direction cosine of angle between the axis of the column and the y axis, as unity, since $\cos\theta_y = \cos 0 = 1$.

Now if v from Equation 3.56 is substituted into Equation 3.62b, we obtain

$$-EA\sum_{j=1}^{4} \int_{y_1}^{y_4} \frac{dN_j}{dy} \cdot \frac{dN_i}{dy} \cdot dy v_j = \int_{y_1}^{y_4} f N_i dy - EA\frac{dv}{dy} N_i \bigg|_{y_1}^{y_4}, \tag{3.63}$$

$$i = 1,2,3,4, \ j = 1,2,3,4.$$

The index notation implies

$$\begin{aligned} \text{set } i &= 1 \text{ and vary } j = 1, 2, 3, 4, \\ i &= 2 \text{ and vary } j = 1, 2, 3, 4, \\ i &= 3 \text{ and vary } j = 1, 2, 3, 4, \\ i &= 4 \text{ and vary } j = 1, 2, 3, 4, \end{aligned}$$

which leads to

$$EA\int_{y_1}^{y_4}\begin{bmatrix} \frac{dN_1}{dy}\cdot\frac{dN_1}{dy} & \frac{dN_2}{dy}\cdot\frac{dN_1}{dy} & \frac{dN_3}{dy}\cdot\frac{dN_1}{dy} & \frac{dN_4}{dy}\cdot\frac{dN_1}{dy} \\ \frac{dN_1}{dy}\cdot\frac{dN_2}{dy} & \frac{dN_2}{dy}\cdot\frac{dN_2}{dy} & \frac{dN_3}{dy}\cdot\frac{dN_2}{dy} & \frac{dN_4}{dy}\cdot\frac{dN_2}{dy} \\ \frac{dN_1}{dy}\cdot\frac{dN_3}{dy} & \frac{dN_2}{dy}\cdot\frac{dN_3}{dy} & \frac{dN_3}{dy}\cdot\frac{dN_3}{dy} & \frac{dN_4}{dy}\cdot\frac{dN_3}{dy} \\ \frac{dN_1}{dy}\cdot\frac{dN_4}{dy} & \frac{dN_2}{dy}\cdot\frac{dN_4}{dy} & \frac{dN_3}{dy}\cdot\frac{dN_4}{dy} & \frac{dN_4}{dy}\cdot\frac{dN_4}{dy} \end{bmatrix} dy \begin{Bmatrix} v_1 \\ v_2 \\ v_3 \\ v_4 \end{Bmatrix}$$

$$= -\int_{y_1}^{y_4}\begin{Bmatrix} fN_1 \\ fN_2 \\ fN_3 \\ fN_4 \end{Bmatrix} dy + EA\begin{Bmatrix} \frac{dv}{dy}N_1\Big|_{y_1}^{y_4} \\ \frac{dv}{dy}N_2\Big|_{y_1}^{y_4} \\ \frac{dv}{dy}N_3\Big|_{y_1}^{y_4} \\ \frac{dv}{dy}N_4\Big|_{y_1}^{y_4} \end{Bmatrix} \tag{3.64a}$$

or

$$[\mathbf{K}]\{\mathbf{r}\} = \{\mathbf{R}\} = \{\mathbf{R}_T\} + \{\mathbf{R}_B\}. \tag{3.64b}$$

It is important to note that the terms in Equation 3.64 are integrated over the total length and that the limits in the boundary term $(dv/dy)\mathrm{N}_i\big|_{y1}^{y4}$, are defined at the ends.

Terms such as

$$\int_{y_1}^{y_4} \frac{dN_3}{dy} \cdot \frac{dN_1}{dy} dy = \int_{y_1}^{y_2} \frac{dN_3}{dy} \cdot \frac{dN_1}{dy} dy + \int_{y_2}^{y_3} \frac{dN_3}{dy} \cdot \frac{dN_1}{dy} dy + \int_{y_3}^{y_4} \frac{dN_3}{dy} \cdot \frac{dN_1}{dy} dy$$

vanish because $N_3 = 0$ in region y_1 to y_2 and $N_i = 0$ in regions y_2 to y_3, and y_3 to y_4. Thus only those terms that have nonzero values in one or two elements contribute to the integration. Hence, the left-hand side reduces to

$$EA\int_{y_1}^{y_4} \begin{bmatrix} \frac{dN_1}{dy} \cdot \frac{dN_1}{dy} & \frac{dN_2}{dy} \cdot \frac{dN_1}{dy} & 0 & 0 \\ \frac{dN_1}{dy} \cdot \frac{dN_2}{dy} & \frac{dN_2}{dy} \cdot \frac{dN_2}{dy} & \frac{dN_3}{dy} \cdot \frac{dN_2}{dy} & 0 \\ 0 & \frac{dN_2}{dy} \cdot \frac{dN_3}{dy} & \frac{dN_3}{dy} \cdot \frac{dN_3}{dy} & \frac{dN_4}{dy} \cdot \frac{dN_3}{dy} \\ 0 & 0 & \frac{dN_3}{dy} \cdot \frac{dN_4}{dy} & \frac{dN_4}{dy} \cdot \frac{dN_4}{dy} \end{bmatrix} dy \begin{Bmatrix} v_1 \\ v_2 \\ v_3 \\ v_4 \end{Bmatrix} \tag{3.65a}$$

or

$$[\mathbf{K}]\{\mathbf{r}\},$$

where

$$[\mathbf{K}] = \tag{3.65b}$$

$$EA\begin{bmatrix} \int_{y_1}^{y_2} \frac{dN_1}{dy} \cdot \frac{dN_1}{dy} dy & \int_{y_1}^{y_2} \frac{dN_2}{dy} \cdot \frac{dN_1}{dy} dy & 0 & 0 \\ \int_{y_1}^{y_2} \frac{dN_1}{dy} \cdot \frac{dN_2}{dy} dy & \int_{y_1}^{y_2} \frac{dN_2}{dy} \cdot \frac{dN_2}{dy} dy + \int_{y_2}^{y_3} \frac{dN_2}{dy} \cdot \frac{dN_2}{dy} dy & \int_{y_2}^{y_3} \frac{dN_3}{dy} \cdot \frac{dN_2}{dy} dy & 0 \\ 0 & \int_{y_2}^{y_3} \frac{dN_2}{dy} \cdot \frac{dN_3}{dy} dy & \int_{y_2}^{y_3} \frac{dN_3}{dy} \cdot \frac{dN_3}{dy} dy + \int_{y_3}^{y_4} \frac{dN_3}{dy} \cdot \frac{dN_3}{dy} dy & \int_{y_3}^{y_4} \frac{dN_4}{dy} \cdot \frac{dN_3}{dy} dy \\ 0 & 0 & \int_{y_3}^{y_4} \frac{dN_3}{dy} \cdot \frac{dN_4}{dy} dy & \int_{y_3}^{y_4} \frac{dN_4}{dy} \cdot \frac{dN_4}{dy} dy \end{bmatrix}.$$

Note that the terms enclosed in the dashed lines represent $[\mathbf{B}]^T[\mathbf{B}]$ for the three elements. This indicates that the contributions of the element stiffness properties are added such that compatibility at common nodes is assured. Integrations for the relevant limits will lead to the assemblage stiffness matrix $[\mathbf{K}]$. For instance,

$$K_{22} = EA\int_{y_1}^{y_2} \frac{dN_2^1}{dy} \cdot \frac{dN_2^1}{dy} dy + EA\int_{y_2}^{y_3} \frac{dN_1^2}{dy} \cdot \frac{dN_1^2}{dy} dy. \tag{3.66a}$$

Assuming elements of equal length l, we have

$$N_2^1 = s_1 = (y - y_1)/l,$$
$$N_1^2 = s_2 = (y_3 - y)/l. \tag{3.66b}$$

Now $ds_1 = dy/l$ and $ds_2 = -dy/l$. By noting that

$$\frac{dN_2^1}{dy} = \frac{dN_2^1}{ds_1} \cdot \frac{ds_1}{dy} = \frac{1}{l}, \text{ and } \frac{dN_1^2}{dy} = \frac{dN_1^2}{ds_2} \cdot \frac{ds_2}{dy} = -\frac{1}{l},$$

we have

$$K_{22} = \frac{AE}{l}\left[\int_0^1 (1)(1)ds_1 + \int_1^0 (-1)(-1)ds_2\right] = \frac{2}{l}, \tag{3.66c}$$

and finally

$$[\mathbf{K}] = \frac{EA}{l}\begin{bmatrix} 1 & -1 & 0 & 0 \\ -1 & 2 & -1 & 0 \\ 0 & -1 & 2 & -1 \\ 0 & 0 & 1 & 1 \end{bmatrix}, \tag{3.67a}$$

which is the same as [**K**] in Equation 3.45 for constant element length and uniform A and E.

The first term on the right-hand side of Equation 3.64a denotes

$$\{\mathbf{R}_T\} = -\begin{Bmatrix} \int_{y_1}^{y_2} f N_1^1 dy \\ \int_{y_1}^{y_2} f N_2^1 dy + \int_{y_2}^{y_3} f N_1^2 dy \\ \int_{y_2}^{y_3} f N_2^2 dy + \int_{y_3}^{y_4} f N_1^3 dy \\ \int_{y_3}^{y_4} f N_2^3 dy \end{Bmatrix} = -\frac{fl}{2}\begin{Bmatrix} 1 \\ 2 \\ 2 \\ 1 \end{Bmatrix} \tag{3.67b}$$

which is the same as the contribution to the assemblage load vector $\{\mathbf{R}\}$ by $\bar{T}_y$, in Equation 3.45.

The second term on the right-hand side in Equation 3.64a denotes essentially the end boundary terms, because N_i have nonzero values only at the end points 1 and 4. For instance,

$$\frac{dv}{dy}N_2\Bigg|_{y_1}^{y_4} = \frac{dv}{dy}N_2^1\Bigg|_{y_1}^{y_4} + \frac{dv}{dy}N_1^2\Bigg|_{y_1}^{y_4} = 0$$

since $N_1^2 = N_2^1 = 0$ at nodes 1 and 4. Hence it reduces to

$$\{\mathbf{R}_B\} = AE\begin{Bmatrix} \left(-\dfrac{dv}{dy}N_1^1\right)_1 \\ 0 \\ 0 \\ \left(\dfrac{dv}{dy}N_2^3\right)_4 \end{Bmatrix}. \tag{3.67c}$$

Since $N_1^1 = N_2^3 = 1$ at nodes 1 and 4, we have

$$\{\mathbf{R}_B\} = \begin{Bmatrix} \left(-AE\dfrac{dv}{dy}\right)_1 \\ 0 \\ 0 \\ \left(-AE\dfrac{dv}{dy}\right)_4 \end{Bmatrix}. \tag{3.67d}$$

Equation 3.67d represents *natural* boundary conditions defined through prescribed values of (normal) derivative of the displacement. The term $AE(dv/dy) = AE\varepsilon_y = A\sigma_y$ denotes (internal) force, and hence

$$\{\mathbf{R}_B\} = \begin{Bmatrix} F_1 \\ 0 \\ 0 \\ F_4 \end{Bmatrix} \tag{3.67e}$$

where F_1 and F_4 are forces at the boundary nodes.

Comment

If an external point load is applied at node 1, F_1 equals the prescribed external load. If there is no external load there, $F_1 = 0$. When the column base has a prescribed boundary condition, say $v_4 = 0$, then the term F_4 will not appear in the final results when the assemblage equations (Equation 3.64) are modified for the given boundary condition. In other words, the natural boundary condition at a point will not have influence on the final results if that point has a prescribed geometric boundary condition.

It is important to understand the foregoing idea that Galerkin's residual procedure is applied to the entire domain and that its application yields boundary conditions which should be properly interpreted. In the subsequent use of Galerkin's method we may consider it to be applied element by element, with a clear understanding of the foregoing implications.

When Galerkin's procedure is applied to an element, we can look upon the interpolation functions for the element to have nonzero values within themselves and zero elsewhere. For instance, in Figure 3.11, N_1^2 and N_2^2 have nonzero values within element 2 but zero values at other locations in the domain.

The foregoing finite element equations from the Galerkin method are essentially the same as those from the variational approach. This happens usually in the case of problems mathematically classified as self-adjoint.[8] In a rather elementary manner self-adjointness can be attributed to the existence of the symmetric operator matrix [**K**]. If the governing equations are nonlinear or the problem is nonself-adjoint, the two results can be different. An example of a nonself-adjoint problem is given in Chapter 8.

Computer Implementation

Although we could perform hand calculations for the one-dimensional column problem, almost all problems solved by using the finite element method involve large matrices, and recourse has to be made to electronic computation.

To understand computer codes in a progressive manner, we plan to introduce a rather simple code (program) DFT/C-1DFE in Chapter 6. This code can handle the problems of axial deformation, one-dimensional flow, and one-dimensional temperature distribution or consolidation.

The reader should study the code in Chapter 6 and use it for solving the problems in Examples 3.1 and 3.2 and some of the subsequent problems.

Other Procedures for Formulation

In this section, we shall introduce alternative methods of formulation by using variational principles such as the principle of stationary complementary energy and the mixed principle.[5,9-11] Although hybrid principles are found to be successful for a number of situations, they will not be covered in this elementary text except for one example for Torsion in Chapter 11.

Comment

For an undergraduate course, the instructor may wish to postpone coverage of the following material for a later stage or for a graduate course.

Complementary Energy Approach

The complementary energy Π_c is defined as the sum of the complementary strain energy U_c and the complementary potential of the external loads W_{pc}

$$\Pi_c = U_c + W_{pc}. \tag{3.68a}$$

Graphical interpretation of U_c is shown in Figure 3.4. According to the principle of stationary (minimum for an elastic body in equilibrium) complementary energy, we have

$$\delta\Pi_c = \delta U_c + \delta W_{pc} = 0$$

or

$$\delta\Pi_c = \delta U_c - \delta W_c = 0, \tag{3.68b}$$

where W_c is the complementary work of external loads. The minus sign occurs because the potential is lost into work. In terms of the components of stresses, strains, and displacements,

$$\Pi_c = \tfrac{1}{2}\iiint_V \{\sigma\}^T[\mathbf{D}]\{\sigma\}dV - \iint_{S_2}\{\mathbf{T}\}^T\{\bar{\mathbf{u}}\}dS, \tag{3.68c}$$

where $[\mathbf{D}]$ is the strain–stress matrix = $[\mathbf{C}]^{-1}$, $\{\bar{\mathbf{u}}\}$ is the vector of prescribed displacements, and S_2 is the part of the boundary on which $\{\bar{\mathbf{u}}\}$ are prescribed. For the column problem, Equation 3.68c can specialize to

$$\Pi_c = \frac{A}{2}\int_{y_1}^{y_2}\sigma_y\frac{1}{E}\sigma_y dy - \int_{y_1}^{y_2}T_y\bar{v}dy. \tag{3.68d}$$

In the case of the complementary energy approach, we assume the stresses as unknowns and express them in terms of nodal stresses, or joint or nodal forces. Hence, the approach is also called the *stress* or *equilibrium* method. For matrix structural analysis, this procedure can specialize to the well-known force method. Advanced study and applications of this approach are beyond the scope of this text; however, we shall illustrate the method by using the rather elementary problem of the axial column.

We express stresses $\{\sigma\}$ in the column element (Figure 3.7) as

$$\{\sigma\} = [\mathbf{N}_\sigma]\{\mathbf{Q}_n\}, \tag{3.69a}$$

where $[\mathbf{N}_\sigma]$ is the matrix of interpolation functions and $\{\mathbf{Q}_n\}$ is the vector of nodal or joint forces related to a statically determinate support system for the element.[5,6,9,12,13]

The surface tractions $\{\mathbf{T}\}$ can be expressed as

$$\{\mathbf{T}\} = [\mathbf{N}_s]\{\mathbf{Q}_n\}, \tag{3.69b}$$

where $[\mathbf{N}_\sigma]$ is a relevant matrix of interpolation functions.

Substitution of Equations 3.69a and 3.69b into Π_c (Equation 3.68d) gives

$$\Pi_c = \frac{1}{2}\iiint_V \{\mathbf{Q}_n\}^T[\mathbf{N}_\sigma]^T[\mathbf{D}][\mathbf{N}_\sigma]\{\mathbf{Q}_n\}dV - \iint_{S_2}\{\mathbf{Q}_n\}^T[\mathbf{N}_s]^T\{\bar{\mathbf{u}}\}dS. \tag{3.70}$$

Differentiation of Π_c with respect to $\{\mathbf{Q}_n\}$and equating to zero leads to

$$[\mathbf{f}]\{\mathbf{Q}_n\} = \{\bar{\mathbf{q}}\} \tag{3.71a}$$

where

$$[\mathbf{f}] = \iiint_V [\mathbf{N}_\sigma]^T[\mathbf{D}][\mathbf{N}_\sigma]dv \tag{3.71b}$$

and

$$\{\bar{\mathbf{q}}\} = \iint_{S_2}[N_s]^T\{\bar{\mathbf{u}}\}dS. \tag{3.71c}$$

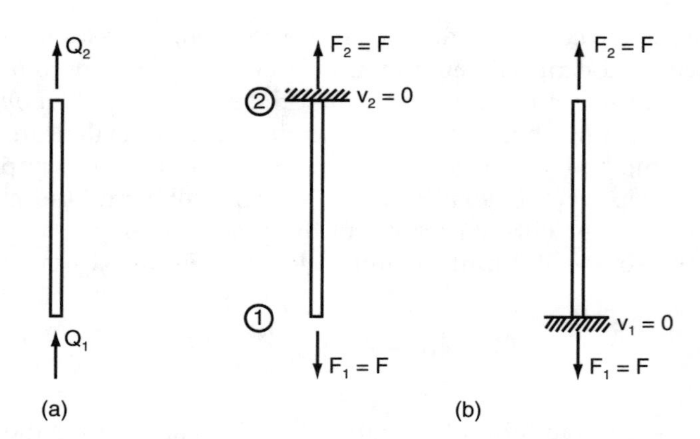

FIGURE 3.13
Force systems: (a) Displacement approach and (b) Stress approach.

[f] is called the element flexibility matrix and $\{\bar{\mathbf{q}}\}$ is the vector of prescribed displacements.

The element flexibility equations can be assembled into global equations and then solved for the unknown forces (stresses). This procedure can be difficult and cumbersome. It is often convenient to transform the element flexibility matrix into the element stiffness matrix.

Example 3.3

Consider the column problem (Figure 3.7). The force systems used in the displacement and stress approaches are shown in Figure 3.13. In the case of the latter, the flexibility matrix is formed by eliminating the degree of freedom corresponding to the rigid-body motion; it is done by constraining one of the ends while the other is loaded, which yields a stable element.

For F acting at node 1, $v_2 = 0$, and for F acting at node 2, $v_1 = 0$. Under these circumstances the axial stress can be expressed as

$$\sigma_y = \frac{1}{A}[1]\{F_1\} = [N_\sigma]\{Q_n\}. \tag{3.72}$$

Then use of Equation 3.71b gives the flexibility matrix as

$$\begin{aligned}[\mathbf{f}] &= A\int_{y_1}^{y_2} \frac{1}{A}\frac{1}{E}\frac{1}{A}dy \\ &= \frac{l}{AE}.\end{aligned} \tag{3.73}$$

The equation relating the reactions is given by

$$F_2 = -F_1 = [-1]F_1 = [\mathbf{G}]F_1. \tag{3.74}$$

From matrix structural analysis, the flexibility relation can be transformed to the stiffness relation as[9,12,13]

$$[\mathbf{k}] = \left[\begin{array}{c:c} [\mathbf{f}]^{-1} & [\mathbf{f}]^T[\mathbf{G}]^T \\ [\mathbf{G}][\mathbf{f}]^{-1} & [\mathbf{G}][\mathbf{f}]^{-1}[\mathbf{G}] \end{array}\right] = \begin{bmatrix} \frac{AE}{l} & \frac{-AE}{l} \\ \frac{-AE}{l} & \frac{AE}{l} \end{bmatrix} \tag{3.75}$$

which is the same matrix $[\mathbf{k}]$ as in Equation 3.41.

Comment

The complementary energy approach can also be used in conjunction with the concept of stress functions. This approach can be easier to implement and understand in comparison to the foregoing procedure. However, it is somewhat difficult to formulate with one-dimensional problems; we have illustrated its application for the two-dimensional torsion (Chapter 11).

Mixed Approach

In the mixed approach, both the element displacements and stresses are assumed to be unknowns. The formulation results in coupled sets of equations in which nodal displacements and stresses appear as unknowns.

Variational Method

In the mixed method, the variational functional, often known as the Hellinger-Reissner principle, is expressed as[5,10]

$$\Pi_R = \iiint_V \left(\{\sigma\}^T\{\varepsilon\} - \tfrac{1}{2}\{\sigma\}^T[\mathbf{D}]\{\sigma\} - \{\mathbf{u}\}^T\{\overline{\mathbf{X}}\}\right)dV$$
$$- \iint_{S_1} \{\mathbf{u}\}^T\{\overline{\mathbf{T}}\}dS_1 - \iint_{S_2} \{\mathbf{u}-\overline{\mathbf{u}}\}^T\{\mathbf{T}\}dS_2 - \{\overline{\mathbf{P}}_{1l}\}^T\{\mathbf{q}_l\}. \tag{3.76}$$

Here $(u - \bar{u})$ denotes the difference between the actual and prescribed displacements along the boundary S_2, and $\{\overline{\mathbf{P}}_i\}$ and $\{\mathbf{q}_i\}$ are the vectors of applied nodal or joint loads and displacements, respectively.

The stationary value of Π_R yields a set of equilibrium and stress–displacement equations. In these equations both the displacements and the stresses simultaneously appear as primary unknowns.

Residual Methods

These methods can be used to formulate finite element equations based on governing differential equations. For example, Galerkin's method can be applied to the equilibrium and stress–displacement (stress–strain) equations, which results in coupled equations in displacements and stresses.

Example 3.4

We illustrate the mixed approach for the column problem (Figure 3.7). As in Equation 3.8d the vertical displacement is assumed to be linear:

$$\begin{aligned} v &= \tfrac{1}{2}\left(1-L\right)v_1 + \tfrac{1}{2}\left(1+L\right)v_2 \\ &= \left[\mathbf{N}_u\right]\{\mathbf{q}\}. \end{aligned} \tag{3.77a}$$

Here $[\mathrm{N}_u]$ is the same as [N] in Equation 3.8d.

The axial stress can also be assumed linear as

$$\begin{aligned} \sigma_y &= \tfrac{1}{2}\left(1-L\right)\sigma_1 + \tfrac{1}{2}\left(1+L\right)\sigma_2 \\ &= \left[\mathbf{N}_\sigma\right]\{\sigma_n\}, \end{aligned} \tag{3.77b}$$

where $\{\sigma_n\}^T = [\sigma_1\ \sigma_2]$ is the vector of nodal stresses and $[N_\sigma]$ is the matrix of interpolation functions. Note that here we have assumed $[N_u] = [N_\sigma] = [\mathrm{N}]$.

Variational Method

Substitution of v, σ_y, and ε_y into the special form of Equation 3.76 leads to

$$\begin{aligned}\Pi_R &= \frac{Al}{2}\int_{-1}^{1}\{\sigma_n\}^T[\mathbf{N}]^T[\mathbf{B}]\{\mathbf{q}\}dL \\ &\quad -\frac{Al}{4}\int_{-1}^{1}\{\sigma_n\}^T[\mathbf{N}]^T\frac{1}{E}[\mathbf{N}]\{\sigma_n\}dL-\left(P_1v_1+P_2v_2\right).\end{aligned} \tag{3.78}$$

Here for convenience we have not included the body force, surface traction, and the term related to difference in displacements. These terms can, however, be included without difficulty.

Now we differentiate Π_R with respect to σ_1, σ_2 and ($\{\sigma_n\}$) and v_1, v_2 ($\{\mathbf{q}\}$) and equate the results to zero:

$$\frac{\partial\Pi_R}{\partial\{\sigma_n\}}=0\Rightarrow\frac{Al}{2}\int_{-1}^{1}[\mathbf{N}]^T[\mathbf{B}]dL\{\mathbf{q}\}-\frac{Al}{4E}\int_{-1}^{1}[\mathbf{N}]^T[\mathbf{N}]dL\{\sigma_n\}=0, \tag{3.79a}$$

$$\frac{\partial\Pi_R}{\partial\{\mathbf{q}\}}=0\Rightarrow\frac{Al}{2}\int_{-1}^{1}[\mathbf{B}]^T[\mathbf{N}]dL\{\sigma_n\}+\{0\}=\{\mathbf{P}_{il}\}, \tag{3.79b}$$

or

$$\begin{aligned}&[\mathbf{k}_{\tau\tau}]\{\sigma_n\}+[\mathbf{k}_{\tau u}]\{\mathbf{q}\}=0,\\ &[\mathbf{k}_{\tau u}]^T\{\sigma_n\}+\{0\}=\{\mathbf{Q}\},\end{aligned} \tag{3.79c}$$

or

$$\begin{bmatrix}[\mathbf{k}]_{\tau\tau} & [\mathbf{k}_{\tau\mathbf{u}}]\\ [\mathbf{k}_{\tau u}]^T & [0]\end{bmatrix}\begin{Bmatrix}\{\sigma_n\}\\ \{\mathbf{q}\}\end{Bmatrix}=\begin{Bmatrix}\{0\}\\ \{\mathbf{Q}\}\end{Bmatrix}, \tag{3.80a}$$

where

$$[\mathbf{k}_{\tau\tau}]=-\frac{Al}{4E}\int_{-1}^{1}[\mathbf{N}]^T[\mathbf{N}]dL=-\begin{bmatrix}\frac{Al}{3E} & \frac{Al}{6E}\\ \frac{Al}{6E} & \frac{Al}{3E}\end{bmatrix}, \tag{3.80b}$$

$$[\mathbf{k}_{\tau u}]=\frac{Al}{2}\int_{-1}^{1}[\mathbf{N}]^T[\mathbf{B}]dL=\begin{bmatrix}\frac{-A}{2} & \frac{A}{2}\\ \frac{-A}{2} & \frac{A}{2}\end{bmatrix}, \tag{3.80c}$$

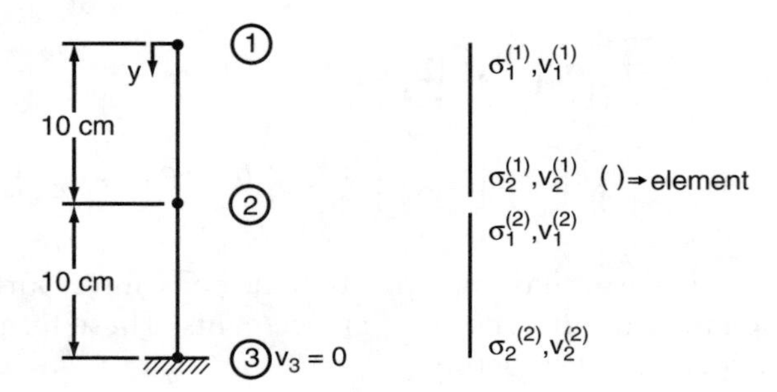

FIGURE 3.14
Column problem for mixed procedure.

and

$$\{\mathbf{Q}\} = \begin{Bmatrix} P_{1l} \\ P_{2l} \end{Bmatrix}. \tag{3.80d}$$

For further illustration, consider the following properties of the column divided in two elements (Figure 3.14):

$A = 1\ \text{cm}^2$,
$P_i = 10\ \text{kg}$,
$l = 10\ \text{cm}$,
$E = 1000\ \text{kg/cm}^2$.

Substitution of these properties into Equations 3.80 gives the element equations as

$$\left[\begin{array}{cc|cc} \frac{-10}{3000} & \frac{-10}{6000} & -\frac{1}{2} & \frac{1}{2} \\ \frac{-10}{6000} & \frac{-10}{3000} & -\frac{1}{2} & \frac{1}{2} \\ \hline -\frac{1}{2} & -\frac{1}{2} & 0 & 0 \\ \frac{1}{2} & \frac{1}{2} & 0 & 0 \end{array}\right] \begin{Bmatrix} \sigma_1 \\ \sigma_2 \\ \hline v_1 \\ v_2 \end{Bmatrix} = \overset{\text{Element 1}}{\begin{Bmatrix} 0 \\ 0 \\ \hline 10 \\ 0 \end{Bmatrix}} \quad \text{or} \quad \overset{\text{Element 2}}{\begin{Bmatrix} 0 \\ 0 \\ \hline 0 \\ 0 \end{Bmatrix}} \tag{3.81}$$

The element equations can be assembled now. For convenience, in the following we have rearranged the nodal unknowns as $[\sigma_1\ v_1\ \sigma_2\ v_2\ \sigma_3\ v_3]$; hence the assembled equations are

$$\begin{bmatrix} \frac{-10}{3000} & -\frac{1}{2} & \frac{-10}{6000} & \frac{1}{2} & 0 & 0 \\ -\frac{1}{2} & 0 & -\frac{1}{2} & 0 & 0 & 0 \\ \frac{-10}{6000} & -\frac{1}{2} & -\frac{20}{3000} & 0 & \frac{-10}{6000} & \frac{1}{2} \\ \frac{1}{2} & 0 & 0 & 0 & -\frac{1}{2} & 0 \\ 0 & 0 & \frac{-10}{6000} & -\frac{1}{2} & \frac{-10}{3000} & \frac{1}{2} \\ 0 & 0 & 0 & 0 & 0 & 1 \end{bmatrix} \begin{Bmatrix} \sigma_1 \\ v_1 \\ \sigma_2 \\ v_2 \\ \sigma_3 \\ v_3 \end{Bmatrix} = \begin{Bmatrix} 0 \\ 10 \\ 0 \\ 0 \\ 0 \\ 0 \end{Bmatrix}. \tag{3.82}$$

Here the boundary condition $v_3 = 0$ has been introduced as described previously. Solution of Equation 3.82 leads to

$$\begin{aligned} \sigma_1 &= -10 \text{ kg/cm}^2, & v_1 &= \frac{2}{10} \text{ cm}, \\ \sigma_2 &= -10, & v_2 &= \frac{1}{10} \text{ cm}, \\ \sigma_3 &= -10, & v_3 &= 0 \text{ (prescribed)}. \end{aligned}$$

These results are similar to those in Example 3.2. We may note that this is rather an elementary problem used simply to illustrate the procedure. Also, the results would have been different if we had used other orders of approximation and different loading and geometry.

Comment

The matrix in Equation 3.82 contains zeros on the main diagonal. For the standard Gaussian elimination, this can cause computational difficulties since the zeros can appear as pivots in the denominator (Appendix 2). Then it is necessary to use the idea of partial and complete pivoting,[7] which involves exchanges of rows and columns during the elimination procedure. Mathematically, the matrix **[K]** in Equation 3.82 is positive-semidefinite in constrast to the positive-definite character of the stiffness matrix in the displacement approach.

Galerkin's Method

With the displacement formulation, we used the governing equation (Equation 3.58b). It was derived on the basis of the equilibrium equation (Equation 3.58a) and the stress–strain or stress–displacement equation

(Equation 3.14a). We can use both the equilibrium and the stress–strain displacement equation for formulating the mixed approach with Galerkin's method. The residuals according to the two equations are

$$R_1 = \frac{d\sigma_y}{dy} - \overline{T}_y \tag{3.83a}$$

and

$$R_2 = -\frac{dv}{dy} + \frac{\sigma_y}{E}. \tag{3.83b}$$

Assuming the approximation models for v and σ_y as in Equation 3.77, we have, according to Galerkin's method,

$$A\int_{y_1}^{y_2} \cdot \left(\frac{d\sigma_y}{dy} - \overline{T}_y \right) N_i dy = 0 \tag{3.84a}$$

and

$$A\int_{y_1}^{y_2} \cdot \left(-\frac{dv}{dy} + \frac{\sigma_y}{E} \right) N_i dy = 0 \tag{3.84b}$$

After proper integration by parts, Equations 3.84a and 3.84b will lead to the same results as in Equation 3.80, obtained by using the variational procedure.

Comment

In this chapter and in Chapters 7 and 11 we have presented formulations by variational and residual methods. In the variational methods, the attention has been given to the displacement, complementary, and mixed procedures, while Galerkin's procedure has been used mainly in the residual methods. A number of useful hybrid procedures[11] are available in the variational methods; Gallagher[9] has presented examples of some of these procedures. As described in Appendix 1, other residual methods are also possible.

In this elementary text, it will be difficult to cover the total range of the methods of formulation. Hence, we have chosen only a few of them, particularly those which can be illustrated easily with simple problems.

Example 3.5

Prove that when the first variation of the functional given in Equation 3.31 is equated to zero then the governing equation and natural boundary conditions are satisfied.

Solution: Equation 3.31) can be written in terms of the primary unknown v and its first derivative,

$$\Pi = \frac{A}{2}\int_{y_1}^{y_2} E\left(\frac{dv}{dy}\right)^2 dy - A\int_{y_1}^{y_2} \bar{y}\, v dy - \int_{y_1}^{y_2} \bar{T}_y v dy - \sum_{i=1}^{2} \bar{P}_{il} v_i.$$

Following the steps described in Chapter 2, we can write

$$\delta\Pi = A\int_{y_1}^{y_2} E\left(\frac{dv}{dy}\right)\Delta\left(\frac{dv}{dy}\right)dy - A\int_{y_1}^{y_2} \bar{y}\,\Delta v dy - \int_{y_1}^{y_2} \bar{T}_y \Delta v dy - \sum_{i=1}^{2} \bar{P}_{il}\Delta v_i$$

or

$$\delta\Pi = AE\left(\frac{dv}{dy}\right)\Delta v\Bigg|_{y_1}^{y_2} - A\int_{y_1}^{y_2} E\left(\frac{d^2v}{dy^2}\right)\Delta v dy - A\int_{y_1}^{y_2} \bar{Y}\Delta v dy - \int_{y_1}^{y_2} \bar{T}_y \Delta v dy - \sum_{i=1}^{2} \bar{P}_{il}\Delta v_i,$$

or

$$\delta\Pi = AE\left(\frac{dv}{dy}\right)\Delta v\Bigg|_{y=y_2} - AE\left(\frac{dv}{dy}\right)\Delta v\Bigg|_{y=y_1} - A\int_{y_1}^{y_2} E\left(\frac{d^2v}{dy^2}\right)\Delta v dy$$

$$-A\int_{y_1}^{y_2} \bar{Y}\Delta v dy - \int_{y_1}^{y_2} \bar{T}_y \Delta v dy - \bar{P}_{il}\Delta v_i\Big|_{y=y_1} - \bar{P}_{il}\Delta v_i\Big|_{y=y_2},$$

or

$$\delta\Pi = \left[\left\{AE\left(\frac{dv}{dy}\right) - \bar{P}_{2l}\right\}\Delta v\right]_{y=y_2} - \left[\left\{AE\left(\frac{dv}{dy}\right) + \bar{P}_{1l}\right\}\Delta v\right]_{y=y_1}$$

$$-\int_{y_1}^{y_2}\left\{AE\left(\frac{d^2v}{dy^2}\right) + A\bar{Y} + \bar{T}_y\right\}\Delta v dy = 0.$$

Clearly for arbitrary Δv the above equation will be satisfied only if

$$AE\left(\frac{d^2v}{dy^2}\right) + A\overline{Y} + \overline{T} = 0, \text{ for y between } y_1 \text{ and } y_2,$$

$$AE\left(\frac{dv}{dy}\right) + \overline{P}_{1l} = 0, \text{ at } y = y_1 \text{ and}$$

$$AE\left(\frac{dv}{dy}\right) - \overline{P}_{2l} = 0, \text{ at } y = y_2.$$

Above equations are the governing equation and natural boundary conditions of the one-dimensional stress–deformation problem.

Bounds

In Chapter 1, we discussed rather qualitatively that different approaches yield different bounds to the exact solution. In the case of the variational procedures, if the physical and mathematical requirements are fulfilled, the potential (displacement) and complementary (stress) energy approaches yield, respectively, lower and upper bounds to the exact solution for displacement.

The approximate (algebraic) value of potential energy with a given approximation function is higher than the exact or minimum value (Figure 3.6). That is, the approximate stiffness is higher than its exact value. Consequently, the approximate displacements u are lower than the exact displacements. This is depicted in Figure 3.15(a), which shows the convergence behavior as the mesh is refined.

The physical explanation is that the displacement approach yields a stiffer or stronger structure than what it really is, because the assumed displacement functions, although continuous within elements, provide an approximate (global) distribution of displacements. Indirectly, this introduces additional supports or constraints in the structure and makes it stiffer.

In the case of the complementary energy approach, the approximate value of complementary energy is higher than the exact or minimum value. The flexibility value from a numerical approximation has a higher value than the corresponding exact value; that is, the stiffness has lower value than the exact stiffness. Consequently, the approximate displacements are higher than the exact displacements, Figure 3.15(b).

The stress approach yields a discontinuous distribution of displacement. Physically, this may be construed to introduce gaps and overlaps in the structure. This makes the structure weaker or less stiff than what it really is.

Thus the two approaches bound the exact solution from below and from above (Figures 3.15(a) and (b)). This property is useful, and often we can bound the exact solution by using both approaches.

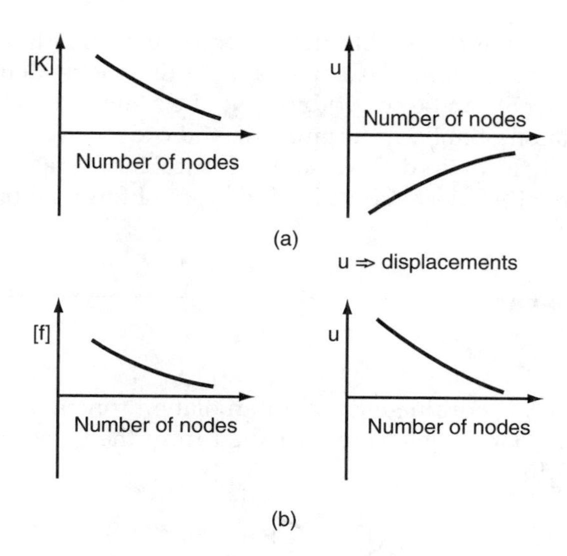

FIGURE 3.15
Symbolic representation of bounds in finite element analysis: (a) Bounds in displacement approach and (b) Bounds in stress approach.

Comment

Although the mixed and hybrid approaches can yield satisfactory and often improved solutions, they do not possess bounds.

Advantages of the Finite Element Method

At this stage, it may be useful to list some of the advantages of the finite element method.

Since the properties of each element are evaluated separately, an obvious advantage is that we can incorporate different material properties for each element. Thus almost any degree of nonhomogeneity can be included. There is no restriction as to the shape of the medium; hence arbitrary and irregular geometries cause no difficulty. Any type of boundary conditions can also be accommodated easily; this can be an important factor.

We have not yet discussed other factors such as nonlinear behavior and dynamic effects. Subsequently, and during study beyond this book, the reader will find that the method can easily handle factors such as nonlinearities, aribitrary loading conditions, and time dependence.

Finally, it is the generality of the method that makes it so appealing and powerful for the solution of problems in a wide range of disciplines in engineering and mathematical physics. This generality allows application of formulations and codes developed for one class of problems to other classes of problems with no or minor modifications. For instance, one-dimensional stress deformation (Chapter 3), flow (Chapter 4), and time-dependent flow of heat or fluids (Chapter 5) can be solved by using the same code (Chapter 6) with minor modifications. Similarly, formulation and code for such field problems as torsion, heat flow, fluid flow, and electrical and magnetic potentials (Chapters 11 and 12) have almost identical bases of formulations and codes.

Problems

3.1. Derive local coordinate L and interpolation functions N_1 and N_2 if the local coordinates were measured from the Quarter point. See Figure 3.16.

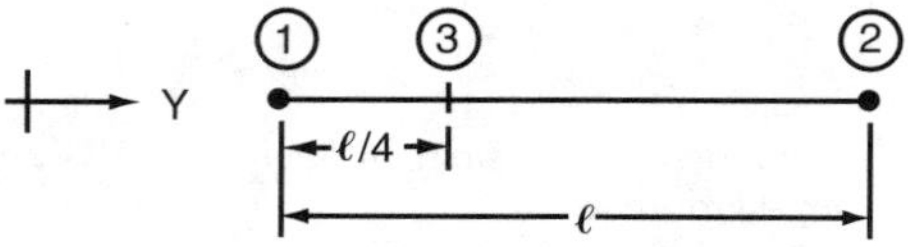

FIGURE 3.16

Solution: (a) For $L = (y - y_3)/(l/4)$, $L(1) = -1$, $L(2) = 3$, and $N_1 = \frac{1}{4}(3 - L)$, $N_2 = \frac{1}{4}(1 + L)$. (b) For $L = (y - y_3)/(3l/4)$, $L(1) = -\frac{1}{3}$, $L(2) = 1$, and $N_1 = \frac{3}{4}(1 - L)$, $N_2 = \frac{1}{4}(1 + 3L)$.

3.2. Derive element equations if the value of the elastic modulus E and the area A vary linearly as

$$E = \begin{bmatrix} \frac{1}{2}(1-L) & \frac{1}{2}(1+L) \end{bmatrix} \begin{Bmatrix} E_1 \\ E_2 \end{Bmatrix}$$

and

$$A = \begin{bmatrix} \frac{1}{2}(1-L) & \frac{1}{2}(1+L) \end{bmatrix} \begin{Bmatrix} A_1 \\ A_2 \end{Bmatrix},$$

where A_1 and A_2 and E_1 and E_2 are the values at nodes 1 and 2, respectively. Answer:

$$[\mathbf{k}] = \lambda \begin{bmatrix} 1 & -1 \\ -1 & 1 \end{bmatrix},$$

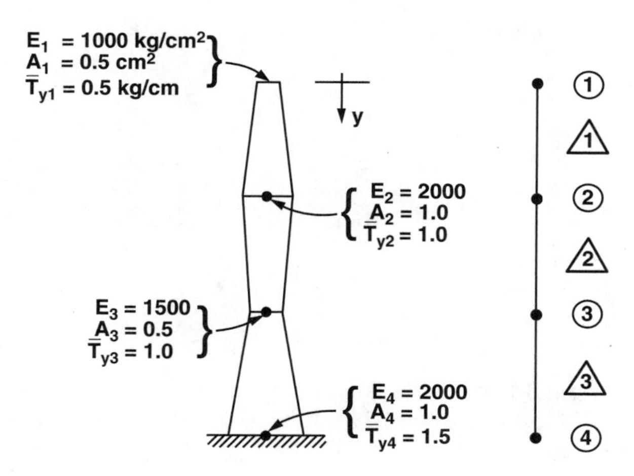

FIGURE 3.17

where the scalar λ is given by

$$\lambda = \frac{1}{6l}\begin{bmatrix} A_1 & E_1 & A_2 & E_2 \end{bmatrix}\begin{bmatrix} 0 & 1 & 0 & \frac{1}{2} \\ 1 & 0 & \frac{1}{2} & 0 \\ 0 & \frac{1}{2} & 0 & 1 \\ \frac{1}{2} & 0 & 1 & 0 \end{bmatrix}\begin{Bmatrix} A_1 \\ E_1 \\ A_2 \\ E_2 \end{Bmatrix}.$$

3.3. Consider the column in Figure 3.17 with linearly varying E, A, and T_y. Length l of each element = 10 cm. Derive element equations, assemble them, introduce boundary condition $v_4 = 0$, and solve for displacements, strains, and stresses. Plot variation of displacements and stresses.

Partial solution:

Equations for element 2,

$$\frac{800}{6}\begin{bmatrix} 1 & -1 \\ -1 & 1 \end{bmatrix}\begin{Bmatrix} v_1 \\ v_2 \end{Bmatrix} = \begin{Bmatrix} 5 \\ 5 \end{Bmatrix}.$$

Assemblage equations,

$$\begin{bmatrix} 7 & -7 & 0 & 0 \\ -7 & 15 & -8 & 0 \\ 0 & -8 & 16 & -8 \\ 0 & 0 & -8 & 8 \end{bmatrix} \begin{Bmatrix} v_1 \\ v_2 \\ v_3 \\ v_4 \end{Bmatrix} = \begin{Bmatrix} 0.20 \\ 0.55 \\ 0.65 \\ 0.40 \end{Bmatrix}.$$

$v_1 = 0.29732$ cm, $v_2 = 0.26875$ cm, $v_3 = 0.17500$ cm, and $v_4 = 0$.

3.4. In Example 3.1, consider surface loading $\bar{T}_y$ to vary linearly from zero at the top to 2 kg/cm² at the base, and compute displacement and stresses.

Hint: Consider

$$\bar{T}_y = \left[\tfrac{1}{2}(1-L) \quad \tfrac{1}{2}(1+L)\right] \begin{Bmatrix} \bar{T}_{y_1} \\ \bar{T}_{y_2} \end{Bmatrix}.$$

Then the load vector is

$$\{\mathbf{Q}\} = \frac{l}{6} \begin{Bmatrix} 2\bar{T}_{y_1} + \bar{T}_{y_2} \\ \bar{T}_{y_1} + 2\bar{T}_{y_2} \end{Bmatrix}.$$

3.5. For Example 3.1, assume that the base settles by 0.1 cm. Find displacements and stresses.

3.6. Use the code in Chapter 6 and obtain computer solutions for Examples 3.1 and 3.2.

3.7. For Example 3.1, study convergence of solutions for displacements and stresses as the number of nodes are increased using the code in Chapter 6. Consider 4, 8, 16, and 32 nodes; plot results, and comment on improvement in accuracy, if any.

Partial solution: See Figure 3.18. Note that in view of the linear model and plane axial problem, the displacement solution is not significantly influenced by the refinement of mesh.

3.8. Sketch the variations of N_i in Equation 3.53) for the element in Figure 3.10, and derive the element stiffness matrix assuming A and E to be constant.

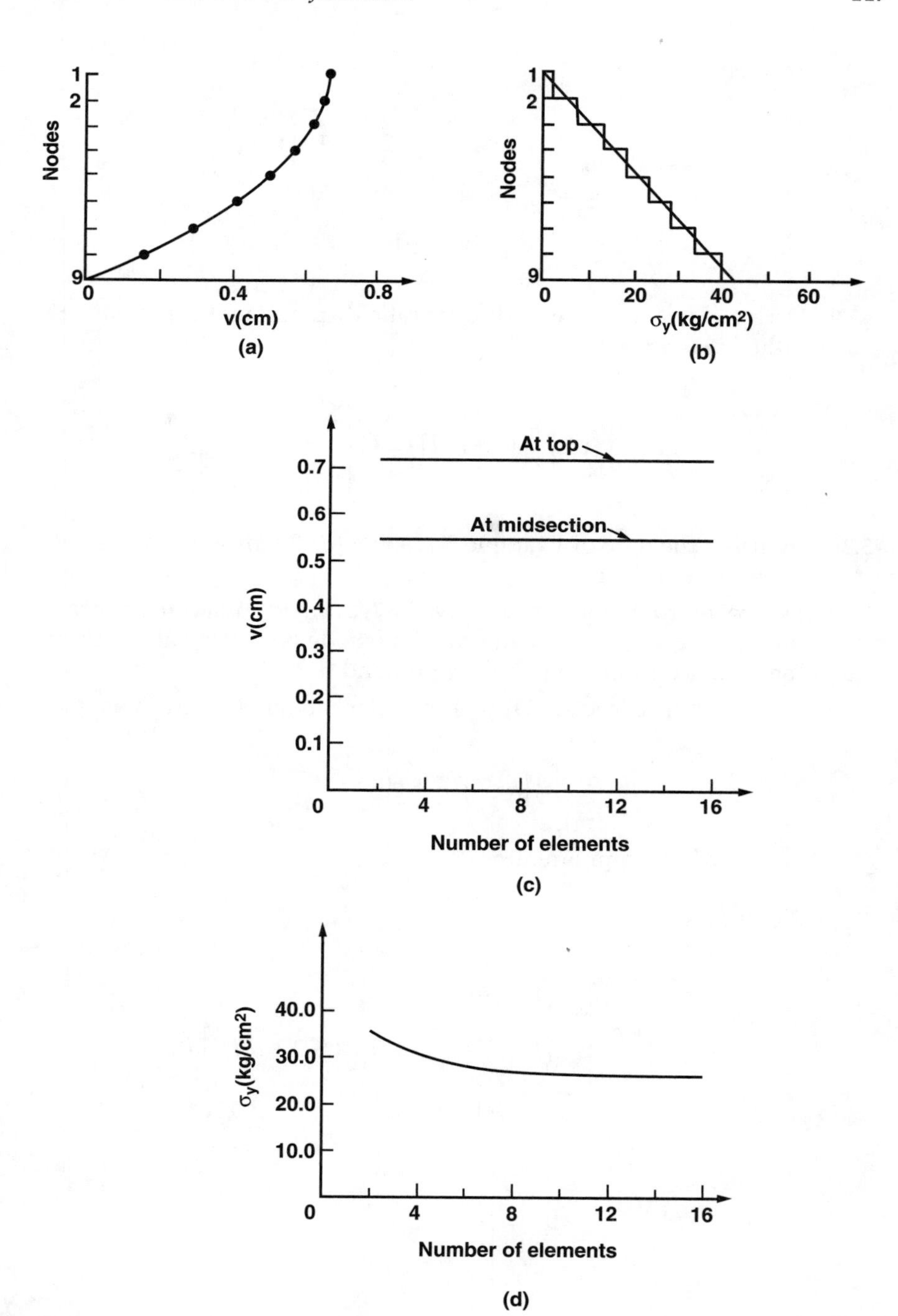

FIGURE 3.18
Results with mesh refinement and convergence: (a) Distribution of displacement v for eight elements, (b) Distribution of stress σ_y for eight elements, (c) Convergence of displacements, and (d) Convergence of stress at midsection.

Solution:

$$[\mathbf{k}] = \frac{AE}{6l}\begin{bmatrix} 14 & 2 & -16 \\ 2 & 14 & -16 \\ -16 & -16 & 32 \end{bmatrix}$$

3.9. Derive element stiffness for quadratic displacement (Problem 3.8) if the area varies linearly as

$$A = \begin{bmatrix} \tfrac{1}{2}(1-L) & \tfrac{1}{2}(1+L) \end{bmatrix}\begin{Bmatrix} A_1 \\ A_2 \end{Bmatrix}.$$

3.10. By using the data of Example 3.1, solve for the displacements and stresses in Problem 3.8.

3.11. Derive interpolation functions N_i, $i = 1, 2, 3$, for a quadratic approximation model when the intermediate node is situated at one third the distance from the left- or right-hand node.

3.12. (a) Derive load vector $\{\mathbf{Q}\}$ due to surface loading T_y given as

$$\overline{T}_y = \tfrac{1}{2}(1-L)\overline{T}_{y_1} + \tfrac{1}{2}(1+L)\overline{T}_{y_2}$$

and quadratic v in Equation 3.53.

Solution:

$$\{\mathbf{Q}\} = \frac{1}{2}\int_{-1}^{1}\begin{Bmatrix} \tfrac{1}{2}L(L-1) \\ \tfrac{1}{2}L(L+1) \\ (1-L^2) \end{Bmatrix}\left[\tfrac{1}{2}(1-L)\overline{T}_{y_1} + \tfrac{1}{2}(1+L)\overline{T}_{y_2}\right]dL$$

$$= \frac{l}{6}\begin{Bmatrix} \overline{T}_{y_1} \\ \overline{T}_{y_2} \\ 2\overline{T}_{y_1} + 2\overline{T}_{y_2} \end{Bmatrix}.$$

(b) Compute the load vector due to gravity given by

$$\overline{Y} = \tfrac{1}{2}(1-L)\overline{Y}_1 + \tfrac{1}{2}(1+L)\overline{Y}_2.$$

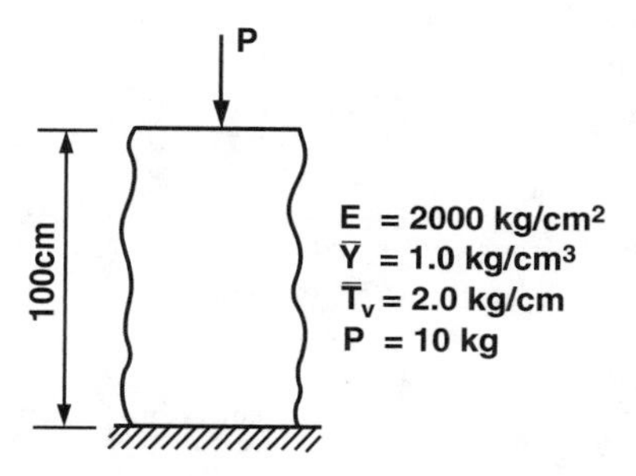

FIGURE 3.19

Assume linear variation for v.

Answer:

$$\{\mathbf{Q}\} = \frac{Al}{6}\begin{Bmatrix} 2\overline{Y}_1 + \overline{Y}_2 \\ \overline{Y}_1 + 2\overline{Y}_2 \end{Bmatrix}.$$

(c) Find the load vector if

$$\overline{T}_y = \tfrac{1}{2}L(L-1)\overline{T}_{y_1} + \tfrac{1}{2}L(L+1)\overline{T}_{y_2} + (1-L^2)\overline{T}_{y_0}.$$

Answer:

$$\{\mathbf{Q}\} = \frac{1}{30}\begin{Bmatrix} 4\overline{T}_{y_1} - \overline{T}_{y_2} + 2\overline{T}_{y_0} \\ -\overline{T}_{y_1} + 4\overline{T}_{y_2} + 2\overline{T}_{y_0} \\ 2\overline{T}_{y_1} + 2\overline{T}_{y_2} + 16\overline{T}_{y_0} \end{Bmatrix}.$$

3.13. Idealize as closely as possible the irregularly shaped column in Figure 3.19; solve for displacements and stresses by using DFT/C-1DFE.

3.14. Consider the coordinate system in Figure 3.2(c) and derive the quadratic interpolation function model (Equation 3.53) from the following generalized coordinate model for v:

$$v = \alpha_1 + \alpha_2 y + \alpha_3 y^2 = [\phi]\{\alpha\}.$$

Solution:

$$\{\alpha\} = [\mathbf{A}]^{-1}\{\mathbf{q}\},$$

$$[\mathbf{A}]^{-1} = \begin{bmatrix} \frac{y_2 y_3}{(y_2 - y_1)(y_3 - y_1)} & \frac{-y_1 y_3}{(y_2 - y_1)(y_3 - y_2)} & \frac{y_1 y_2}{(y_3 - y_2)(y_3 - y_1)} \\ \frac{-(y_3 + y_2)}{(y_2 - y_1)(y_3 - y_1)} & \frac{y_3 + y_1}{(y_2 - y_1)(y_3 - y_2)} & \frac{-(y_2 + y_1)}{(y_3 - y_2)(y_3 - y_1)} \\ \frac{1}{(y_2 - y_1)(y_3 - y_1)} & \frac{-1}{(y_2 - y_1)(y_3 - y_2)} & \frac{1}{(y_3 - y_2)(y_3 - y_1)} \end{bmatrix},$$

$$[\mathbf{N}] = [\phi][\mathbf{A}]^{-1}$$
$$= \left[\frac{(y - y_2)(y - y_3)}{(y_2 - y_1)(y_3 - y_1)} \quad \frac{-(y - y_1)(y - y_3)}{(y_2 - y_1)(y_3 - y_2)} \quad \frac{(y - y_1)(y - y_2)}{(y_3 - y_2)(y_3 - y_1)} \right].$$

Because $y_3 - y_1 = l/2$, $y_3 - y_3 = -l/2$, $y_2 - y_1 = l$, and $(y - y_3)/(l/2) = L$, we finally have

$$[\mathbf{N}] = \left[\tfrac{1}{2}L(L-1) \quad \tfrac{1}{2}L(L+1) \quad 1 - L^2 \right].$$

References

1. Zienkiewicz, O. C., Irons, B. M., Ergatoudis, J., Ahmad, S., and Scott, F. C. 1969. "Isoparametric and associated element families for two- and three-dimensional analysis," in *Finite Element Methods in Stress Analysis,* Holland, I. and Bell, K., Eds., Tech. Univ. of Norway, Trondheim.
2. Bazeley, G. P., CHEUNG, Y. K., Irons, B. M., and Zienkiewicz, O. C. 1965. "Triangular elements in plate bending-conforming and nonconforming solutions," in *Proc. Conf. on Matrix Methods of Struct. Mechanics,* Wright Patterson Air Force Base. Dayton, OH.
3. Strang, G. and Fix, G. J. 1973. *An Analysis of the Finite Element Method,* Prentice-Hall, Englewood Cliffs, NJ.
4. Bathe, K. J. and Wilson, E. L. 1976. *Numerical Methods in Finite Element Analysis,* Prentice-Hall, Englewood Cliffs, NJ.
5. Desai, C. S. and Abel, J. F. 1972. *Introduction to the Finite Element Method,* Van Nostrand Reinhold, New York.
6. Turner, M. J., Clough, R. W., Martin, H. C., and Topp, L. C., 1956. "Stiffness and Deflection Analysis of Complex Structures," *J. Aeronaut. Sci.,* 23, 9, 805-823.
7. Fox, L. 1965. *An Introduction to Numerical Linear Algebra,* Oxford University Press, New York.
8. Prenter, P. M. 1975. *Splines and Variational Methods,* John Wiley & Sons, New York.

9. Gallagher, R. H. 1975. *Finite Element Analysis Fundamentals,* Prentice-Hall, Englewood Cliffs, NJ.
10. Washizu, K. 1968. *Variational Methods in Elasticity and Plasticity,* Pergamon Press, Elmsford, NY.
11. Pian, T. H. H. and Tong, P. 1972. "Finite element methods in continuum mechanics," in *Advances in Applied Mechanics,* Vol. 12, Academic Press, New York.
12. Przemieniecki, J. S. 1968. *Theory of Matrix Structural Analysis,* McGraw-Hill, New York.
13. Beaufait, F. W., Rowan, W. H., Hoadley, P. O., and Hackett, R. M. 1960. *Computer Methods of Structural Analysis,* Prentice-Hall, Englewood Cliffs, NJ.

4

One-Dimensional Flow

Flow of heat or fluid through solids is a problem that is frequently encountered in engineering. In general, such a flow occurs in three spatial dimensions. For some problems, however, we can assume the flow to occur essentially in one dimension. Examples of these problems are heat flow through uniform bars, fluid flow through pipes of uniform cross section, vertical flow in a medium of large extent, and flow in and out of vertical banks and cuts and toward vertical retaining structures.

Theory and Formulation

Governing Equation

The schematic of one-dimensional flow through a pipe is shown in Figure 4.1. The governing differential equation for the one-dimensional steady-state fluid flow in the x-direction can be expressed as

$$Ak_x \frac{d^2\phi^*}{dx^2} + \bar{q}(x) = 0 \tag{4.1}$$

where $\bar{q}(x)$ is the fluid influx per unit length, k_x is the permeability coefficient in the x-direction, ϕ^* is the fluid head, and A is the cross-sectional area of the flow region.

A comparison between Equations 3.28 and 4.1 shows that the two equations have the same form; AE of Equation 3.28 has been replaced by Ak_x in Equation 4.1, force per unit length F_y of Equation 3.28 has been replaced by the fluid influx $\bar{q}(x)$ per unit length in Equation 4.1. The primary unknowns are displacement (v) in Equation 3.28 and fluid head (ϕ^*) in Equation 4.1. This indicates that the phenomenon of deformations in a column and the one-dimensional flow follow similar natural behavior. We can even consider an interpretation that the rate of change of deformation due the (axial) load

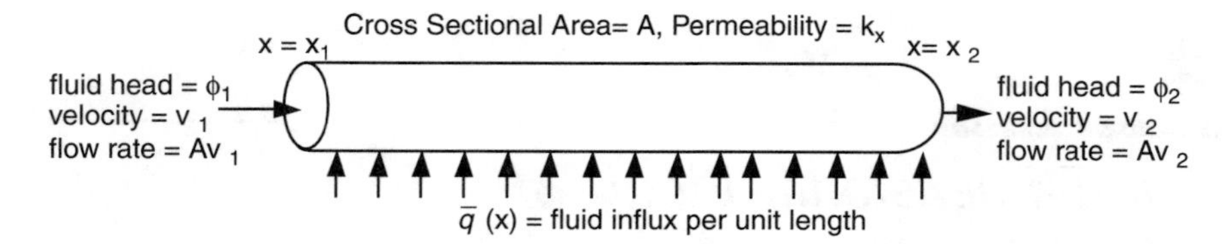

FIGURE 4.1
Schematic of one-dimensional flow through a pipe of cross-sectional area A.

and the flow or rate of change of potential or temperature due to applied fluid head or temperature are analogous.

Equation 4.1 governs steady or time-independent flow of heat and fluid through one-dimensional media; in fact there are a number of other phenomena such as one-dimensional moisture migration that are also governed by similar equations. The finite element formulation for Equation 4.1 can therefore be used for all these problems with only minor modifications. In Chapter 5, we shall derive equations for transient flow by using the problems of heat flow and consolidation, while in this chapter we use steady fluid flow through rigid media as the example problem. We may note, however, that the finite element formulation for all these problems will essentially be the same except for different relevance for material properties and meanings of the unknowns such as temperature and fluid head.

Now, we follow the steps (Chapter 2) required in the finite element formulation.

Finite Element Formulation

As in Chapter 3, the flow domain (Figure 4.1) can be idealized as a one-dimensional line as shown in Figure 4.2. Then we can follow the steps (Chapter 2) required in the finite element formulation.

Step 1. Choose Element Configuration

As in Chapter 3, we use the one-dimensional line element (Figure 3.2(c)).

Step 2. Choose Approximation Function

We choose a linear approximation model for fluid potential as

$$\varphi = \alpha_1 + \alpha_2 x \tag{4.2a}$$

or

$$\varphi = N_1\varphi_1 + N_2\varphi_2 = [\mathbf{N}]\{Q_n\}, \tag{4.2b}$$

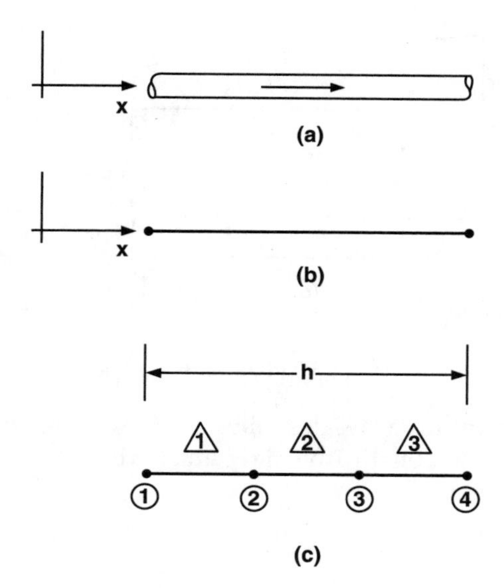

FIGURE 4.2
Idealization for flow pipe: (a) Flow through pipe, (b) One dimensional idealization, and (c) Discretization in three elements.

where $N_1 = \frac{1}{2}(1 - L)$, $N_2 = \frac{1}{2}(1 + L)$, *and* $\{\varphi_n\}^T = [\varphi_1 \ \varphi_2]$ is the vector of elemental nodal fluid heads. Equation 4.2 is similar to Equation 3.7 except that here φ replaces v.

The requirements for choosing the approximation function for φ are similar to those in Chapter 3. The linear approximation yields *continuous* variation of φ within the element. For simple plane and linear flow, the interelement compatibility is required only for the nodal values of φ. In other words, since the highest order of derivation in the energy function, given subsequently in Equation 4.6, is 1, the interelement compatibility is required only up to order equal to $1 - 1 = 0$, that is, for nodal heads. The completeness requirement is satisfied since the linear function, Equation 4.2, allows for rigid body motion and for constant state of gradient g_x, Equation 4.3a.

Step 3. Define Gradient-Potential Relation and Constitutive Law

The relation between the gradient of φ and φ can be defined in a manner analogous to the definition of strain ε_y in Equation 3.12a; hence

$$g_x = \frac{\partial \varphi}{\partial x}, \tag{4.3a}$$

where g_x is the gradient of φ with respect to x.

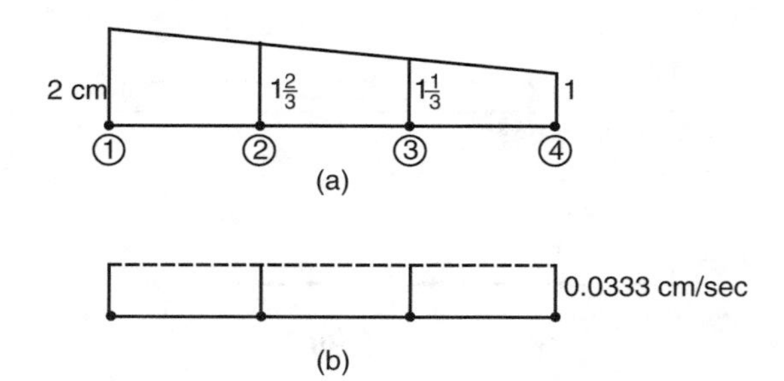

FIGURE 4.3
Plots of finite element computations: (a) Fluid heads and (b) Velocity.

The relevant constitutive law describes the flow behavior through porous media. The simplest constitutive law that we can use is Darcy's law (Figure 4.3) given by

$$v_x = -k_x \frac{\partial \varphi}{\partial x} = -k_x g_x, \tag{4.3b}$$

where v_x = velocity in the x direction and g_x = hydraulic gradient. The negative sign occurs because the velocity is in the negative gradient direction. The value of the gradient can be obtained by differentiating φ in Equation 4.2 with respect to x as

$$\frac{\partial \varphi}{\partial x} = \frac{\partial}{\partial x}\left[\frac{1}{2}(1-L)\varphi_1 + \frac{1}{2}(1+L)\varphi_2\right]. \tag{4.4a}$$

Use of the chain rule of differentiation leads to

$$\begin{aligned} \frac{\partial \varphi}{\partial x} &= \frac{1}{l}[-1 \quad 1]\begin{Bmatrix}\varphi_1 \\ \varphi_2\end{Bmatrix} \\ &= [\mathbf{B}]\{\varphi_n\}. \end{aligned} \tag{4.4b}$$

As before (Equation 3.13) [**B**] is the gradient-potential transformation matrix. Substitution of Equation 4.4b into Equation 4.3 gives

$$v_x = -[\mathbf{R}][\mathbf{B}]\{\varphi_n\}, \tag{4.5}$$

where [**R**] = matrix of coefficients of permeability; for the one-dimensional case it is simply the scalar k_x. It is interesting to note the similarity between Equations 4.4b and 3.13 and between Equations 4.5 and 3.15; the gradient and velocity terms here are, respectively, analogous to the strain and stress terms in the stress–deformation problem.

Step 4. Derive Element Equations

The finite element equations can be derived by using either a variational principle or a residual procedure. For problems with certain mathematical properties such as self-adjointness, valid variational principles are available. For certain other problems, it may not be possible to establish a mathematically consistent variational principle. Residual methods such as Galerkin's procedure can be used to derive finite element equations for such problems; in fact, the residual methods can be used as general procedures.

For the problem governed by Equation 4.1 we shall first use a variational principle. Later in this chapter, we shall illustrate the use of Galerkin's method.

Variational Approach

In Chapter 3 we have seen that in the variational method one needs the right functional to obtain the element equations. For stress–deformation problems this functional is the total potential energy of the system. What is the appropriate functional expression for the flow problem? The functional for solving the one-dimensional flow problem is obtained from its governing equation and natural boundary conditions, as discussed in Chapter 2. From Chapters 2 and 3, the functional that should be minimized to satisfy the governing Equation 4.1 in absence of a natural boundary condition is given by

$$\Omega = \frac{1}{2}\int_{x_1}^{x_2} Ak_x\left(\frac{d\phi}{dx}\right)^2 dx - \int_{x_1}^{x_2} \bar{q}\phi dx \tag{4.6}$$

Note that Equation 4.6 is identical to Equation 3.29.

Substitution of $\partial\varphi/\partial x$ and φ into Equation 4.6 leads to

$$\Omega = A\int_{x_1}^{x_2} \frac{1}{2}\{\varphi_n\}^T[\mathbf{B}]^T k_x[\mathbf{B}]\{\varphi_n\}dx - \int_{x_1}^{x_2} \bar{q}[\mathbf{N}]\{\varphi_n\}dx. \tag{4.7}$$

Differentiation of Ω with respect to φ_1 and φ_2 yields

$$\delta\Omega = 0 \Rightarrow \begin{cases} \dfrac{\partial\Omega}{\partial\varphi_1} = 0, \\ \dfrac{\partial\Omega}{\partial\varphi_2} = 0, \end{cases} \tag{4.8}$$

or

$$A\int_{x_1}^{x_2}[\mathbf{B}]^T k_x[\mathbf{B}]dx\{\varphi_n\} = \int_{x_1}^{x_2}\bar{q}[\mathbf{N}]^T dx, \tag{4.9a}$$

or

$$[\mathbf{k}]\{\varphi_n\} = \{\mathbf{Q}\}, \tag{4.9b}$$

where $[\mathbf{k}]$ = element property (permeability) matrix,

$$\begin{aligned}[\mathbf{k}] &= A\int_{x_1}^{x_2}[\mathbf{B}]^T k_x[\mathbf{B}]dx \\ &= \frac{Al}{2}\int_{-1}^{1}[\mathbf{B}]^T k_x[\mathbf{B}]dL,\end{aligned} \tag{4.9c}$$

and $\{\mathbf{Q}\}$ = element nodal forcing parameter vector,

$$\begin{aligned}\{\mathbf{Q}\} &= \int_{x_1}^{x_2}\bar{q}[\mathbf{N}]^T dx \\ &= \frac{1}{2}\int_{-1}^{1}\bar{q}[\mathbf{N}]^T dL.\end{aligned} \tag{4.9d}$$

Evaluation of [k] and {Q}

Substitution for $[\mathbf{B}]$ and $[\mathbf{N}]$ in Equations 4.9c and 4.9d and integrations lead to

$$\begin{aligned}[\mathbf{k}] &= \frac{Ak_x l}{2}\int_{-1}^{1}[\mathbf{B}]^T[\mathbf{B}]dL \\ &= \frac{Ak_x}{l}\begin{bmatrix}1 & -1 \\ -1 & 1\end{bmatrix},\end{aligned} \tag{4.10}$$

$$\{\mathbf{Q}\} = \frac{\bar{q}l}{2}\int\begin{Bmatrix}\frac{1}{2}(1-L) \\ \frac{1}{2}(1+L)\end{Bmatrix}dL = \frac{\bar{q}l}{2}\begin{Bmatrix}1 \\ 1\end{Bmatrix}. \tag{4.11}$$

Notice that Equation 4.10 is similar to Equation 3.28 except for the fact that k_x has replaced E. The lumping of half the applied flux at each node is the result of the linear approximation model, and for higher-order models, the forcing vector can be different.

The element equations can thus be expressed as

$$\frac{Ak_x}{l}\begin{bmatrix} 1 & -1 \\ -1 & 1 \end{bmatrix}\begin{Bmatrix} \varphi_1 \\ \varphi_2 \end{Bmatrix} = \begin{Bmatrix} \bar{q}l/2 \\ \bar{q}l/2 \end{Bmatrix}. \tag{4.12}$$

Step 5. Assemble

As explained in Chapter 3, we can differentiate the total of Ω^t for all elements and obtain the assemblage equations. The procedure is essentially similar to the direct stiffness approach and is based on the physical requirement that the fluid heads at a common node between two elements are equal. This is analogous to the interelement compatibility for displacements in Chapter 3. Then for a domain discretized in three elements (Figure 4.2) with uniform area of cross section, we have

$$\frac{Ak_x}{l}\begin{bmatrix} 1 & -1 & 0 & 0 \\ -1 & 2 & -1 & 0 \\ 0 & -1 & 2 & -1 \\ 0 & 0 & -1 & 1 \end{bmatrix}\begin{Bmatrix} \varphi_1 \\ \varphi_2 \\ \varphi_3 \\ \varphi_4 \end{Bmatrix} = \frac{\bar{q}l}{2}\begin{Bmatrix} 1 \\ 2 \\ 2 \\ 1 \end{Bmatrix} \tag{4.13a}$$

or

$$[\mathbf{K}]\{\mathbf{r}\} = \{\mathbf{R}\}, \tag{4.13b}$$

where $[\mathbf{K}]$ = assemblage property (permeability) matrix, $\{\mathbf{r}\}$ = assemblage vector of nodal heads, and $\{\mathbf{R}\}$ = assemblage vector of nodal forcing parameters.

Step 6. Solve for Potentials

Solution of Equation 4.13 after introduction of the prescribed boundary conditions allows computations of nodal heads. To illustrate this step, assume the following properties:

Example 4.1

$$A = 1.0 \text{ cm}^2,$$

$$k_x = 1 \text{ cm/sec},$$

$$q = 0.0 \text{ cm}^2/\text{sec},$$

$$l = 10 \text{ cm}.$$

Boundary conditions:

$$\varphi(x=0)=2.0 \text{ cm},$$

$$\varphi(x=h)=1.0 \text{ cm}.$$

We use Equation 4.12 to find the element equations. For instance, for element number 1 we have

$$\frac{1\times 1}{10}\begin{bmatrix} 1 & -1 \\ -1 & 1 \end{bmatrix}\begin{Bmatrix} \varphi_1 \\ \varphi_2 \end{Bmatrix}=\begin{Bmatrix} 0 \\ 0 \end{Bmatrix}$$

and so on for the other elements. The assembled equations are

$$\frac{1}{10}\begin{bmatrix} 1 & -1 & 0 & 0 \\ -1 & 2 & -1 & 0 \\ 0 & -1 & 2 & -1 \\ 0 & 0 & -1 & 1 \end{bmatrix}\begin{Bmatrix} \varphi_1 \\ \varphi_2 \\ \varphi_3 \\ \varphi_4 \end{Bmatrix}=\begin{Bmatrix} 0 \\ 0 \\ 0 \\ 0 \end{Bmatrix}.$$

The boundary conditions are introduced as follows:

$$\text{set } k_{11}=1 \text{ and } k_{1j}=0, \quad j=2, 3, 4,$$

$$k_{44}=1 \text{ and } k_{4i}=0, \quad i=1, 2, 3,$$

$$R_1=2, \quad R_4=1.$$

Therefore

$$\begin{bmatrix} 1 & 0 & 0 & 0 \\ -\frac{1}{10} & \frac{2}{10} & -\frac{1}{10} & 0 \\ 0 & -\frac{1}{10} & \frac{2}{10} & -\frac{1}{10} \\ 0 & 0 & 0 & 1 \end{bmatrix}\begin{Bmatrix} \varphi_1 \\ \varphi_2 \\ \varphi_3 \\ \varphi_4 \end{Bmatrix}=\begin{Bmatrix} 2 \\ 0 \\ 0 \\ 1 \end{Bmatrix}.$$

From the first equation we have φ_1 = 1. From the second equation,

$$-\frac{\varphi_1}{10}+\frac{2\varphi_2}{10}-\frac{\varphi_3}{10}=0$$

or

$$-\varphi_1+2\varphi_2-\varphi_3=0,$$

but $\varphi_1 = 2$; therefore

$$2\varphi_2 - \varphi_3 = 2. \tag{i}$$

From the third equation

$$-\varphi_2 + 2\varphi_3 - \varphi_4 = 0, \tag{ii}$$

but $\varphi_4 = 1$. Therefore,

$$-\varphi_2 + 2\varphi_3 = 1.$$

Equations i and ii lead to

$$\begin{array}{r} 2\varphi_2 - \varphi_3 = 2 \\ -2\varphi_2 + 4\varphi_3 = 2 \\ \hline 3\varphi_3 = 4 \end{array}$$

Hence

$$\varphi_3 = \frac{4}{3}$$

and

$$2\varphi_2 - \frac{4}{3} = 2.$$

Hence

$$\varphi_2 = \frac{5}{3}.$$

Step 7. Secondary Quantities

The secondary quantities for the fluid flow problem are the velocities and quantity of flow. The knowledge of φ from Step 6 now permits evaluation of these quantities.

Velocities. Element 1:

$$\begin{aligned} \text{Eq. } (4.5) \rightarrow v_{x_1} &= -k_x[\mathbf{B}]\{\varphi_n\} \\ &= -\frac{1}{10}[-1 \quad 1]\begin{Bmatrix} 2 \\ \frac{5}{3} \end{Bmatrix} \\ &= 0.0333 \text{ cm/sec}. \end{aligned}$$

Similarly for elements 2 and 3:

$$v_{x_2} = -\frac{1}{10}[-1 \quad 1]\begin{Bmatrix} \frac{5}{3} \\ \frac{4}{3} \end{Bmatrix}$$

$$= 0.0333 \text{ cm/sec}.$$

$$v_{x_3} = -\frac{1}{10}[-1 \quad 1]\begin{Bmatrix} \frac{4}{3} \\ 1 \end{Bmatrix}$$

$$= 0.0333 \text{ cm/sec}.$$

Quantity of flow:

$$Q_f = v \times A = 0.0333 \times 1 = 0.0333 \text{ cc/sec},$$

where v is the velocity normal to the cross-sectional area A, and the subscript in v_{x1} denotes element 1 and so on.

Step 8. Interpret and Plot Results

The computed values of φ and v_x are plotted in Figure 4.3. We can see that the finite element computations give exact solutions both for fluid heads and velocities. This is because the cross section and material properties are uniform, and constitutive (Darcy's) law is linear. As indicated in Chapter 3, nonuniform cross section and nonlinearity can yield nonlinear distributions for φ and v.

Formulation by Galerkin's Method

The residual corresponding to Equation 4.1 is

$$R(\varphi) = Ak_x \frac{\partial^2 \varphi}{\partial x^2} + \bar{q} \tag{4.14}$$

The details of derivation are almost identical to those for the column problem in Chapter 3. The final results for the three-element system (Figure 4.1) will be

$$A\frac{k_x}{l}\begin{bmatrix} 1 & -1 & 0 & 0 \\ -1 & 2 & -1 & 0 \\ 0 & -1 & 2 & -1 \\ 0 & 0 & -1 & 1 \end{bmatrix}\begin{Bmatrix} \varphi_1 \\ \varphi_2 \\ \varphi_3 \\ \varphi_4 \end{Bmatrix} = -\frac{\bar{q}l}{2}\begin{Bmatrix} 1 \\ 2 \\ 2 \\ 1 \end{Bmatrix} + \begin{Bmatrix} \left(-Ak_x \frac{\partial \varphi}{\partial x}\right)_1 \\ 0 \\ 0 \\ \left(Ak_x \frac{\partial \varphi}{\partial x}\right)_2 \end{Bmatrix}, \tag{4.15a}$$

$$[\mathbf{K}]\{\mathbf{r}\} = \{\mathbf{R}_T\} + \{\mathbf{R}_B\} = \{\mathbf{R}\}. \tag{4.15b}$$

The boundary terms $\{\mathbf{R}_B\}$ include flux, $k_x(\partial\varphi/\partial x)$, or Neumann-type boundary condition at end nodes 1 and 4. They are prescribed by given values of the (normal) derivative $\partial\varphi/\partial x$. If the boundary is impervious, $\partial\varphi/\partial x = 0$. For the node with geometric or Dirichlet boundary conditions, that is, prescribed φ, the term is removed while introducing the condition in the assemblage equations.

Forced and Natural Boundary Conditions for Flow Problems

At the boundary points, $x = x_1$ and $x = x_2$ (Figure 4.1), either fluid head or fluid velocity (or flow rate, that is equal to the velocity multiplied by the cross-sectional area) should be specified. If fluid heads are specified then the boundary conditions are of the first kind, known as forced boundary conditions. Forced boundary conditions do not change the functional expressions given in Equation 4.6. However, if the fluid velocity (or flow rate) is specified at a boundary point then it becomes a natural boundary condition. In this case the functional, Equation 4.6 should be modified to incorporate the natural boundary conditions.

The natural boundary conditions at the two boundary points can be written as

$$\begin{aligned} -k_x \left.\frac{d\phi}{dx}\right|_{x=x_1} &= v_1, \\ -k_x \left.\frac{d\phi}{dx}\right|_{x=x_2} &= v_2. \end{aligned} \tag{4.16}$$

To incorporate these two boundary conditions two additional terms in the functional must be included, these are $-Av_1\phi_1$ and $Av_2\phi_2$. With these two extra terms the complete expression of the functional becomes

$$\Omega = \frac{1}{2}\int_{x_1}^{x_2} Ak_x\left(\frac{d\phi}{dx}\right)^2 dx - \int_{x_1}^{x_2} \bar{q}\phi dx - A\left(v\phi|_{x=x_1} - v\phi|_{x=x_2}\right) \tag{4.17}$$

Note that although the fluid velocity direction is the same at both boundary points, the fluid is entering into the pipe causing an influx of fluid at $x = x_1$ and the fluid is leaving the pipe causing an outflux of fluid at $x = x_2$. As a result, the signs associated with these two terms in the functional expression are opposite.

Example 4.2

Prove that when the first variation of the functional Equation 4.17 is equated to zero then the governing Equation 4.1 and natural boundary conditions Equation 4.16 are satisfied.

Solution: Consider the functional in Equation 4.17,

$$\Omega = \frac{1}{2}\int_{x_1}^{x_2} Ak_x\left(\frac{d\phi}{dx}\right)^2 dx - \int_{x_1}^{x_2} \bar{q}\phi dx - A\left(v_1\phi\big|_{x=x_1} - v_2\phi\big|_{x=x_2}\right).$$

Following the steps described in Chapter 2 we can write

$$\delta\Omega = \int_{x_1}^{x_2} Ak_x\left(\frac{d\phi}{dx}\right)\Delta\left(\frac{d\phi}{dx}\right)dx - \int_{x_1}^{x_2} \bar{q}\Delta\phi dx - A\left(v_1\Delta\phi\big|_{x=x_1} - v_2\Delta\phi\big|_{x=x_2}\right),$$

or

$$\delta\Omega = Ak_x\left(\frac{d\phi}{dx}\right)\Delta\phi\Bigg|_{x_1}^{x_2} - \int_{x_1}^{x_2} Ak_x\left(\frac{d^2\phi}{dx^2}\right)\Delta\varphi dx - \int_{x_1}^{x_2} \bar{q}\Delta\phi dx - A\left(v_1\Delta\phi\big|_{x=x_1} - v_2\Delta\phi\big|_{x=x_2}\right),$$

or

$$\delta\Omega = A\left\{k_x\left(\frac{d\phi}{dx}\right) + v_2\right\}\Delta\phi\Bigg|_{x_2} - A\left\{k_x\left(\frac{d\phi}{dx}\right) + v_1\right\}\Delta\phi\Bigg|_{x_1} - \int_{x_1}^{x_2}\left[Ak_x\left(\frac{d^2\phi}{dx^2}\right) + \bar{q}\right]\Delta\varphi dx = 0.$$

Clearly for arbitrary $\Delta\phi$ the above equation will be satisfied only if

$$Ak_x\left(\frac{d^2\phi}{dx^2}\right) + \bar{q} = 0, \text{ for x between } x_1 \text{ and } x_2,$$

$$k_x\left(\frac{d\phi}{dx}\right) + v_1 = 0, \text{ at } x = x_1,$$

$$k_x\left(\frac{d\phi}{dx}\right) + v_2 = 0, \text{ at } x = x_2.$$

The above equations are the governing Equation 4.1 and natural boundary conditions Equation 4.17 for the one-dimensional fluid flow problem.

Problems

4.1. Solve Example 4.1 by considering a nonzero value of $\bar{q}$.

4.2. Use code DFT/C-1DFE and verify the results in Example 4.1.

4.3. Use code DFT/C-1DFE and compute the steady-state temperature in a bar 30 cm long with the following properties,

$$k_x = 1.0,$$

$$\bar{q} = 0,$$

$$l = 10 \text{ cm}$$

and temperatures at the ends,

$$T(0) = 100 \text{ degrees},$$

$$T(30 \text{ cm}) = 0 \text{ degrees}.$$

4.4. Two students A and B solved two different fluid flow problems using finite elements. Both of them considered fluid flow through a 40-cm long pipe with global node numbers as shown in Figure 4.4 below. Student A considered 4 elements, each having 2 nodes, and student B considered 2 elements, each having 3 nodes. Both of them obtained the following fluid head values, $\phi_1 = 1$ cm, $\phi_2 = 2$ cm, $\phi_3 = 4$ cm, $\phi_4 = 7$ cm, $\phi_5 = 12$ cm. What will be the magnitude and direction of fluid velocities computed by the two students at point P? P is the center point between nodes 1 and 2. Assume $k_x = 1$ cm/s.

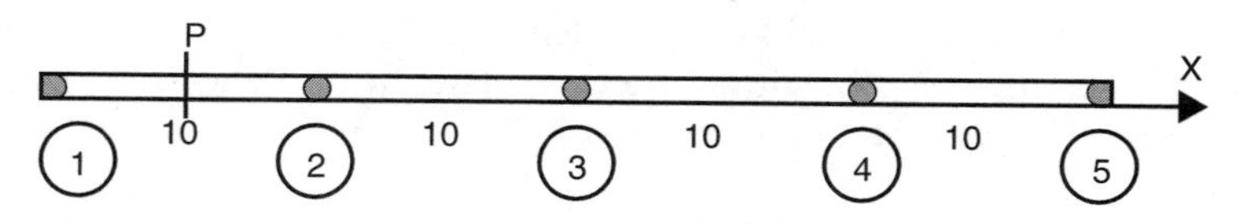

FIGURE 4.4

4.5. In a fluid-flow problem the fluid influx varies linearly from 0 to 10 units per unit length along the length of the flow tube (Figure 4.5). If the tube is discretized into (1) 2 linear (2-node) elements and (2) 1 quadratic (3-node) element, derive the load vector for both cases.

4.6. The 4.node element mesh is shown in the following Figure (4.6). In a finite element formulation the following relations are obtained:

$$\phi = [N]\{q\}, \qquad v_x = -k_x[B]\{q\}, \qquad [k]\{q\} = \{Q\}$$

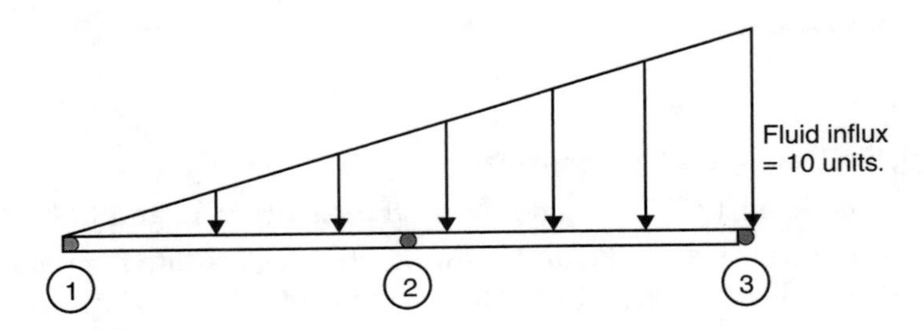

FIGURE 4.5

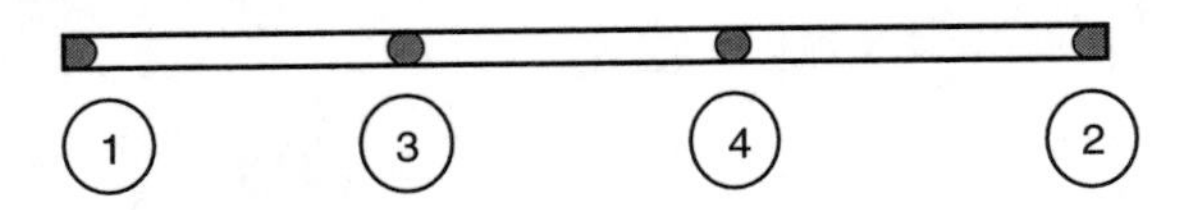

FIGURE 4.6

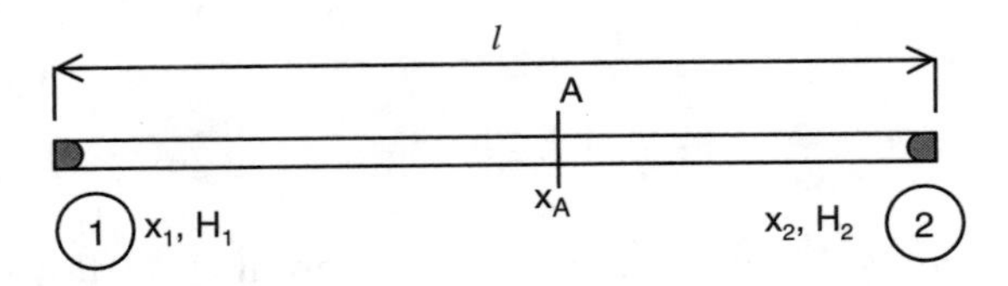

FIGURE 4.7

where ϕ is the fluid head, v_x is the fluid velocity, and {q} is the vector of the nodal degrees of freedom.

(a) What are the dimensions of [N], [B], [k], {q}, and {Q}?

(b) What should be the order of interpolation functions?

(c) How (constant, linear, parabolic, cubic, etc.) should fluid head (ϕ) and fluid velocity (v_x) vary between nodes 1 and 3?

4.7. For a one-dimensional flow problem the fluid head H is assumed to be a linearly varying function. The fluid head at point A (Figure 4.7) can be expressed as

$$H_A = f(x_1, x_2, x_A, H_1, H_2)/l.$$

Find the function *f*.

4.8. Derive element equations for the steady-state flow problem assuming quadratic variation of fluid head ϕ and linear variation of cross-sectional area A.

$$\phi = N_1\phi_1 + N_2\phi_2 + N_3\phi_3 \qquad (N_1,\ N_2 \text{ and } N_3 \text{ are quadratic})$$
$$A = N_{1L}A_1 + N_{2L}A_2 \qquad (N_{1L} \text{ and } N_{2L} \text{ are linear})$$

The fluid influx $\bar{q}$ is constant over the element.

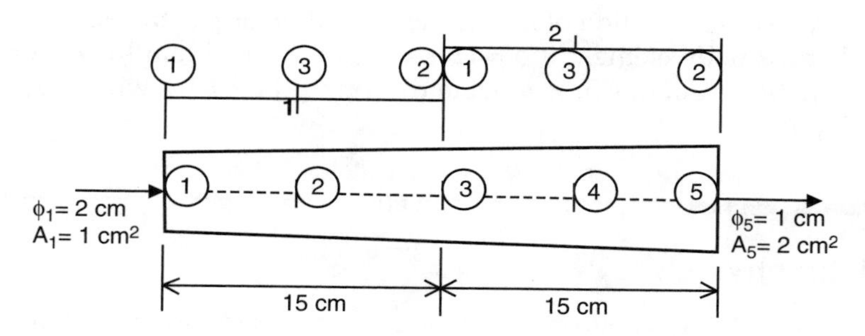

FIGURE 4.8

4.9. Consider fluid through a tapered tube. The cross section of the tube varies linearly from 1 cm^2 at the left end to 2 cm^2 at the right end as shown in Figure 4.8.

For this problem k_x = 1 cm/s, $\bar{q}$ = 0. Discretize the problem geometry into two 3-node elements (local and global node numberings are shown in the figure). Assume quadratic variation of the fluid head within an element. (1) Calculate fluid heads at global nodes 2, 3, and 4. (2) Plot variations of the fluid head along the tube length. (3) Calculate fluid velocities at nodes 2 and 4.

4.10. (a) Consider a 5-node element (nodes are equally spaced). For this element write down the expression of the interpolation function N_3 corresponding to the middle node 3. Express N_3 in terms of the local coordinate s, where s varies linearly from 0 at the left boundary to 1 at the right boundary of the element.

(b) For what types of variations (constant, linear, quadratic, etc.) of the primary unknown would you obtain the exact solution from the finite element analysis using these 5-node elements?

4.11. Consider the following functional

$$\Omega = \int_{-1}^{+1} \left[\phi^2 + 2\phi\left(\frac{\partial \phi}{\partial x}\right) + \left(\frac{\partial \phi}{\partial x}\right)^2 - \phi f_1 - \frac{\partial \phi}{\partial x} f_2 - f_3 \right] dL$$

where $\phi = [N]\{q\}$, $\frac{\partial \phi}{\partial x} = [B]\{q\}$, and f_1, f_2, and f_3 are scalar functions of L.

From this functional obtain the element equation in the form $[k]\{q\} = \{Q\}$. Express [k] and {Q} in terms of $[N]$, $[B]$, f_1, f_2, and f_3.

4.12. Consider the functional

$$\Omega(\phi, \phi_x) = \int_0^1 (\phi_x^2 - 4\phi)dx$$

where ϕ is a function of x, ϕ_x is the derivative of ϕ with respect to x. For a 2-node element if $\phi = (1 - x) \cdot \phi_1 + x \cdot \phi_2$ obtain the element equation from this functional in the form $[k]\{\phi_n\} = \{r\}$, where $\{\phi_n\}^T = [\phi_1 \;\; \phi_2]$

Bibliography

Desai, C. S. 1973. "An approximate solution for unconfined seepage," *J. Irrigation Drain Div.*, ASCE, Vol. 99, No. IR1.

Desai, C. S. 1975. "Finite element methods for flow in porous media," in *Finite Elements in Fluids*, Vol. 1, Gallagher, R. H., Oden, J. T., Taylor, C., and Zienkiewicz, O. C., Eds., John Wiley & Sons, New York, Chap. 8.

Desai, C. S. 1977. "Seepage in porous media," in *Numerical Methods in Geotechnical Engineering*, Desai, C. S. and Christian, J. T., Eds., McGraw-Hill, New York, Chap. 14.

Pinder, G. F. and Gray, W. G. 1977. *Finite Element Simulation in Surface and Subsurface Hydrology*, Academic Press, New York.

Remson, I., Hornberger, G. M., and Molz, F. J. 1971. *Numerical Methods in Subsurface Hydrology*, Wiley-Interscience, New York.

5

One-Dimensional Time Dependent Flow: Introduction to Uncoupled and Coupled Problems

In this chapter we treat problems in which temperature or fluid pressures act in addition to external loading. These effects can occur in two ways. For the case when the magnitudes of temperature or fluid pressures are known, it is relatively easier to include them in the finite element formulations because the effects can be superimposed or considered as uncoupled. Some examples of these are known temperature distribution in a structure and known fluid pressure in a porous body. The general case occurs when temperature is unknown just like the displacement. Then we need to consider interaction or coupling between deformation and thermal effects.

Uncoupled Case

As an illustration, let us assume that the effect of a known change in temperature T is to cause a known strain ε_{y0} in the column (Figure 5.1(a)),

$$\varepsilon_{y_0} = \alpha' T, \tag{5.1}$$

where α' = coefficient of thermal expansion. If the total strain is denoted by ε_y, the effective elastic strain ε_{yn} is given by

$$\varepsilon_{y_n} = \varepsilon_y - \varepsilon_{y_0}. \tag{5.2}$$

The stress–strain relation can then be written as

$$\sigma_y = E\varepsilon_{y_n} = E\left(\varepsilon_y - \varepsilon_{y_0}\right). \tag{5.3}$$

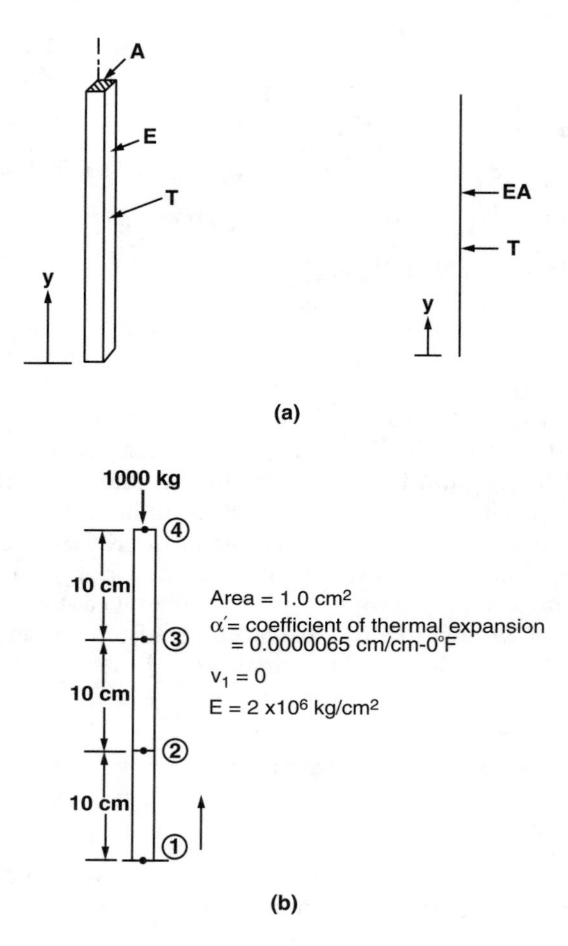

FIGURE 5.1
Thermal effects in a bar idealized as one-dimensional: (a) Bar and one-dimensional idealization and (b) Finite element mesh and properties.

For deriving finite element equations, we need to modify the potential energy function (Equation 3.21), as

$$\Pi_p = \frac{A}{2}\int_{y_1}^{y_2} \sigma_y \varepsilon_{y_n} dy - A\int_{y_1}^{y_2} \bar{Y} v dy - \int_{y_1}^{y_2} \bar{T}_y v dy - \sum P_{il} v_i. \tag{5.4}$$

Note that the terms related to the external forces are the same as before. The first term related to the strain energy can be written by using Equation 5.3 as

$$U' = \frac{A}{2}\int_{y_1}^{y_2} E\left(\varepsilon_y - \varepsilon_{y_0}\right)\left(\varepsilon_y - \varepsilon_{y_0}\right) dy \tag{5.5a}$$

$$= \frac{A}{2}\int_{y_1}^{y_2} E\varepsilon_y^2 dy - A\int_{y_1}^{y_2} E\varepsilon_y \varepsilon_{y_0} dy + \frac{A}{2}\int_{y_1}^{y_2} E\varepsilon_{y_0}^2 dy.$$

$$= U_1 + U_2$$

We can drop the last term of Equation 5.5a, since, being a constant, it will not contribute to element equations when Π_p is differentiated. Thus, compared to the problem in Chapter 3 (Equation 3.21), the only new term that enters is the second term in Equation 5.5a, which after substitution for ε_y from Equation 3.13 can be expanded as

$$U_2 = A\int_{y_1}^{y_2} [v_1 \quad v_2]\frac{1}{l}\begin{Bmatrix} -1 \\ 1 \end{Bmatrix} E\varepsilon_{y_0} dy = A\{\mathbf{q}\}^T \int_{y_1}^{y_2} [\mathbf{B}]^T[\mathbf{C}]\{\varepsilon_{y_0}\} dy$$

$$= \frac{AE\varepsilon_{y_0}}{l}\int_{y_1}^{y_2} (-v_1 + v_2) dy$$

$$= \frac{AE\varepsilon_{y_0}}{l}\frac{l}{2}\int_{-1}^{1} (-v_1 + v_2) dL \tag{5.5b}$$

$$= \frac{AE\varepsilon_{y_0}}{2}\left[(-v_1 L)\Big|_{-1}^{1} + (v_2 L)\Big|_{-1}^{1}\right]$$

$$= AE\varepsilon_{y_0}(-v_1 + v_2).$$

Here U_2 is the part of the strain energy relevant to ε_{y0}; the other part U_1 is the same as U in Equation 3.21 and corresponds to the strain due to the external loads. Now differentiation of Π_p with respect to v_1 and v_2 gives

$$\frac{\partial U_2}{\partial v_1} = AE\varepsilon_{y_0}(-1), \tag{5.6a}$$

$$\frac{\partial U_2}{\partial v_2} = AE\varepsilon_{y_0}(1), \tag{5.6b}$$

or, in matrix notation,

$$\{\mathbf{Q}_0\} = AE\varepsilon_{y_0}\begin{Bmatrix} -1 \\ 1 \end{Bmatrix} = A\sigma_{y_0}\begin{Bmatrix} -1 \\ 1 \end{Bmatrix}. \tag{5.7a}$$

In general terms,

$$\{\mathbf{Q}_0\} = A\int_{y_1}^{y_2} [\mathbf{B}]^T [\mathbf{C}]\{\varepsilon_{y_0}\}dy. \tag{5.7b}$$

Finally, if we add $\{\mathbf{Q}_0\}$ in Equation 5.7 to Equation 3.28, the modified element equations are

$$[\mathbf{k}]\{\mathbf{q}\} = \{\mathbf{Q}\} + \{\mathbf{Q}_0\}, \tag{5.8}$$

where all terms except $\{\mathbf{Q}_0\}$have the same meanings as before. We call $\{\mathbf{Q}_0\}$ an additional, extra, correction, initial, or residual load vector.

Thus the known strains due to temperature changes become an additional load which is superimposed on the external load $\{\mathbf{Q}\}$. It appears on the right-hand side of Equation 5.8 and does not contribute to the left-hand side, which contains the unknown displacements.

Initial Stress

As can be seen from Equation 5.7, the known temperature or fluid pressure can act as initial stress or pressure. For instance, we may know from the solutions of the Laplace equation (Chapter 12) the distribution of seepage pressures below a dam; such pressures can be converted as additional load vectors by using Equation 5.7.

Residual Stresses

The concept of initial strain or stress, which can be called by a common term residual, is important and useful for incorporation of residual conditions caused by temperature, fluid pressure, creep, lack of fit, and other phenomena. This concept has also been used widely in a number of techniques for nonlinear analysis.

Example 5.1

We now illustrate the concept of initial load vector by using the foregoing formulation. The details are shown in Figure 5.1(b). It is required to derive the element equations to include the effect of a change (increase) in temperature of 100°F and then to solve for displacements and stresses for combined (uncoupled) superimposed thermal effects and external load = 1000 kg at the top of the bar.

According to Equation 5.1, the initial strain is

$$\varepsilon_{y_0} = \alpha' T = 0.0000065 \times 100 = 0.00065 \text{ cm/cm}.$$

Hence, the resulting stress (Equation 5.3) is

$$E\varepsilon_{y_0} = 2 \times 10^6 \times 0.00065 = 1300 \text{ kg/cm}^2,$$

and the initial load (Equation 5.7) is

$$\{\mathbf{Q}_0\} = 1 \times 1300 \begin{Bmatrix} -1 \\ 1 \end{Bmatrix}.$$

The assembled initial load vector $\{\mathbf{R}\}$ for the three elements is

$$\{\mathbf{R}_0\} = 1300 \begin{Bmatrix} -1 \\ 1-1 \\ 1-1 \\ 1 \end{Bmatrix} = \begin{Bmatrix} -1300 \\ 0 \\ 0 \\ 1300 \end{Bmatrix}.$$

By following the procedure in Chapter 3 and by adding $\{\mathbf{R}_0\}$ to the external load vector, we obtain

$$\frac{2 \times 10^6 \times 1}{10} \begin{bmatrix} 1 & -1 & 0 & 0 \\ -1 & 2 & -1 & 0 \\ 0 & -1 & 2 & -1 \\ 0 & 0 & -1 & 1 \end{bmatrix} \begin{Bmatrix} v_1 \\ v_2 \\ v_3 \\ v_4 \end{Bmatrix} = \begin{Bmatrix} 0 \\ 0 \\ 0 \\ -1000 \end{Bmatrix} + \begin{Bmatrix} -1300 \\ 0 \\ 0 \\ 1300 \end{Bmatrix}$$

$$= \begin{Bmatrix} -1300 \\ 0 \\ 0 \\ 300 \end{Bmatrix}.$$

Note that here (Figure 5.1) we have measured y positive in the upward direction.

The boundary condition $v_1 = 0$ is now introduced, and the solution of the resulting equations yield

$$v_1 = 0.000 \text{ cm (specified)},$$

$$v_2 = 0.0015,$$

$$v_3 = 0.0030,$$

$$v_4 = 0.0045.$$

The total strains due to the effects of both the external loads and thermal effects are found as

$$\varepsilon_y = [\mathbf{B}]\{\mathbf{q}\} = \frac{1}{l}\begin{bmatrix}-1 & 1\end{bmatrix}\begin{Bmatrix}v_1\\v_2\end{Bmatrix},$$

$$\varepsilon_{y_1} = +0.00015, \ \text{cm/cm},$$

$$\varepsilon_{y_2} = +0.00015,$$

$$\varepsilon_{y_3} = +0.00015.$$

These strains are equal to the sum of the strain due to the external load, $\varepsilon_y = -0.00050$, and that due to the thermal effect, $\varepsilon_{yo} =$ 0.00065.

Comment

This example is included mainly as an elementary illustration of the concept of initial load vector. It does not necessarily represent a practical situation. Moreover, unless the ends are constrained, the temperature change may not influence the stresses in the bar.

Time-Dependent Problems

As illustrations, we shall consider two problems governed essentially by the same equation. They are time-dependent temperature effects[1] and time-dependent deformations with expulsion of fluid from the pores (of a homogeneous medium); in geomechanics, the latter is called consolidation.[2]

Governing Equation

Figure 5.2 shows the schematic of the time dependent heat flow through a rod. The heat energy is entering the rod at the left end ($x = x_1$) at a rate of Q per unit area, and is leaving at the right end ($x = x_2$) at a rate of ($Q + \Delta Q$) per unit area. Heat energy also enters along the length of the rod as q(x) per unit length.

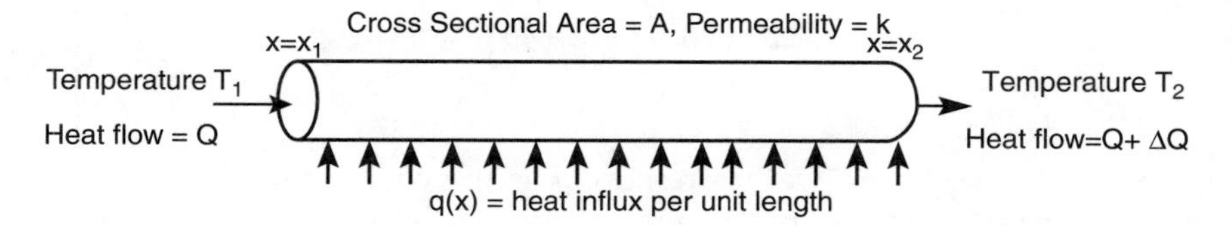

FIGURE 5.2
Schematic of one-dimensional heat flow along rod of cross-sectional area A.

The difference between the amount of heat energy entering the control volume and the amount leaving the control volume should contribute to the increase in temperature of the rod. From conservation of heat energy one can write

$$A \cdot Q + q\left(x_2 - x_1\right) - A\left(Q + \Delta Q\right) = \rho c \frac{\partial T^*}{\partial t} V \tag{5.9a}$$

where ρ is the density (lbm/ft^3 or g/cm^3) and c is the specific heat (Btu/lbm-°F or J/kg-K) of the rod material, V is the volume of the rod, and T and t represent temperature and time, respectively. If we take an elemental length of the rod, e.g., $x_2 - x_1 = dx$, then the above equation can be written as

$$A \cdot Q + q \cdot dx - A\left(Q + \frac{\partial Q}{\partial x} dx\right) = \rho c \frac{\partial T^*}{\partial t} A \cdot dx \tag{5.9b}$$

Equation 5.9b can be further simplified by substituting the relation $Q = -k\frac{\partial T}{\partial x}$ between the heat flow rate Q and the temperature gradient $\frac{\partial T}{\partial x}$, where k is the thermal conductivity,

$$q + Ak \frac{\partial^2 T^*}{\partial x^2} = \rho c \frac{\partial T^*}{\partial t} A, \tag{5.9c}$$

or

$$A\alpha \frac{\partial^2 T^*}{\partial x^2} + \bar{q} = A \frac{\partial T^*}{\partial t}, \tag{5.9d}$$

where $\alpha (= k/\rho c)$ is the thermal diffusivity (Btu/hr-ft-°F or W/m · K). The term $\bar{q}\left(=\frac{q}{\rho c}\right)$ drops out when the heat influx along the length of the rod is zero.

For the consolidation problem[2-4] the governing differential equation has the following form,

$$\frac{k}{\gamma_w}\frac{\partial^2 p^*}{\partial x^2} = m_v \frac{\partial p^*}{\partial t}. \tag{5.9e}$$

Here p^* = (excess) pore water pressure, k is the coefficient of permeability (ft/s or cm/s), γ_w is the unit weight of water (lb/ft^3 or g/cm^3), and m_v = coefficient of volume compressibility (ft^2/lb or cm^2/g).

The constraint conditions associated with Equation 5.9 can be expressed as initial conditions,

$$(\text{i})\ T^*(x,0) = \bar{T}_0^*(x),\quad 0 \le x \le h,\quad t \le 0 \tag{5.10}$$

and boundary conditions,

$$(\text{ii})\ T^*(0,t) = \bar{T}_0^*(t),\quad t > 0, \tag{5.11a}$$

$$(\text{iii})\ T^*(h,t) = \bar{T}_h(t),\quad t > 0, \tag{5.11b}$$

where h is the total length of the domain.

We note that for the time-dependent problem extra boundary conditions occur as initial conditions (Equation 5.10) to define the initial or starting conditions of the body. In the case of static or time-independent problems in Chapters 3 and 4 such initial conditions are not required.

We shall first follow the steps for the thermal problem and then illustrate the case of consolidation.

Step 1. Discretize and Choose Element Configuration

The one-dimensional medium is divided into line elements, as in Figure 3.2.

Step 2. Choose Approximation Model

We choose a linear model to express temperature within an element as

$$\begin{aligned} T &= \tfrac{1}{2}(1-L)T_1(t) + \tfrac{1}{2}(1+L)T_2(t) \\ &= [\mathbf{N}(x)]\{\mathbf{T}_n(t)\}, \end{aligned} \tag{5.12}$$

where $\{\mathbf{T}_n\}^T = [T_1\ T_2]$, the vector of unknown temperatures at nodes 1 and 2, respectively. Here T_1 and T_2 are time dependent, and $[\mathbf{N}]$ is dependent on the spatial coordinate x. As stated in Chapter 4, this function satisfies the requirements for approximation function for T for the one-dimensional flow.

Step 3. Define Gradient-Temperature and Constitutive Relation

The following rate equation can be assumed to describe the mechanism of heat flow and is used as the constitutive relation:

$$q' = -kA\frac{\partial T}{\partial x}, \tag{5.13}$$

where q' = rate of heat flow in the x direction (Btu/hr or W), k = thermal conductivity (Btu/hr-ft-°F or W/m · K), and A = area normal to x direction (ft^2 or m^2).

Step 4. Derive Element Equations

From Example 2.12 one can write that the appropriate functional that should be minimized for satisfying the governing Equation 5.9d is[5,6]

$$\Omega = A\int_{x_1}^{x_2}\left[\frac{1}{2}\alpha\left(\frac{\partial T}{\partial x}\right)^2 + \frac{\partial T}{\partial t}T\right]dx - \int_{x_1}^{x_2}\bar{q}T dx. \tag{5.14}$$

Example 5.2

Prove that the minimization of the functional given in Equation 5.14 guarantees the satisfaction of the governing Equation 5.9d.

Solution: Minimizing the functional given in Equation 5.14 means that its first variation must be equal to zero. Thus,

$$\delta\Omega = A\int_{x_1}^{x_2}\left[\alpha\left(\frac{\partial T}{\partial x}\right)\cdot\Delta\left(\frac{\partial T}{\partial x}\right) + \frac{\partial T}{\partial t}\Delta T\right]dx - \int_{x_1}^{x_2}\bar{q}\cdot\Delta T dx = 0$$

or

$$\delta\Omega = A\int_{x_1}^{x_2}\alpha\left(\frac{\partial T}{\partial x}\right)\cdot\Delta\left(\frac{\partial T}{\partial x}\right)dx + \int_{x_1}^{x_2}\left(A\frac{\partial T}{\partial t} - \bar{q}\right)\cdot\Delta T dx$$

$$= \left[A\alpha\left(\frac{\partial T}{\partial x}\right)\cdot\Delta T\right]_{x=x_1}^{x=x_2} - A\int_{x_1}^{x_2}\alpha\left(\frac{\partial^2 T}{\partial x^2}\right)\cdot\Delta T\cdot dx + \int_{x_1}^{x_2}\left(A\frac{\partial T}{\partial t} - \bar{q}\right)\cdot\Delta T dx$$

$$= \left[A\alpha\left(\frac{\partial T}{\partial x}\right)\cdot\Delta T\right]_{x=x_1}^{x=x_2} - \int_{x_1}^{x_2}\left[A\alpha\left(\frac{\partial^2 T}{\partial x^2}\right) + \bar{q} - A\frac{\partial T}{\partial t}\right]\cdot\Delta T\cdot dx = 0.$$

If the above equation is satisfied for arbitrary ΔT between x = x_1 and x = x_2 then $A\alpha\frac{\partial^2 T}{\partial x^2} + \bar{q} - A\frac{\partial T}{\partial t}$ must be equal to zero between x = x_1 and x = x_2. Thus the governing Equation 5.9d can be satisfied by minimizing the functional of Equation 5.14.

By using Equation 5.12, we have

$$\begin{aligned} \frac{\partial T}{\partial x} &= \frac{\partial}{\partial x}\left[\tfrac{1}{2}(1-L)T_1 + \tfrac{1}{2}(1+L)T_2\right] \\ &= \frac{1}{l}\begin{bmatrix} -1 & 1 \end{bmatrix}\begin{Bmatrix} T_1 \\ T_2 \end{Bmatrix} \\ &= [\mathbf{B}]\{\mathbf{T}_n\} \end{aligned} \tag{5.15a}$$

and

$$\begin{aligned} \frac{\partial T}{\partial t} &= \dot{T} = \frac{\partial}{\partial t}\left[\tfrac{1}{2}(1-L)T_1 + \tfrac{1}{2}(1+L)T_2\right] \\ &= [\mathbf{N}]\begin{Bmatrix} \frac{\partial T_1}{\partial t} \\ \frac{\partial T_2}{\partial t} \end{Bmatrix} = [\mathbf{N}]\begin{Bmatrix} \dot{T}_1 \\ \dot{T}_2 \end{Bmatrix} \\ &= [\mathbf{N}]\{\dot{\mathbf{T}}_n\}, \end{aligned} \tag{5.15b}$$

where $\{\dot{\mathbf{T}}_n\} = [\dot{T}_1 \ \ \dot{T}_2]$ and the overdot denotes derivative with respect to time. Note that [**N**] is a function of space coordinates and hence constant for the time derivative. Substitutions from Equations 5.15a and 5.15b into Equation 5.14 yield

$$\begin{aligned} \Omega &= \frac{A}{2}\frac{l}{2}\int_{-1}^{1}\{\mathbf{T}_n\}^T[\mathbf{B}]^T\alpha[\mathbf{B}]\{\mathbf{T}_n\}dL \\ &+ \frac{Al}{2}\int_{-1}^{1}\{\dot{\mathbf{T}}_n\}^T[\mathbf{N}]^T[\mathbf{N}]\{\mathbf{T}_n\}dL \\ &- \frac{\bar{q}l}{2}\int_{-1}^{1}[\mathbf{N}]\{\mathbf{T}_n\}dL. \end{aligned} \tag{5.16}$$

Expansion of Equation 5.16 leads to (assuming α to be constant)

$$\Omega = \frac{A\alpha l}{4}\int_{-1}^{1}\begin{bmatrix} T_1 & T_2 \end{bmatrix}\frac{1}{l}\begin{Bmatrix} -1 \\ 1 \end{Bmatrix}\frac{1}{l}\begin{bmatrix} -1 & 1 \end{bmatrix}\begin{Bmatrix} T_1 \\ T_2 \end{Bmatrix} dL$$

$$+\frac{Al}{2}\int_{-1}^{1}\begin{bmatrix} \dot{T}_1 & \dot{T}_2 \end{bmatrix}\begin{Bmatrix} N_1 \\ N_2 \end{Bmatrix}\begin{bmatrix} N_1 & N_2 \end{bmatrix}\begin{Bmatrix} T_1 \\ T_2 \end{Bmatrix} dL \tag{5.17a}$$

$$-\frac{\bar{q}l}{2}\int_{-1}^{1}\begin{bmatrix} N_1 & N_2 \end{bmatrix}\begin{Bmatrix} T_1 \\ T_2 \end{Bmatrix} dL$$

$$= \frac{A\alpha}{4l}\int_{-1}^{1}\begin{bmatrix} T_1 & T_2 \end{bmatrix}\begin{bmatrix} 1 & -1 \\ -1 & 1 \end{bmatrix}\begin{Bmatrix} T_1 \\ T_2 \end{Bmatrix} dL$$

$$+\frac{Al}{2}\int_{-1}^{1}\begin{bmatrix} \dot{T}_1 & \dot{T}_2 \end{bmatrix}\begin{bmatrix} N_1^2 & N_1N_2 \\ N_1N_2 & N_2^2 \end{bmatrix}\begin{Bmatrix} T_1 \\ T_2 \end{Bmatrix} dL \tag{5.17b}$$

$$-\frac{\bar{q}l}{2}\int_{-1}^{1}\left(N_1T_1 + N_2T_2\right) dL.$$

Further expansion of the matrix terms leads to

$$\Omega = \frac{A\alpha}{4l}\int_{-1}^{1}\left(T_1^2 - 2T_1T_2 + T_2^2\right) dL$$

$$+\frac{Al}{2}\int_{-1}^{1}\left(N_1^2\dot{T}_1T_1 + N_1N_2\dot{T}_1T_2 + N_1N_2\dot{T}_2T_1 + N_2^2\dot{T}_2T_2\right) dL \tag{5.18}$$

$$-\frac{\bar{q}l}{2}\int_{-1}^{1}\left(N_1T_1 + N_2T_2\right) dL.$$

We now take differentiation (variation) of Ω with respect to T_1 and T_2. An important difference between this variation and the variation in Chapters 3 and 4 may be noted. Here, the functional involves time derivatives, $\dot{T}_1$ and $\dot{T}_2$. During the variations with respect to T_1 and T_2, we make an assumption that $\dot{T}_1$ and $\dot{T}_2$ remain constant. This assumption makes the variational approach mathematically less rigorous. The differentiation yields

$$\frac{\partial \Omega}{\partial T_1} = \frac{A\alpha}{4l}\int_{-1}^{1}(2T_1 - 2T_2) dL$$

$$+\frac{Al}{2}\int_{-1}^{1}\left(N_1^2\dot{T}_1 + N_1N_2\dot{T}_2\right) dL - \frac{\bar{q}l}{2}\int_{-1}^{1}N_1 dL = 0, \tag{5.19a}$$

$$\frac{\partial\Omega}{\partial T_2} = \frac{A\alpha}{4l}\int_{-1}^{1}\left(-2T_1 + 2T_2\right)dL$$
$$+\frac{Al}{2}\int_{-1}^{1}\left(N_1N_2\dot{T}_1 + N_2^2\dot{T}_2\right)dL \tag{5.19b}$$
$$-\frac{\bar{q}l}{2}\int_{-1}^{1}N_2 dL = 0.$$

Therefore,

$$\frac{\partial\Omega}{\partial T_1} = \frac{A\alpha}{2l}\int_{-1}^{1}\left(T_1 - T_2\right)dL$$
$$+\frac{Al}{2}\int_{-1}^{1}\left(N_1^2\dot{T}_1 + N_1N_2\dot{T}_2\right)dL - \frac{\bar{q}l}{2}\left(1\right) = 0, \tag{5.19c}$$

$$\frac{\partial\Omega}{\partial T_2} = \frac{A\alpha}{2l}\int_{-1}^{1}\left(-T_1 + T_2\right)dL + \frac{Al}{2}\int_{-1}^{1}\left(N_1N_2\dot{T}_1 + N_2^2\dot{T}_2\right)dL$$
$$-\frac{\bar{q}l}{2}\left(1\right) = 0. \tag{5.19d}$$

Integration of the terms gives

$$\frac{\partial\Omega}{\partial T_1} = A\frac{\alpha}{l}\left(T_1 - T_2\right) + \frac{Al}{6}\left(2\dot{T}_1 + \dot{T}_2\right) - \frac{\bar{q}l}{2} = 0, \tag{5.20a}$$

$$\frac{\partial\Omega}{\partial T_2} = \frac{A\alpha}{l}\left(-T_1 + T_2\right) + \frac{Al}{6}\left(\dot{T}_1 + 2\dot{T}_2\right) - \frac{\bar{q}l}{2} = 0. \tag{5.20b}$$

Rearranging in matrix notation leads to

$$\frac{A\alpha}{l}\begin{bmatrix} 1 & -1 \\ -1 & 1 \end{bmatrix}\begin{Bmatrix} T_1 \\ T_2 \end{Bmatrix} + \frac{Al}{6}\begin{bmatrix} 2 & 1 \\ 1 & 2 \end{bmatrix}\begin{Bmatrix} \dot{T}_1 = \frac{\partial T_1}{\partial t} \\ \dot{T}_2 = \frac{\partial T_2}{\partial t} \end{Bmatrix} = \frac{\bar{q}l}{2}\begin{Bmatrix} 1 \\ 1 \end{Bmatrix} \tag{5.21a}$$

or

$$\left[\mathbf{k}_\alpha\right]\left\{\mathbf{T}_n\right\} + \left[\mathbf{k}_t\right]\left\{\dot{\mathbf{T}}_n\right\} = \left\{\mathbf{Q}(t)\right\}. \tag{5.21b}$$

Layered Media

In the case of layered media, the formulation based on Equation 5.9c or 5.9e will lead to the following element equations:

$$\lambda_1 \begin{bmatrix} 1 & -1 \\ -1 & 1 \end{bmatrix} \begin{Bmatrix} T_1 \\ T_2 \end{Bmatrix} + \lambda_2 \begin{bmatrix} 2 & 1 \\ 1 & 2 \end{bmatrix} \begin{Bmatrix} \dot{T}_1 \\ \dot{T}_2 \end{Bmatrix} = \frac{\bar{q}l}{2} \begin{Bmatrix} 1 \\ 1 \end{Bmatrix}, \tag{5.21c}$$

where $\lambda_1 = Ak/l$ and $\lambda_2 = A\rho cl/6$ or $\lambda_1 = Ak/\gamma_w l$ and $\lambda_2 = Am_v l/6$, corresponding to Equation 5.9c for the thermal problem and Equation 5.9e for the consolidation problem. In the case of the latter, temperature is replaced by pore water pressure. Although the same symbol $\bar{q}$ is used for prescribed flux in Equations 5.21a and 5.21c, their units will be different and should be defined carefully. Here $[\mathbf{k}_\alpha]$ = element thermal diffusivity matrix, $[\mathbf{k}_t]$ = element matrix related to time dependence, and $\{\mathbf{Q}(t)\}$ = element nodal vector of forcing (flux) parameters, which can be time dependent. The matrices in Equation 5.21 can be expressed as

$$[\mathbf{k}_\alpha] = A\int_{x_1}^{x_2} [\mathbf{B}]^T \alpha [\mathbf{B}] dx, \tag{5.22a}$$

$$[\mathbf{k}_t] = A\int_{x_1}^{x_2} [\mathbf{N}]^T [\mathbf{N}] dx, \tag{5.22b}$$

$$\{\mathbf{Q}(t)\} = \bar{q}\int_{x_1}^{x_2} [\mathbf{N}]^T dx. \tag{5.22c}$$

Solution in Time

Additional derivations are required in the time-dependent problems because of the appearance of the second term on the left-hand side of Equation 5.21. Up to Equation 5.21 we have discretized the domain only in the spatial direction, x. Now we need to obtain solutions in time. This can be done by using the finite difference[7] type of discretization for the derivatives with respect to time. The first derivative $\partial T/\partial t$ can be written approximately as (Figure 5.3(b))

$$\left(\frac{\partial T}{\partial t}\right)_{t+\Delta t} \cong \frac{T(t+\Delta t) - T(t)}{\Delta t} \tag{5.23}$$

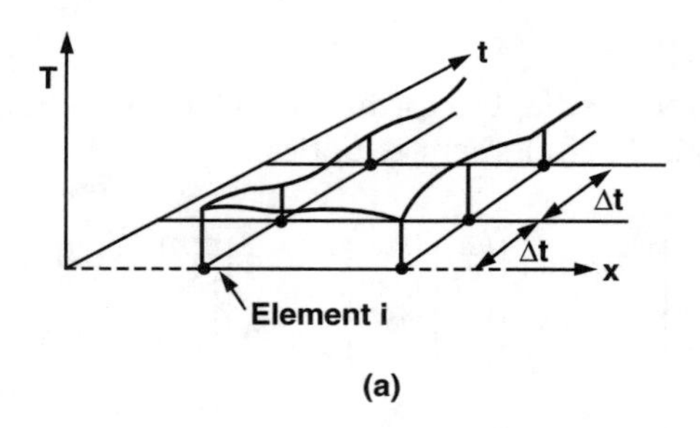

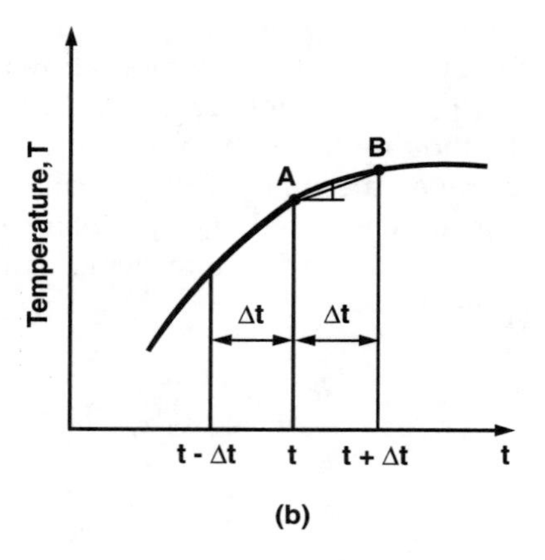

FIGURE 5.3
Solution in time domain: (a) Time variation of temperature in a typical element and (b) Finite difference approximation for first derivative.

where Δt = time increment. Equation 5.23 essentially gives the slope of the chord joining points A and B as an approximation to the continuous derivative $\partial T/\partial t$. By using Equation 5.23, we can now write approximations to the time derivative at the two nodes of an element (Figure 5.3(b)):

$$\frac{\partial T_1}{\partial t} \simeq \frac{T_1(t+\Delta t) - T_1(t)}{\Delta t}, \tag{5.24a}$$

$$\frac{\partial T_2}{\partial t} \simeq \frac{T_2(t+\Delta t) - T_2(t)}{\Delta t}. \tag{5.24b}$$

The time integration scheme (Equation 5.23) is the Euler-type procedure.[7] We use this simple scheme only for the sake of introduction. In fact, a number of improved and mathematically superior schemes such as the Crank–Nicholson procedure are commonly used in finite element applications.

Substitution of Equation 5.24 into Equation 5.21 leads to

$$[\mathbf{k}_\alpha]\begin{Bmatrix} T_1(t+\Delta t) \\ T_2(t+\Delta t) \end{Bmatrix} + \frac{1}{\Delta t}[\mathbf{k}_t]\begin{Bmatrix} T_1(t+\Delta t) - T_1(t) \\ T_2(t+\Delta t) - T_2(t) \end{Bmatrix} = \{\mathbf{Q}(t+\Delta t)\}, \tag{5.25}$$

or

$$\left([\mathbf{k}_\alpha] + \frac{1}{\Delta t}[\mathbf{k}_t]\right)\begin{Bmatrix} T_1(t+\Delta t) \\ T_2(t+\Delta t) \end{Bmatrix} = \{\mathbf{Q}(t+\Delta t)\} + \frac{1}{\Delta t}[\mathbf{k}_t]\begin{Bmatrix} T_1(t) \\ T_2(t) \end{Bmatrix}, \tag{5.26a}$$

or

$$[\bar{\mathbf{k}}]\begin{Bmatrix} T_1(t+\Delta t) \\ T_2(t+\Delta t) \end{Bmatrix} = \{\bar{\mathbf{Q}}\}, \tag{5.26b}$$

where

$$[\bar{\mathbf{k}}] = [\mathbf{k}_a] + \frac{1}{\Delta t}[\mathbf{k}_t],$$

$$\{\bar{\mathbf{Q}}\} = \{\mathbf{Q}(t+\Delta t)\} + \frac{1}{\Delta t}[\mathbf{k}_t]\begin{Bmatrix} T_1(t) \\ T_2(t) \end{Bmatrix}.$$

At any time t, the terms on the right-hand side in Equation 5.26 are usually known, $\{\mathbf{Q}(t + \Delta t)\}$ from the specified forcing functions and the second term from the known values of T at the previous time level. Because the initial conditions (Equation 5.10) are given, we know the initial values of T at any point at time $t = 0$. Hence, we can solve Equation 5.26 for T at $0 + \Delta t$, since the T at the previous time level are known.

The solution process for the time-dependent problem thus involves two steps: (1) discretization in space and then (2) propagation or marching in time at various time levels. The foregoing problem belongs to the class of problems called initial value problems, since we start from an initially known state.

Step 4. Derivation by Galerkin's Method

We now consider formulation of finite element equations (Equation 5.21) by using Galerkin's residual method. Equation 5.9d can be written as (for $q = 0$),

$$\alpha\frac{\partial^2 T^*}{\partial x^2} - \frac{\partial T^*}{\partial t} = 0, \tag{5.27a}$$

or

$$\left(\alpha\frac{\partial^2}{\partial x^2}-\frac{\partial}{\partial t}\right)T^*=0, \tag{5.27b}$$

or

$$LT^*=0 \tag{5.27c}$$

where L is the differential operator. Denoting the approximate solution by T, we have the residual R,

$$R(x)=\alpha\frac{\partial^2 T}{\partial x^2}-\frac{\partial T}{\partial t}. \tag{5.28}$$

On the basis of the explanation in Chapter 3, we consider a generic element for the Galerkin formulation. Assuming linear approximation for an element as in Equation 5.12, we have

$$\begin{aligned} T &= \tfrac{1}{2}(1-L)T_1(t)+\tfrac{1}{2}(1+L)T_2(T) \\ &= \sum_{i=1}^{2} N_i T_i. \end{aligned} \tag{5.29}$$

Here the interpolation functions N_i are functions of x, and the nodal temperatures T_i are the functions of time. We note here that for a given time t, Equation 5.29 gives variation of the temperature along the length of the element. At this stage, the time dependence is included in the temperatures at the nodes only, $T_1(t),T_2(t)$ (Figure 5.3(a)).

Now according to Galerkin's method,

$$\int_{x_1}^{x_2}\left(\alpha\frac{\partial^2 T}{\partial x^2}-\frac{\partial T}{\partial t}\right)N_i dx=0, \quad i=1,2. \tag{5.30a}$$

Integration by parts of the first term leads to

$$\int_{x_1}^{x_2}\alpha\frac{\partial T}{\partial x}\frac{\partial N_i}{\partial x}dx-\alpha\frac{\partial T}{\partial x}N_i\Bigg|_{x_1}^{x_2}+\int_{x_1}^{x_2}\frac{\partial T}{\partial t}N_i dx=0 \tag{5.30b}$$

or

$$\int_{x_1}^{x_2}\alpha\frac{\partial T}{\partial x}\frac{\partial N_i}{\partial x}dx+\int_{x_1}^{x_2}\frac{\partial T}{\partial t}N_i dx=\alpha\frac{\partial T}{\partial x}N_i\Bigg|_{x_1}^{x_2}. \tag{5.30c}$$

Substitution of $T = \sum N_i T_i$ into Equation 5.30c leads to

$$\int_{x_1}^{x_2} \alpha \frac{\partial N_j}{\partial x} \frac{\partial N_i}{\partial x} T_i dx + \int_{x_1}^{x_2} N_j N_i \frac{\partial T_i}{\partial t} dx = \alpha \frac{\partial T}{\partial x} N_i \Big|_{x_1}^{x_2}, \quad i = 1, 2, j = 1, 2. \quad (5.31a)$$

The index notation has an implication similar to that explained in Chapter 3. In matrix notation, we have

$$\int_{x_1}^{x_2} \alpha \begin{bmatrix} \frac{\partial N_1}{\partial x} \cdot \frac{\partial N_1}{\partial x} & \frac{\partial N_2}{\partial x} \cdot \frac{\partial N_1}{\partial x} \\ \frac{\partial N_1}{\partial x} \cdot \frac{\partial N_2}{\partial x} & \frac{\partial N_2}{\partial x} \cdot \frac{\partial N_2}{\partial x} \end{bmatrix} \begin{Bmatrix} T_1 \\ T_2 \end{Bmatrix} dx + \int_{x_1}^{x_2} \begin{bmatrix} N_1^2 & N_2 N_1 \\ N_1 N_2 & N_2^2 \end{bmatrix} \begin{Bmatrix} \frac{\partial T_1}{\partial t} \\ \frac{\partial T_2}{\partial t} \end{Bmatrix} dx$$

$$= \begin{Bmatrix} \alpha \frac{\partial T}{\partial x} N_1 \Big|_{x_1}^{x_2} \\ \alpha \frac{\partial T}{\partial x} N_2 \Big|_{x_1}^{x_2} \end{Bmatrix}. \quad (5.31b)$$

The two terms on the left-hand side yield the same element equations $[\mathbf{k}_\alpha]$ and $[\mathbf{k}_t]$ as in Equation 5.21.

Because $N_1 = 0$ at x_2 and $N_2 = 0$ at x_1, and $N_1 = 1$ at x_1 and $N_2 = 1$ at x_1, the left-hand side yields

$$\begin{Bmatrix} -\alpha \left(\frac{\partial T}{\partial x} \right)_1 \\ \alpha \left(\frac{\partial T}{\partial x} \right)_2 \end{Bmatrix} = \begin{Bmatrix} -q_1 \\ q_2 \end{Bmatrix}, \quad (5.32)$$

which denotes (joint) fluid fluxes at nodes 1 and 2, respectively. As explained in Chapter 3, when we assemble the element equations, the terms in Equation 5.32 can vanish, except those at the ends, and they can constitute conditions specified at the ends.

Step 5. Assembly for Global Equations

As an illustration, consider a one-dimensional bar of metal (Figure 5.2a) through which heat flow occurs. The bar is divided into three elements (Figure 5.4) with four nodes; the elements are assumed to be of uniform length and cross section. Use of Equation 5.21 allows generation of element equations for the three elements. By observing the fact that temperatures at

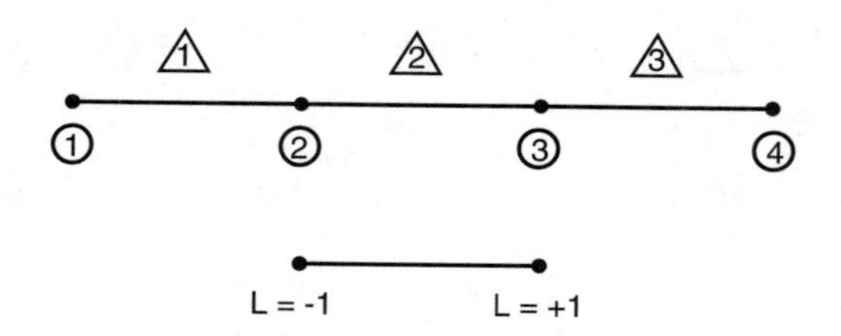

FIGURE 5.4
Discretization of bar.

adjacent nodes are continuous, we can add the element equations to obtain global equations as follows:

$$\frac{A\alpha}{l}\begin{bmatrix} 1 & -1 & 0 & 0 \\ -1 & 2 & -1 & 0 \\ 0 & -1 & 2 & -1 \\ 0 & 0 & -1 & 1 \end{bmatrix}\begin{Bmatrix} T_1 \\ T_2 \\ T_3 \\ T_4 \end{Bmatrix}_{t+\Delta t} + \frac{Al}{6\Delta t}\begin{bmatrix} 2 & 1 & 0 & 0 \\ 1 & 4 & 1 & 0 \\ 0 & 1 & 4 & 1 \\ 0 & 0 & 1 & 2 \end{bmatrix}\begin{Bmatrix} T_1 \\ T_2 \\ T_3 \\ T_4 \end{Bmatrix}_{t+\Delta t}$$

$$= \begin{Bmatrix} R_1 \\ R_2 \\ R_3 \\ R_4 \end{Bmatrix}_{t+\Delta t} + \frac{Al}{6\Delta t}\begin{bmatrix} 2 & 1 & 0 & 0 \\ 1 & 4 & 1 & 0 \\ 0 & 1 & 4 & 1 \\ 0 & 0 & 1 & 2 \end{bmatrix}\begin{Bmatrix} T_1 \\ T_2 \\ T_3 \\ T_4 \end{Bmatrix}_{t}, \tag{5.33a}$$

or in matrix notation,

$$\frac{A\alpha}{l}\left[\mathbf{K}_{\alpha}^{*}\right]\{\mathbf{r}\}_{t+\Delta t} + \frac{Al}{6\Delta t}\left[\mathbf{K}_{t}^{*}\right]\{\mathbf{r}\}_{t+\Delta t} = \{\mathbf{R}\}_{t+\Delta t} + \frac{Al}{6\Delta t}\left[\mathbf{K}_{t}^{*}\right]\{\mathbf{r}\}_{t}, \tag{5.33b}$$

Here $[K_{\alpha}^{*}]$ and $[K_{t}^{*}]$ are the assemblage matrices and $\{r\}$ and $[R]$ are the assemblage nodal unknown (temperature) and known forcing function vectors, respectively, and the superscript asterisk denotes assemblage matrices in Equation 5.33a. Note that we can take out α/l and $l/6\Delta t$ as common if the material is assumed to be homogeneous; then Equation 5.33b can be written as

$$\Delta T\left[\mathbf{K}_{\alpha}^{*}\right]\{\mathbf{r}\}_{t+\Delta t} + \frac{1}{6}\left[\mathbf{K}_{t}^{*}\right]\{\mathbf{r}\}_{t+\Delta t} = \frac{\Delta t}{Al}\{\mathbf{R}\}_{t+\Delta t} + \frac{1}{6}\left[\mathbf{K}_{t}^{*}\right]\{\mathbf{r}\}_{t} \tag{5.33c}$$

where $\Delta T = \alpha\Delta t/l^2$ is a nondimensional time increment (with area A assumed to be unity):

$$\alpha = \frac{k}{\rho c} = \frac{\text{Btu}}{\text{hr-ft-°F}} \frac{\text{lbm-°F}}{\text{Btu}} = \frac{\text{ft}^2}{\text{hr}}.$$

Therefore,

$$\Delta T = \frac{\text{ft}^2}{\text{hr}} \cdot \frac{\text{hr}}{\text{ft}^2} = \text{nondimensional increment of time factor; T here denotes time factor and not temperature.}$$

Boundary Conditions

At $t = 0$, we have the initial condition (Equation 5.10). We choose to prescribe values of T_1, T_2, T_3, and T_4 at $t \leq 0$ as uniform $= \bar{T}_0$. Hence Equation 5.33 at $0 + \Delta t$ is

$$\frac{\alpha}{l}\left[\mathbf{K}_\alpha^*\right]\left\{\mathbf{r}\right\}_{0+\Delta t} + \frac{1}{6\Delta t}\left[\mathbf{K}_t^*\right]\left\{\mathbf{r}\right\}_{0+\Delta t} = \left\{\mathbf{R}\right\}_{0+\Delta t} + \frac{1}{6\Delta t}\left[\mathbf{K}_t\right]\left\{\mathbf{r}\right\}_0. \tag{5.34}$$

The initial conditions are applied at $t = 0$.

Now we can introduce end (geometric) boundary conditions

$$\begin{aligned} T_1 &= T_1(0,t) = \delta_1, \quad t > 0, \\ T_4 &= T_4(h,t) = \delta_4, \quad t > 0. \end{aligned} \tag{5.35}$$

A procedure similar to the one described in Chapter 3 can be used. Here δ_1 and δ_4 are the known values applied at nodes 1 and 4, respectively, and can be assumed to remain constant with time.

For clarity of illustration let us assume that $\Delta T = 1$, $T(x, 0) = T_0$, and $\{\mathbf{R}\} = \{0\}$; then Equation 5.34 can be expressed as

$$\begin{aligned} &\begin{bmatrix} \frac{4}{3} & -\frac{5}{6} & 0 & 0 \\ & \frac{8}{3} & -\frac{5}{6} & 0 \\ \text{sym.} & & \frac{8}{3} & -\frac{5}{6} \\ & & & \frac{4}{3} \end{bmatrix} \begin{Bmatrix} T_1 \\ T_2 \\ T_3 \\ T_4 \end{Bmatrix}_{\Delta t} = \begin{bmatrix} \frac{1}{3} & \frac{1}{6} & 0 & 0 \\ & \frac{2}{3} & \frac{1}{6} & 0 \\ \text{sym.} & & \frac{2}{3} & \frac{1}{6} \\ & & & \frac{1}{3} \end{bmatrix} \begin{Bmatrix} T_0 \\ T_0 \\ T_0 \\ T_0 \end{Bmatrix}_0 \\ &= \begin{Bmatrix} \frac{T_0}{3} + \frac{T_0}{6} \\ \frac{T_0}{6} + \frac{2T_0}{3} + \frac{T_0}{6} \\ \frac{T_0}{6} + \frac{2T_0}{3} + \frac{T_0}{6} \\ \frac{T_0}{6} + \frac{T_0}{3} \end{Bmatrix}_0 = \begin{Bmatrix} \frac{T_0}{2} \\ T_0 \\ T_0 \\ \frac{T_0}{2} \end{Bmatrix}_0. \end{aligned} \tag{5.36}$$

It is important to note that the right-hand side essentially yields an equivalent or residual load vector $\{\mathbf{R}_0\}$ corresponding to the known temperatures at $t = 0$.

Equation 5.36 can now be modified to include boundary conditions (Equation 5.35). The modification can be performed by setting

$$
\begin{aligned}
&K_{11} = 1, && K_{44} = 1,\\
&K_{1j} = 0, && j = 2\ 4,\\
&K_{4j} = 0, && j = 1, 2, 3,\\
&R_1 = \delta_1, &&\\
&R_4 = \delta_4, &&
\end{aligned}
\tag{5.37}
$$

which leads to

$$
\begin{bmatrix} 1 & 0 & 0 & 0 \\ -\frac{5}{6} & \frac{8}{3} & -\frac{5}{6} & 0 \\ 0 & -\frac{5}{6} & \frac{8}{3} & -\frac{5}{6} \\ 0 & 0 & 0 & 1 \end{bmatrix} \begin{Bmatrix} T_1 \\ T_2 \\ T_3 \\ T_4 \end{Bmatrix}_{\Delta t} = \begin{Bmatrix} \delta_1 \\ \overline{T_0} \\ \overline{T_0} \\ \delta_4 \end{Bmatrix}_0 . \tag{5.38}
$$

Note that Equation 5.38 lost its symmetric nature. Often, it is economical and convenient from the computational viewpoint to restore the symmetry of the matrix. It can be achieved, as described in Chapter 3, by modifying further the other equations as

$$
\begin{aligned}
&K_{j1} = 0, && j = 2, 3, 4,\\
&K_{j4} = 0, && j = 1, 2, 3,
\end{aligned}
\tag{5.39a}
$$

and the right-hand term as

$$
\begin{bmatrix} 1 & 0 & 0 & 0 \\ 0 & \frac{8}{3} & -\frac{5}{6} & 0 \\ 0 & -\frac{5}{6} & \frac{8}{3} & 0 \\ 0 & 0 & 0 & 1 \end{bmatrix} \begin{Bmatrix} T_1 \\ T_2 \\ T_3 \\ T_4 \end{Bmatrix}_{\Delta t} = \begin{Bmatrix} \delta_1 \\ \overline{T_0} + \frac{5}{6}\delta_1 \\ \overline{T_0} + \frac{5}{6}\delta_4 \\ \delta_4 \end{Bmatrix}_0 . \tag{5.39b}
$$

Step 6. Solve for Primary Unknowns

Example 5.3

To illustrate the solution of Equation 5.39b, let us adopt quantitative values for $\overline{T}_0$, δ_1, and δ_4 as follows:

$$\overline{T}_0 = 0.0, \qquad 0 \le x \le h, \;\; t \le 0,$$
$$\left.\begin{aligned} \delta_1 &= 10 \text{ degrees} \\ \delta_4 &= 20 \text{ degrees} \end{aligned}\right\} \quad t \ge 0. \tag{5.40}$$

Then Equation 5.39b for the first time increment ($t = 0 + \Delta t$) becomes

$$\begin{bmatrix} 1 & 0 & 0 & 0 \\ 0 & \frac{8}{3} & -\frac{5}{6} & 0 \\ 0 & -\frac{5}{6} & \frac{8}{3} & 0 \\ 0 & 0 & 0 & 1 \end{bmatrix} \begin{Bmatrix} T_1 \\ T_2 \\ T_3 \\ T_4 \end{Bmatrix}_{\Delta t} = \begin{Bmatrix} 10 \\ \frac{50}{6} \\ \frac{100}{6} \\ 20 \end{Bmatrix}_0 . \tag{5.41a}$$

Therefore,

$$\tfrac{8}{3} T_2 - \tfrac{5}{6} T_3 = \frac{50}{6},$$
$$-\tfrac{5}{6} T_2 + \tfrac{8}{3} T_3 = \tfrac{100}{6},$$

or

$$16T_2 - 5T_3 = 50,$$
$$-5T_2 + 16T_3 = 100,$$

or by multiplication by 5 and 16, respectively, of the two equations, we have

$$\begin{array}{r} 80T_2 - 25T_3 = 250 \\ -80T_2 + 256T_3 = 1600 \\ \hline 231T_3 = 1850 \Rightarrow T_3 = 8.01 \end{array}$$

and

$$16T_2 - \frac{5 \times 1850}{231} = 50,$$
$$16T_2 - 40.04 = 50,$$
$$T_2 = 5.63.$$

Second Time Increment

For the next time increment ($t = 0 + \Delta t + \Delta t = 0 + 2\Delta t$), we have

$$\begin{bmatrix} \frac{4}{3} & -\frac{5}{6} & 0 & 0 \\ -\frac{5}{6} & \frac{8}{3} & -\frac{5}{6} & 0 \\ 0 & -\frac{5}{6} & \frac{8}{3} & -\frac{5}{6} \\ 0 & 0 & -\frac{5}{6} & \frac{4}{3} \end{bmatrix} \begin{Bmatrix} T_1 \\ T_2 \\ T_3 \\ T_4 \end{Bmatrix}_{2\Delta t} = \begin{bmatrix} \frac{1}{3} & \frac{1}{6} & 0 & 0 \\ \frac{1}{6} & \frac{2}{3} & \frac{1}{6} & 0 \\ 0 & \frac{1}{6} & \frac{2}{3} & \frac{1}{6} \\ 0 & 0 & \frac{1}{6} & \frac{1}{3} \end{bmatrix} \begin{Bmatrix} 10 \\ 5.63 \\ 8.01 \\ 20 \end{Bmatrix}_{\Delta t} = \frac{1}{6} \begin{Bmatrix} 25.63 \\ 40.53 \\ 57.67 \\ 48.01 \end{Bmatrix}_{\Delta t}. \tag{5.41b}$$

Introduction of boundary conditions, $\delta_1 = 10$ and $\delta_4 = 20$, yields

$$\begin{bmatrix} 1 & 0 & 0 & 0 \\ 0 & \frac{8}{3} & -\frac{5}{6} & 0 \\ 0 & -\frac{5}{6} & \frac{8}{3} & 0 \\ 0 & 0 & 0 & 1 \end{bmatrix} \begin{Bmatrix} T_1 \\ T_2 \\ T_3 \\ T_4 \end{Bmatrix}_{2\Delta t} = \begin{Bmatrix} 10 \\ \frac{40.53}{6} + \frac{5}{6} \times 10 \\ \frac{57.67}{6} + \frac{5}{6} \times 20 \\ 20 \end{Bmatrix} = \begin{Bmatrix} 10 \\ 90.43/6 \\ 157.67/6 \\ 20 \end{Bmatrix}. \tag{5.41c}$$

Therefore

$$\frac{8}{3} T_2 - \frac{5}{6} T_3 = \frac{90.43}{6},$$
$$-\frac{5}{6} T_2 + \frac{8}{3} T_3 = \frac{157.67}{6},$$

or

$$16T_2 - 5T_3 = 90.43,$$
$$-5T_2 + 16T_3 = 157.67,$$

or

$$80T_2 - 25T_3 = 452.15,$$
$$-80T_2 + 256T_3 = 2522.72,$$

or

$$231T_3 = 2974.87,$$
$$T_3 = 12.878,$$

and

$$16T_2 - 5 \times 12.878 = 90.43,$$
$$16T_2 = 154.821,$$
$$T_2 = 9.676,$$

and so on for other time increments.

Step 7. Compute the Derived or Secondary Quantities

The derived quantities can be rate of flow and quantity of flow. For example, from Equation 5.13 rate of heat flow q' for an element can be written as

$$\begin{aligned} q' &= -kA\frac{\partial T}{\partial x} \\ &= -kA[\mathbf{B}]\{\mathbf{T}_n\} \qquad (5.42a) \\ &= \frac{-kA}{l}\begin{bmatrix} -1 & 1 \end{bmatrix}\begin{Bmatrix} T_1 \\ T_2 \end{Bmatrix}. \end{aligned}$$

Since we know T_1 and T_2 for each element, computation of q' is straightforward. For instance, for element 2 at time Δt, $T_1 = 10$ and $T_2 = 5.63$.

Therefore

$$q'(2) = \frac{-kA}{l}(-10+5.63)$$
$$= 4.37\frac{kA}{l}\text{Btu/unit time.} \tag{5.42b}$$

One-Dimensional Consolidation[3,8,9]

Consolidation is the phenomenon which describes time-dependent deformation in a saturated porous medium such as soil under applied (external) loading. The material deforms with time while the liquid or water in the pores gradually squeezes or diffuses out. This phenomenon involves coupling or interaction between deformation and pressure in the fluid.

Under a number of assumptions,[2] it is possible to approximate the phenomenon to occur only in one (vertical) direction; then the stress-deformation behavior of the skeleton of the medium and the behavior of fluid can be treated separately. The stress–strain behavior is expressed through an effective stress concept given by

$$\sigma = \sigma' + p, \tag{5.43}$$

where σ = total applied stress (Figure 5.5), σ' = effective stress carried by the soil skeleton, and p = pore fluid (water) pressure. The stress–strain or constitutive relation for the deformation behavior of the skeleton can be expressed as (see Ref. 2 or other undergraduate texts on geomechanics)

$$\Delta\sigma' = -\frac{1}{a_v}\Delta e, \tag{5.44}$$

where a_v = coefficient of compressibility; e = void ratio, proportional to axial or vertical strain, and Δ denotes change due to a load increment.

Under the above assumptions, the governing equation for one-dimensional consolidation is given by Equation 5.9c; for homogeneous media, it can be expressed as

$$c_v\frac{\partial^2 p}{\partial y^2} = \frac{\partial p}{\partial t}, \tag{5.45}$$

where c_v = coefficient of consolidation = $k(1+e_0)/\gamma_w a_v$, k = coefficient of permeability, γ_w = unit weight of water, y = vertical coordinate, and e_0 = initial void ratio. The constitutive law for the flow behavior is Darcy's law:

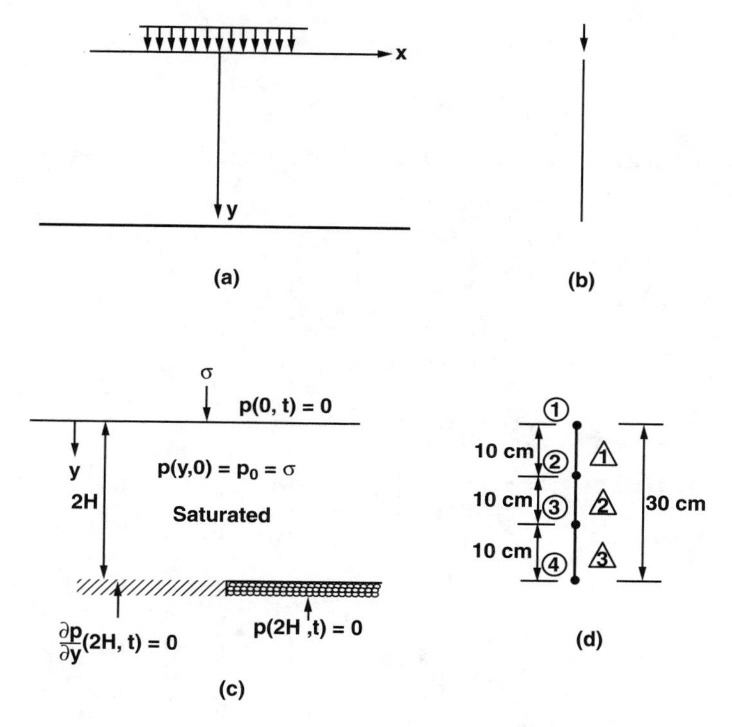

FIGURE 5.5
One-dimensional consolidation: (a) Consolidating mass, (b) One-dimensional idealization, (c) Boundary conditions, and (d) Discretization.

$$v = -\frac{k}{\gamma_w}\frac{\partial p}{\partial y}. \tag{5.46}$$

Equation 5.45 is of the same form as Equation 5.9 for heat flow, and all steps of the finite element formulations are essentially the same.

We shall therefore detail an illustrative problem for consolidation by using the results derived previously.

Consider a consolidating mass (Figure 5.5) divided into three elements (Figure 5.5(d)). Assume the following properties:

Length of element, $l = 10$ cm;
Coefficient of consolidation, $c_v = 1$ cm²/sec;
Applied vertical load, $\sigma = 1$ g/cm².

By using Equation 5.21a, we obtain, for each element (with $A = 1$),

$$\frac{1}{10}\begin{bmatrix} 1 & -1 \\ -1 & 1 \end{bmatrix}\begin{Bmatrix} p_1 \\ p_2 \end{Bmatrix} + \frac{10}{6}\begin{bmatrix} 2 & 1 \\ 1 & 2 \end{bmatrix}\begin{Bmatrix} \dot{p}_1 \\ \dot{p}_2 \end{Bmatrix} = \begin{Bmatrix} Q_1(t) \\ Q_2(t) \end{Bmatrix} \tag{5.47}$$

or

$$[\mathbf{k}_\alpha]\{\mathbf{p}_n\}+[\mathbf{k}_t]\{\dot{\mathbf{p}}_n\}=\{\mathbf{Q}(t)\}.$$

The simple backward difference Euler-type integration process will lead to

$$\left([\mathbf{k}_\alpha]+\frac{1}{\Delta t}[\mathbf{k}_t]\right)\begin{Bmatrix}p_1\\p_2\end{Bmatrix}_{t+\Delta t}=\{\mathbf{Q}(t)\}_{t+\Delta t}+\frac{1}{\Delta t}[\mathbf{k}_t]\begin{Bmatrix}p_1\\p_2\end{Bmatrix}_t. \tag{5.48}$$

Assembly of the three element equations gives

$$\frac{1}{10}\begin{bmatrix}1&-1&0&0\\-1&2&-1&0\\0&-1&2&-1\\0&0&-1&1\end{bmatrix}\begin{Bmatrix}p_1\\p_2\\p_3\\p_4\end{Bmatrix}_{t+\Delta t}+\frac{10}{6}\begin{bmatrix}2&1&0&0\\1&4&1&0\\0&1&4&1\\0&0&1&2\end{bmatrix}\begin{Bmatrix}p_1\\p_2\\p_3\\p_4\end{Bmatrix}_{t+\Delta t}$$
$$=\begin{Bmatrix}R_1\\R_2\\R_3\\R_4\end{Bmatrix}_{t+\Delta t}+\frac{10}{6}\begin{bmatrix}2&1&0&0\\1&4&1&0\\0&1&4&1\\0&0&1&2\end{bmatrix}\begin{Bmatrix}p_1\\p_2\\p_3\\p_4\end{Bmatrix}_t. \tag{5.49}$$

We assumed homogeneous medium and $\Delta T = 1$.

Assume initial and boundary conditions (Figure 5.5) as

$$\begin{aligned}p_0=p(y,0)&=1.0\text{ g/cm}^2, && 0\le x\le h,\ t\le 0,\\ p(0,t)&=0.0, && t>0,\\ p(2H,t)&=0.0, && t>0.\end{aligned} \tag{5.50a}$$

Here we assume that the top and bottom of the consolidating mass are pervious, and hence the pore water pressure there is essentially zero at all times; this constitutes the first or Dirichlet-type boundary condition. It is possible to incorporate an impervious boundary, that is, the second or Neumann-type condition. For instance, we can have impervious base, and then

$$\frac{\partial p}{\partial y}(2H,t)=0,\quad t>0. \tag{5.50b}$$

In the finite element formulation herein, it is not necessary to make any special modification for this boundary condition. The condition of no flow across the base implied in Equation 5.50b is achieved simply by leaving the boundary node as it is and treating it just like other nodes where p is assumed as the unknown and obtained through the finite element computations.

The procedure for introduction of the boundary conditions (Equation 5.50a) is essentially the same as for the temperature problem.

An important observation can be made at this stage. The problem of consolidation is a coupled problem involving interaction between the deformation of the skeleton of the medium and the pressure in the pores of the medium. However, it is only because of one-dimensional idealization and the accompanying assumptions[2,8,9] that we are able to consider separately the deformation (Equation 5.43) and pore pressures (Equation 5.45). In reality and in a mathematical sense, the effects of the two phenomena, deformation and fluid pressure, are coupled.

Computer Code

The code DFT/C-1DFE mentioned in Chapters 3 and 4 and described in Chapter 6 can be used to solve the problems of transient heat flow and consolidation by specifying the required option. At this stage, the reader may study the portion of the code relevant to these problems. In the following are described some results from application of this code.

Example 5.4

Figure 5.6 shows a one-dimensional idealization of a homogeneous medium divided into 10 elements and 11 nodes. The following properties are given:

Initial conditions: $T(x, 0)$ or $p(x, 0) = 100$ units,

Boundary conditions: $T(0, t)$ or $p(0, t) = 0$,
$T(h, t)$ or $p(2H, t) = 0$;
α or $c_v = 1$ unit,
$l = 1$ unit,
$t = 0.1$ unit.

Temperature Problem. In the case of heat flow, this problem represents time-dependent cooling of a bar initially at a temperature of 100°F whose ends are cooled to 0°F and kept at that temperature for all subsequent time levels.

A pictorial distribution of computed temperatures at various time steps along the bar is shown in Figure 5.7(a). Distribution along the bar at various times is shown in Figure 5.7(b).

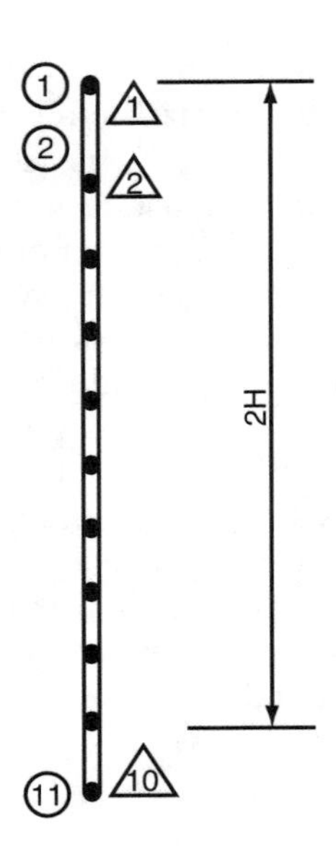

FIGURE 5.6
Finite element mesh.

Thus, it is possible to compute distribution of temperatures in a one-dimensional medium by using the finite element procedure. The temperatures and quantity of heat flow can be found at any point in the bar and at any time level, giving the entire thermal history under cooling or heating due to a given change in temperature. It is possible to include different geometrical and material properties for each element; hence, nonhomogeneous media can be easily accommodated. It is also possible to include natural boundary conditions, that is, boundaries that are insulated against heat. For complete insulation, it can be done just by leaving the node at that boundary free, that is, by allowing the finite element procedure to compute temperature at the node. This is possible because the formulation procedure includes the natural or gradient boundary condition automatically, in an integrated sense.

Consolidation. In this case, the problem represents transient deformations of a porous saturated medium such as a soil foundation subjected to a load of 100 units. Because the medium is saturated, at $t \leq 0$, the fluid carries all the applied load (pressure) and the initial conditions are $p(y, 0) = 100$ and $\sigma'(y, 0) = 0$. As time elapses, the pressure dissipates and the load is gradually transferred to the soil skeleton and σ' increases; at $t = \infty$, $\sigma'(y, \infty) = 100$ units. If the top and bottom boundaries are pervious, we can assume that $p(0, t) = p(2H, t) = 0.0$ at all times.

The time-dependent distribution of the $p(y, \text{t})$ is similar to the distribution of temperature (Figure 5.7). Often, in geomechanics, it is convenient to define degree of consolidation U as[2]

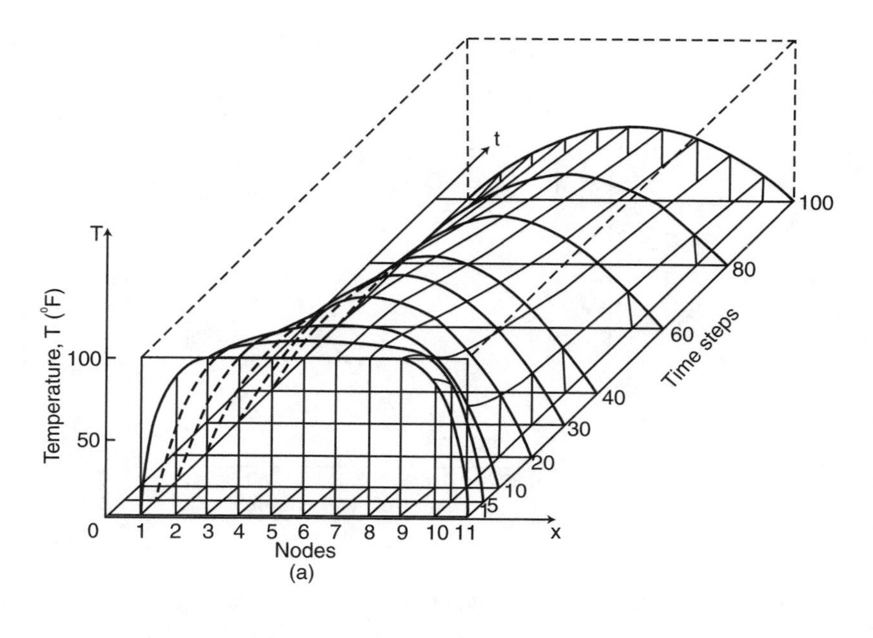

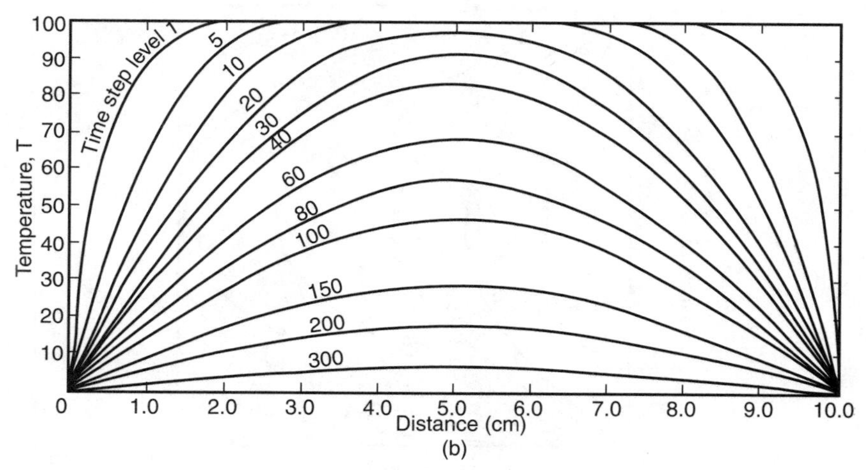

FIGURE 5.7
Finite element solution of heat flow problem: (a) Distribution of temperature at various time levels and (b) Distribution of temperature along bar at various time levels.

$$U = 1 - \frac{\int_0^{2H} p dy}{\int_0^{2H} p_0 dy} = 1 - \sum_{i=0}^{\infty} \frac{2}{M^2} e^{-M^2 T_v}, \tag{5.51}$$

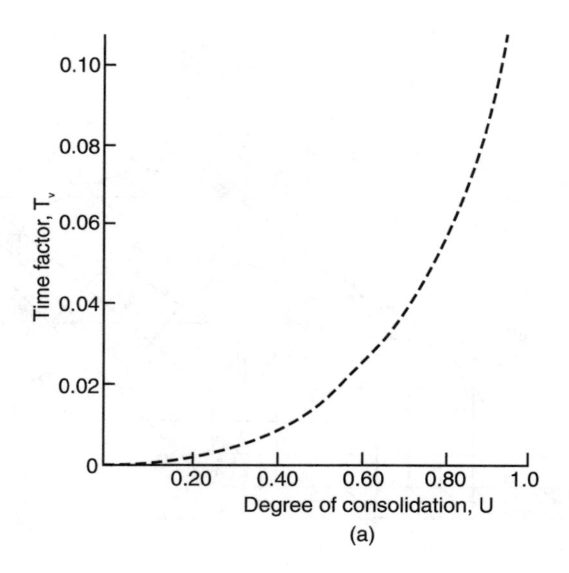

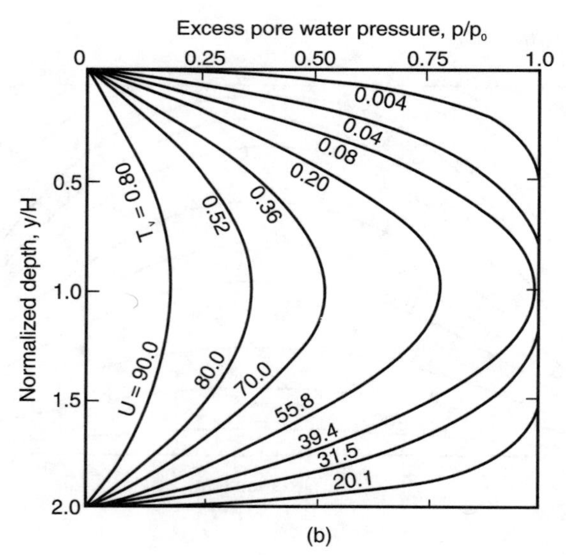

FIGURE 5.8
Finite element results for consolidation problem: (a) Time factor vs. degree of consolidation and (b) Pore water pressure vs. depth.

where p_0 is the initial pressure, T_v = nondimensional time factor = $c_v t/H^2$, and $M = (\pi/2)(2i + 1)$.

Figure 5.8(a) shows the distribution of U vs. time factor T_v. Figure 5.8(b) shows values of excess pore water pressure along the depth at various values of U and T_v.

The foregoing results indicate that the finite element procedure can be used for consolidation or settlement computations of foundations idealized as one-dimensional and subjected to (vertical) structural loads. The procedure can yield history of settlements and pore water pressures with time and the final settlements under a given load increment.

Once the pore pressure at any time under a given load increment is found, Equation 5.43 can be used to compute change in σ', since σ, the applied load, is known. From the knowledge of the change in σ' and the material property a_v, it is then possible to find the change in strain void ratio (Δe) from Equation 5.44. The settlement can be found by multiplying the strain by the total length of the medium:

$$\Delta v(t) = \Delta e(t) 2H, \tag{5.52}$$

where $\Delta v(t)$ is the vertical settlement at time t.

It is also possible to include layered systems with horizontal stratification. An example of such a system follows.

Example 5.5. Consolidation of Layered Clay Media

Figure 5.9(a) shows a layered system of soil with four layers.[10] The material properties are shown in the figure;[10] any convenient units can be assigned to the dimensions and to the properties. Program DFT/C-1DFE was used to solve the problem.

Figure 5.9(b) shows distribution of excess pore water pressures at various time levels during consolidation. The problem of temperature distribution in layered media can be solved almost identically.

It can be seen that the distribution of pore water pressure at the interfaces between the layers is not continuous. If the magnitudes of material properties between two layers differ widely, the discontinuity can cause difficulties. Then it may be necessary to derive criteria for restricting spatial and timewise meshes[8,9,11] or to use alternative formulation in terms of velocities or stream functions as unknowns.

As explained before, once the pore water pressures are known, it is easy to find the deformations at any time level.

Problems

5.1. Derive an expression for the initial load vector for a line element with linear approximation for v and due to given fluid pressure p_0. For $p_0 = 10 \text{ kg/cm}^2$, compute $\{\mathbf{Q}_0\}$; solve the problem in Figure 5.2(b) with an external load of 1000 kg at the top by including the effect of p_0. Hint:

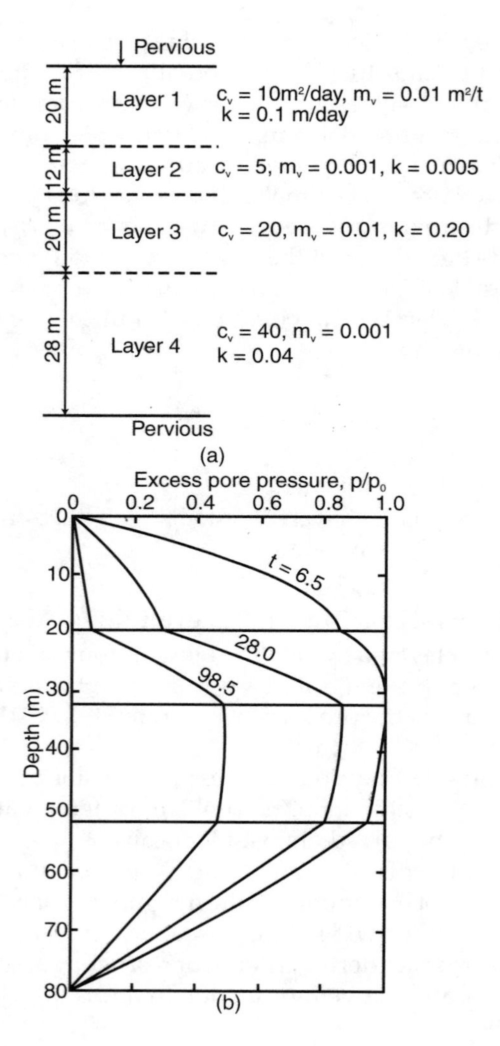

FIGURE 5.9
Consolidation in layered system: (a) Details of layered system[10] and (b) Dissipation of pore water pressure with time.

$$\{\mathbf{Q}_0\} = -A\int_{y_1}^{y_2} [\mathbf{B}]^T p_0 dy.$$

5.2. Formulate the finite element equations for the temperature (or consolidation) problem with linear variation of areas and material properties as

$$A = \tfrac{1}{2}\left(1 - L\right)A_1 + \tfrac{1}{2}\left(1 + L\right)A_2,$$
$$\alpha = \tfrac{1}{2}\left(1 - L\right)\alpha_1 + \tfrac{1}{2}\left(1 + L\right)\alpha_2.$$

5.3. Solve Example 5.3 for two time steps with the properties as before except the initial conditions, which vary as

$$T\left(0,0\right) = 10,$$
$$T\left(h/2,0\right) = 20,$$
$$T\left(h,0\right) = 10.$$

5.4. Derive equations for the three elements in Figure 5.4 with the following properties:

Element 1: $k, \rho, c;$

Element 2: $2k, \rho, c;$

Element 3: $3k, \rho, c.$

Assemble the equations and solve for the conditions that

$$T\left(x, 0\right) = 0.0, \qquad 0 \le x \le h, \ t \le 0,$$
$$T\left(0,t\right) = 10, \qquad t > 0,$$
$$T\left(h,t\right) = 20, \qquad t > 0.$$

Assume that A_1, α_1, and A_2, α_2 are the values at the two end nodes of an element and that they vary linearly.

5.5. Consider gradually refined mesh for both space and time, and by using DFT/C-1DFE, study the behavior of the numerical solution for one-dimensional heat flow (or consolidation); assume the boundary conditions in Problem 5.3. Use 2, 4, 8, 16, and 32 elements and Δt = 0.05, 0.1, 0.2, 0.25, and 0.5; 1.0, 2.0,10.0. Make various combinations of these values and obtain numerical solutions. Plot error vs. $\Delta t = \alpha\Delta t/l^2$. Define error as the difference between the numerical solution (p) and the exact solution $T^*(p^*)$:

$$\text{error} = T^* - T \ \text{ or } \ p^* - p,$$

where

$$p^* = \sum_{n=1}^{\infty} \frac{2p_0}{n\pi}\left(1-\cos n\pi\right)\left(\sin\frac{n\pi y}{2H}\right)e^{-\left(\frac{1}{4}\right)n^2\pi^2 T_v},$$

where p_0 is the uniform initial pressure, for the consolidation problem. Hint: Note that with increasing value of Δt, say beyond 2.0, the solution will become less and less accurate.[8,9]

5.6. Solve Example 5.3 with forcing fluxes given as $\bar{q} = 0.1$ unit/unit length.

5.7. By using DFT/C-1DFE, solve Example 5.4 by considering only half the depth with the boundary condition at midsection as $(\partial p/\partial y)(H,t) = 0.0$. Hint: Since the problem is symmetrical about the midsection, $\partial p/\partial y = 0$.

5.8. By using DFT/C-1DFE, solve Example 5.4 for an impervious bottom boundary, that is,

$$\frac{\partial p}{\partial y}\left(2H, t\right) = 0.$$

5.9. Formulate the consolidation problem if the soil were deposited gradually with time from 1 unit of depth to 10 units of depth in 10 years.[12] Assume linear variation of the deposition. Assume required parameters.

5.10. Let f' denote the true derivative of $f(t)$ at $t = t_1$; f'_F, f'_B, and f'_C denote derivatives of $f(t)$ at $t = t_1$, computed by forward, backward and central difference formulae, respectively. Similar notations are used for two other functions $g(t)$ and $h(t)$. These three functions are shown in Figure 5.10. Fill up the following blanks using any one of the three symbols, = (equal to), > (greater than), or < (less than).

$$\text{(a)}\ f'_F - f'_B,\ f'_B - f'_C,\ f'_C - f'$$

$$\text{(b)}\ g'_F - g'_B,\ g'_B - g'_C,\ g'_C - g'$$

$$\text{(c)}\ h'_F \_ h'_B,\ h'_B \_ h'_C,\ h'_C \_ h'$$

5.11. In a transient heat flow problem formulation one student started with a wrong functional given by

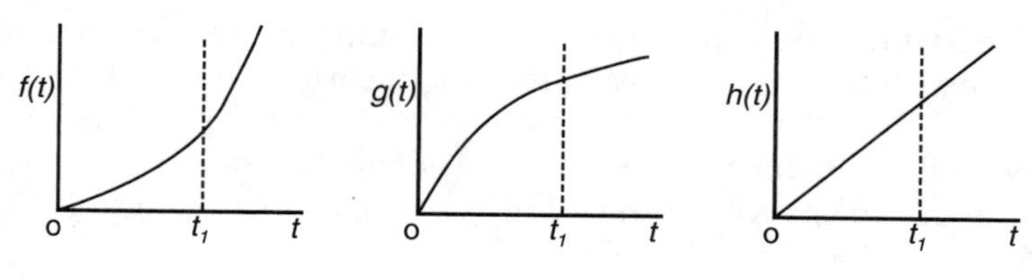

FIGURE 5.10

$$\Omega = \frac{\alpha l}{4}\int_{-1}^{+1}\left[\left(\frac{\partial T}{\partial t}\right)^2 + T^2\right]dL - \frac{l}{2}\int_{-1}^{+1}\bar{q}TdL$$

What element equation will he obtain from this functional? Take $T = [N]\{T_n\}$ and $\frac{\partial T}{\partial t} = [N]\{\dot{T}_n\}$.

5.12. Consider the function $f(t) = t^2$. Calculate its derivatives at $t = 1$ numerically by (1) forward difference formula, (2) backward difference formula, and (3) central difference formula by taking $\Delta t = 0.5$. Compare these numerical values with the exact value of the derivative at $t = 1$ and offer your comments to justify the results.

5.13. The global equation for the one-dimensional transient heat flow problem with zero heat influx (see Figure 5.11) is given by

$$\frac{\alpha(\Delta t)}{l^2}\begin{bmatrix} +\frac{4}{3} & -\frac{5}{6} & 0 & 0 \\ -\frac{5}{6} & +\frac{8}{3} & -\frac{5}{6} & 0 \\ 0 & -\frac{5}{6} & +\frac{8}{3} & -\frac{5}{6} \\ 0 & 0 & -\frac{5}{6} & +\frac{4}{3} \end{bmatrix}\begin{Bmatrix} T_1 \\ T_2 \\ T_3 \\ T_4 \end{Bmatrix}_t = \begin{Bmatrix} \frac{T_1}{3} + \frac{T_2}{6} \\ \frac{T_1}{6} + \frac{2T_2}{3} + \frac{T_3}{6} \\ \frac{T_2}{6} + \frac{2T_3}{3} + \frac{T_4}{6} \\ \frac{T_3}{6} + \frac{T_4}{3} \end{Bmatrix}_{t-\Delta t}$$

(a) If the bar is initially at 0°C and then the temperature field is increased as $T = 10t^2$ $(t \geq 0)$ at the two boundaries [nodes 1 and 4] calculate T_2 and T_3 at t = 1 s and 2 s; assume that $\alpha = l^2$.

(b) If $\alpha = 0.5\, l^2$, and you want to obtain T_2 and T_3 at t = 1 s after only one iteration, what should be the value of Δt? What is the nondimensional time increment ΔT corresponding to this Δt?

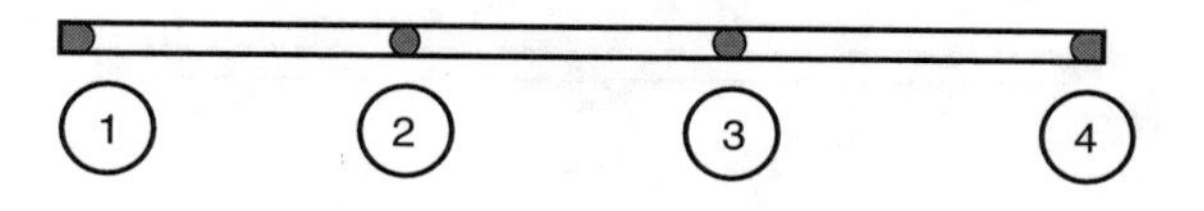

FIGURE 5.11

5.14. The 30 unit long wire (Figure 5.11) is initially at 100°F. Its boundary temperatures are then decreased nonlinearly as a function of time $T_1 = T_4 = 100(1 - t^2)$. Calculate and plot the temperature profiles along the wire for two time steps. Assume that the nondimensional time increment $\Delta T = 1$, heat influx $\bar{q} = 0$, element length $l = 10$, A = 1, and $\alpha = 50$.

5.15. Solve Problem 5.14, but now assume that the heat influx per unit length $\bar{q} = 10$ unit instead of 0.

5.16. Consider the same problem geometry as in Problem 5.14. Assume that the initial temperature of the wire is 100°F; then the boundary temperatures T_1 and T_4 are suddenly dropped to 50°F and 80°F, respectively (i.e., for $t > 0$, $T_1 = 50°F$, $T_4 = 80°F$), and are kept constant at these two temperatures. Calculate and plot the temperature profiles along the wire for two time steps. Assume that the nondimensional time increment $\Delta T = 1$, heat influx $\bar{q} = 0$, element length $l = 10$, A = 1, and $\alpha = 50$.

5.17. A rod of length 20 units has a constant cross-sectional area (A) of 1 unit, and thermal diffusivity $\alpha = 10$ units. It is discretized into two identical finite elements. Initially the rod is at 100°F temperature. Its boundary temperatures (for both boundaries) are then decreased to 0°F. Calculate the temperature at the center node at time t = 2 s, by taking the time increment (1) $\Delta t = 1$ s and (2) $\Delta t = 2$ s. The heat influx $\bar{q} = 0$.

5.18. The element equation for the transient heat flow problem is given by

$$\frac{A\alpha}{L}\begin{bmatrix} 1 & -1 \\ -1 & 1 \end{bmatrix}\begin{Bmatrix} T_1 \\ T_2 \end{Bmatrix}_t + \frac{Al}{6\Delta t}\begin{bmatrix} 2 & 1 \\ 1 & 2 \end{bmatrix}\begin{Bmatrix} T_1 \\ T_2 \end{Bmatrix}_t = \begin{Bmatrix} R_1 \\ R_2 \end{Bmatrix}_t + \frac{Al}{6\Delta t}\begin{bmatrix} 2 & 1 \\ 1 & 2 \end{bmatrix}\begin{Bmatrix} T_1 \\ T_2 \end{Bmatrix}_{t-\Delta t}$$

Assume $\frac{A\alpha}{l} = 1$, $\frac{Al}{6} = 1$ and heat influx along the length of the rod is zero for the rod shown in Figure 5.12.

(a) If you want to solve this problem by the time-marching method (using backward difference formula), derive global equation. Take $\Delta t = 1$.

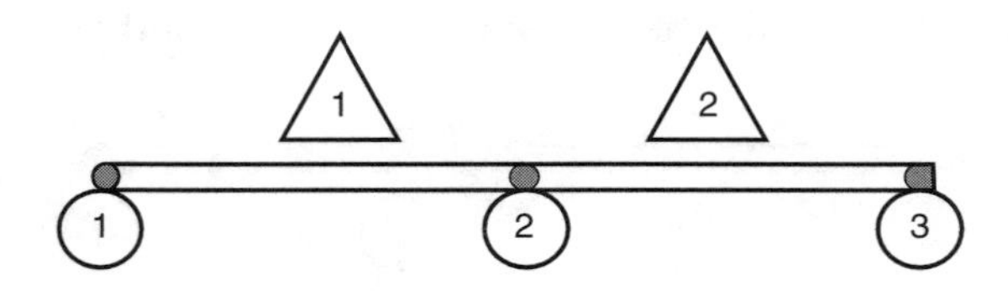

FIGURE 5.12

(b) In general the temperature T_2 of the middle node can be expressed in the following form,

$$T_2(t) = f(T_1(t), T_3(t)) + g(T_1(t-\Delta t), T_2(t-\Delta t), T_3(t-\Delta t))$$

Starting from the global equation, derived in part (a), obtain the functions $f(T_1,T_3)$ and $g(T_1,T_2,T_3)$ for (1) $\Delta t = 1$ and (2) $\Delta t = 0.5$.

(c) Consider the initial condition $T(x,t) = 1.0$, at $t = 0.0$, and the boundary conditions $T_1(t) = 1 + t$, $T_3(t) = 1 + 3t$, then obtain $T_2(t)$ at $t = 1$, 2 and 3; take $\Delta t = 1$ for your calculation.

References

1. Carslaw, H. S. and Jaeger, J. C. 1959. *Conduction of Heat in Solids,* Clarendon Press, Oxford.
2. Terzaghi, K. and Peck, R. B. 1955. *Soil Mechanics in Engineering Practice,* Wiley, New York.
3. Desai, C. S. and Johnson, L. D. 1972. "Some numerical procedures for analysis of one-dimensional consolidation," in *Proc. Symp. on Appl. of Finite Element Methods in Geotech. Eng.,* C. S. Desai, Ed., Waterways Expt. Station, Vicksburg.
4. Schiffman, R. L., private communication.
5. Courant, R. and Hilbert, D. 1965. *Methods of Mathematical Physics,* Wiley-Interscience, New York.
6. Desai, C. S. and Abel, J. F. 1972. *Introduction to the Finite Element Method,* Van Nostrand Reinhold, New York.
7. Crandall, S. H. 1956. *Engineering Analysis,* McGraw-Hill, New York.
8. Desai, C. S. and Johnson, L. D. 1972. "Evaluation of two finite element formulations for one-dimensional consolidation," *Comput. Struct.* 2, 469-486.
9. Desai, C. S. and Johnson, L. D. 1973. "Evaluation of some numerical schemes for consolidation," *Int. J. Num. Methods Eng.* 7, 243-254.
10. Schiffman, R. L., Stein, J. R., and Jones, R. A. 1971. "PROGRS-I, a computer program to calculate the progress of ground settlement — reference manual," Report No. 71-4, Computing Center, University of Colorado, Boulder.
11. Desai, C. S. and Saxena, S. K. 1977. "Consolidation analysis of layered anisotropic foundations," *Int. J. Num. Anal. Methods Geomech.* 1, No. 1, 5-23.
12. Koutsoftas, D. C. and Desai, C. S. 1976. "One-dimensional consolidation by finite elements: solution of some practical problems," Report No. VPI-E-76-17, Dept. of Civil Eng., VPI & SU, Blacksburg, Va.

6

Finite Element Codes: One- and Two-Dimensional Problems

By now the reader must have realized the general nature of the finite element method. A comparison of the derivations and the results in Chapters 3, 4, and 5 indicates that there is significant similarity in the theory and formulation of the three categories of problems. The physical phenomena of deformation (Equation 3.58), steady flow (Equation 4.1), and transient flow (Equation 5.9) are governed by analogous differential equations. The effect of a forcing function, either a load or a flux, is transmitted or diffused through a medium in similar manners, thus indicating the close relationship between these phenomena.

All but very trivial problems that are solved by using the finite element method require use of the computer. The amount of information to be digested and processed is so great for most problems that it is not possible to perform the calculations manually. Hence, a knowledge of and exposure to the development and use of computer programs or codes become necessary.

One-Dimensional Code

Deformation, Flow, and Temperature/Consolidation Problems

To introduce this topic gradually, first we present in this chapter details of a code for the three one-dimensional problems covered in Chapters 3, 4, and 5. It is shown here that the generality of the method permits use of essentially the same program for all three problems. The program is called Deformation Flow, Temperature/Consolidation, One-Dimensional Finite Element ≡ DFT/C-1DFE. A number of example problems and documentation are provided in this chapter. In the following, details of various stages of the code are described.

Philosophy of Codes

A computer code should be concise, but at the same time, it should be properly documented so that the user can understand and employ it without spending an undue amount of time. This is particularly true of a code written for a beginner, as in this book. To achieve this purpose, and to make the code relevant to academic aims, we have provided sufficient commentary in the code and in the following.

Stages

The explanation of the code can be divided into a number of stages, which include input, computer implementation, and output. These stages are described in the subsequent listing of the code.

Stage 1. Input Quantities

This step deals with the input quantities such as the title, common parameters, material properties, and nodal and element data,

Input Set 1. The first card gives details of the number of the problem and the title of the problem; a number of problems can be executed in the same run. If the problem number NPROB* is set equal to zero (a blank line), the program will automatically exit from the computer. The second line includes information on number of nodes (NNP), number of materials (NMAT), number of surface or boundary traction lines (NSLC), option for whether body force is applied or not (NBODY), and option for choosing the category of problem: stress deformation, flow, or temperature/consolidation (NOPT); semibandwidth (IBAND), which in the case of the linear approximation used for the code is 2; and number of selected time levels at which output is desired (NTIME). The third line or lines (as many as number of materials) give the properties relevant to the category; that is, deformation, steady flow, or time-dependent flow. Table 6.1 shows the meaning of the terms for the categories. In the case of body weight DENS denotes material density (unit weight). For consolidation of nonhomogeneous media γ_{wi} = unit weight of water is input in place of RO.

Input Set 2. This set gives the data on the nodal points. Here M denotes the node point number and KODE designates the boundary condition at the node. Y denotes the y coordinate (or any one-dimensional coordinate) of the node point. VLY gives the specified value of the boundary condition; it is

* Explanations of symbols are given later.

tied in with the value of KODE. If KODE = 0, the node is free, and we can input VLY to be a concentrated load or forcing function; if KODE = 1, then VLY = specified value of displacement, fluid head, temperature, or pore water pressure. Note that if there is a natural boundary condition, i.e., if gradient or slope is zero, we input KODE = 0. In general, these data should be provided for all nodes, except in the case of the following.

Within input set 2, we have made a provision for automatic generation of nodal data. This is applicable only if the nodes are equidistant, and if the generated nodes are free, that is, KODE = 0; then nodal data are provided only for the first and last node in the region where these conditions are satisfied. Nodal data should be provided whenever there is a change in the length and properties of the element and in the boundary condition.

We also compute the length l (ALL) of the element and store it for subsequent computations.

Input Set 3. The total number of elements NEL is set equal to NNP – 1. An element is identified by the two end node (global) numbers and a number corresponding to the type of material of that element. Hence, for each element, four quantities are input: element number M, node numbers IE(M, 1) and IE(M, 2), and material type number IE(M, 3). The value of the latter varies from 1 to NMAT.

As explained in the user's guide (given later in this chapter), it is possible to generate automatically the element properties if certain conditions are fulfilled.

Three different variations of (element) areas can be assigned. In the case of uniform area for all elements, IAREA = 1, and only one value of AREA(1) is input. If IAREA = 2, areas are input for the first and the last element referred to their midsections, and the computer assigns intermediate area by assuming a linear variation. If all elements have different areas, IAREA = 3, and areas for all elements (at their midsections) are input.

TABLE 6.1

Material Properties for Various Problems

Terms	PROP	AMV	RO	DENS	Remarks
Stress deformation					
Homogeneous	E	1.0	1.0	γ	Equation 3.14
Nonhomogeneous	E_i^*	1.0	1.0	γ_i	
Steady flow					
Homogeneous	k or α	1.0	1.0	γ	Equation 4.3b
Nonhomogeneous	k_i or α_i[a]	1.0	1.0	γ_i	Equation 5.9d
Time-dependent temperature					
Homogeneous	α	1.0	1.0	γ	Equation 5.9d
Nonhomogeneous	k_i	c_i	ρ_i	γ_i	Equation 5.9c
Time-dependent consolidation					
Homogeneous	c_v	1.0	1.0	γ	Equation 5.45
Nonhomogeneous	k_i[a]	m_{vi}	γ_{wi}	γ_i	Equation 5.46

[a] $i = 1, 2, \ldots, N$ = number of different materials.

Input Set 4. Here NSLC denotes the number of surface traction loading ($\bar{T}_y$) cards, and TY denotes the value of surface loading. For instance, if in Example 6.2, Figure 6.2, $\bar{T}_y$ is applied on two of the elements, NSLC = 2. The computer converts this loading to lumped loads at the two surrounding nodes according to Equation 3.41. KEL denotes the element on which TY is applied. If no TY is specified, set NSLC = 0; then no data are required for surface tractions.

Input Set 5. This set contains three subsets of data for the time-dependent temperature or consolidation problems. The first subset contains the time increment (DT), total time (TOTIM) desired for the time-dependent solutions, and option (INOPT) for specifying uniform, linear, or arbitrary variation for the initial conditions.

In the second subset, input the specific values of time levels TIM(I) at which output is desired; the number of these values is equal to NTIME.

The third subset contains data on initial conditions $T(x, 0)$ (Equation 5.10). Here, according to INOPT, data are input as uniform, linearly varying, or arbitrary values of initial temperature at the nodes. If INOPT = 1, input simply one value UINIT(l) of initial temperature of pore pressure. If INOPT = 2, input the values at the first and the last nodes, and the computer will provide intermediate values by linear interpolation. If INOPT = 3, input temperature or pore pressure for each node.

Note: The information in input set 5 is not required for stress-deformation and steady flow problems, that is, when NOPT = 1 or 2.

Stage 2. Initialize

Various quantities are initialized here. TIME = 0.0 and NCT = 0 are set; NCT denotes the number of time steps and can be printed out at each time step. H(I) is set equal to UINIT(I), the initial conditions. Matrices QK(I,J) and QP(I,J) are the element matrices $[\mathbf{k}_\alpha]$ and $[\mathbf{k}_t]$ in Equations 5.22a and 5.22b, respectively. Vectors Q(I) (Equations 3.41, 4.11, and 5.22c) and R(I) (Equations 3.45, 4.13, and 5.33b) are the element and assemblage forcing vectors, respectively. Matrix A(I,J) stores the assemblage matrices. Matrix AK(I,J) is used repeatedly for the time solution and is modified at each step during the Gaussian elimination process; its value is set equal to A(I,J) before starting a new time step. For the stress and steady flow problems, A(I,J) and AK(I,J) are computed once, and no such resetting is required. AP(I,J) is the assemblage matrix corresponding to $[\mathbf{K}_t^*]$ (Equation 5.33b).

Intermediate quantities required for computation of the time factor in the consolidation problem are evaluated here.

Stage 3. Compute Element Matrices

Element matrix $[\mathbf{k}]$ and load vector $\{\mathbf{Q}\}$ due to surface traction or flux are computed for all elements one by one. The contributions at nodes due to the gravity load are computed according to Equation 3.41 and are added to the element load vector {**Q**}.

Stage 4. Assemble

At this stage, we add the contributions of the element matrix and load vector to the appropriate locations in the global matrix [**K**] and load vector {**R**} (Equation 3.45). This is done by identifying the global nodes corresponding to the local nodes 1 and 2 for each element.

Stage 5. Concentrated Forces

The concentrated forces VLY(I) are added to the corresponding locations in vector {**R**}.

Stage 6. Boundary Conditions

The boundary conditions are introduced here to modify the assemblage matrices and load vectors (Equations 3.45, 4.13, and 5.33). In view of the fact that we have taken advantage of bandedness and symmetry, special logic is used to make this modification.

Stage 7. Time Integration

This is operational only for the time-dependent problem (NOPT = 3 and 4) and pertains to Equation 5.33. Here the equivalent load vector $[\mathbf{K}_t]\{\mathbf{r}\}_t$ is computed and added to the assemblage vector {**R**}.

Stage 8. Solve Equations

The final equations are solved by performing the Gauss–Doolittle elimination procedure. For NOPT = 1 or 2, this is performed only once. For NOPT = 3 and 4, it is performed as many number of times as the time steps = TOTIM/DT.

The space of vector R(I) is used here to store the values of computed displacements, fluid heads, temperatures, or pore water pressures.

Stage 9. Set $\{\mathbf{R}\}_t = \{\mathbf{H}\} = \{\mathbf{R}\}_{t+\Delta t}$

This is relevant only for NOPT = 3 and 4. Here the values of R(I) just computed are stored in H(I) as the value $\{\mathbf{R}\}_t$ at the previous time stage.

Stage 10. Output Quantities

Depending on the value of NOPT, computed results are printed out as follows:

NOPT = 1: results for stress-deformation problem,
= 2: results for flow problem,
= 3: results for time-dependent temperature problem,
= 4: results for consolidation problem.

In the case of consolidation, additional quantities such as time factor and degree of consolidation are also printed out.

Explanation of Major Symbols and Arrays

A = Assemblage matrix which is set equal to AK; only once for stress deformation and steady flow but at every time step for transient flow.

AK = Assemblage matrix which is computed only once. This matrix corresponds to [**K**] (Equations 3.45, 4.13, or 5.33).

ALL = Length of each element.

AMV = Coefficient of volume compressibility.

AP = Assemblage matrix corresponds to $[\mathbf{K}^*_t]$ (Equation 5.33).

AREAEL = Area of elements. If IAREA = 1, input one value of uniform area, = 2 input first and last values, = 3 input all values.

DENS = Density (unit weight) for different materials.

DT = Time increment (Equations 5.23 and 5.33).

GWT = Contribution of gravity or body force at an element node.

H = Vector that stores initial values of temperature (or pore pressure) and also $R_{t+\Delta t}$ computed at $t + \Delta t$ time level for use as R_t, for the subsequent time level.

IAREA = Option for specifying areas of elements:
= 1: for uniform area over all elements.
= 2: for linearly varying areas; input values for only the first and last elements.
= 3: for arbitrary variation, that is, input area for each element.

IBAND = Semibandwidth. Set 2 for the one-dimensional problem with linear approximation.

IE(M,1) = Node 1 of element M.

IE(M,2) = Node 2 of element M.

IE(M,3) = Material type of element M.

INOPT = Option for initial conditions:
= 1: uniform temperature or pressure at all nodes.
= 2: linearly varying temperature or pressure; input values only for the first and the last nodes.
= 3: arbitrary; input values for all nodes.

KEL = Element(s) on which surface tractions (TY) are applied.

KODE = Code for boundary conditions:
= 1: specified displacement, potential, temperature, or pore pressure.
= 0: free node where concentrated force can be applied.

M = Element number.

NBODY = 0 for no body force, = 1 if body force specified.

NCT = Step of time increment; can be printed out if desired.

NMAT = Number of materials.

NNP = Number of nodes.

NOPT = 1 stress deformation, = 2 steady flow, = 3 transient temperature, = 4 consolidation.

N_P_NO = Nodal point number.

NPROB = Problem number; if = 0, program exits from the computer.

NSLC = Number of surface traction cards.

NTIME = Number of time levels at which output is desired.

PROP = Material property:
= E, elastic modulus for stress deformation.
= k, coefficient of permeability for flow.
= α, thermal diffusivity for temperature (for homogeneous medium).
= c_v, coefficient of consolidation (for homogeneous medium). For layered media, see Table 6.1.

QK = Element matrix $[\mathbf{k}]$ in Equations 3.41 and 4.9 and $[\mathbf{k}_\alpha]$ in Equation 5.21.

QP = Element matrix $[\mathbf{k}_t]$ in Equation 5.21.

R = Assemblage forcing vector $\{\mathbf{R}\}$ in Equations 3.45, 4.13, and 5.33. Computed nodal displacements, potential, temperature, or pressure are stored in this vector after Gaussian elimination.

RO = Mass density of material.

TF = Time factor.

TIM = Selected time levels at which output is desired.

TIME = Elapsed time.

TITLE = Title and description of problem.

TUINIT = Total of initial values = (sum of initial pore pressures) × (number of nodes).

TY = Applied surface traction load.

UAV = Average degree of consolidation at a given time.

UINIT = Values of initial temperature or pore pressure at time ≤ 0.

USUM = Average dissipation of pore pressure at a given time.

VLY = Value of applied boundary condition. If KODE = 1, it implies specified displacement, potential, temperature, or pressure; if KODE = 0, it implies free node where forcing parameters (concentrated load) can be specified.

Y = y coordinate for stress deformation (of vertical column and strut) and for consolidation.
= x coordinate for flow and temperature.

User's Guide for Code DFT/C-1DFE*

Computer code for Finite Element Analysis of 1-Dimensional Deformation, Flow Temperature and Consolidation problems

NOTE The input data can be provided in free format.

Problem number and title:

NPROB		TITLE

Basic parameters:

NNP	NMAT	NSLC	NBODY	NOPT	IBAND	NTIME	

Material properties: Input as many sets as NMAT

PROP	AMV	RO	DENS	

Nodal point data:

N_P_NO	KODE	Y-COORD	VLY	

See *Note 1* on page 185 for automatic generation of nodal point data and explanation of KODE.

Element data:

M	IE(M,1)	IE(M,2)	IE(M,3)	

See *Note 2* on page 185 for automatic generation of element data.

Areas of elements:
(Option for Variation)

IAREA	

= 1 for constant areas
= 2 for linearly varying areas
= 3 for different areas for each element

Values of areas:

AREAEL(1)	AREAEL(2)	...

If area is constant, input only one value AREAEL(1).
If areas vary linearly, input values for the first and the last elements.
If areas have different values in each element, input all values.
Note: Area is constant for each element; linear variation relates to change in areas from element to element for entire body.

*The code is available on the web, www.crcpress.com.

Surface Tractions:

KEL	TY	

Input as many sets as NSLC.

For time-dependent problems enter the following input data:

Time increment, total time, and option for variation of initial conditions:

DT	TOTIM	INOPT	

INOPT= 1 for constant values of initial conditions
= 2 for linear variation
= 3 for different values at each node

Selected output time levels:

TIM(1)	TIM(2)	...

Input as many as NTIME values

Initial Conditions:

UINIT(1)	UINIT(2)	UINIT(3)	...

If initial values are constant, input one value UINIT(1).
If linear variation, input values for the first and the last node.
If values are different at each node, input all values.

Blank line

Each run can contain more than one problem, NPROB. Then input all data for each problem consecutively.

Note 1:

1. If the nodal points are equidistant, data are required only for the first and the last node. The data for intermediate nodes are generated by the computer. If a boundary condition or force is applied at a node, data must be input for such nodes.
2. KODE = 0: Node is free, and concentrated force can be applied.

   KODE = 1: Boundary condition in terms of displacement, potential, temperature, or pore pressure is specified.

   KODE is set equal to zero for the nodes that are generated by the computer.

Note 2: If the elements have the same material properties, PROP, AMV, RO, DENS, data are required only for the first and the last element; data for intermediate elements are generated by the computer, and their PROP, AMV, RO, and DENS are set equal to those for the previous element. If there is a change in material property, data must be separately provided. For instance, for 11 elements, if there is a change in material property at element 6, then data for elements 1 and 5 and 6 and 11 should be provided.

Example 6.1*

See Figure 6.1. By using DFT/C-1DFE, compute displacements and axial stresses. Plot the results.

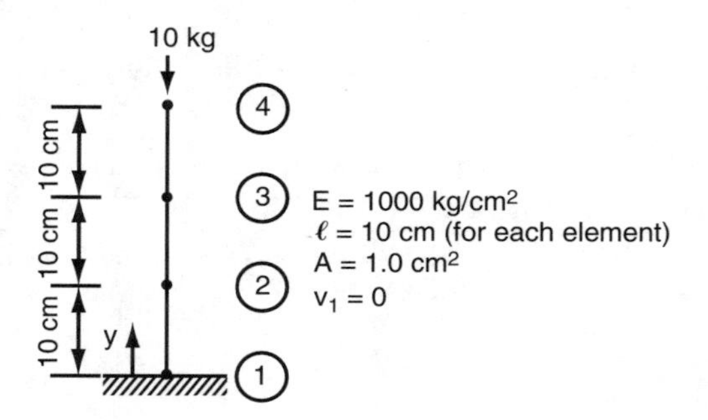

FIGURE 6.1

Example 6.2

See Figure 6.2. By using DFT/C-1DFE, compute and plot displacements and stresses.

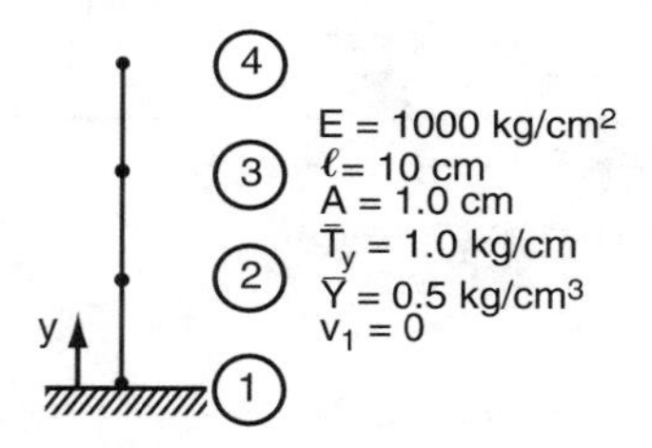

FIGURE 6.2

Example 6.3

The pipe in Figure 6.3 is subjected to potentials $\varphi = 2$ and 1 at the two end points 1 and 4, respectively. Use DFT/C-1DFE, and find and plot distributions of potentials and velocities.

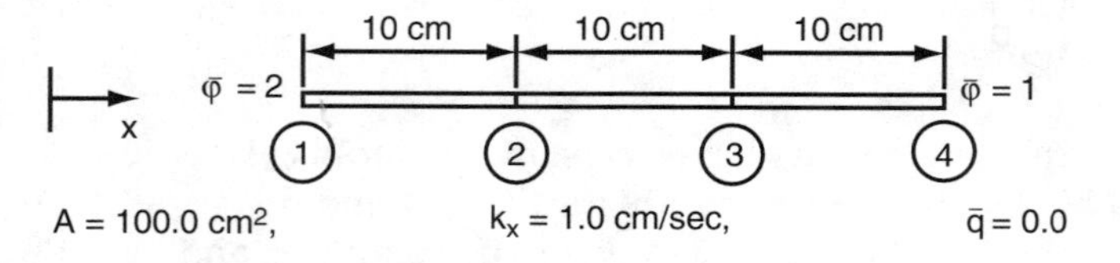

FIGURE 6.3

* Solutions of these examples are available on the web, www.crcpress.com.

Example 6.4

The bar in Figure 6.4 is initially at a temperature of 100°F; that is $T(x, 0) =$ 100°F. The boundary conditions are

$$T(0,t) = T(10,t) = 0,\ t \geq 0.$$

The coefficient of diffusivity $\alpha = 1$. Divide the bar into 10 elements, that is, 11 nodes, and adopt $\Delta t = 0.1$ unit.

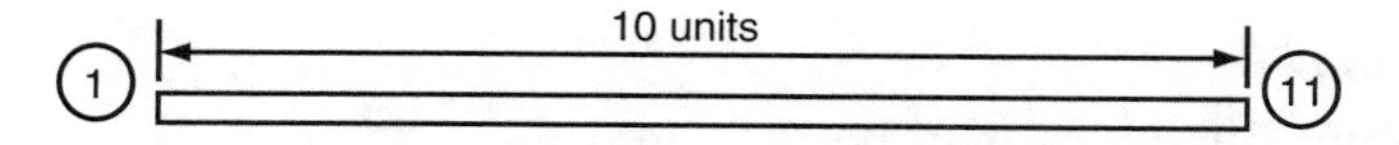

FIGURE 6.4

Use DFT/C-1DFE, and compute the distribution of temperature at various time levels as the bar cools.

Two-Dimensional Code

Stress Deformation and Field Problems

The theory and development of finite element analysis for various two-dimensional problems are presented in Chapters 10–13. A number of two- and three-dimensional finite element procedures have been developed.[1] They include linear and nonlinear finite element codes for static, dynamic, and repetitive thermomechanical loading; a number of constitutive models have been implemented in the codes: elasticity, plasticity, elastoviscoplasticity with the unified disturbed state concept (DSC) for microcracking, degradation (damage), and softening.[1] A list of the codes is provided in Appendix 3.

For the introductory academic levels, the simple two-dimensional codes are considered to be appropriate. Three such codes are included as a part of this text: (1) DFT/C-1DFE: one-dimensional; (2) PLANE-2D: two-dimensional linear with plane strain, plane stress, and axisymmetric idealizations (Chapter 13); and (3) FIELD-2D: two-dimensional for torsion, potential flow, seepage, and heat flow (Chapters 11 and 12). These codes including source listing, user's manual, and sample problems are available on the web, www.crcpress.com. The conditions and cautions for using these codes are also available on the web; the user agrees to abide by these conditions.

User's Guide for Plane-2D

Program Input to be Prepared by User

General Comments

Finite element analyses generally consist of large amounts of input data. The preparation of this data should be as easy as possible and should leave little room for error. In keeping with these concerns, the input to PLANE-2D is free-form according to the conventions described below.

Conventions

1. Input quantities may be delimited with either blanks or <TAB> characters. Whitespace must separate each number and comma separation is not allowed.
2. Real (floating point) numbers may be specified in either a conventional decimal format or by using an "E" format. For example, the number 1234.56 may be specified in any of the following ways:

1234.56	0.123456E+04
1.23456E+03	12.3456e2

   A real number must have a digit appearing before and after the decimal point.
3. Comment lines may be used to enhance the readability of the input data file. A comment line must contain the character "1" in column one. The comment may be up to 80 characters long after the "1" (on the same line). Comments are optional and are not required.
4. More than one data set can be included within one Input file by separating each data set by a blank line.

A better understanding of the correct use of the above conventions can be attained by studying the sample data files presented in this guide.

Units

It is the responsibility of the user to maintain consistent units. The units used to describe the material properties must be consistent with those used to describe the geometry of the body, and the node point specifications. The solution will be expressed in the same units as the input.

0. Title Record

Any information (up to 80 characters) that identifies and describes the problem can be given here. The first character should be the exclamation symbol (!).

Additional comment lines can be given anywhere in the data file by prefacing the line with an "!".

1. Problem Parameters Record (Input Set 1)

a. Total number of nodes	NNP
b. Total number of elements	NEL
c. Number of different materials	NMAT
d. Number of surface loading pain	NSLC
e. Type of idealization	NOPT
1. plane strain	
2. plane stress	
3. axisymmetric	
f. Option for body (weight) force	NBODY
• 0: no body force desired.	
• 1: apply gravity body force.	

2. Material Property Specifications (Input Set 2)

All materials in this program are assumed to be isotropic and linear elastic. Each element is assigned a material number in the Element Specification below. The following Information must be supplied for each distinct material used (repeat NMAT times):

a. Young's modulus	E(MATNUM)
b. Poisson's ratio	PR(MATNUM)
c. Weight density	GAM(MATNUM)
d. Material thickness (3rd dimension)	THICK(MATNUM)

3. Nodal Point Specifications (Input Set 3)

In this section, nodal coordinates are supplied along with the node quantities at these node points. The known quantities are applied displacements or applied loads which are specified in the appropriate direction. A boundary condition code KODEBC(NODNUM) is used to indicate whether X- or Y-direction loads or displacements are applied at each node as follows:

- 0: prescribed (concentrated) forces are specified in both the X- and Y- directions; ULXBC and VLYBC denote the magnitudes of the X and Y forces, respectively

- 1: prescribed displacement is specified in the X-direction and prescribed force in the Y-direction; ULXBC denotes the magnitude of displacement and VLYBC denotes the magnitude of the force
- 2: prescribed force is specified in the X-direction and prescribed displacement in the Y-direction; ULXBC denotes the magnitude of force and VLYBC denotes the magnitude of the displacement
- 3: prescribed displacements are specified in both the X- and Y-directions; ULXBC and VLYBC denote the magnitudes of the displacements

For the axisymmetric idealization, the R-direction is substituted for the X-direction and the Z-direction is substituted for the Y-direction. If for the axisymmetric case a load is applied in the Z-direction, remember to divide the load by 2π.

Automatic Node Generation

The information below must be supplied for each node point of the mesh unless automatic node generation is used. If any nodes lie along the same line and are equidistant, only the first and last node points need to have information specified, as the intermediate nodal data will be generated automatically. For nodes generated automatically, KODEBC(NODNUM) is set to zero.

a. Node number	NODNUM
b. Code for boundary conditions	KODEBC(NODNUM)
c. X-coordinate (R for axisymmetric)	XCOOR(NODNUM)
d. Y-coordinate (Z for axisymmetric)	YCOOR(NODNUM)
e. Boundary condition value in X-dir	ULXBC(NODNUM)
f. Boundary condition value in Y-dir	VLYBC(NODNUM)

4. Element Specification (Input Set 4)

For each element in the mesh the program needs to know the nodal connectivity and the material type of the element in question. The nodal connectivity consists of a list of node point numbers which define the quadrilateral or triangular elements (entered in either a CCW or CW sense around the element). Four node points are specified; for a triangular element the third node is repeated again as the fourth node. See Figure 6.5.

Automatic Element Generation

The information below must be supplied for each clement of the mesh unless automatic element generation is used. If any row of elements along the

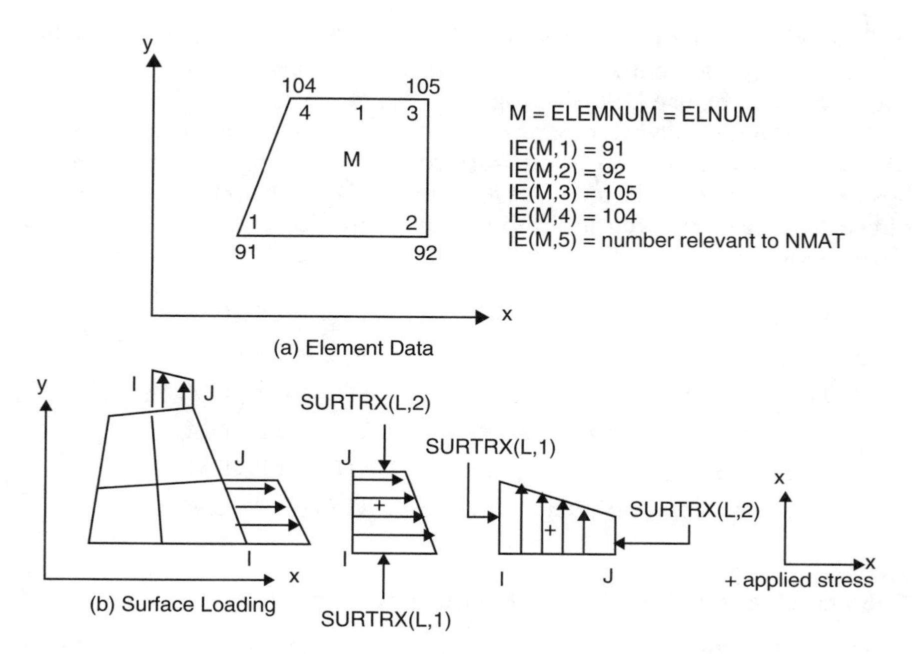

FIGURE 6.5
Input for elements and surface loading.

direction of node numbering all possess the same material properties then only the first and last elements in the row need to have information specified (the intermediate element data will be generated automatically). In this case the four node numbers just have to be incremented by one as one moves from element to element down the row. For elements generated automatically, the material number IE(ELNUM,5) is set equal to that of the previous element.

a. Element number	ELNUM
b. First node of element	IE(ELNUM,1)
c. Second node of element	IE(ELNUM, 2)
d. Third node of element	IE(ELNUM, 3)
e. Fourth node of element	IE(ELNUM, 4)
f. Material number of element	IE(ELNUM, 5)

5. Surface Loading (Pressure) Specifications (Input Set 5)

A uniform or linearly varying pressure may be defined along a boundary segment or interior line by specifying tractions (pressures) on the element edges which compose the boundary segment or line. The pressure on element edges is specified by indicating the nodal pair which define the edge and the magnitude of pressures at each node. This applied pressure distribution on each element edge is automatically transformed into equivalent nodal forces to be used by the program.

The information below must be supplied for each element edge along which the user desires to apply surface loading (repeal NSLP times). A positive magnitude entry indicates tension loading in the positive coordinate direction. Negative loads indicate compression. See Figure 6.5 for further clarification.

a. Node I	ISURND(SLPNUM)
b. Node J	JSURND(SLPNUM)
c. Pressure in the X-direction at I	SURTRX(SLPNUM, 1)
d. Pressure in the X-direction at J	SURTRX(SLPNUM, 2)
e. Pressure in the Y-direction at I	SURTRY(SLPNUM, 1)
f. Pressure in the Y-direction at J	SURTRY(SLPNUM, 2)

6. Blank Line (required if more than one data set)

If several different problems are included in one data file (a stack), then a blank line is needed to separate the input data of each problem. For a given problem the input data should include the same Input Data Sets defined above.

Sample Problems for Plane-2D

Plane Stress

Figure 6.6 shows finite element mesh for a quarter of the beam (plate) subjected to surface loading varying linearly from zero at node 5 to 1000 kg/m^2 at node 25.

Results

Finite element results for displacements at points A and B and stresses (σ_x) at centroids of element 12 and 16 are shown below:

Displacement (10^{-4}m)

	Closed Form		Finite Element	
Point	u	v	u	v
Node 25 (A)	1.500	–1.275	1.467	–1.249
Node 13 (B)	0.375	–0.318	0.369	–0.314

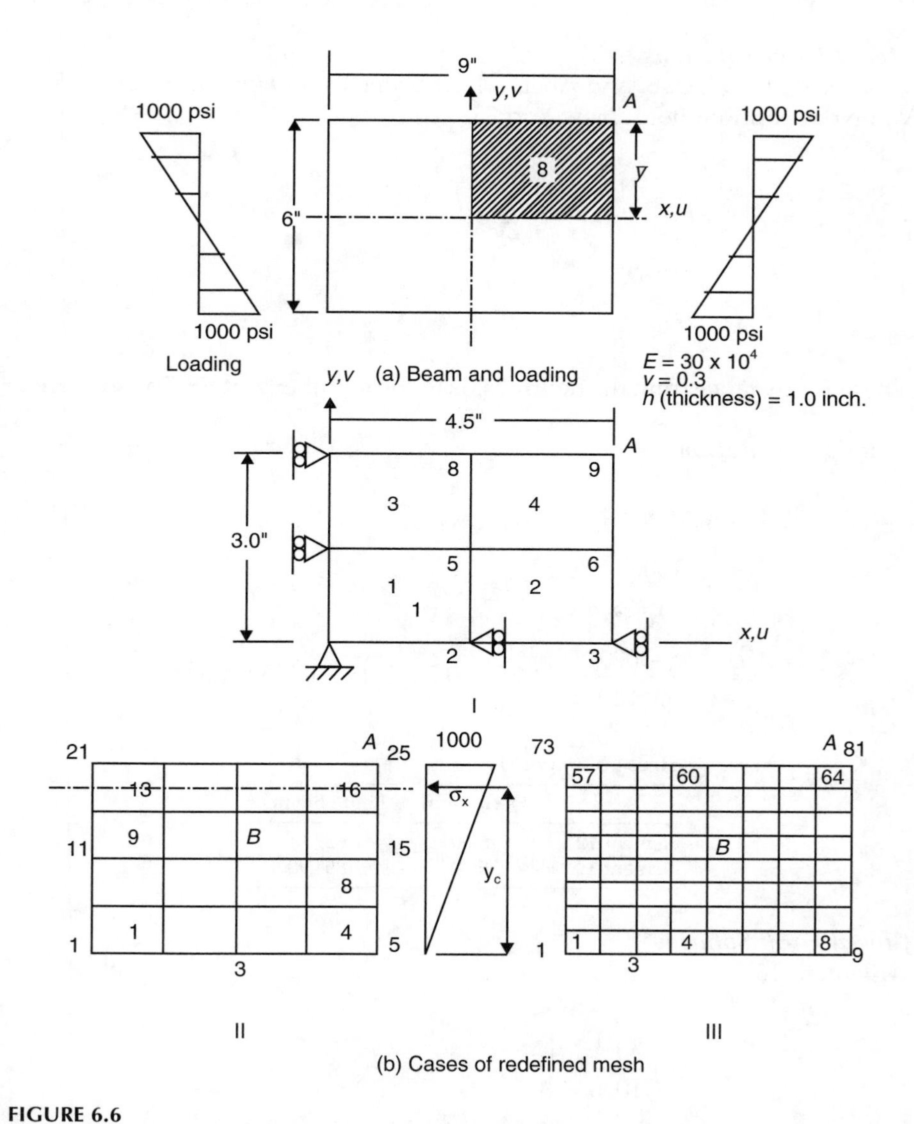

FIGURE 6.6
Example of a two-dimensional stress-deformation problem: (a) Beam and loading and (b) Cases of redefined mesh.

Closed Form Solutions

(See, e.g., Desai, C. S. and Abel, J. F., *Introduction to Finite Element Method*, Van Nostrand Reinhold, New York, 1972.)

$$u = \frac{p}{E\bar{y}} xy$$

$$v = \frac{-p}{2E\bar{y}}\left(x^2 + vy^2\right)$$

where $\bar{y}$ = distance from the neutral axis to the top fiber and v = Poisson ratio.

Sample Calculation

$$\text{Point A:}\ \ x = 4.5,\ y = 3,\ \bar{y} = 3$$

$$u = \frac{1000}{30 \times 10^6 \times 3} 4.5 \times 3 = 1.5 \times 10^{-4}\, m$$

$$v = \frac{-1000}{2 \times 30 \times 10^6 \times 3}\left(4.5^2 + 0.3 \times 3^2\right) = -1.275 \times 10^{-4}\, m$$

Stress σ_x (Kg/m²)

	Closed Form	Finite Element
Element 12	625.00	623.64
Element 16	875.00	854.98

Closed Form Solution

Element 16

$$\frac{\sigma_x}{1000} = \frac{y_c}{3}$$

$$\sigma_x = \frac{2.625 \times 1000}{3} = 875.0 kg/m^2$$

User's Guide for Field-2D

Input Set 1

NOTE: Input data should be in free format except on the title line.

(i) Title (MAIN PROGRAM)

This data identifies and gives problem description and can permit implementation of multiple problems in one run.

(ii) Problem Parameters (MAIN PROGRAM)

	Description	Symbol
(a)	Total no. of nodal points	NNP
(b)	Total no. of elements	NEL
(c)	Number of different materials	NMAT
(d)	Problem option: 1 = torsion, 2 = flow; velocity potential, 3 = flow; stream function, 4 = seepage, 5 = heat flow	NTYPE
(e)	Total no. of nodes at which boundary conditions are specified	NUMPOT
(f)	Flow quantity option: 0 = no flow calculated, 1 = flow calculated at specified cross sections	NQYES

(iii) Material Properties (MAIN PROGRAM)

	Description	Symbol
(a)	For torsion, the shear modulus; for potential flow problems set equal to 1; for seepage, it is the permeability in the x-direction; for heat flow it is the coefficient of conductivity in the x-direction	PERX(I)
(b)	For torsion, the angle of twist per unit length; for flow problems, set equal to 1; for seepage it is the permeability in the y-direction; for heat flow, it is the coefficient of heat conduction in the y-direction	PERY(I)
(c)	Thickness or length (in general, set equal to 1)	TH(I)

Note: Input as many sets as the number of materials (NMAT), up to 10.

Input Set 2

(i) Nodal Point Data (MAIN PROGRAM)

Description	Symbol
Node number	M
Blank	
x-coordinate	X(M)
y-coordinate	Y(M)

Note: Input as many sets as total number of nodes (NNP). If the nodes are along the same line and equidistant, input only the first and the last nodal data, then the intermediate nodal data are generated automatically. Input for the last node is always required.

Input Set 3

(i) Element Data (MAIN PROGRAM)

Description	Symbol
Element No.	M
Node number 1	IE(M,1)
Node number 2	IE(M,2)
Node number 3	IE(M,3)
Node number 4	IE(M,4)
Material number for element	IE(M,5)

Note: Input as many sets as total number of elements (NEL). Input node numbers counterclockwise around element. If the elements are in the same material properties, data are required for only the first and last elements in that sequence. Data for the last element are always required.

Input Set 4

(i) Boundary Conditions (MAIN PROGRAM)

Description	Symbol
Node no. at which boundary condition is specified	NPPOT(I)
Value of specified boundary condition	H(I)

Note: Input as many lines as number of nodes of specified boundary conditions (NUMPOT).

Input Set 5 (Not Required if NQYES = 0)

(i) Computation of' "Flow2" (Figure 6.7) (MAIN PROGRAM)

	Description	Symbol
(a)	No. of cross sections at which flow is desired	NQLINE
(b)	Number of elements along each cross section (should start in a new line)	NFEL(I) I = 1,2…NQLINE
(c)	Specific numbers of elements along each cross section (should start in a new line)	NELQ(I,J) I = 1,2…NOLINE J = 1,2…NFEL(I)

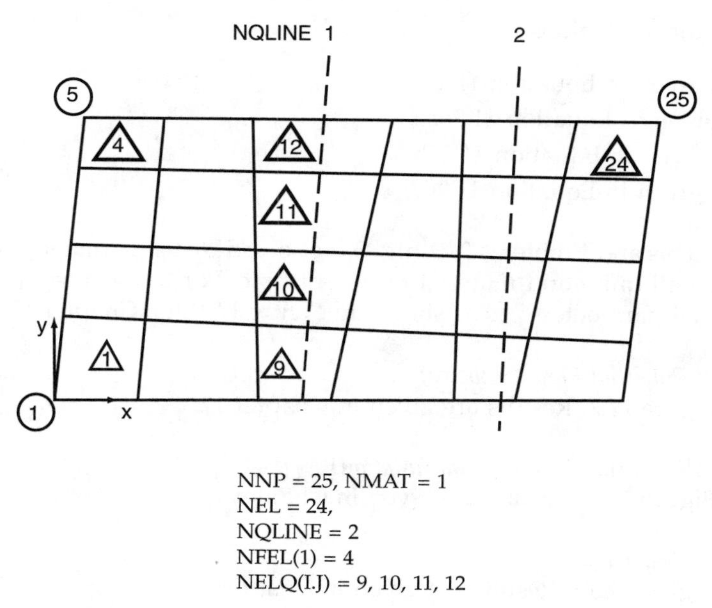

FIGURE 6.7
Schematic representation of input. (*Note*: See also the descriptions in Chapters 10, 11 and 12.)

Sample Problems for FIELD-2D

Problem 1: Torsion

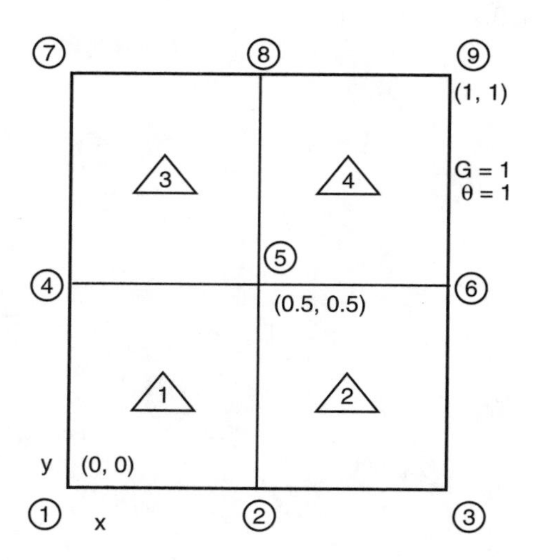

Closed form solutions for

M_t is given in Equation 11.25
ϕ is given in Equation 11.26a
τ_{xz} is given in Equation 11.26b
τ_{yz} is given in Equation 11.26c

Note 1: This and Problems 2, 3, and 5 are solved by using material parameters to be all unity and material to be isotropic. For the seepage problem, the material parameters are as shown in Figure 12.12(c), Chapter 12.

Problem 2: Potential Flow (velocity)
See Figure 12.7. Results are given in Chapter 12.

Problem 3: Potential Flow (stream function)
See Figure 12.7. Results are given in Chapter 12.

Problem 4: Seepage
See Figure 12.12. Results are given in Chapter 12.

Problem 5: Heat Flow
See Figure 12.11. Results are given in Chapter 12.

References

1. Desai, C.S. 2001. *Mechanics of Materials and Interfaces: The Disturbed State Concept*, CRC Press, Boca Raton, FL.

7

Beam Bending and Beam-Column

Introduction

Problems of beam bending and beam-column analysis using one-dimensional idealization are considered in this chapter. Figure 7.1(a) shows a beam-column subjected to the transverse load $p(x)$ and axial load $\bar{P}$. We first treat the case of bending only, without the load $\bar{P}$. Under the usual assumptions of beam bending theory,[1] the governing differential equation can be written as

$$\frac{d^2}{dx^2}\left(F(x)\frac{d^2w^*}{dx^2}\right) = p(x), \tag{7.1a}$$

where w^* = transverse displacement; $F(x) = EI(x)$, flexural rigidity; and x = coordinate along the centroidal axis of the beam. If $F(x)$ is assumed uniform along the beam, Equation 7.1a specializes to

$$F\frac{d^4w^*}{dx^4} = p(x). \tag{7.1b}$$

Here we have used z as the vertical coordinate (Figure 7.1(a)) and w as the corresponding transverse displacement.

We now follow the various steps of finite element formulation.

Step 1. Discretize and Choose Element Configuration

With the one-dimensional idealization the beam can be replaced by a line (Figure 7.l(b)), with the rigidity F lumped at the line. The idealized beam is now discretized into one-dimensional line elements (Figure 7.l(c)). A generic element is shown in Figure 7.l(d).

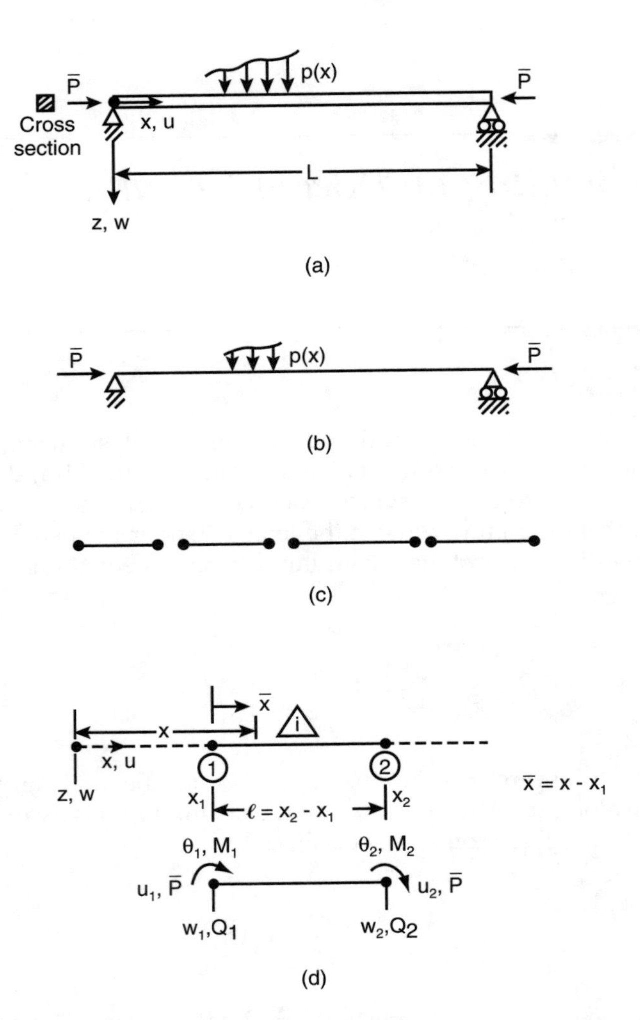

FIGURE 7.1
Beam bending and beam-column: (a) Beam with transverse and axial loads, (b) One-dimensional idealization, (c) Discretized beam, and (d) Generic element.

Step 2. Choose Approximation Model

In the case of (one-dimensional) column deformation (Chapter 3), we dealt only with plane deformations. To satisfy the physical condition of continuity of such structures, it was necessary to satisfy interelement compatibility at least with respect to nodal displacement (Figure 7.2(a)). It was therefore possible to fulfill the physical and mathematical requirements of the problem by using only first-order (linear) approximation. In contrast to the plane deformations, for realistic approximation of the physical conditions in the

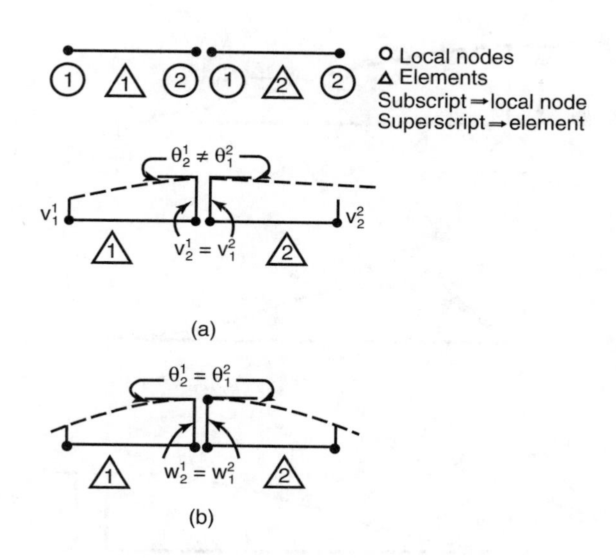

FIGURE 7.2
Requirements of interelement compatibility: (a) Interelement compatibility for axial deformation (Chapter 3) and (b) Interelement compatibility for beam bending.

case of bending, it is necessary to satisfy interelement compatibility with respect to both the displacements and slopes, that is, first derivative (gradient) of displacement (Figure 7.2(b)). As a consequence, it becomes necessary to use higher-order approximation for the displacement in the case of bending. Since it is necessary to provide for interelement compatibility for slopes also, we can add slope at the node as an additional unknown. This leads to two primary unknowns, displacement w and slope $\theta = dw/dx$, at each node; hence, for an element there are a total of four degrees of freedom: w_1, θ_1 at node 1 and w_2, θ_2 at node 2 (Figure 7.1(d)).

The commonly used interpolation approximation model for displacement w at any point $s = \bar{x}/l = (x - x_1)/l$, where s = local coordinate, x = global coordinate of any point, x_1 = global coordinate of point 1, and l = length of the element, is given by

$$w(x) = N_1 w_1 + N_2 \theta_1 + N_3 w_2 + N_4 \theta_2 \tag{7.2a}$$

or

$$w(x) = [\mathbf{N}]\{\mathbf{q}\}. \tag{7.2b}$$

Here $\{\mathbf{q}\}^T = [w_1\ \theta_1\ w_2\ \theta_2]$ and $[\mathbf{N}] = [N_1\ N_2\ N_3\ N_4]$ is the matrix of interpolation functions N_i, $i = 1, 2, 3, 4,$

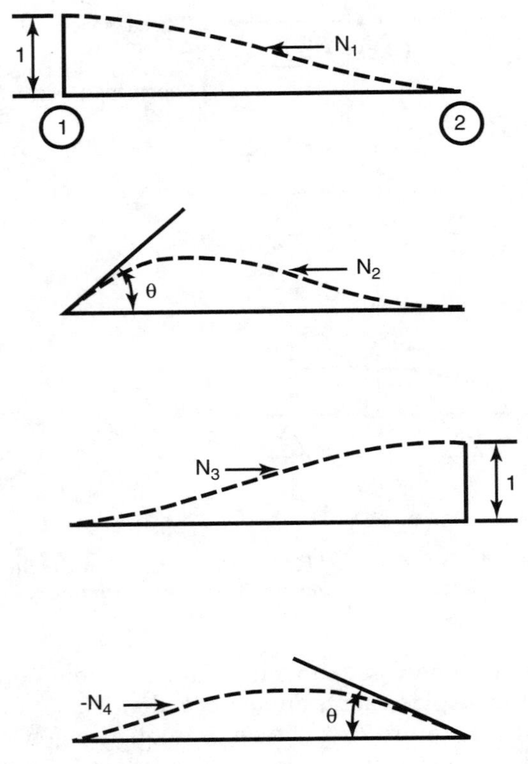

FIGURE 7.3
Plots of N_i, $i = 1, 2, 3, 4$ (Equation 7.3)

$$
\begin{aligned}
N_1 &= 1 - 3s^2 + 2s^3, \\
N_2 &= ls\left(1 - 2s + s^2\right), \\
N_3 &= s^2\left(3 - 2s\right), \\
N_4 &= ls^2\left(s - 1\right).
\end{aligned}
\tag{7.3}
$$

In Figure 7.3, we have shown plots of N_i, $i = 1, 2, 3, 4$. In mathematical literature the functions N_i are called Hermitian functions.[2,3]

It is worthwhile here to illustrate that the matrix [**N**] in Equation 7.2 can be derived by following the procedure outlined in Chapter 3, where we started with polynomial functions. Here we can use the following cubic polynomial:

$$
w(x) = \alpha_1 + \alpha_2 x + \alpha_3 x^2 + \alpha_4 x^3, \tag{7.4}
$$

where α_i = generalized coordinates. Equation 7.4 can be expressed in matrix notation as

$$w(x) = \begin{bmatrix} 1 & x & x^2 & x^3 \end{bmatrix} \begin{Bmatrix} \alpha_1 \\ \alpha_2 \\ \alpha_3 \\ \alpha_4 \end{Bmatrix} \tag{7.5}$$

$$= [\Phi]\{t\alpha\mathbf{a}\}.$$

Differentiation of $w(x)$ with respect to x leads to

$$\begin{aligned} \frac{dw}{dx} = \theta &= \alpha_2 + 2\alpha_3 x + 3\alpha_4 x^2 \\ &= \begin{bmatrix} 0 & 1 & 2x & 3x^2 \end{bmatrix}\{\alpha\} \\ &= [\Phi']\{\alpha\}. \end{aligned} \tag{7.6}$$

Here the prime denotes derivative with respect to x. To express w(x) in term of nodal values of w and θ, we use Equations 7.5 and 7.6 to first evaluate their values at nodes 1 and 2 as

$$\{\mathbf{q}\} = \begin{Bmatrix} w_1(x=0) \\ \theta_1(x=0) \\ w_2(x=l) \\ \theta_2(x=l) \end{Bmatrix} = \begin{bmatrix} 1 & 0 & 0 & 0 \\ 0 & 1 & 0 & 0 \\ 1 & l & l^2 & l^3 \\ 0 & 1 & 2l & 3l^2 \end{bmatrix} \begin{Bmatrix} \alpha_1 \\ \alpha_2 \\ \alpha_3 \\ \alpha_4 \end{Bmatrix} \tag{7.7}$$

or

$$\{\mathbf{q}\} = [\mathbf{A}]\{\alpha\}. \tag{7.8a}$$

Therefore,

$$\{\alpha\} = [\mathbf{A}]^{-1}\{\mathbf{q}\}. \tag{7.8b}$$

Finally, substitution of $\{\alpha\}$ into Equation 7.5 leads to

$$w(x) = [\Phi][\mathbf{A}]^{-1}\{\mathbf{q}\}. \tag{7.9}$$

where

$$[\mathbf{A}]^{-1} = \begin{bmatrix} 1 & 0 & 0 & 0 \\ 0 & 1 & 0 & 0 \\ -3/l^2 & -2/l & 3/l^2 & -1/l \\ 2/l^3 & 1/l^2 & -2/l^3 & 1/l^2 \end{bmatrix}.$$

If we perform the multiplication $[\Phi][\mathbf{A}]^{-1}$, we obtain $[\mathbf{N}]$ in Equation 7.2.

Comment on Requirements for Approximation Function

The approximation function (Equation 7.2) is conformable since it provides for interelement compatibility up to $n - 1 = 2 - 1 = 1$ derivative of w, that is, for both w and its first derivative (Figure 7.2(b)), where $n = 2$ is the highest order of derivative in the potential energy function in Equation 7.14a below. It also satisfies the completeness criterion since it allows for rigid body motion because of α_1 in Equation 7.4 and for constant states of strain (slope), Equation 7.6). Notice that the complete expansion (Equation 3.3d) up to n = 3 is retained in Equation 7.4.

Step 3. Define Strain-Displacement and Stress-Strain Relationships

From beam bending theory,[1] the relevant strain-displacement relation is

$$\varepsilon(x,z) = \frac{du}{dx} = -z\frac{d^2w}{dx^2} = -zw'', \tag{7.10}$$

where u = axial displacement and the superscript (two) primes denote second derivative. In bending problems the constitutive law is commonly expressed through the moment-curvature relationship

$$M(x) = F(x)w''(x). \tag{7.11}$$

Differentiation of $w(x)$ in Equation 7.2 twice with respect to x, through use of the chain rule of differentiation,

$$\frac{d}{dx} = \frac{1}{l}\frac{d}{ds} \quad \text{and} \quad \frac{d^2}{dx^2} = \frac{1}{l^2}\frac{d^2}{ds^2}, \tag{7.12}$$

leads to

$$w''(x) = \frac{d^2w}{dx^2} = \frac{1}{l^2}\frac{d^2w}{ds^2} = \frac{1}{l^2}\frac{d^2}{ds^2}[\mathbf{N}]\{\mathbf{q}\}, \tag{7.13a}$$

$$= [\mathbf{B}]\{\mathbf{q}\}, \tag{7.13b}$$

where $[\mathbf{B}]$ = transformation matrix; its coefficients are obtained by proper differentiation. For example,

$$\frac{dw}{dx} = \frac{1}{l}\frac{d}{ds}\left(N_1 w_1 + N_2\theta_1 + N_3 w_2 + N_4\theta_2\right)$$

$$= \frac{1}{l}\left[-6s + 6s^2 \quad l\left(1 - 4s + 3s^2\right) \quad 6s - s^2 \quad l\left(3s^2 - 2s\right)\right]\begin{Bmatrix} w_1 \\ \theta_1 \\ w_2 \\ \theta_2 \end{Bmatrix} \tag{7.13c}$$

and

$$w'' = \frac{d^2w}{dx^2} = \frac{1}{l}\frac{d}{ds}\left(\frac{dw}{ds}\right)$$

$$= \frac{1}{l^2}\left[-6 + 12s \quad -4l + 6ls \quad 6 - 12s \quad 6ls - 2l\right]\begin{Bmatrix} w_1 \\ \theta_1 \\ w_2 \\ \theta_2 \end{Bmatrix}. \tag{7.13d}$$

Step 4. Derive Element Equations

Energy Approach

We now use the principle of stationary (minimum) potential energy to derive element equations. The potential energy for the beam element assuming only surface traction loading $p(x)$ is expressed as[4,5]

$$\Pi_p = \int_{x_1}^{x_2} \tfrac{1}{2} F\left(w''\right)^2 dx - \int_{x_1}^{x_2} pw\, dx. \tag{7.14a}$$

Substitution of w and w'' from Equations 7.2 and 7.13d into Equation 7.14a leads to

$$\Pi_p = \tfrac{1}{2} Fl \int_0^1 \{\mathbf{q}\}^T [\mathbf{B}]^T [\mathbf{B}] \{\mathbf{q}\} ds - l \int_0^1 [\mathbf{N}] p \{\mathbf{q}\} ds, \tag{7.14b}$$

where F is assumed uniform, and from $s = (x - x_1)/l$, we have substituted the transformation $dx = l ds$.

Differentiating Π_p with respect to w_1, θ_1, w_2, and θ_2; and equating each term to zero, we obtain four element equations expressed in matrix notation as

$$Fl \int_0^1 [\mathbf{B}]^T [\mathbf{B}] ds \{\mathbf{q}\} = l \int_0^1 [\mathbf{N}]^T p ds \tag{7.15a}$$

or

$$[\mathbf{k}]\{\mathbf{q}\} = \{\mathbf{Q}\}, \tag{7.15b}$$

where

$$[\mathbf{k}] = Fl \int_0^1 [\mathbf{B}]^T [\mathbf{B}] ds$$

and

$$\{\mathbf{Q}\} = l \int_0^1 [\mathbf{N}]^T p ds,$$

which in the expanded form is

$$[\mathbf{k}] = Fl \int \begin{Bmatrix} 6(2s-1)/l^2 \\ 2(3s-2)/l \\ -6(2s-1)/l^2 \\ 2(3s-1)/l \end{Bmatrix} \begin{bmatrix} \dfrac{6(2s-1)}{l^2} & \dfrac{2(3s-2)}{l} & \dfrac{-6(2s-1)}{l^2} & \dfrac{2(3s-1)}{l} \end{bmatrix} ds. \tag{7.16a}$$

After the integrations,

$$[\mathbf{k}] = \frac{F}{l^3} \begin{bmatrix} 12 & 6l & -12 & 6l \\ & 4l^2 & -6l & 2l^2 \\ & \text{sym.} & 12 & -6l \\ & & & 4l^2 \end{bmatrix} \tag{7.16b}$$

and, assuming the surface traction p varies linearly as

$$p = (1-s)p_1 + sp_2 \tag{7.16c}$$

where p_1 and p_2 are values of loading at nodes 1 and 2, respectively,

$$\{Q\} = l\int_0^1 \begin{Bmatrix} 1-3s^2+2s^3 \\ ls(1-2s+s^2) \\ s^2(3-2s) \\ ls^2(s-1) \end{Bmatrix} [1-s \quad s] \begin{Bmatrix} p_1 \\ p_2 \end{Bmatrix} ds \tag{7.16d}$$

or

$$\{Q\} = \frac{l}{20}\begin{Bmatrix} 7p_1+3p_2 \\ \frac{l}{3}(3p_1+2p_2) \\ 3p_1+7p_2 \\ -\frac{l}{3}(2p_1+3p_2) \end{Bmatrix}. \tag{7.16e}$$

Derivation Using Galerkin's Method

On the basis of the explanation in Chapter 3, we consider one generic element. The residual is given by

$$R(x) = F\frac{d^4w}{dx^4} - p. \tag{7.17}$$

Therefore, use of the Galerkin method gives

$$\int_{x_1}^{x_2}\left(F\frac{d^4w}{dx^4} - p\right)N_i dx = 0, \quad i = 1, 2, 3, 4. \tag{7.18}$$

Integration by parts twice of the first term leads to

$$F\int_{x_1}^{x_2}\frac{d^2w}{dx^2}\frac{d^2N_i}{dx^2}dx - \int_{x_1}^{x_2} pN_i dx + \frac{d^3w}{dx^3}N_i\Bigg|_{x_1}^{x_2} - \frac{d^2w}{dx^2}\frac{dN_i}{dx}\Bigg|_{x_1}^{x_2} = 0. \tag{7.19a}$$

Substitution of w from Equation 7.2 leads to

$$F\int_{x_1}^{x_2}\frac{d^2N_j}{dx^2}\frac{d^2N_i}{dx^2}dxq_i-\int_{x_1}^{x_2}pN_idx+F\frac{d^3w}{dx^3}N_i\Bigg|_{x_1}^{x_2}-F\frac{d^2w}{dx^2}\frac{dN_i}{dx}\Bigg|_{x_1}^{x_2}=0, \tag{7.19b}$$

$$i=1,2,3,4,\quad j=1,2,3,4$$

or

$$F\int_{x_1}^{x_2}\frac{d^2N_j}{dx^2}\frac{d^2N_i}{dx^2}dxq_j=\int_{x_1}^{x_2}pN_idx-F\frac{d^3w}{dx^3}N_i\Bigg|_{x_1}^{x_2}+F\frac{d^2w}{dx^2}\frac{dN_i}{dx}\Bigg|_{x_1}^{x_2}. \tag{7.19c}$$

Expansion of the term on the left-hand side leads to

$$F\int_{x_1}^{x_2}\begin{bmatrix} N_1''^2 & N_2''N_1'' & N_3''N_1'' & N_4''N_1'' \\ & N_2''^2 & N_3''N_2'' & N_4''N_2'' \\ & \text{sym.} & N_3''^2 & N_4''N_3'' \\ & & & N_4''^2 \end{bmatrix}\begin{Bmatrix} w_1 \\ \theta_1 \\ w_2 \\ \theta_2 \end{Bmatrix}dx. \tag{7.20a}$$

Here the double prime indicates second derivative with respect to x. After proper transformations for differentiations and integrations, this expression will yield the same results in Equation 7.16b.

The first term on the right-hand-side will lead to the load vector in Equation 7.16e. The remaining two terms on the right-hand side yield internal joint shear and moment forces. Let us consider

$$F\frac{d^3w}{dx^3}N_i\Bigg|_{x_1}^{x_2}=\begin{Bmatrix} F\dfrac{d^3w}{dx^3}N_1\Bigg|_{x_1}^{x_2} \\ F\dfrac{d^3w}{dx^3}N_2\Bigg|_{x_1}^{x_2} \\ F\dfrac{d^3w}{dx^3}N_3\Bigg|_{x_1}^{x_2} \\ F\dfrac{d^3w}{dx^4}N_4\Bigg|_{x_1}^{x_2} \end{Bmatrix}. \tag{7.20b}$$

Noting the properties of N_i (Figure 7.3), we find that this reduces to

$$\begin{Bmatrix} \left(F\dfrac{d^3w}{dx^3}\right)_1 \\ 0 \\ -\left(F\dfrac{d^3w}{dx^3}\right)_2 \\ 0 \end{Bmatrix}, \tag{7.20c}$$

which gives joint shear forces at the two nodes. Similarly,

$$F\frac{d^2w}{dx^2}\frac{dN_i}{dx}\bigg|_{x_1}^{x_2}$$

leads to

$$\begin{Bmatrix} \left(F\dfrac{d^2w}{dx^2}\dfrac{dN_1}{dx}\right)\bigg|_{x_1}^{x_2} \\ \left(F\dfrac{d^2w}{dx^2}\dfrac{dN_2}{dx}\right)\bigg|_{x_1}^{x_2} \\ \left(F\dfrac{d^2w}{dx^2}\dfrac{dN_3}{dx}\right)\bigg|_{x_1}^{x_2} \\ \left(F\dfrac{d^2w}{dx^2}\dfrac{dN_4}{dx}\right)\bigg|_{x_1}^{x_2} \end{Bmatrix} \tag{7.20d}$$

or

$$\begin{Bmatrix} 0 \\ \left(F\dfrac{d^2w}{dx^2}\right)_1 \\ 0 \\ \left(-F\dfrac{d^2w}{dx^2}\right)_2 \end{Bmatrix} \tag{7.20e}$$

because

$$\frac{dN_1}{dx} = \frac{dN_3}{dx} = 0 \qquad \text{at } x_1 \text{ and } x_2,$$

$$\frac{dN_2}{dx} = \frac{dN_4}{dx} = 1 \qquad \text{at } x_1 \text{ and } x_2,$$

and

$$\frac{dN_2}{dx} = 0 \quad \text{at } x_2,$$

$$\frac{dN_4}{dx} = 0 \quad \text{at } x_1,$$

which yield joint moment forces at the two nodes. It may be noted that if there were no externally applied joint forces, when joint load vectors (Equations 7.20c and 7.20e) are assembled for all elements in the discretized body, only the terms related to the two ends of the entire beam will remain, whereas the other terms will be zero due to alternating plus and minus signs. The nonzero end terms will denote the boundary conditions; for instance, for a simply supported beam the end term in Equation 7.20e, which denotes moment, will vanish. Note that the element equations from both the energy and residual procedures will be essentially the same. Further details of applications of the Galerkin and other residual methods are given in Appendix 1.

Steps 5 to 8

We shall illustrate the steps of assembly and computation of primary and secondary quantities by using an example. Figure 7.4 shows a beam with length $L = 20$ cm and cross-sectional area = 2 cm^2 (2 cm deep × 1 cm wide) and $E = 10^6$ kg/cm^2. It is subjected to a uniform surface traction $p(x) = 100$ kg/cm. The beam is divided into two elements of length $l = 10$ cm each.

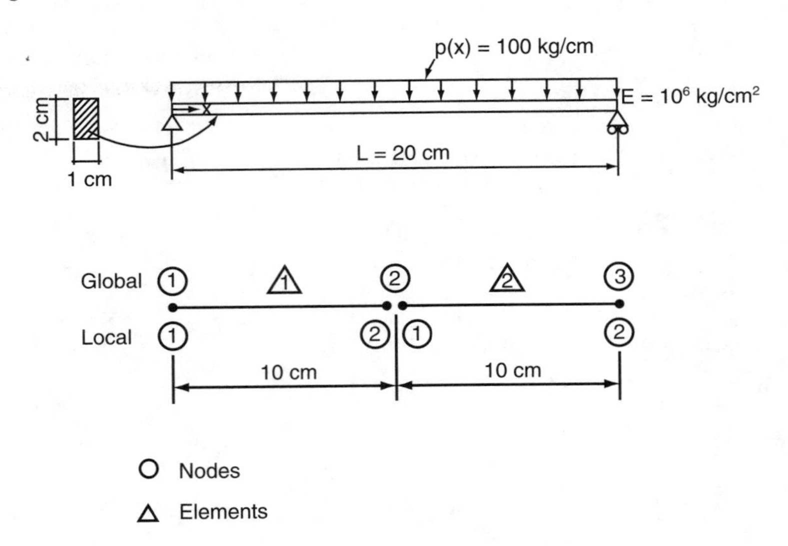

FIGURE 7.4
Example for beam bending.

Use of Equations 7.16b and 7.16e leads to the following (local) stiffness relationship for both elements (with $\boldsymbol{I} = \frac{2}{3}$ cm^4):

$$\frac{16\times 10^3}{6}\begin{bmatrix} 3 & 15 & -3 & 15 \\ 15 & 100 & -15 & 50 \\ -3 & -15 & 3 & -15 \\ 15 & 50 & -15 & 100 \end{bmatrix}\begin{Bmatrix} w_1 \\ \theta_1 \\ w_2 \\ \theta_2 \end{Bmatrix} = \frac{1000}{6}\begin{Bmatrix} 3 \\ 5 \\ 3 \\ -5 \end{Bmatrix}. \tag{7.21}$$

By following the assembly procedure as described in Chapter 3, we obtain the following global stiffness relation for the two elements:

$$16\begin{bmatrix} 3 & 15 & -3 & 15 & 0 & 0 \\ 15 & 100 & -15 & 50 & 0 & 0 \\ -3 & -15 & 6 & 0 & -3 & 15 \\ 15 & 50 & 0 & 200 & -15 & 50 \\ 0 & 0 & -3 & -15 & 3 & -15 \\ 0 & 0 & 15 & 50 & -15 & 100 \end{bmatrix}\begin{Bmatrix} w_1 \\ \theta_1 \\ w_2 \\ \theta_2 \\ w_3 \\ \theta_3 \end{Bmatrix} = \begin{Bmatrix} 3 \\ 5 \\ 6 \\ 0 \\ 3 \\ -5 \end{Bmatrix}. \tag{7.22a}$$

The beam (Figure 7.4) is supported at two unyielding ends; therefore, the boundary conditions are $w_1 = w_3 = 0$. Introduction of these constraints into Equation 7.22a leads to modified global equations as

$$16\begin{bmatrix} 1 & 0 & 0 & 0 & 0 & 0 \\ 15 & 100 & -15 & 50 & 0 & 0 \\ -3 & -15 & 6 & 0 & -3 & 15 \\ 15 & 50 & 0 & 200 & -15 & 50 \\ 0 & 0 & 0 & 0 & 1 & 0 \\ 0 & 0 & 15 & 50 & -15 & 100 \end{bmatrix}\begin{Bmatrix} w_1 \\ \theta_1 \\ w_2 \\ \theta_2 \\ w_3 \\ \theta_3 \end{Bmatrix} = \begin{Bmatrix} 0 \\ 5 \\ 6 \\ 0 \\ 0 \\ -5 \end{Bmatrix}. \tag{7.22b}$$

Solution of the four simultaneous equations gives

$$\begin{aligned} &\omega_1 = 0.0000\ (\text{given}), && w_2 = 0.3125\ \text{cm}, && w_3 = 0.0000\ (\text{given}), \\ &\theta_1 = 0.0500\ \text{radian}, && \theta_2 = 0.0000\ \text{rad.}, && \theta_3 = -0.0500\ \text{rad.} \end{aligned} \tag{7.22c}$$

Closed Form Solutions

From beam bending theory[1] the closed form or exact solutions are

$$w(x) = \frac{px}{24F}\left(L^3 - 2Lx^2 + x^3\right), \tag{7.23}$$

$$\frac{dw}{dx} = \frac{p}{24F}\left(L^3 - 6x^2 + 4x^3\right), \tag{7.24}$$

which yield the same results as those in Equation 7.22c from the finite element computations. That is, with the cubic approximation model, the finite element method yields the same results as the closed form solutions insofar as the displacements and slopes are concerned.

Secondary Quantities

Let us now consider the secondary quantities: moments (M) and shear forces (V). To find moment M we substitute relevant nodal displacements into

$$M = F\frac{d^2w}{dx^2} \tag{7.25}$$

where d^2w/dx^2 is defined in Equation 7.13d.

For element 1,

$$M\ (\text{at } s = 0) = \frac{F}{l^2}\begin{bmatrix}-6 & -4l & 6 & -2l\end{bmatrix}\begin{Bmatrix}0.0000\\0.0500\\0.3125\\0.0000\end{Bmatrix}$$

$$= \frac{-0.125F}{l^2} = 0.08334 \times \frac{10^6}{10^2} = -833.4 \text{ kg-cm.}$$

Similarly,

$$M\ (\text{at } s = 0.5) = \frac{-0.50F}{l^2} = -3333.3 \text{ kg-cm,}$$

$$\text{M}\ (\text{at } s = 1) = \frac{-0.875F}{l^2} = -5833.4 \text{ kg-cm.}$$

For element 2,

$$M\ (\text{at } s=0)=\frac{F}{l^2}[-6\quad -4l\quad 6\quad -2l]\begin{Bmatrix} 0.3125 \\ 0.0000 \\ 0.0000 \\ -0.0500 \end{Bmatrix}$$

$$= -5833.4 \text{ kg-cm.}$$

Similarly,

$$M\ (\text{at } s=0.5)=\frac{-0.05F}{l^2}=-3333.3 \text{ kg-cm,}$$

$$\text{M}\ (\text{at } s=1)=\frac{-0.125F}{l^2}=-833.4 \text{ kg-cm.}$$

The closed form solution for moment is[1]

$$M=Fw''(x)=\frac{12p}{24}\left(-Lx+x^2\right), \tag{7.26}$$

which at $x = l/4$ and $x = l/2$ gives

$$M\left(x=\frac{l}{4}\right)=-3750.0 \text{ kg-cm,}$$

$$M\left(x=\frac{l}{2}\right)=-5000 \text{ kg-cm.}$$

Figure 7.5(a) shows the bending moment diagrams from the finite element computations and from the closed form solution.

Next we compute shear force V, which is given in closed form as

$$V=F\frac{d^3w}{dx^3}, \tag{7.27}$$

which yields

$$V(x=0)=\frac{-pL}{2}=-1000.0 \text{ kg,}$$

$$V\left(x=\frac{L}{2}\right)=0.0 \text{ kg,}$$

$$V(x=L)=1000 \text{ kg.}$$

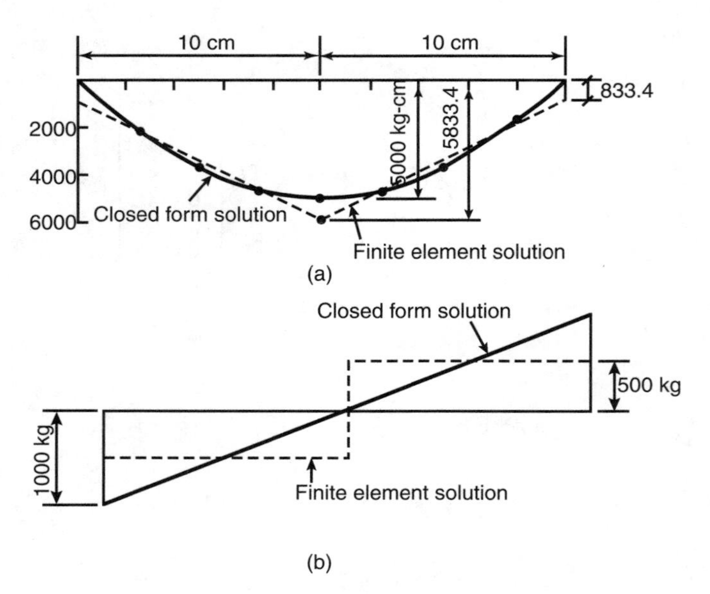

FIGURE 7.5
Comparisons for bending moments and shear forces (two elements): (a) Bending moment diagrams and (b) Shear force diagrams.

From the finite element computations,

$$V = F\frac{d}{dx}(w'') = \frac{F}{l}\frac{dw''}{ds}$$

$$= \frac{1}{l^3}[12 \quad 6l \quad -12 \quad 6l]\begin{Bmatrix} w_1 \\ \theta_1 \\ w_2 \\ \theta_2 \end{Bmatrix}. \tag{7.28}$$

Therefore, *for element 1*,

$$V = \frac{F}{l^3}[12 \quad 6l \quad -12 \quad 6l]\begin{Bmatrix} 0.0000 \\ 0.0500 \\ 0.3125 \\ 0.0000 \end{Bmatrix} = -500.25 \text{ kg},$$

and for element 2,

$$V = \frac{0.75\,F}{l^3} = +500.25 \text{ kg}.$$

Figure 7.5(b) shows plots of shear force from closed form and finite element computations.

Comment

As noted previously, both methods yield the same values of displacements and slopes. However, for moments, the finite element method yields a bilinear distribution (Figure 7.5(a)), whereas the closed form solution shows continuous distribution. Also, introduction of only $w_1 = w_3 = 0$ in the finite element procedure does not yield zero moments at the ends, which is required for the assumption of simple support in the closed form solution. Overall, the magnitude of the moments from the finite element computations is satisfactory.

In the case of the shear forces, however, the finite element computations show a wide disparity as compared with the closed form results (Figure 7.5(b)). At the ends the difference between the two results is high.

As discussed in Chapter 3, there are two possible methods by which we can improve computations of bending moments and shear forces: (1) refine mesh and/or (2) choose different (higher-order) approximation models.

Mesh Refinement

The beam is now divided into four elements of l = 5 cm, as shown in Figure 7.6. Substitution of E, I, l, and p into Equations 7.16b and 7.16e leads to the following general element equation:

$$\frac{10^6 \times 2}{125 \times 3}\begin{bmatrix} 12 & 30 & -12 & 30 \\ & 100 & -30 & 50 \\ & & 12 & -30 \\ & \text{sym.} & & 100 \end{bmatrix}\begin{Bmatrix} w_1 \\ \theta_1 \\ w_2 \\ \theta_2 \end{Bmatrix} = \begin{Bmatrix} 250 \\ 2500/12 \\ 250 \\ -2500/12 \end{Bmatrix}.$$

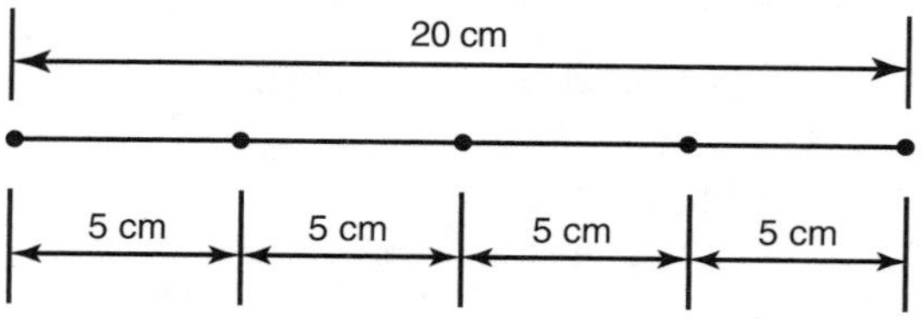

FIGURE 7.6
Mesh refinement for beam bending.

Assembly of the four element equations and introduction of the boundary conditions $w_1 = w_5 = 0$ lead to the following modified assemblage equations:

$$5333.33\begin{bmatrix} 1 & 0 & 0 & 0 & 0 & 0 & 0 & 0 & 0 & 0 \\ 30 & 100 & -30 & 50 & 0 & 0 & 0 & 0 & 0 & 0 \\ -12 & -30 & 24 & 0 & -12 & 30 & 0 & 0 & 0 & 0 \\ 30 & 50 & 0 & 200 & -30 & 50 & 0 & 0 & 0 & 0 \\ 0 & 0 & -12 & -30 & 24 & 0 & -12 & 30 & 0 & 0 \\ 0 & 0 & 30 & 50 & 0 & 200 & -30 & 50 & 0 & 0 \\ 0 & 0 & 0 & 0 & -12 & -30 & 24 & 0 & -12 & 30 \\ 0 & 0 & 0 & 0 & 30 & 50 & 0 & 200 & -30 & 50 \\ 0 & 0 & 0 & 0 & 0 & 0 & 0 & 0 & 1 & 0 \\ 0 & 0 & 0 & 0 & 0 & 0 & 30 & 50 & -30 & 100 \end{bmatrix}\begin{Bmatrix} w_1 \\ \theta_1 \\ w_2 \\ \theta_2 \\ w_3 \\ \theta_3 \\ w_4 \\ \theta_4 \\ w_5 \\ \theta_5 \end{Bmatrix} = \begin{Bmatrix} 0.00 \\ 208.33 \\ 500.00 \\ 0.00 \\ 500.00 \\ 0.00 \\ 500.00 \\ 0.00 \\ 0.00 \\ -208.33 \end{Bmatrix}.$$

Solution of these equations gives

$$w_1 = 0.0000, \quad w_2 = 0.2227, \quad w_3 = 0.3125, \quad w_4 = 0.2227,$$

$$w_5 = 0.0000,$$

$$\theta_1 = 0.0500, \quad \theta_2 = 0.0344, \quad \theta_3 = 0.0000, \quad \theta_4 = -0.0344,$$

$$\theta_5 = 0.0500.$$

Figure 7.7 shows comparisons between bending moments and shear forces from the finite element analysis and those from the closed form solution. It can be seen that the computed values are now closer to the exact solutions than the comparisons in Figure 7.5 for the two-element approximation.

Higher-Order Approximation

As the next step from the cubic approximation, we can adopt a fifth-order approximation as follows:

$$\begin{aligned} w(x) &= \alpha_1 + \alpha_2 x + \alpha_3 x^2 + \alpha_4 x^3 + \alpha_5 x^4 + \alpha_6 x^5 \\ &= \begin{bmatrix} 1 & x & x^2 & x^3 & x^4 & x^5 \end{bmatrix}\{\alpha\} \\ &= [\Phi]\{\alpha\}, \end{aligned} \tag{7.29}$$

where

$$\{\alpha\}^T = \begin{bmatrix} \alpha_1 & \alpha_2 & \alpha_3 & \alpha_4 & \alpha_5 & \alpha_6 \end{bmatrix}.$$

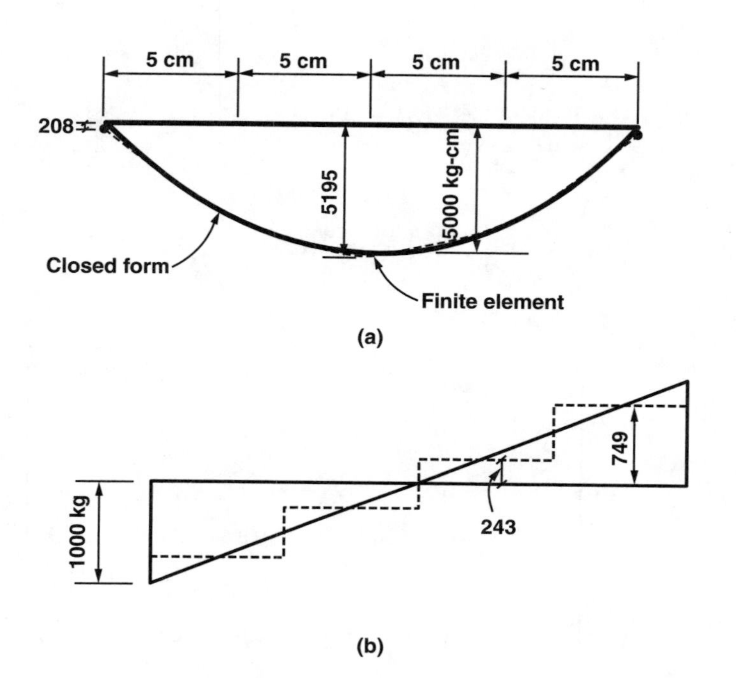

FIGURE 7.7
Comparisons for bending moments and shear forces (four elements): (a) Bending moment diagrams and (b) Shear force diagrams.

The first and second derivatives are obtained as

$$\theta = \frac{dw}{dx} = \begin{bmatrix} 0 & 1 & 2x & 3x^2 & 4x^3 & 5x^4 \end{bmatrix} \{\alpha\} \tag{7.30}$$

and

$$w'' = \frac{d^2w}{dx^2} = \begin{bmatrix} 0 & 0 & 2 & 6x & 12x^2 & 20x^3 \end{bmatrix} \{\alpha\}, \tag{7.31}$$

where w'' is the second derivation or curvature. Evaluation of w, θ, and w'' at the two nodes leads to

$$\{\mathbf{q}\} = \begin{Bmatrix} w_1 \\ \theta_1 \\ w_1'' \\ w_2 \\ \theta_2 \\ w_2'' \end{Bmatrix} = \begin{bmatrix} 1 & 0 & 0 & 0 & 0 & 0 \\ 0 & 1 & 0 & 0 & 0 & 0 \\ 0 & 0 & 2 & 0 & 0 & 0 \\ 1 & l & l^2 & l^3 & l^4 & l^5 \\ 0 & 1 & 2l & 3l^2 & 4l^3 & 5l^4 \\ 0 & 0 & 2 & 6l & 12l^2 & 20l^3 \end{bmatrix} \tag{7.32a}$$

or

$$\{\mathbf{q}\} = [\mathbf{A}]\{\alpha\}. \tag{7.32b}$$

Therefore

$$\{\alpha\} = [\mathbf{A}]^{-1}\{\mathbf{q}\}, \tag{7.33}$$

where

$$[\mathbf{A}]^{-1} = \frac{1}{l^9}\begin{bmatrix} l^9 & 0 & 0 & 0 & 0 & 0 \\ 0 & l^9 & 0 & 0 & 0 & 0 \\ 0 & 0 & l^9/2 & 0 & 0 & 0 \\ -10l^6 & -6l^7 & -3l^8/2 & 10l^6 & -4l^7 & l^8/2 \\ 15l^5 & 8l^6 & 3l^7/2 & -15l^5 & 7l^6 & -l^7 \\ -6l^4 & -3l^5 & -l^6/2 & 6l^4 & -3l^5 & -l^6/2 \end{bmatrix}.$$

Hence,

$$w(x) = [\Phi][\mathbf{A}]^{-1}\{\mathbf{q}\} = [\mathbf{N}]\{\mathbf{q}\}. \tag{7.34}$$

As before, multiplication of $[\Phi][\mathbf{A}]^{-1}$ leads to the interpolation functions N_i, $i = 1, 2,\ldots,6$, given by

$$\begin{aligned} N_1 &= 1 - 10s^3 + 15s^4 - 6s^5, \\ N_2 &= l\left(s - 6s^3 + 8s^4 - 3s^5\right), \\ N_3 &= \frac{l^2}{2}\left(s^2 - 3s^3 + 3s^4 - s^5\right), \\ N_4 &= 10s^3 - 15s^4 + 6s^5, \\ N_5 &= l\left(-4s^3 + 7s^4 - 3s^5\right), \\ N_6 &= l^2\left(\frac{s^3}{2} - s^4 + \frac{s^5}{2}\right). \end{aligned} \tag{7.35}$$

Use of these interpolation functions will enforce interelement compatibility of displacement w, slope θ, and curvature w''. This would provide an improvement in the approximation to the continuity of the deformed beam. The shape functions are plotted in Figure 7.8.

Problem 7.9 refers to derivation of equations for the beam bending problem with the fifth-order approximation. We have given partial results and comments with Problem 7.9.

Beam-Column

If in addition to the lateral load the beam is subjected to a constant axial load $\bar{P}$ (Figure 7.1(a)), then the beam also acts as a column, and the combined effect is called a beam-column.

To simplify the problem, we assume that the axial load is small in comparison with the lateral load and that the problem is linear. Under these assumptions, we can superimpose the bending and axial load effects.

Step 1

As before, the beam is divided into one-dimensional line elements.

Step 2

Now with the cubic model we have three degrees of freedom at each end of the element: w, θ for bending (Equation 7.2), and u for the axial deformation (Figure 7.1(a)). Thus the total number of degrees of freedom for the element (Figure 7.1(d)) is six (w_1, θ_1, u_1; w_2, θ_2, u_2). As in Chapter 3, we choose linear approximation for axial deformation (denoted here as u instead of as v in Chapter 3); accordingly, in terms of local coordinate s,

$$u(x) = [1-s \quad s]\begin{Bmatrix} u_1 \\ u_2 \end{Bmatrix} = [\mathbf{N}_a]\{\mathbf{q}_a\}. \tag{7.36}$$

For the bending part, we use the cubic function (Equation 7.2). The combined interpolation model can be now written as

$$\begin{Bmatrix} u(x) \\ w(x) \end{Bmatrix} = \begin{bmatrix} \underset{(1\times 2)}{[\mathbf{N}_a]} & \underset{(1\times 4)}{[0]} \\ \underset{(1\times 2)}{[0]} & \underset{(1\times 4)}{[\mathbf{N}_b]} \end{bmatrix} \begin{Bmatrix} \{q_a\} \\ \{q_b\} \end{Bmatrix}, \tag{7.37}$$

where $[\mathbf{N}_b]$ and $\{\mathbf{q}_b\}$ are the same as $[\mathbf{N}]$ and $\{\mathbf{q}\}$ in Equation 7.2; here we have used the subscript b to denote bending.

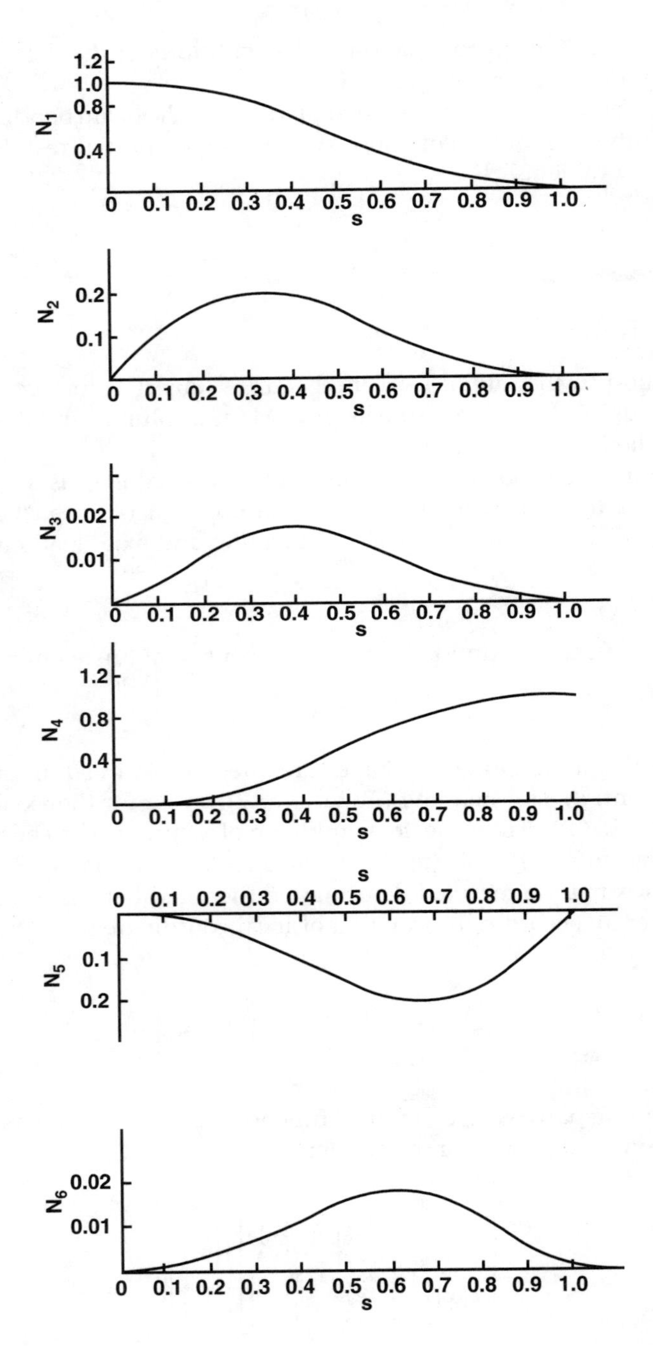

FIGURE 7.8
Interpolation or shape functions for fifth-order model.

Steps 3 to 5

The strain-displacement and stress–strain relations for the axial and the bending problem are given in Equations 3.12a and 3.15 and in Equations 7.10 and 7.11, respectively.

The potential energy is given as essentially the sum of the energies due to bending (Equation 7.14) and axial deformation (Equation 3.21). Hence

$$\Pi_p = l\int_0^1 \frac{1}{2} F(w'')^2 ds + \frac{Al}{2}\int_0^1 E\left(\frac{du}{dx}\right) ds - l\int_0^1 wpds - \bar{P}u \tag{7.38}$$

Here we assume $\bar{P}$ is a concentrated uniform axial load; the term $\bar{P}\,u$ denotes the potential of load $\bar{P}$.

Substitution for w (Equation 7.2), w'' (Equation 7.13), u (Equation 7.36), and du/dx (Equation 3.12) leads to

$$\begin{aligned}\Pi_p = {} & \frac{Fl}{2}\int_0^1 \{\mathbf{q}_b\}^T[\mathbf{B}_b]^T[\mathbf{B}_b]ds\{\mathbf{q}_b\} + \frac{AEL}{2}\int_0^1 \{\mathbf{q}_a\}^T[\mathbf{B}_a]^T[\mathbf{B}_a]ds\{\mathbf{q}_a\} \\ & - l\int_0^1 [\mathbf{N}_b]\{\mathbf{q}_b\}ds - \bar{P}[\mathbf{N}_a]\{\mathbf{q}_a\}\end{aligned} \tag{7.39}$$

Here the subscripts a and b denote axial and bending nodes, respectively.

Now we differentiate Π_p with respect to w_1, θ_1, u_1, w_2, θ_2, and u_2 and equate the results to zero to obtain a set of six linear algebraic equations. After proper integrations and rearrangement of terms, and assuming p to be uniform in Equation 7.16e, the resulting element equations can be expressed as

$$\begin{bmatrix} \frac{12EI}{l^3} & \frac{6EI}{l^2} & -\frac{12EI}{l^3} & \frac{6EI}{l^2} & 0 & 0 \\ & \frac{4EI}{l} & -\frac{6EI}{l^2} & \frac{2EI}{l} & 0 & 0 \\ & & \frac{12EI}{l^3} & -\frac{6EI}{l^2} & 0 & 0 \\ & & & \frac{4EI}{l} & 0 & 0 \\ & \text{sym.} & & & \frac{EA}{l} & -\frac{EA}{l} \\ & & & & & \frac{EA}{l} \end{bmatrix} \begin{Bmatrix} w_1 \\ \theta_1 \\ w_2 \\ \theta_2 \\ u_1 \\ u_2 \end{Bmatrix} = \begin{Bmatrix} \frac{pl}{2} \\ \frac{pl^2}{12} \\ \frac{pl}{2} \\ -\frac{pl^2}{12} \\ \bar{P} \\ \bar{P} \end{Bmatrix}. \tag{7.40}$$

We have arranged the element relations such that the superimposition of the bending and axial effects can be seen clearly. The top relation is the same as in Equation 7.15 and the bottom relation as in Equation 3.28.

The remaining steps of assembly and introduction of boundary conditions are essentially the same as before. The reader should at this stage undertake solution of Problem 7.8.

Comment

What happens if the load $\bar{P}$ is not small and if the coupling or interaction between the bending and axial effects needs to be considered? For instance, in the foregoing, we ignored the bending effects that may be caused by $\bar{P}$. Under these circumstances, the problem can no longer be considered as linear, and the principle of superimposition usually does not hold. One of the manifestations of such a nonlinear problem is buckling. This type of nonlinearity falls under the category of geometrical nonlinearity and is beyond the scope of this text.

Other Procedures of Formulation

In this section we shall discuss the use of the complementary energy and mixed principles for the derivation of finite element equations. Only a simple introduction to these procedures is included; for advanced study the reader may consult texts and publications referenced elsewhere in the text, e.g., in Chapters 3 and 15.

Complementary Energy Approach

For generating a stable structural element for beam bending, we need to eliminate rigid-body degrees of freedom. For the beam element (Figure 7.1(d)), such degrees of freedom are those relevant to the conditions of displacement and rotations at one of the nodes. This leads to an element like a built-in beam or cantilever (Figure 7.9). Then the complementary energy is expressed as a special case of Equation 3.55c as

$$\Pi_c = \frac{1}{2}\int_{x_1}^{x_2} M\frac{1}{F}M dx - [Q_1 \quad M_1]\begin{Bmatrix} w_1 \\ \theta_1 \end{Bmatrix}. \tag{7.41}$$

The variation of moment M along x can be expressed as

$$\begin{aligned} M &= xQ_1 - M_1 \\ &= [x \quad -1]\begin{Bmatrix} Q_1 \\ M_1 \end{Bmatrix}. \end{aligned} \tag{7.42}$$

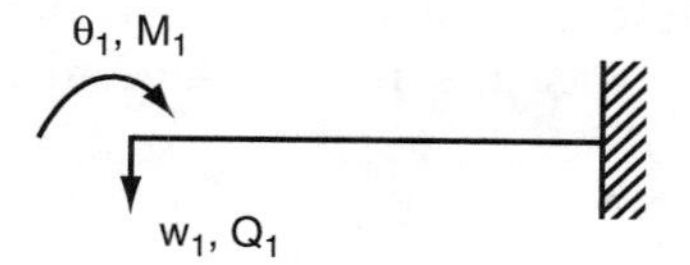

FIGURE 7.9
Stable element for bending.

Note that the moment replaces stress as the primary unknown. The flexibility matrix according to Equation 3.58a is

$$[f]=\frac{1}{F}\int_{x_1}^{x_2}\begin{Bmatrix} x \\ -1 \end{Bmatrix}[x \quad -1]dx \tag{7.43a}$$

$$=\frac{1}{6F}\begin{bmatrix} 2l^2 & -3l \\ -3l & 6 \end{bmatrix}. \tag{7.43b}$$

The relation between the end nodal forces is given by

$$Q_2=-Q_1, \tag{7.44a}$$

$$M_2=lQ_1-M_1, \tag{7.44b}$$

or

$$\begin{Bmatrix} Q_2 \\ M_2 \end{Bmatrix}=\begin{bmatrix} -1 & 0 \\ 1 & -1 \end{bmatrix}\begin{Bmatrix} Q_1 \\ M_1 \end{Bmatrix} \tag{7.44c}$$

$$=[\mathbf{G}]\begin{Bmatrix} Q_1 \\ M_1 \end{Bmatrix}. \tag{7.44d}$$

Now the inverse of [**f**] is

$$[\mathbf{f}]^{-1}=\frac{2F}{l^3}\begin{Bmatrix} 6 & 3l \\ 3l & 2l^2 \end{Bmatrix}. \tag{7.45}$$

By using Equation 3.62, we obtain the element stiffness matrix [**k**], which will be the same as that in Equation 7.16b. If the variation of moment were different from linear, we would obtain a different stiffness matrix.

As stated earlier (Chapter 3), use of the stress function approach can be more straightforward and easier. It is illustrated in Chapter 11.

Mixed Approach

Here the formulation involves two primary unknowns: displacement w and moment (stress) M. We assume the same cubic function (Equation 7.2) for w and linear variation for M as

$$\begin{aligned} M &= (1-s)M_1 + sM_2 \\ &= [\mathbf{N}_m]\{\mathbf{M}_n\}, \end{aligned} \tag{7.46}$$

where M is the moment at any point, $\{\mathbf{M}_n\}^T = [M_1\ M_2]$ is the vector of nodal moments, and $[\mathbf{N}_m]$ is the matrix of interpolation functions.

A special form of Equation 3.63 for beam bending can be written as

$$\Pi_R = -\frac{1}{2}\int_{x_1}^{x_2} M\frac{1}{F}M dx + \int_{x_1}^{x_2} M\frac{d^2w}{dx^2}dx - \{\mathbf{Q}\}^T\{\mathbf{q}\}, \tag{7.47}$$

where

$$\{\mathbf{Q}\}^T = [Q_1 \quad M_1 \quad Q_2 \quad M_2]$$

and

$$\{\mathbf{q}\}^T = [w_1 \quad \theta_1 \quad w_2 \quad \theta_2].$$

Substitution of M and w'' from Equations 7.46 and 7.13d, respectively, into Π_R leads to

$$\begin{aligned} \Pi_R = &-\frac{\{\mathbf{M}_n\}^T}{2F}\int_{x_1}^{x_2}[\mathbf{N}_m]^T[\mathbf{N}_m]dx\{\mathbf{M}_n\} \\ &+\{\mathbf{M}_n\}^T\int_{x_1}^{x_2}[\mathbf{N}_m]^T[\mathbf{N}'']dx\{\mathbf{q}\} \\ &-\{\mathbf{Q}\}^T\{\bar{\mathbf{q}}\}. \end{aligned} \tag{7.48}$$

For a stationary value, we take derivations of Π_R with respect to $(M_1\ M_2)$ and $(w_1, \theta_1, w_2, \theta_2)$. Thus

$$\frac{\partial \Pi_R}{\partial\{\mathbf{M}_n\}} = -\frac{1}{2F}\int_{x_1}^{x_2}[\mathbf{N}_m]^T[\mathbf{N}_m]dx\{\mathbf{M}_n\} + \int_{x_1}^{x_2}[\mathbf{N}_m]^T[\mathbf{N}'']dx\{\mathbf{q}\} = 0, \tag{7.49a}$$

$$\frac{\partial \Pi_R}{\partial\{\mathbf{q}\}} = 0 + \int_{x_1}^{x_2}[\mathbf{N}_m][\mathbf{N}'']^T dx\{\mathbf{M}_n\} - \{\mathbf{Q}\} = 0, \tag{7.49b}$$

or in matrix notation,

$$\begin{bmatrix} [\mathbf{k}_{\tau\tau}] & [\mathbf{k}_{\tau u}] \\ [\mathbf{k}_{\tau u}]^T & [0] \end{bmatrix} \begin{Bmatrix} M_1 \\ M_2 \\ w_1 \\ \theta_1 \\ w_2 \\ \theta_2 \end{Bmatrix} = \begin{Bmatrix} 0 \\ 0 \\ Q_1 \\ M_1 \\ Q_2 \\ M_2 \end{Bmatrix}, \tag{7.49c}$$

where

$$\begin{aligned}
[\mathbf{k}_{\tau\tau}] &= -\frac{l}{2F}\int_{x_1}^{x_2}[\mathbf{N}_m]^T[\mathbf{N}_m]dx \\
&= -\frac{l}{2F}\int_0^1 \begin{bmatrix} (1-s)^2 & s(1-s) \\ s(1-s) & s^2 \end{bmatrix} ds \\
&= -\frac{l}{6F}\begin{bmatrix} 2 & 1 \\ 1 & 2 \end{bmatrix},
\end{aligned} \tag{7.49d}$$

$$\begin{aligned}
[\mathbf{k}_{\tau\tau}] &= \int_{x_1}^{x_2}[\mathbf{N}_m]^T[\mathbf{N}'']dx \\
&= \frac{l}{l^2}\int_0^1 \begin{Bmatrix} 1-s \\ s \end{Bmatrix}[-6+2s \quad -4l+6ls \quad 6-12s \quad 6ls-2l]ds \\
&= \begin{bmatrix} -1/l & -1 & 1/l & 0 \\ 1/l & 0 & -1/l & 1 \end{bmatrix}.
\end{aligned} \tag{7.49e}$$

Hence the element equations are

$$\begin{bmatrix} -\frac{l}{3F} & -\frac{l}{6F} & -\frac{1}{l} & -1 & \frac{1}{l} & 0 \\ -\frac{l}{6F} & -\frac{l}{3F} & \frac{1}{l} & 0 & -\frac{1}{l} & 1 \\ -\frac{1}{l} & \frac{1}{l} & 0 & 0 & 0 & 0 \\ -1 & 0 & 0 & 0 & 0 & 0 \\ \frac{1}{l} & -\frac{1}{l} & 0 & 0 & 0 & 0 \\ 0 & \frac{1}{l} & 0 & 0 & 0 & 0 \end{bmatrix} \begin{Bmatrix} M_1 \\ M_2 \\ w_1 \\ \theta_1 \\ w_2 \\ \theta_2 \end{Bmatrix} = \begin{Bmatrix} 0 \\ 0 \\ Q_1 \\ M_1 \\ Q_2 \\ M_2 \end{Bmatrix} \tag{7.50}$$

As an illustration, we consider the beam in Figure 7.4, divided in two elements. The assemblage relation for the two elements is obtained by observing compatibility of nodal moments, displacements, and rotations; this gives

$$\begin{bmatrix} -\frac{l}{3F} & -\frac{1}{l} & -1 & -\frac{l}{6F} & \frac{1}{l} & 0 & 0 & 0 & 0 \\ -\frac{1}{l} & 0 & 0 & \frac{1}{l} & 0 & 0 & 0 & 0 & 0 \\ -1 & 0 & 0 & 0 & 0 & 0 & 0 & 0 & 0 \\ -\frac{l}{6F} & \frac{1}{l} & 0 & -\frac{2l}{3F} & -\frac{2}{l} & 0 & -\frac{l}{6F} & \frac{1}{l} & 0 \\ \frac{1}{l} & 0 & 0 & 0 & 0 & 0 & \frac{1}{l} & 0 & 0 \\ 0 & 0 & 0 & 0 & 0 & 0 & 0 & 0 & 0 \\ 0 & 0 & 0 & -\frac{l}{6F} & \frac{1}{l} & 0 & -\frac{l}{3F} & -\frac{1}{l} & 1 \\ 0 & 0 & 0 & \frac{1}{l} & 0 & 0 & -\frac{1}{l} & 0 & 0 \\ 0 & 0 & 0 & 0 & 0 & 0 & 1 & 0 & 0 \end{bmatrix} \begin{Bmatrix} M_1 \\ w_1 \\ \theta_1 \\ M_2 \\ w_2 \\ \theta_2 \\ M_3 \\ w_3 \\ \theta_3 \end{Bmatrix} = \begin{Bmatrix} 0 \\ Q_1^{(1)} \\ M_1^{(1)} \\ 0 \\ Q_2^{(1)} + Q_1^{(2)} \\ M_2^{(1)} + M_1^{(2)} \\ 0 \\ Q_2^{(2)} \\ M_2^{(2)} \end{Bmatrix} \tag{7.51a}$$

or

$$[\mathbf{K}]\{\mathbf{r}\} = \{\mathbf{R}\}. \tag{7.51b}$$

Here the superscripts (1) and (2) denote element numbers.

By substitution of the numerical values as in the example of Figure 7.4, we have, for uniform loading,

$$\begin{bmatrix} -\frac{5}{10^6} & -\frac{1}{10} & -1 & -\frac{2.5}{10^6} & \frac{1}{10} & 0 & 0 & 0 & 0 \\ -\frac{1}{10} & 0 & 0 & \frac{1}{10} & 0 & 0 & 0 & 0 & 0 \\ -1 & 0 & 0 & 0 & 0 & 0 & 0 & 0 & 0 \\ -\frac{2.5}{10^6} & \frac{1}{10} & 0 & -\frac{10}{10^6} & -\frac{2}{10} & 0 & -\frac{2.5}{10^6} & \frac{1}{10} & 0 \\ \frac{1}{10} & 0 & 0 & -\frac{2}{10} & 0 & 0 & \frac{1}{10} & 0 & 0 \\ 0 & 0 & 0 & 0 & 0 & 0 & 0 & 0 & 0 \\ 0 & 0 & 0 & -\frac{2.5}{10^6} & \frac{1}{10} & 0 & -\frac{5}{10^6} & -\frac{1}{10} & 1 \\ 0 & 0 & 0 & \frac{1}{10} & 0 & 0 & -\frac{1}{10} & 0 & 0 \\ 0 & 0 & 0 & 0 & 0 & 0 & 1 & 0 & 0 \end{bmatrix} \begin{Bmatrix} M_1 \\ w_1 \\ \theta_1 \\ M_2 \\ w_2 \\ \theta_2 \\ M_3 \\ w_3 \\ \theta_3 \end{Bmatrix} = \begin{Bmatrix} 0 \\ \frac{1}{12} \\ \frac{10}{12} \\ 0 \\ 1 \\ 0 \\ 0 \\ \frac{1}{2} \\ -\frac{10}{12} \end{Bmatrix} \quad (7.51c)$$

We can see that at the midsection the slope $\theta_2 = 0$. Introduction of the boundary conditions $w_1 = w_3 = 0$ into Equation 7.51b leads to

$$\begin{bmatrix} -\frac{5}{10^6} & -\frac{1}{10} & -1 & -\frac{2.5}{10^6} & \frac{1}{10} & 0 & 0 & 0 & 0 \\ 0 & 1 & 0 & 0 & 0 & 0 & 0 & 0 & 0 \\ -1 & 0 & 0 & 0 & 0 & 0 & 0 & 0 & 0 \\ -\frac{2.5}{10^6} & \frac{1}{10} & 0 & -\frac{10}{10^6} & -\frac{2}{10} & 0 & -\frac{2.5}{10^6} & \frac{1}{10} & 0 \\ \frac{1}{10} & 0 & 0 & -\frac{2}{10} & 0 & 0 & \frac{1}{10} & 0 & 0 \\ 0 & 0 & 0 & 0 & 0 & 0 & 0 & 0 & 0 \\ 0 & 0 & 0 & -\frac{2.5}{10^6} & \frac{1}{10} & 0 & -\frac{5}{10^6} & -\frac{1}{10} & 1 \\ 0 & 0 & 0 & 0 & 0 & 0 & 0 & 1 & 0 \\ 0 & 0 & 0 & 0 & 0 & 0 & 1 & 0 & 0 \end{bmatrix} \begin{Bmatrix} M_1 \\ w_1 \\ \theta_1 \\ M_2 \\ w_2 \\ \theta_2 \\ M_3 \\ w_3 \\ \theta_3 \end{Bmatrix} = 1000 \begin{Bmatrix} 0 \\ 0 \\ \frac{10}{12} \\ 0 \\ 1 \\ 0 \\ 0 \\ 0 \\ -\frac{10}{12} \end{Bmatrix} \quad (7.51d)$$

Note that the assemblage matrix in Equation 7.51 has zeros on the diagonal. If such a system is solved by using Gaussian elimination, as discussed in Chapter 3, we shall need to resort to partial or complete pivoting.

Solution of Equation 7.51d leads to

$$\begin{array}{lll} M_1 = -833.3 \text{ kg-cm}, & \mathrm{M}_2 = -5833.33 \text{ mg-cm}, & \mathrm{M}_3 = -833.33 \text{ kg-cm}, \\ \mathrm{w}_1 = 0.00 \text{ cm (given)}, & w_2 = 0.312495 \text{ cm}, & \mathrm{w}_3 = 0.000 \text{ (given)}, \\ \theta_1 = 0.04999 \text{ rad}, & \theta_2 = 0.0000 \text{ rad}, & \theta_3 = -0.05100 \text{ rad}. \end{array}$$

These results are the same as in the case of the displacement method. The difference is that here the moments are primary unknowns, whereas in the displacement approach they were derived from computed displacements and slopes. The results are the same because we assume linear variation for moments. This is a rather elementary problem included simply to illustrate the mixed procedure.

Problems

7.1. (a) Derive stiffness matrix [**k**] for the beam bending problem if $EI = F$ varies linearly within the element as

$$F = N_1 F_1 + N_2 F_2,$$

where $N_1 = 1 - s$ and $N_2 = s$, $s = (x - x_1)/l$. (b) Derive [**k**] if the area of the beam varies linearly within the element as

$$A = A_1 N_1 + A_2 N_2.$$

7.2. Derive the load vector due to uniform body force $\bar{Z}$ acting in the z direction, with the approximation function for w as in Equation 7.2.

Solution:

$$\{\mathbf{Q}\}^T = Al\bar{Z}\begin{bmatrix} \frac{1}{2} & \frac{l}{12} & \frac{1}{2} & \frac{-l}{12} \end{bmatrix}.$$

7.3. Derive stiffness matrix [**k**] for the beam bending problem (with uniform F) if the beam is supported on an elastic foundation which can be represented by a series of linear elastic springs with uniform spring constant k_f (F/L) (Figure 7.10).

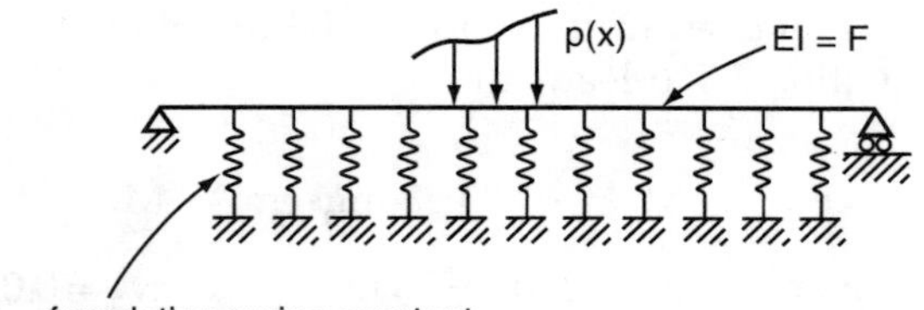

FIGURE 7.10
Beam on elastic foundation.

Partial solution: The additional term in Π_p (Equation 7.14a) due to the foundation spring support is given by

$$\Pi_{pf} = l\int_0^1 \frac{1}{2}\left(k_f w\right) w ds$$
$$= \frac{1}{2}\int_0^1 \{\mathbf{q}\}^T [\mathbf{N}]^T k_f [\mathbf{N}]\{\mathbf{q}\} ds,$$

which will lead to the following contribution by the spring support to the stiffness matrix:

$$[\mathbf{k}]_f = lk_f \int_0^1 \begin{Bmatrix} 1-3s^2+2s^3 \\ ls\left(1-2s+s^2\right) \\ s^2(3-2s) \\ ls^2(s-1) \end{Bmatrix} \times$$
$$\left[1-3s^2+2s^3 \quad ls\left(1-2s+s^2\right) \quad s^2(3-2s) \quad ls^2(s-1)\right] ds$$

$$= k_f \begin{bmatrix} \frac{13}{35} & \frac{11l}{210} & \frac{9}{70} & -\frac{13l}{420} \\ & \frac{l^2}{105} & \frac{13l}{420} & -\frac{l^2}{140} \\ & \text{sym.} & \frac{13}{35} & -\frac{11l}{210} \\ & & & \frac{l^2}{105} \end{bmatrix}$$

7.4. Multiply $[\Phi][\mathbf{A}]^{-1}$ in Equation 7.9 and show that the result yields N_i, $i = 1, 2, 3, 4$, in Equation 7.3.

7.5. Perform all steps of multiplications and integrations for going from Equation 7.16a to Equation 7.16b and from Equation 7.16c to 7.16e.

7.6. Derive $\{\mathbf{Q}\}$ if the load varies as

$$p(s) = N_1 p_1 + N_2 p_2 + N_0 p_0,$$

where

$$N_1 = 2s^2 - 3s + 1,$$
$$N_2 = s(2s-1),$$
$$N_0 = 4s(1-s).$$

Subscript zero denotes the node at midpoint of the element.

7.7. Solve the example problem in Figure 7.4 with the following conditions: (1) One end fixed against movement and rotation and the other end free; (2) One end settles by 0.05 cm.

7.8. By using two elements (Figure 7.4) and with constant axial load $\bar{P} =$ 50 kg, perform the entire process of formulation, assembly, introduction of boundary conditions, and solution of equations. Assume boundary conditions $w_1 = w_3 = 0$.

7.9. Compute the stiffness matrix and load vectors due to body force $\bar{P}$ and surface traction p with the fifth-order approximations (Equation 7.34).

Solution:

$$[\mathbf{k}] = Fl\int_0^1 [\mathbf{B}]^T[\mathbf{B}]ds$$

where

$$[B]^T = \frac{1}{l^2}\begin{Bmatrix} -60S + 180s^2 - 120s^3 \\ l(-36s + 96s^2 - 60s^3) \\ l^2(1 - 9s + 18s^2 - 10s^3) \\ 60s - 180s^2 + 120s^3 \\ l(-24s + 84s^2 - 60s^3) \\ l^2(3s - 12s^2 + 10s^3) \end{Bmatrix}.$$

Partial results:

$$k_{11} = 17.41\ F/l^3,\ k_{12} = 175.24\ F/l^2$$

$$k_{54} = 8.57\ F/l^2,\ K_{33} = 0.088\ F/l.$$

For constant body force $\bar{Z}$ and surface load p, the load vector is

$$\{\mathbf{Q}\} = A\bar{Z}\begin{Bmatrix} l/2 \\ l^2/10 \\ l^3/120 \\ l/2 \\ -l^2/10 \\ l^3/120 \end{Bmatrix} + p\begin{Bmatrix} l/2 \\ l^2/10 \\ l^3/120 \\ l/2 \\ -l^2/10 \\ l^3/120 \end{Bmatrix}.$$

7.10. Compute the load vector for the beam bending element with the interpolation model (Equation 7.34) and surface loading varying linearly as

$$p(s) = (1 - s)p_1 + sp_2.$$

Solution:

$$\{\mathbf{Q}\} = \begin{Bmatrix} l\left(\dfrac{p_1}{2} + \dfrac{p_2 - p_1}{2}\right) \\ l^2\left[\dfrac{p_1}{10} + \dfrac{4}{105}(p_2 - p_1)\right] \\ l^3\left(\dfrac{p_1}{120} + \dfrac{p_2 - p_1}{280}\right) \\ l\left[\dfrac{p_1}{2} + \dfrac{5}{14}(p_2 - p_1)\right] \\ l^2\left[-\dfrac{p_1}{10} - \dfrac{13}{210}(p_2 - p_1)\right] \\ l^3\left(\dfrac{p_1}{120} + \dfrac{p_2 - p_1}{210}\right) \end{Bmatrix}.$$

7.11. Consider two elements (Figure 7.4) with different values of E:

Element 1: $E = 10^6 \text{ kg/cm}^2$,

Element 2: $E = 2 \times 10^6 \text{ kg/cm}^2$.

Assemble and solve for displacements, slopes, moments, and shear forces for the other properties as for the example of Figure 7.4. Comment on the distribution of these quantities around the junction of the two elements.

7.12. Assume the area of the beam element to vary linearly as, see Problem 7.l(b),

$$A = N_1 A_1 + N_2 A_2$$

and derive element equations for uniform loading. Assuming the area to vary linearly as shown in Figure 7.11, compute displacements, slopes, movement, and shear forces for the loading and properties of the example in Figure 7.4.

Note: Since area varies, corresponding variation in moment of inertia should also be considered.

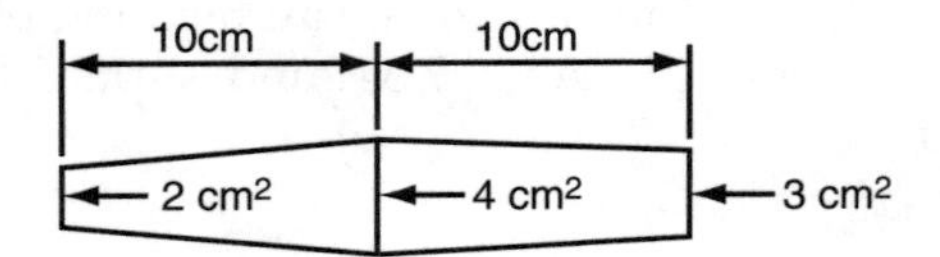

FIGURE 7.11

7.13. If the stiffness matrix for the uncoupled beam-column element of Figure 7.12(a) is given by

$$[k]=\begin{bmatrix} a_1 & a_2 & a_3 & a_4 & 0 & 0 \\ & a_5 & a_6 & a_7 & 0 & 0 \\ & & a_8 & a_9 & 0 & 0 \\ & & & a_{10} & 0 & 0 \\ & & & & b & -b \\ & & & & & b \end{bmatrix}$$

(since the [k] matrix is symmetric only the upper triangular part is shown), what should be the stiffness matrices for elements in Figures 12(b) and (c)? Note that the three elements are identical but the directions and numbering of the degrees of freedom are different in three figures.

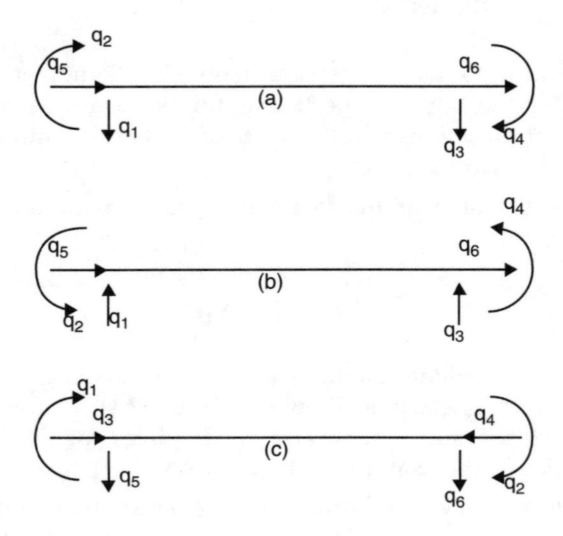

FIGURE 7.12

7.14. In a beam bending problem a three-node beam element is considered with minimum degrees of freedom at each node necessary to satisfy the interelement compatibility of the displacement and slope.

(a) For this element how many interpolation functions do you expect in the expression

$$w = \sum_{i=1}^{M} N_i q_i$$

where w is the transverse displacement, N_i is the interpolation function, and q_i are the nodal degrees of freedom?

(b) What order of polynomial is N_i?

(c) Sketch all your interpolation functions showing their important characteristics.

(d) How many degrees of freedom should each node have?

(e) Answer parts (a) through (d) for a three-node element used for solving the uncoupled beam-column problem.

7.15. Can we solve a beam problem with the three-node beam elements, with only one nodal degree of freedom (transverse displacement) at each node? What difficulty, if any, should we encounter if we try to do it?

7.16. For a three-node beam element the six degrees of freedom are as shown in Figure 7.13. Plot N_4 in the following 4 coordinates, shown in Figures 7.14(a), (b), (c), and (d). In all four cases give values of the inclination angles of the tangents to N_4 at three node positions.

7.17. The structure shown in Figure 7.15 is to be analyzed by discretizing it into two beam-column elements. The global numbering of nodal degrees of freedom is shown in the figure. For each element numerical values of E (Young's modulus), A (cross-sectional area), I (modulus of rigidity), and l (length) are unity.

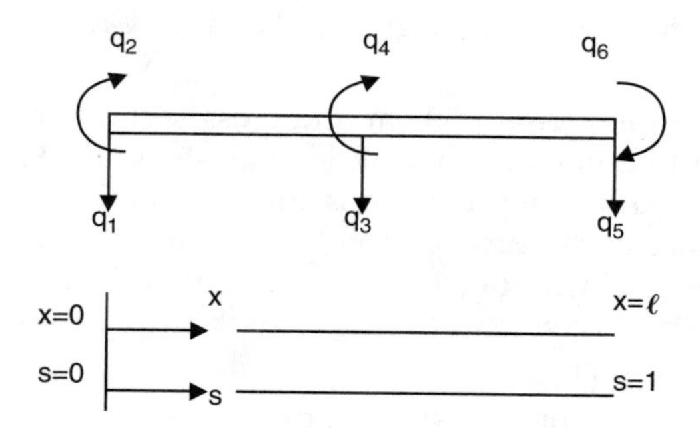

FIGURE 7.13

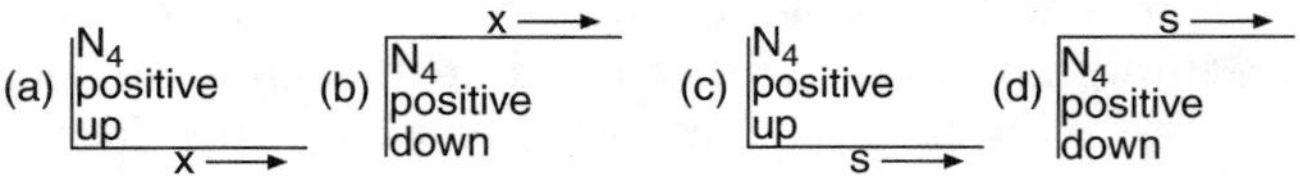

(a)
N4
positive
up
x
(b)
x
N4
positive
down
(c)
N4
positive
up
s
(d)
s
N4
positive
down

FIGURE 7.14

q8
P
3
q9
q7
Element 2
q2
q5
Element 1
q3
q6
q1
1
2
q4

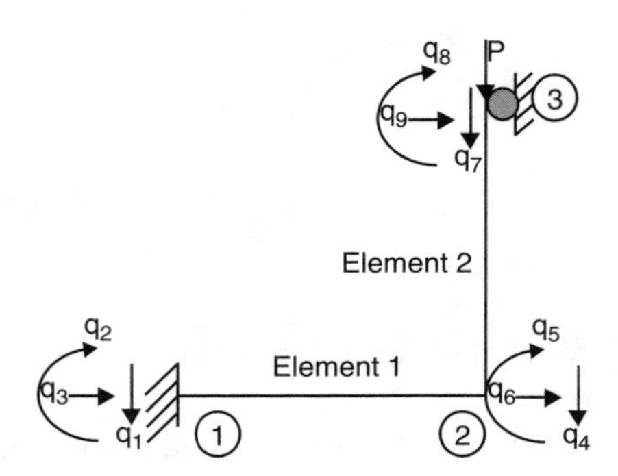

FIGURE 7.15

(a) Draw two elements separately and number the nodal degrees of freedom at the element level for each element.

(b) Construct [k] matrix for each element consistent with the nodal degrees of freedom numbering in (a).

(c) Assemble two element stiffness matrices to construct the global stiffness matrix [K] consistent with the global numbering shown in the figure.

(d) Apply boundary conditions and obtain the global equations in the following form:

$$\left[\bar{K}\right]\{\bar{r}\} = \left\{\bar{R}\right\}$$

where $[\bar{K}]$ is the nonsingular global stiffness matrix and $\{\bar{r}\}$ is the vector of nonzero degrees of freedom. Give all entries of $[\bar{K}]$, $\{\bar{r}\}$, and $\{\bar{R}\}$.

7.18. A beam of total length 30 cm and cross section area 2 × 1 cm^2 is subjected to a linearly varying load as shown in Figure 7.16. The Young's modulus of the beam is 10^6 kg/cm^2. Calculate θ_B, θ_C, and θ_D (rotations at the support points B, C, and D, also mention if the rotation is clockwise or counterclockwise) by discretizing the beam into three beam elements of equal length and carrying out the finite element analysis.

7.19. A 30-inch long continuously supported beam, shown in Figure 7.17, is subjected to a parabolic load distribution with a peak load of

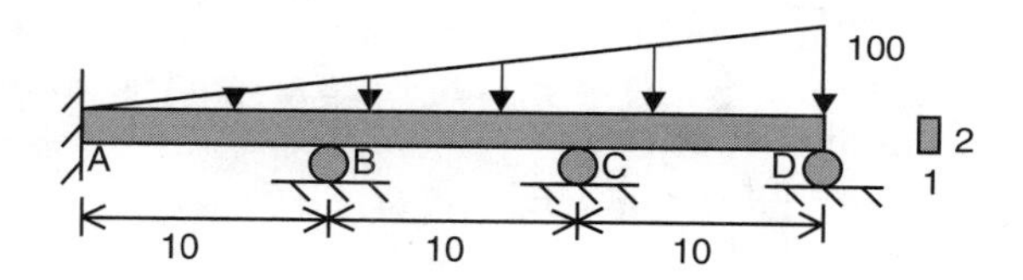

FIGURE 7.16

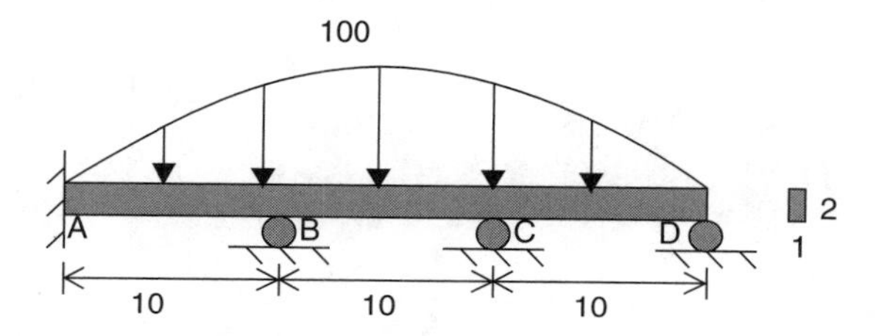

FIGURE 7.17

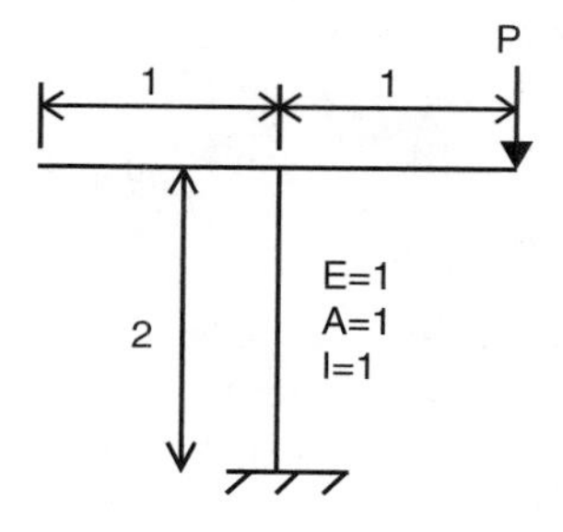

FIGURE 7.18

100 kg/cm acting at the midpoint. The Young's modulus of the beam is 10^6 kg/cm^2. Calculate θ_B, θ_C, and θ_D (rotations at the support points B, C, and D; also mention if the rotation is clockwise or counterclockwise) by discretizing the beam into three beam elements of equal length and carrying out the finite element analysis.

7.20. Two concrete beams are joined to form a T-shaped structure as shown in Figure 7.18. The joint is firm and no relative rotation between the beams is permitted.

(a) To obtain the displacement under the load P how would you discretize the structure so that you need the minimum number of two-node beam-column elements?

(b) In each element show all degrees of freedom and number them such that you can use the stiffness matrix given in Equation 7.40.

(c) Assign global numbers to all degrees of freedom; then assemble the global stiffness matrix.

(d) Introduce the boundary conditions and obtain the global equation in the form $[\bar{K}]\{\bar{r}\} = \{\bar{R}\}$ after the geometric boundary conditions are incorporated.

7.21. Let us consider the functional,

$$\Omega = \int_0^1 \left\{ \phi \left[\frac{d\phi}{ds} + \frac{d^2\phi}{ds^2} + \frac{d^3\phi}{ds^3} \right] - \left[\frac{d^2\phi}{ds^2} + \frac{d^3\phi}{ds^3} + \frac{d^4\phi}{ds^4} \right] \right\} ds,$$

where $\phi = \left[(1 - 3s + 2s^2) \ (2s^2 - s) \ (4s - 4s^2) \right] \begin{Bmatrix} \phi_1 \\ \phi_2 \\ \phi_3 \end{Bmatrix} = [N]\{q\}.$

(a) What will stiffness matrix [k] and load vector {Q} be, if you derive the finite element equation [k]{q} = {Q} from the above functional? Give the expressions in terms of integrals, without carrying out any matrix multiplication or integration.

(b) If $\phi = [(1-s) \ s] \begin{Bmatrix} \phi_1 \\ \phi_2 \end{Bmatrix} [N]\{q\}$, derive [k] matrix and {Q} vector from the above functional.

7.22. For the frame structure shown below the two columns, AB and CD have very high axial stiffnesses; hence, their axial displacement may be assumed to be zero. The beam resting on these two columns has low axial stiffness; hence it can be treated as a beam- column. The element stiffness matrices for the beam element and the beam-column element are given below. Directions of the nodal degrees of freedom of the beam element and the beam-column element are shown in Figure 7.19.

$$[k]_{beam} = \begin{bmatrix} A & B & C & D \\ & E & F & G \\ & & H & I \\ & & & J \end{bmatrix}, \quad [k]_{beam-column} = \begin{bmatrix} a & b & c & d & 0 & 0 \\ & e & f & g & 0 & 0 \\ & & h & i & 0 & 0 \\ & & & j & 0 & 0 \\ & & & & k & -k \\ & & & & & k \end{bmatrix}$$

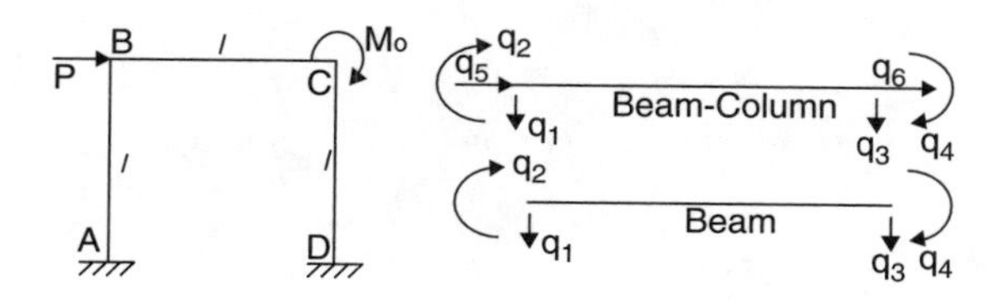

FIGURE 7.19

(a) Show all nonzero nodal degrees of freedom on the structure.
(b) Write the global equation in the form [K]{r} = {R}, after incorporating the boundary conditions, such that {r} contains only the nonzero nodal degrees of freedom.
(c) How (constant, linear, quadratic, cubic, etc.) should axial stress, curvature, bending moment, and shear force vary between AB and BC?

References

1. Timoshenko, S. 1956. *Strength of Materials,* Van Nostrand Reinhold, New York.
2. Strang, G. and Fix, G. J. 1973. *An Analysis of the Finite Element Method,* Prentice-Hall, Englewood Cliffs, NJ.
3. Prenter, P. M. 1975. *Splines and Variational Methods,* John Wiley & Sons, New York.
4. Desai, C. S. and Abel, J. F. 1972. *Introduction to the Finite Element Method,* Van Nostrand Reinhold, New York.
5. Crandall, S. H. 1956. *Engineering Analysis,* McGraw-Hill, New York.

8

One-Dimensional Mass Transport

Introduction

A number of problems in various disciplines of engineering involve the phenomenon of mass transport. This can include transport through diffusion and convection of chemicals, pollutants, contaminants, and dissolved salts in water. In this chapter, we shall treat mass transport for simple one-dimensional idealizations.

The differential equation governing one-dimensional mass transport can be stated as[1]

$$\frac{\partial}{\partial x}\left(D_x \frac{\partial c^*}{\partial x}\right) - \frac{\partial}{\partial x}(v_x c^*) - W = \frac{\partial c^*}{\partial t}, \tag{8.1}$$

where D_x is the dispersion coefficient, c^* is the unknown or state variable (concentration of pollutant or dissolved salt), v_x is the velocity or convection parameter, W is the applied source or sink, and x and t are the space and time coordinates. The first term is essentially the same as the first term in the flow equation (Equation 5.9) and represents the phenomenon of diffusion. The second term denotes the process of transport by convection.

Finite Element Formulation

In this and the following chapters, we shall often detail and label only the main steps, while some of the common steps will be discussed without explicit labels.

The domain of mass transport (Figure 8.1) is idealized as one-dimensional. Use of the line element and the linear approximation gives the approximation for c as

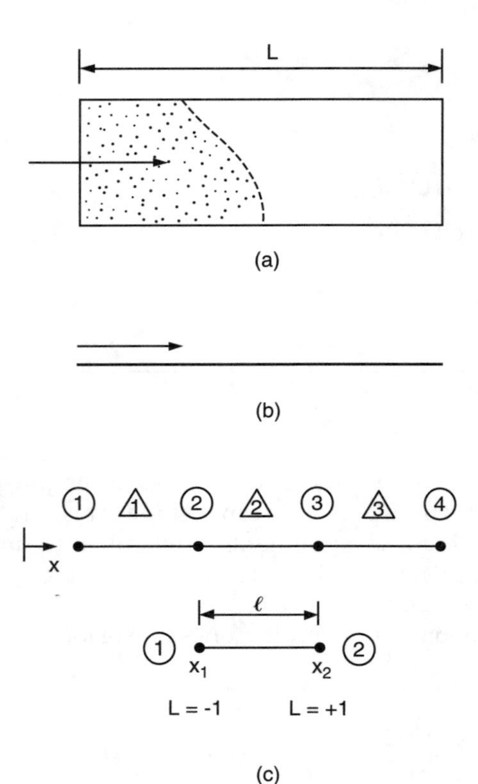

FIGURE 8.1
One-dimensional mass transport; (a) Mass transport by diffusion and convection, (b) One-dimensional idealization, and (c) Discretization.

$$c = \tfrac{1}{2}(1 - L)c_1 + \tfrac{1}{2}(1 + L)c_2 \tag{8.2a}$$

or

$$c = [\mathbf{N}]\{\mathbf{q}\} = \sum N_i c_i, \quad i = 1, 2, \tag{8.2b}$$

where$\{\mathbf{q}\}^T = [c_1\ c_2]$, and c is the approximation to the concentration c^* and L denotes that local coordinate; in Figure 8.1, it also denotes length of the medium.

Step 4. Derivation of Element Equations

For derivation by Galerkin's method, the residual R is

$$R(c) = \frac{\partial}{\partial x}\left(D_x \frac{\partial c}{\partial x}\right) - \frac{\partial}{\partial x}(v_x c) - W - \frac{\partial c}{\partial t}. \tag{8.3}$$

Weighing R with respect to N_1 yields

$$\int_{x_1}^{x_2}\left\{\frac{\partial}{\partial x}\left[D_x\frac{\partial}{\partial x}\left(\sum N_j c_j\right)\right]-\frac{\partial}{\partial x}\left[v_x\left(\sum N_j c_j\right)\right]-W-\left[\frac{\partial}{\partial t}\left(\sum N_j c_j\right)\right]\right\}N_i dx=0. \quad (8.4)$$

As explained in Chapter 3, we shall treat the elements one by one. However, it is understood that the procedure is applied for the entire medium. Equation 8.4 is expanded as

$$\int_{x_1}^{x_2}\left\{\frac{\partial}{\partial x}\left[D_x\frac{\partial}{\partial x}\left(\sum N_j c_j\right)\right]-\frac{\partial v_x}{\partial x}\left(\sum N_j c_j\right)-v_x\frac{\partial\left(\sum N_j c_j\right)}{\partial x}\right.$$

$$\left.-W-\frac{\partial c_j}{\partial t}\left(\sum N_j\right)\right\}N_i dx=0. \quad (8.5)$$

Integrating by parts the first term in Equation 8.5 and assuming D_x to be constant, we have

$$\int_{x_1}^{x_2}\left\{\frac{\partial}{\partial x}\left[D_x\frac{\partial}{\partial x}\left(\sum N_j c_j\right)\right]\right\}N_i dx=D_x N_i\frac{\partial c}{\partial n}\bigg|_{x_1 i}^{x_2}-\int_{x_1}^{x_2}D_x\frac{\partial N_j}{\partial x}\frac{\partial N_i}{\partial x}c_j dx. \quad (8.6)$$

The first term on the right-hand side in Equation 8.6 denotes Neumann-type boundary conditions and can be specified as a known value of the normal derivative $\partial c/\partial n$. The term $D_x(\partial c/\partial n)\,|_{x_1}^{x_2}$ is the (known) flux on the end boundaries.

Substitution of Equation 8.6 into Equation 8.5 leads to element equations in matrix form as

$$[\mathbf{E}]\{\mathbf{q}\}+[\mathbf{E}_t]\{\dot{\mathbf{q}}\}=\{\mathbf{Q}\}, \quad (8.7)$$

where $\{\dot{\mathbf{q}}\}^T=[\partial c_1/\partial t\ \partial c_2/\partial t]$ $[\mathbf{E}]$ and $[\mathbf{E}_t]$ are the element property matrices, and $\{\mathbf{Q}\}$ is the element forcing parameter vector. These matrices are defined as follows:

$$[\mathbf{E}]=\int_{x_1}^{x_2}\left([\mathbf{B}]^T D_x[\mathbf{B}]dx+[\mathbf{N}]^T v_x[\mathbf{B}]dx\right) \quad (8.8a)$$

$$[\mathbf{E}_t]=\int_{x_1}^{x_2}[\mathbf{N}]^T[\mathbf{N}]dx, \quad (8.8b)$$

and

$$\{\mathbf{Q}\} = \int_{x_1}^{x_2} [\mathbf{N}]^T W dx - D_x N_i \frac{\partial c}{\partial n}\bigg|_{x_1}^{x_2}. \tag{8.8c}$$

Here **[B]** is the usual transformation matrix (Equations 3.13 and 4.4) obtained by taking the proper derivative of c (Equation 8.2). In writing Equation 8.8a we assumed, only for simplicity, that v_x is constant with respect to x; hence the second term in Equation 8.5 is not included. We note that it is not necessary to make this assumption because that term can be easily included.

The evaluation of **[E]**, **[E$_t$]**, and **{Q}** is as follows:

$$\begin{aligned}
[\mathbf{E}] &= \frac{D_x l}{2}\int_{-1}^{1} \frac{1}{l^2}\begin{Bmatrix} -1 \\ 1 \end{Bmatrix}[-1 \quad 1]\, dL \\
&+ \frac{v_x l}{2}\int \frac{1}{2}\begin{Bmatrix} 1-L \\ 1+L \end{Bmatrix}\frac{1}{l}[-1 \quad 1]\, dL \\
&= \frac{D_x}{l}\begin{bmatrix} 1 & -1 \\ -1 & 1 \end{bmatrix} + \frac{v_x}{2}\begin{bmatrix} -1 & +1 \\ 1 & 1 \end{bmatrix} \\
&= \begin{bmatrix} D_x/l - v_x/2 & -D_x/l + v_x/2 \\ -D_x/l - v_x/2 & D_x/l + v_x/2 \end{bmatrix}.
\end{aligned} \tag{8.9}$$

We note here the important characteristic that the element matrix **[E]** is nonsymmetric; this property is contributed by the convection part of the mass transport. Now

$$\begin{aligned}
[\mathbf{E}_t] &= \frac{l}{2}\int_{-1}^{1} \frac{1}{2}\begin{Bmatrix} 1-L \\ 1+L \end{Bmatrix}\frac{1}{2}[1-L \quad 1+L]\, dL \\
&= \frac{l}{6}\begin{bmatrix} 2 & 1 \\ 1 & 2 \end{bmatrix}.
\end{aligned} \tag{8.10}$$

This matrix is similar to the matrix arising from the time-dependent term (Equation 5.21).

$$\{\mathbf{Q}\} = \frac{Wl}{2}\int_{-1}^{1} \frac{1}{2}\begin{Bmatrix} 1-L \\ 1+L \end{Bmatrix} dL - \left(\frac{D_x}{2l}\begin{Bmatrix} 1-L \\ 1+L \end{Bmatrix}[-1 \quad 1]\begin{Bmatrix} c_1 \\ c_2 \end{Bmatrix}\right)\Bigg|_{x_1}^{x_2} \tag{8.11}$$

$$= \left. \frac{Wl}{2}\begin{Bmatrix}1\\1\end{Bmatrix} - \left(\frac{D_x}{2l}\begin{Bmatrix}1-L\\1+L\end{Bmatrix}[-1 \quad 1]\begin{Bmatrix}c_1\\c_2\end{Bmatrix}\right)\right|_{x_1}^{x_2}.$$

The first term in Equation 8.11 indicates that applied sink or source quantity is lumped equally at the two nodes. As discussed in previous chapters, only the terms at the end nodes remain when the second part of {**Q**} is considered.

Step 5. Assembly

Consider the three-element mesh (Figure 8.1). Combination of element matrices for the three elements leads to the assemblage equations as

$$\begin{bmatrix} D_x/l - v_x/2 & -D_x/l + v_x/2 & & \\ -D_x/l - v_x/2 & 2D_x/l + 0 & -D_x/l + v_x/2 & \\ & -D_x/l - v_x/2 & 2D_x/l + 0 & -D_x/l + v_x/2 \\ & & -D_x/l - v_x/2 & D_x/l + v_x/2 \end{bmatrix}\begin{Bmatrix}c_1\\c_2\\c_3\\c_4\end{Bmatrix}$$

$$+\frac{l}{6}\begin{bmatrix}2 & 1 & 0 & 0\\1 & 4 & 1 & 0\\0 & 1 & 4 & 1\\0 & 0 & 1 & 2\end{bmatrix}\begin{Bmatrix}\partial c_1/\partial t\\ \partial c_2/\partial t\\ \partial c_3/\partial t\\ \partial c_4/\partial t\end{Bmatrix} = \frac{Wl}{2}\begin{Bmatrix}1\\2\\2\\1\end{Bmatrix}. \tag{8.12a}$$

The contribution of the boundary terms is not included because we assume that

$$\frac{\partial c}{\partial x}(L, t) = 0 \tag{8.13a}$$

and

$$c(0, t) = \bar{c}. \tag{8.13b}$$

In matrix notation, Equation 8.12a becomes

$$[\mathbf{K}]\{\mathbf{r}\} + [\mathbf{K}_t]\{\dot{\mathbf{r}}\} = \{\mathbf{R}\}, \tag{8.12b}$$

where [**K**] and [$\mathbf{K}_t$] are the assemblage property matrices, {**R**} is the assemblage forcing parameter vector, and {**r**} is the vector of assemblage nodal unknowns.

Solution in Time

A number of integration schemes can be used to solve the matrix equations in the time domain. For instance, Equation 8.12b can be expressed in finite difference form as[2]

$$[\mathbf{K}]\left(\theta\{\mathbf{r}\}_{t+\Delta t}+(1-\theta)\{\mathbf{r}\}_t\right)$$
$$+[\mathbf{K}_t]\left(\theta\left\{\frac{\partial\{\mathbf{r}\}}{\partial t}\right\}_{t+\Delta t}+(1-\theta)\left\{\frac{\partial\{\mathbf{r}\}}{\partial t}\right\}_t\right)=\{\mathbf{R}\}_{t+\Delta t}, \tag{8.14}$$

where θ is a scalar. We obtain various implicit and explicit methods depending on the value of θ. If $\theta = 1$, the fully implicit method is obtained, and if $\theta = \frac{1}{2}$, we obtain the Crank–Nicholson scheme. These schemes possess different mathematical properties such as convergence and stability which can influence the quality of the numerical solution. Detailed discussion of these aspects is beyond the scope of this text.

If we use the following scheme to approximate the first derivatives in Equation 8.14,

$$\frac{1}{2}\left(\left\{\frac{\partial(\mathbf{r})}{\partial t}\right\}_t+\left\{\frac{\partial\{\mathbf{r}\}}{\partial t}\right\}_{t+\Delta t}\right)\simeq\frac{\{\mathbf{r}\}_{t+\Delta t}-\{\mathbf{r}\}_t}{\Delta t}, \tag{8.15}$$

then Equation 8.14 becomes (for $\theta = \frac{1}{2}$)

$$\left([\mathbf{K}]+\frac{2}{\Delta t}[\mathbf{K}_t]\right)\{\mathbf{r}\}_{t+\Delta t}\simeq 2\{\mathbf{R}\}_{t+\Delta t}-\left([\mathbf{K}]-\frac{2}{\Delta t}[\mathbf{K}_t]\right)\{\mathbf{r}\}_t. \tag{8.16}$$

Since the values of $\{\mathbf{R}\}_{t+\Delta t}$ and $\{\mathbf{r}\}$ are known, we can compute $\{\mathbf{r}\}_{t+\Delta t}$ at the next time step by solving Equation 8.16. The procedure starts from the first step when the values of $\{\mathbf{r}\}_0$ at t = 0 are prescribed as initial conditions.

The boundary conditions are prescribed as known values of $\{\mathbf{r}\}$ at given nodes. Equation 8.16 is modified for these conditions before the solutions are obtained at various time levels.

Convection Parameter v_x

For the solution of Equation 8.16, the value of the convection velocity v_x should be known. It can often be obtained by using available data or formulas.

One of the ways to compute it is to first solve the flow or seepage equations (Chapter 4), which allow computation of v_x.

Comment

In most engineering problems the assemblage matrices are symmetric and banded. In the case of mass transport, the matrix **[K]** in Equation 8.16 is banded but not symmetric. Consequently, as discussed in Appendix 2, we have to store in the computer core the coefficients on the entire band (2B-1). This aspect will involve relatively more computer time in solving the equations as compared to the time required for symmetric and banded systems. The nonsymmetric property of **[K]** can be considered to be a characteristic of the nonself-adjoint nature of the problem.

The behavior of the numerical solution can be influenced significantly due to the existence of the convection term. If the magnitude of the convection term is relatively large, the system (Equation 8.1) is predominantly convective. This can render the solutions more susceptible to numerical instability. On the other hand, if the diffusion term predominates, the numerical solution can be well behaved.

Example 8.1

To illustrate a few time steps in the solution of Equation 8.16, we adopt the following properties:

$$
\begin{aligned}
D_x &= 1 \quad \left(L^2/T\right),\\
v_x &= 1 \quad \left(L/T\right),\\
l &= 1 \quad (L),\\
W &= 0 \quad \left(M/TL^3\right),\\
\Delta t &= 1 \quad (T).
\end{aligned}
$$

Here L denotes length, M mass, and T time. These properties are chosen simply to illustrate the procedure, and may not necessarily represent a field situation.

Boundary conditions:

$$c(0,\, t) = 1.0, \quad \left(M/L^3\right) \text{ at node 1, } t > 0, \ \text{(Figure 8.1)}.$$

Initial conditions:

$$c(x,\, 0) = 0.0,\ t \le 0.$$

At time $t = 0 + \Delta t$, from Equation 8.16, we have

$$\left(\begin{bmatrix} \frac{1}{2} & -\frac{1}{2} & 0 & 0 \\ -\frac{3}{2} & 2 & -\frac{1}{2} & 0 \\ 0 & -\frac{3}{2} & 2 & -\frac{1}{2} \\ 0 & 0 & -\frac{3}{2} & \frac{3}{2} \end{bmatrix} + \frac{1}{3}\begin{bmatrix} 2 & 1 & 0 & 0 \\ 1 & 4 & 1 & 0 \\ 0 & 1 & 4 & 1 \\ 0 & 0 & 1 & 2 \end{bmatrix}\right)\begin{Bmatrix} c_1 \\ c_2 \\ c_3 \\ c_4 \end{Bmatrix}_{\Delta t}$$

$$= \left(\begin{bmatrix} \frac{1}{2} & -\frac{1}{2} & 0 & 0 \\ -\frac{3}{2} & 2 & -\frac{1}{2} & 0 \\ 0 & -\frac{3}{2} & 2 & -\frac{1}{2} \\ 0 & 0 & -\frac{1}{2} & \frac{3}{2} \end{bmatrix} - \frac{2 \times 1}{1 \times 6}\begin{bmatrix} 2 & 1 & 0 & 0 \\ 1 & 4 & 1 & 0 \\ 0 & 1 & 4 & 1 \\ 0 & 0 & 1 & 2 \end{bmatrix}\right)\begin{Bmatrix} c_1 \\ c_2 \\ c_3 \\ c_4 \end{Bmatrix}_{0}.$$

Because $c_1 = c_2 = c_3 = c_4 = 0$ at $t = 0$,

$$\begin{bmatrix} \frac{7}{6} & -\frac{7}{6} & 0 & 0 \\ -\frac{1}{6} & \frac{20}{6} & -\frac{7}{6} & 0 \\ 0 & -\frac{1}{6} & \frac{20}{6} & -\frac{7}{6} \\ 0 & 0 & -\frac{1}{6} & \frac{13}{6} \end{bmatrix}\begin{Bmatrix} c_1 \\ c_2 \\ c_3 \\ c_4 \end{Bmatrix}_{\Delta t} = \begin{Bmatrix} 0 \\ 0 \\ 0 \\ 0 \end{Bmatrix}.$$

Introduction of the boundary condition $c_1 = 1$ gives

$$\begin{bmatrix} 1 & 0 & 0 & 0 \\ -\frac{1}{6} & \frac{20}{6} & -\frac{1}{6} & 0 \\ 0 & -\frac{1}{6} & \frac{20}{6} & -\frac{1}{6} \\ 0 & 0 & -\frac{7}{6} & \frac{13}{6} \end{bmatrix}\begin{Bmatrix} c_1 \\ c_2 \\ c_3 \\ c_4 \end{Bmatrix}_{\Delta t} = \begin{Bmatrix} 1 \\ 0 \\ 0 \\ 0 \end{Bmatrix}.$$

Solution by Gaussian elimination leads to

$$c_1 = 1.00 \text{ (given)},$$

$$c_2 = 3564 \times 10^{-4},$$

$$c_3 = 1282 \times 10^{-4},$$

$$c_4 = 690 \times 10^{-4},$$

The next stage, $t = t + \Delta t = 2\Delta t$ can now be performed by using the values of c computed at the end of $t = \Delta t$ as initial conditions and so on for other time steps.

Example 8.2

A solution to Equation 8.1 was obtained by Guymon[3] using a variational procedure. We present here his results, which compare numerical predictions with a closed form solution for a problem idealized as one-dimensional. The results are presented in terms of the nondimensional quantities x/L, c/c_0, $v_x t/L$, and $v_x l/D_x$, where L is the total length of the one-dimensional medium.

The initial conditions are

$$c(x, 0) = 0,$$

and the boundary condition is

$$c(0, t) = c_0.$$

That is,

$$\frac{c}{c_0}(0, t) = 1.$$

Table 8.1 shows a comparison between the numerical solutions and the closed form solution. The former are obtained for two conditions: $v_x l/D_x = 0.5$ and $v_x l/D_x = 0.25$ at a time level $v_x t/L = 0.5$. The closed form solution is obtained from[3]

$$\frac{c}{c_0} = \frac{1}{2}\left\{\operatorname{erfc}\left[\frac{x - v_x t}{(4D_x t)^{1/2}}\right] + \exp\left(\frac{x v_x}{D_x}\right)\operatorname{erfc}\left[\frac{x + v_x t}{(4D_x t)^{1/2}}\right]\right\},$$

where erfc denotes error function.

Comment

The results indicate that as the magnitude of the convective term $v_x l/D_x$ increases, the numerical solutions are less accurate. This term has significant influence on numerical solutions of the one-, two-, and three-dimensional problems involving diffusion-convection phenomena.

TABLE 8.1

Comparison Between Finite Element Predictions and Closed Form Solution at $v_x t/L = 0.5$[3]

x/L	Closed Form Solution	Finite Element Solution $v_x l/D_x = 0.25$	Finite Element Solution $v_x l/D_x = 0.50$
0	1.000	1.000	1.000
0.1	0.976	0.975	0.972
0.2	0.928	0.927	0.923
0.3	0.851	0.850	0.845
0.4	0.745	0.743	0.738
0.5	0.616	0.614	0.606
0.6	0.478	0.475	0.463
0.7	0.345	0.341	0.324
0.8	0.230	0.228	0.209
0.9	0.142	0.145	0.132
1.0	0.080	0.080	0.080

References

1. Bear, J. 1972. *Dynamics of Fluids in Porous Media*, American Elsevier, New York.
2. Richtmeyer, R. D. and Morton, K. W. 1957. *Difference Methods for Initial-Value Problems*, Wiley-Interscience, New York.
3. Guymon, G. L. 1970. "A finite element solution of the one-dimensional diffusion-convection equations," *Water Resour. Res.* 6(1), 204-210.

Bibliography

Amend, J. H., Contractor, D. N., and Desai, C. S. 1976. "Oxygen depletion and sulfate production in strip mine spoil dams," in *Proc. 2nd Int. Conf. on Numerical Methods in Geomechanics*, Blacksburg, VA, C. S. Desai, Ed., 1976, ASCE, New York.

Cheng, R. T. 1974. "On the study of convective dispersion equations," in *Proc. Int. Symp. on Finite Element Methods in Flow Problems*, Swansea, U.K., 1974, University of Alabama Press, Huntsville.

Desai, C. S. and Contractor, D. N. 1976. "Finite element analysis of flow, diffusion and salt water intrusion in porous media," in *Proc. U.S.-Germany Symposium on Theory and Algorithms in Finite Element Analysis*, Bathe, K. J., Oden, J. T., and Wunderlich, W., Eds., MIT, Cambridge, MA.

Guymon, G. L., Scott, V. H., and Hermann, C. R. 1970. "A general numerical solution of two-dimensional diffusion-convection equation by the finite element method," *Water Resour. Res.* 6(6), 1611-1617.

Gray, W. G. and Pinder, G. F., Eds. 1977. *Proceedings, First International Conference on Finite Elements in Water Resources*, Pentech Press, London.

Segol, G., Pinder, G. F., and Gray, W. G. 1975. "A Galerkin finite element technique for calculating the transient position of the salt water front," *Water Resour. Res.* 11(2), 343-347.

Smith, I. M., Farraday, R. V., and O'connor, B. A. 1973. "Rayleigh-Ritz and Galerkin finite elements for diffusion-convection problems," *Water Resour. Res.* 9(3), 593-606.
Wu, T. H., Desai, C. S., and Contractor, D. N. 1976. "Finite element procedure for salt water intrusion in coastal aquifers," *Report No. VPI-E-76-23,* Department of Civil Engineering, Virginia Polytechnic Institute and State University, Blacksburg.

9

One-Dimensional Stress Wave Propagation

Introduction

In Chapters 5 and 8 time-dependent problems of heat and fluid flow and mass transport were considered. Although these problems were represented by different mathematical equations, the finite element solutions in time involved first-order derivatives with respect to time. Now we consider a different kind of time-dependent problem which involves second-order time derivatives in the finite element equations.

When a time-dependent force caused by factors such as an impact, blast, and earthquake loading impinges on a medium, it is transmitted through the medium as a (stress) wave. Generally such waves propagate in all the three spatial directions. Under certain circumstances and assumptions, it is possible to idealize the medium as one-dimensional.

Consider a homogeneous bar of uniform cross section (Figure 9.1). A time-dependent force $P_x(t)$ acting on the bar causes vibrations in the bar, and a (stress) wave propagates to and fro in the bar. The governing differential equation for the one-dimensional case, often known as the wave equation, is given by[1,2]

$$\frac{\partial \sigma_x}{\partial x} = \rho \frac{\partial^2 u}{\partial t^2} + P_x(t), \tag{9.1a}$$

where σ_x is the axial stress in the x direction, ρ is the mass density of the material of the bar, and u is the axial displacement. Equation 9.1a is a statement of the dynamic equilibrium at an instant of time and can be derived by using Newton's second law. The left-hand side denotes internal force; the term $\rho(\partial^2 u/\partial t^2) = p\ddot{u}$ denotes inertia force, where $\ddot{u}$ is the acceleration and $P_x(t)$ is the external force. If the material is assumed to be linearly elastic, the stress–strain law is

$$\sigma_x = E\varepsilon_x = E\frac{\partial u}{\partial x}. \tag{9.2a}$$

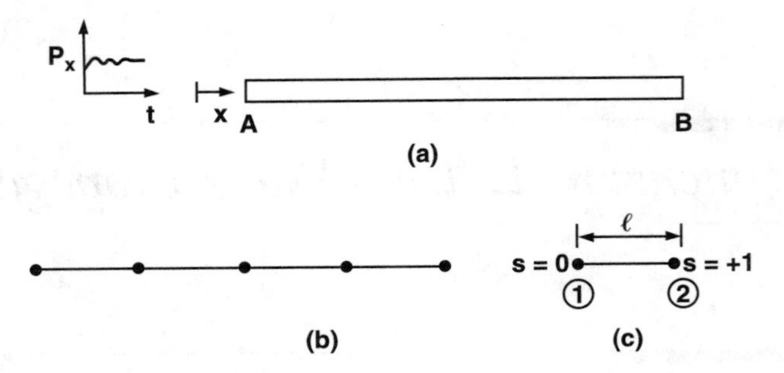

FIGURE 9.1
One-dimensional wave propagation; (a) Bar subjected to time-dependent load, (b) Idealization and discretization, and (c) Generic element.

Taking first derivatives, we obtain

$$\frac{\partial \sigma_x}{\partial x} = E\frac{\partial^2 u}{\partial x^2}, \tag{9.2b}$$

where E is the elastic modulus and ε_x is the gradient of u or axial strain. Substitution of Equation 9.2b into Equation 9.1a gives

$$E\frac{\partial^2 u}{\partial x^2} = \rho\frac{\partial^2 u}{\partial t^2} + P_x(t). \tag{9.1b}$$

The quantity $\sqrt{E/\rho} = c_x$ is called the velocity of elastic wave propagation.

A number of other physical phenomena such as propagation of waves of sound and vibrations of strings are also governed by Equation 9.1, which in mathematical terms is classified as a hyperbolic equation.[3,4]

Finite Element Formulation

For the body idealized as one-dimensional, we use line elements (Figure 9.1) and linear approximation as

$$u = N_1 u_1 + N_2 u_2 = [\mathbf{N}]\{\mathbf{q}\}, \tag{9.3}$$

where $N_1 = 1 - s$, $N_2 = s$, and $s = (x - x_1)/l$.

Step 4. Derive Element Equations

This step can be achieved by using the variational (Hamilton's) principle and the principle of virtual work;[4,5] these two are essentially statements of the same phenomenon. Here we shall illustrate the use of the virtual work principle,[4] which is somewhat easier to understand for the introductory treatment of the dynamics problem.

Equilibrium of the dynamical system at an instant of time requires satisfaction of the virtual work equation:[4,5]

$$\iiint_V \{\sigma_x\}^T \delta\{\varepsilon_x\} dV = \iiint_V \{\mathbf{F}_x\}^T \delta\{\mathbf{u}\} dV + \iiint_V \{\mathbf{P}_x\}^T \delta\{\mathbf{u}\} dV, \tag{9.4a}$$

where δ denotes a small virtual change or perturbation and F_x is the equivalent body force per unit volume caused by the inertial effect,

$$F_x = -\rho \frac{\partial^2 u}{\partial t^2}. \tag{9.4b}$$

The quantities required in Equation 9.4a can be evaluated as

$$\varepsilon_x = \frac{\partial u}{\partial x} = \frac{1}{l}[-1 \quad 1]\begin{Bmatrix} u_1 \\ u_2 \end{Bmatrix} \tag{9.5a}$$

or

$$\{\varepsilon\} = [\mathbf{B}]\{\mathbf{q}\},$$

$$\{\dot{\mathbf{u}}\} = \left\{\frac{\partial u}{\partial t}\right\} = [\mathbf{N}]\begin{Bmatrix} \dfrac{\partial u_1}{\partial t} \\ \dfrac{\partial u_2}{\partial t} \end{Bmatrix} \tag{9.5b}$$

$$= [\mathbf{N}]\{\dot{\mathbf{q}}\},$$

and

$$\{\ddot{\mathbf{u}}\} = \left\{\frac{\partial^2 u}{\partial t^2}\right\} = [\mathbf{N}]\begin{Bmatrix} \dfrac{\partial^2 u_1}{\partial t^2} \\ \dfrac{\partial^2 u_2}{\partial t^2} \end{Bmatrix} \tag{9.5c}$$

$$= [\mathbf{N}]\{\ddot{\mathbf{q}}\};$$

since the N_i are functions of space coordinates only; the time derivatives apply only to the nodal displacements.

Substitution of $\{\varepsilon\}$, $\{\mathbf{u}\}$, and $\{\ddot{\mathbf{u}}\}$ into Equation 9.4a leads to

$$\{\delta\mathbf{q}\}^T\left(\iiint_V [\mathbf{B}]^T[\mathbf{C}][\mathbf{B}]dV\right)\{\mathbf{q}\} = -\{\delta\mathbf{q}\}^T\left(\iiint_V \rho[\mathbf{N}]^T[\mathbf{N}]dV\right)\{\ddot{\mathbf{q}}\} + \{\delta\mathbf{q}\}^T\iiint_V [\mathbf{N}]\{\mathbf{P}_x\}dV. \tag{9.6}$$

Since $[\delta\mathbf{q}]$ represents arbitrary virtual changes, we have

$$\iiint_V [\mathbf{B}]^T[\mathbf{C}][\mathbf{B}]dV\{\mathbf{q}\} = -\iiint_V \rho[\mathbf{N}]^T[\mathbf{N}]dV\,\{\ddot{\mathbf{q}}\} + \iiint_V [\mathbf{N}]^T\{\mathbf{P}_x\}dV \tag{9.7a}$$

or

$$[\mathbf{k}]\{\mathbf{q}\} + [\mathbf{m}]\{\ddot{\mathbf{q}}\} = \{\mathbf{Q}(t)\}. \tag{9.7b}$$

We note that we could have obtained these results by using Hamilton's principle[5] and by properly differentiating the associated variational function with respect to the components of $\{\mathbf{q}\}$ and equating the results to zero. The terms $[\mathbf{k}]$ and $\{\mathbf{Q}\}$ have the same meaning as before (Equation 3.28). The additional matrix $[\mathbf{m}]$ is called the element mass matrix. For the line element with uniform area A and constant P_x at nodes, the matrices are

$$\{\mathbf{k}\} = \frac{AE}{l}\begin{bmatrix} 1 & -1 \\ -1 & 1 \end{bmatrix}, \tag{9.8a}$$

$$\{\mathbf{Q}(t)\} = \frac{AlP_x(t)}{2}\begin{Bmatrix} 1 \\ 1 \end{Bmatrix}, \tag{9.8b}$$

and

$$\begin{aligned}[\mathbf{m}] &= \rho Al\int_0^1 \begin{Bmatrix} 1-s \\ s \end{Bmatrix}[1-s \quad s]ds \\ &= \frac{\rho Al}{6}\begin{bmatrix} 2 & 1 \\ 1 & 2 \end{bmatrix}.\end{aligned} \tag{9.8c}$$

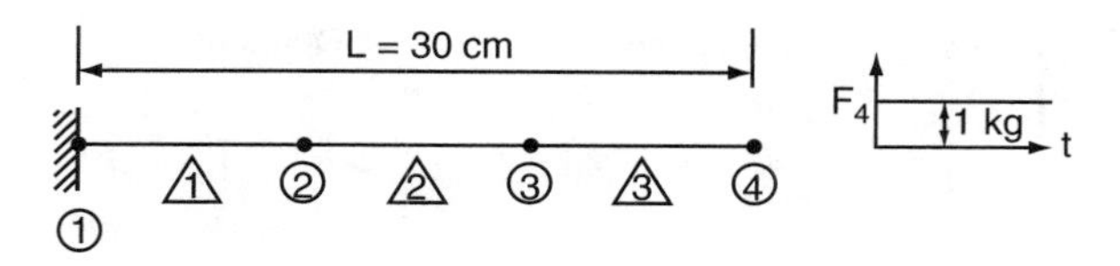

FIGURE 9.2
Mesh for one-dimensional medium.

The matrix in Equation 9.8c is called consistent because it is derived from the consistent (variational) principle; in other words, the mass is distributed to the nodes consistent with the first term on the right-hand side of Equation 9.7a. Often, it may be convenient to approximate it as lumped, where the total mass $A\rho l$ is divided equally among the two nodes:

$$[\mathbf{m}] = \frac{A\rho l}{2}\begin{bmatrix} 1 & 0 \\ 0 & 1 \end{bmatrix}. \tag{9.8d}$$

Because the lumped matrix is diagonal, it can offer computational advantages. On the other hand, the consistent matrix can be more accurate for mathematical analyses. Detailed consideration of this aspect is beyond the scope of this text.

Step 5. Assemble Element Equations

The element equations can now be added such that interelement continuity of displacements (and accelerations) are ensured at common nodes. Then for the three-element discretization (Figure 9.2) we have

$$\underset{(4\times 4)}{[\mathbf{K}]}\;\underset{(4\times 1)}{\{\mathbf{r}\}} + \underset{(4\times 4)}{[\mathbf{M}]}\;\underset{(4\times 1)}{\{\ddot{\mathbf{r}}\}} = \underset{(4\times 1)}{\{\mathbf{R}(t)\}}, \tag{9.9}$$

where $[\mathbf{M}]$ is the assemblage mass matrix and the other terms have the same meanings as in Equation 3.45.

Equation 9.9 represents a set of (matrix) partial differential equations and is the result of the discretization of physical space in the first phase. The time dependence is contained in $\{\ddot{\mathbf{r}}\} = \partial^2\{\mathbf{r}\}/\partial t^2$, and the next phase involves discretization in time in order to find solution in time.

As in the case of Equations 5.21 and 8.7, Equation 9.9 represents a time-dependent phenomenon. The difference is that here we need to consider second derivatives instead of the first derivative in the other two equations.

Time Integration

As a simple approximation, we assume that the acceleration $\{\mathbf{r}\}$ varies linearly between a time step Δt from time level t to $t + \Delta t$ (Figure 9.3(a)). As a

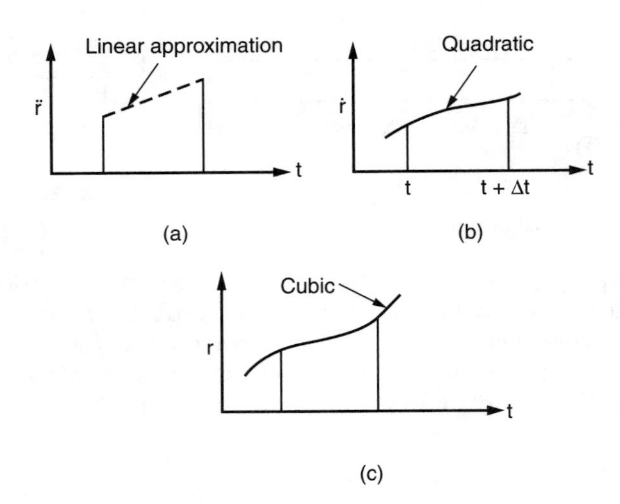

FIGURE 9.3
Approximations for time integration; (a) Acceleration, (b) Velocity, and (c) Displacement.

result, the velocities will be approximated as quadratic and the displacements as cubic (Figures 9.3(b) and (c)). As presented in References 5 and 6, this assumption leads to finite difference discretization of Equation 9.9 as

$$[\overline{\mathbf{K}}]\{\mathbf{r}\}_{t+\Delta t} = \{\overline{\mathbf{R}}\}_{t+\Delta t}, \tag{9.10a}$$

where

$$[\overline{\mathbf{K}}] = [\mathbf{K}] + \frac{6}{(\Delta t)^2}[\mathbf{M}], \tag{9.10b}$$

$$\{\overline{\mathbf{R}}\}_{t+\Delta t} = \{\mathbf{R}\}_{t+\Delta t} + \frac{6}{(\Delta t)^2}[\mathbf{M}]\left(\{\mathbf{r}\}_t + \Delta t(\dot{\mathbf{r}})_t + \frac{1}{3}(\Delta t)^2\{\ddot{\mathbf{r}}\}_t\right). \tag{9.10c}$$

Because the time integration starts from time $t = 0$, $\{\mathbf{r}\}_0$, $\{\dot{\mathbf{r}}\}_0$, and $\{\ddot{\mathbf{r}}\}_0$ are known from the given boundary (displacement) and initial (velocity and acceleration) conditions. Thus the vector $\{\bar{\mathbf{R}}\}_{\Delta t}$ for the first time step $t = 0 + \Delta t = \Delta t$ is known since $\{\mathbf{R}\}$ is known at all time levels. Thus we can solve Equation 9.10a at $t = \Delta t$ and compute displacements at Δt as $\{\mathbf{r}\}_{\Delta t}$. Once the $\{\mathbf{r}\}_{\Delta t}$ are known, velocities and accelerations at $t = \Delta t$ are found from[5,6]

$$\{\dot{\mathbf{r}}\}_{\Delta t} = \frac{3}{\Delta t}\left(\{\mathbf{r}\}_{\Delta t} - \{\mathbf{r}\}_0\right) - 2\{\dot{\mathbf{r}}\}_0 - \frac{\Delta t}{2}\{\ddot{\mathbf{r}}\}_0, \tag{9.11a}$$

$$\{\ddot{\mathbf{r}}\}_{\Delta t} = \frac{6}{(\Delta t)^2}\left(\{\mathbf{r}\}_{\Delta t} - \{\mathbf{r}\}_0\right) - \frac{6}{\Delta t}\{\dot{\mathbf{r}}\}_0 - 2\{\ddot{\mathbf{r}}\}_0. \tag{9.11b}$$

The procedure can now be continued for subsequent time steps $2\Delta t$, $3\Delta t$, ..., $n\Delta t$, where $n\Delta t$ is the total time period for which the solution is desired.

The assumption of linear acceleration between t and $t + \Delta t$ may not be the best procedure for the time integration. We have used it mainly for a simple illustration. Alternative and mathematically superior schemes are available and used for practical problems (Reference 4 and the bibliography).

Example 9.1

Consider the propagation of an elastic wave in a bar discretized in three elements (Figure 9.2). Assume the following properties:

Element length, $l = 10$ cm;
Element area, $A = 1\ \text{cm}^2$;
Elastic modulus, $E = 1000\ \text{kg/cm}^2$;
Density, $\rho = 10^{-11}\ \text{kg/sec}^2/\text{cm}^4$;
Wave velocity, $c_x = \sqrt{\dfrac{E_x}{\rho}} = 10^7$ cm/s.

These simple properties are chosen only to illustrate the procedure; they may not necessarily represent field situations.

An approximate size of the time step can be found as

$$\Delta t = \frac{l}{c_x} = \frac{10}{10^7} = 10^{-6}\ \text{s}. \tag{9.12}$$

This size is often called the characteristic time step. Substitution of these numbers into Equations 9.8a and 9.8b and an assemblage (Equations 9.9 and 9.10) leads to

$$[\mathbf{K}] = \frac{1 \times 1000}{10}\begin{bmatrix} 1 & -1 & 0 & 0 \\ -1 & 2 & -1 & 0 \\ 0 & -1 & 2 & -1 \\ 0 & 0 & -1 & 1 \end{bmatrix},$$

$$[\mathbf{M}] = \frac{1 \times 10^{-11} \times 10}{6}\begin{bmatrix} 2 & 1 & 0 & 0 \\ 1 & 4 & 1 & 0 \\ 0 & 1 & 4 & 1 \\ 0 & 0 & 1 & 2 \end{bmatrix},$$

and hence

$$
\left[\bar{\mathbf{K}}\right] = 10^2 \begin{bmatrix} 1 & -1 & 0 & 0 \\ -1 & 2 & -1 & 0 \\ 0 & -1 & 2 & -1 \\ 0 & 0 & -1 & 1 \end{bmatrix} + \frac{6 \times 10^{-11} \times 10}{10^{-6} \times 10^{-6} \times 6} \begin{bmatrix} 2 & 1 & 0 & 0 \\ 1 & 4 & 1 & 0 \\ 0 & 1 & 4 & 1 \\ 0 & 0 & 1 & 2 \end{bmatrix}
$$

$$
= 10^2 \begin{bmatrix} 1 & -1 & 0 & 0 \\ -1 & 2 & -1 & 0 \\ 0 & -1 & 2 & -1 \\ 0 & 0 & -1 & 1 \end{bmatrix} + 10^2 \begin{bmatrix} 2 & 1 & 0 & 0 \\ 1 & 4 & 1 & 0 \\ 0 & 1 & 4 & 1 \\ 0 & 0 & 1 & 2 \end{bmatrix}. \tag{9.13}
$$

Boundary and Initial Conditions

The boundary condition is that the displacement at the fixed end of node 1 (Figure 9.2) is zero at all times; that is,

$$
u_1(0, t) = 0. \tag{9.14a}
$$

The initial conditions are that displacements, velocities, and acceleration are zero at $t = 0$:

$$
\begin{aligned} u(x, 0) &= 0, \\ \dot{u}(x, 0) &= 0, \\ \ddot{u}(x, 0) &= 0. \end{aligned} \tag{9.14b}
$$

An initial condition defines the state of the body at the start or initiation of loading in terms of the displacement and/or its derivatives.

An external force is assumed to be applied at the free end, node 4, as a constant force:

$$
F_4(t) = 1\ \text{kg}. \tag{9.14c}
$$

With these conditions, the load vector $\{\bar{\mathbf{R}}\}$ is

$$\{\overline{\mathbf{R}}\}_{t+\Delta t} = \{\mathbf{R}\}_{t+\Delta t} + \frac{6\times10^{-11}\times10}{10^{-6}\times10^{-6}\times6}\begin{bmatrix}2&1&0&0\\1&4&1&0\\0&1&4&1\\0&0&1&2\end{bmatrix} \tag{9.15}$$

$$\left(\{\mathbf{r}\}_t + 10^{-6}\{\dot{\mathbf{r}}\}_t + \frac{1}{3}\times10^{-12}\{\ddot{\mathbf{r}}\}_t\right).$$

Using Equations 9.13 to 9.15, we can write Equation 9.10 at $t = 0 + \Delta t = \Delta t$ as

$$\left(10^2\begin{bmatrix}1&-1&0&0\\-1&2&-1&0\\0&-1&2&-1\\0&0&-1&1\end{bmatrix} + 10^2\begin{bmatrix}2&1&0&0\\1&4&1&0\\0&1&4&1\\0&0&1&2\end{bmatrix}\right)\begin{Bmatrix}u_1\\u_2\\u_3\\u_4\end{Bmatrix}_{\Delta t} = \begin{Bmatrix}0\\0\\0\\1\end{Bmatrix}_{\Delta t}. \tag{9.16}$$

Now we introduce the boundary condition

$$u_1(\Delta t) = 0$$

in Equation 9.16, which leads to

$$\begin{bmatrix}600&0&0\\0&600&0\\0&0&300\end{bmatrix}\begin{Bmatrix}u_2\\u_3\\u_4\end{Bmatrix}_{\Delta t} = \begin{Bmatrix}0\\0\\1\end{Bmatrix}_{\Delta t}. \tag{9.17}$$

Solution of Equation 9.17 by Gaussian elimination (with u_1 prescribed) gives

$$\begin{Bmatrix}u_1\\u_2\\u_3\\u_4\end{Bmatrix}_{\Delta t} = \begin{Bmatrix}0\\0\\0\\\frac{1}{300}\end{Bmatrix}. \tag{9.18a}$$

Substitution of Equation 9.18a and the initial conditions (Equation 9.14b) in Equations 9.11a and 9.11b allows computation of velocities and accelerations at $t = \Delta t$ as

$$\begin{Bmatrix} \dfrac{\partial u_1}{\partial t} \\ \dfrac{\partial u_2}{\partial t} \\ \dfrac{\partial u_3}{\partial t} \\ \dfrac{\partial u_4}{\partial t} \end{Bmatrix}_{\Delta t} = \frac{3}{10^{-6}} \left(\begin{Bmatrix} 0 \\ 0 \\ 0 \\ \frac{1}{300} \end{Bmatrix}_{\Delta t} - \begin{Bmatrix} 0 \\ 0 \\ 0 \\ 0 \end{Bmatrix}_0 \right) - 2 \begin{Bmatrix} 0 \\ 0 \\ 0 \\ 0 \end{Bmatrix}_0 - \frac{10^{-6}}{2} \begin{Bmatrix} 0 \\ 0 \\ 0 \\ 0 \end{Bmatrix}_0 \tag{9.18b}$$

and

$$\begin{Bmatrix} \dfrac{\partial^2 u_1}{\partial t^2} \\ \dfrac{\partial^2 u_2}{\partial t^2} \\ \dfrac{\partial^2 u_3}{\partial t^2} \\ \dfrac{\partial^2 u_4}{\partial t^2} \end{Bmatrix}_{\Delta t} = \frac{6}{10^{-6} \times 10^{-6}} \left(\begin{Bmatrix} 0 \\ 0 \\ 0 \\ \frac{1}{300} \end{Bmatrix}_{\Delta t} - \begin{Bmatrix} 0 \\ 0 \\ 0 \\ 0 \end{Bmatrix}_0 \right) - \frac{6}{10^{-6}} \begin{Bmatrix} 0 \\ 0 \\ 0 \\ 0 \end{Bmatrix}_0 - 2 \begin{Bmatrix} 0 \\ 0 \\ 0 \\ 0 \end{Bmatrix}_0 . \tag{9.18c}$$

Equation 9.10a can now be solved for time level $t = 2\Delta t$, $3\Delta t$, and so on. We note here that the foregoing numerical calculations are meant only as an illustration of the procedure and not necessarily as acceptable solutions. For the latter, one needs to program the procedure and select optimum spatial and temporal meshes.

Example 9.2

Now we present results from a problem in one-dimensional wave propagation solved by Yamada and Nagai.[7] The properties of the problem (Figure 9.4) are as follows:

Total length of bar,	L or h = 500 mm;
Number of nodes	=51;
Number of elements	= 50;
Length of each element,	l = 10 mm;
Area of cross section assumed arbitrarily	=1 mm^2;
Modulus of elasticity,	E = 20,000 kg/mm^2;
Density,	ρ=0.008 kg ms^2/mm^4;
Wave velocity,	c_x = 5000 mm/ms;
Time increment,	$\Delta t = l/c_x = 0.002$ ms.

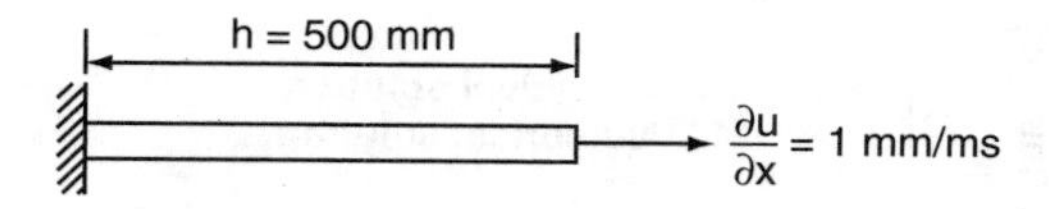

FIGURE 9.4
One-dimensional wave propagation.[7]

Boundary Conditions

$$u(0,\ t) = 0,$$

$$\frac{\partial u}{\partial x}(h,\ t) = 1\ \text{mm/ms},$$

or

$$u(h,\ t) = \frac{\partial u}{\partial x}(h,\ t) \times t.$$

The results in terms of particle velocity at time $t = 0.08$ ms (for $\Delta t = l/c_x = 10/5000 = 0.002$ ms), that is, after 40 time steps, are compared with closed form analytical solutions (Figures 9.5(a) and (b)) for consistent and lumped mass approximations, respectively. It can be seen that for the characteristic time step, the consistent mass yields exact solution, whereas the lumped mass formulation is not that accurate.

The size of the time step can have significant influence on the numerical solutions. This is shown in Figures 9.5(c) and (d) for both approximations for $\Delta t = 0.5\ l/c_x$. Again the lumped mass formulation is not close to the analytical solution. On the other hand, the consistent mass formulation shows oscillations and inaccuracies, particularly in the vicinity of the wave front. The aspect of numerical stability and comparison of lumped and consistent masses is wide in scope, and the reader may consult the references and the bibliography.

Damping

Most natural systems possess damping. It can be easily included in the dynamic equations; the resulting element equations will then be

$$[\mathbf{k}]\{\mathbf{q}\} + [\mathbf{c}]\{\dot{\mathbf{q}}\} + [\mathbf{m}]\{\ddot{\mathbf{q}}\} = \{\mathbf{Q}(t)\}, \tag{9.7c}$$

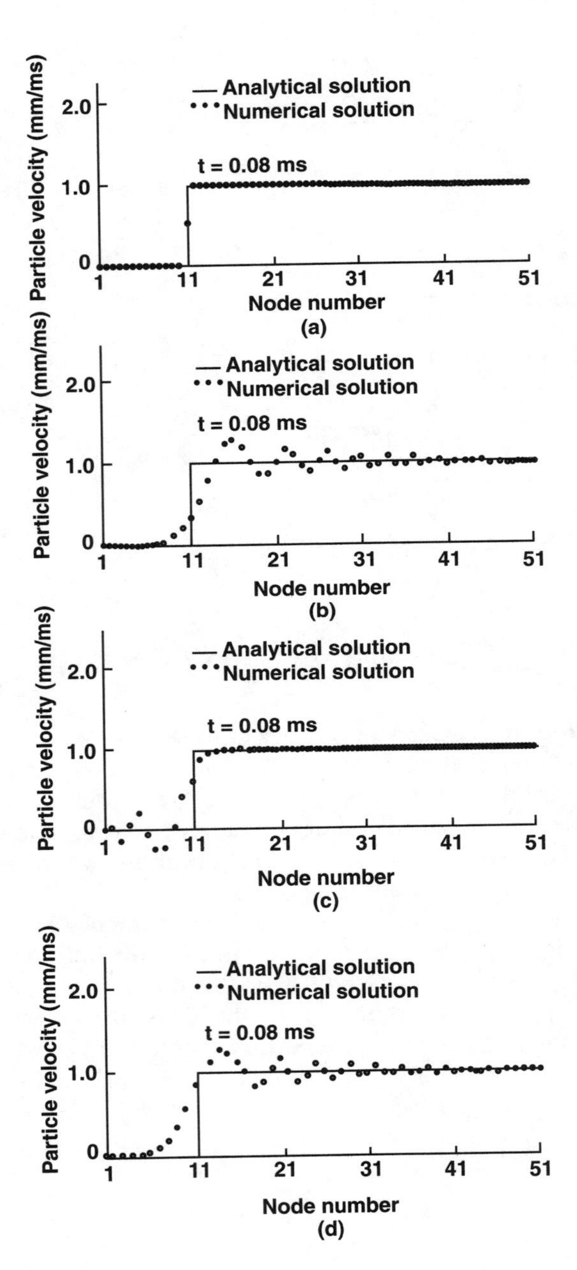

FIGURE 9.5
Results for wave propagation in bar:[7] (a) Consistent mass: $\Delta t = l/c_x$, (b) Lumped mass: $\Delta t = l/c_x$, (c) Consistent mass: $\Delta t = 0.5\, l/c_x$, and (d) Lumped mass: $\Delta t = 0.5\, l/c_x$.

where **[c]** is the damping matrix. Determination of damping properties of material is a very important topic and is discussed in various publications (Reference 4; the bibliography).

Problems

9.1. Compute the element consistent mass matrix for the quadratic approximations [Equation 3.40)]:

$$u = \tfrac{1}{2}L(L-1)u_1 + \left(1-L^2\right)u_0 + \tfrac{1}{2}L(L+1)u_2,$$

where 0 denotes the middle node.

Solution:

$$[\mathbf{m}] = \frac{A\rho l}{4}\int_{-1}^{1}\begin{Bmatrix} L(L-1) \\ \left(1-L^2\right) \\ L(L+1) \end{Bmatrix}\left[L(L-1) \quad \left(1-L^2\right) \quad L(L+1)\right]dL$$

$$= \frac{A\rho l}{120}\begin{bmatrix} 16 & 8 & -4 \\ 8 & 64 & 8 \\ -4 & 8 & 16 \end{bmatrix}.$$

9.2. Compute the element consistent mass matrix for cubic (Hermitian function) u:

$$u = \left(1-3s^2+2s^3\right)u_1 + ls(s-1)^2\frac{\partial u_1}{\partial x}$$

$$+ s^2(3-2s)u_2 + ls^2(s-1)\frac{\partial u_2}{\partial x}.$$

Solution:

$$[\mathbf{m}] = \frac{A\rho l}{420}\begin{bmatrix} 156 & 22 & 54 & -13 \\ 22 & 4 & 13 & -3 \\ 54 & 13 & 156 & -22 \\ -13 & -3 & -22 & 4 \end{bmatrix}.$$

9.3. Perform computations for three time steps in Example 9.1.

9.4. Consider a single-element discretization which can be replaced by a spring-mass system. It is subjected to a forcing function P_x as shown in Figure 9.6. Specialize Equation 9.10a for this single element and obtain solutions for two time steps. Assume properties as in Example 9.1. Compare numerical results with the closed form solution from

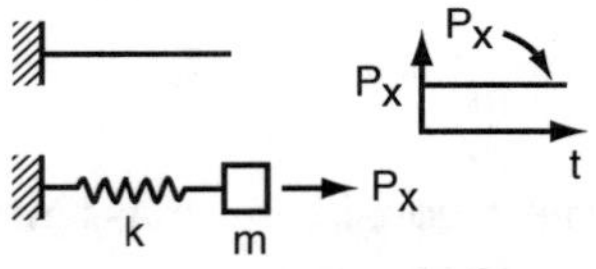

FIGURE 9.6

$$u(t) = \frac{P_x}{k}\left(1 - \cos pt\right),$$

where $p^2 = k/m$, where k and m are the stiffness and mass of the spring, respectively. The initial conditions are $u(x, 0) = \dot{u}\ (x, 0) = 0$.

9.5. Prepare a computer problem based on Equation 9.10 and solve Problems 9.3 and 9.4.

References

1. Love, A. E. H. 1944. *A Treatise on Mathematical Theory of Elasticity,* Dover, New York.
2. Timoshenko, S. and Goodier, J. N. 1970. *Theory of Elasticity,* McGraw-Hill, New York.
3. Carnahan, B., Luther, H. A., and Wilkes, J. O. 1969. *Applied Numerical Methods,* John Wiley & Sons, New York.
4. Desai, C. S., and Christian, J. T., Eds. 1977. *Numerical Methods in Geotechnical Engineering,* McGraw-Hill, New York.
5. Desai, C. S. and Abel, J. F. 1972. *Introduction to the Finite Element Method,* Van Nostrand Reinhold, New York.
6. Wilson, E. L. 1968. "A computer program for the dynamic stress analysis of underground structures," Report No. 68-1, University of California, Berkeley, CA.
7. Yamada, Y. and Nagai, Y. May 1971. "Analysis of one-dimensional stress wave by the finite element method," *Seisan Kenkyu, J. Ind. Sci.,* University of Tokyo, Japan, 23(5), 186-189.

Bibliography

Archer, J. S. Aug. 1963. "Consistent mass matrix for distributed mass systems," *J. Struct. Div. ASCE* 39, No. ST4.

Clough, R. W. 1971. "Analysis of structural vibrations and dynamic response," in *Proc. U.S. Japan Seminar,* Tokyo, 1969, Gallagher, R. H., Yamada, Y., and Oden, J. T., Eds., University of Alabama Press, Huntsville.

Desai, C. S., Ed. 1972. *Proc. Symp. on Appl. of Finite Element Methods in Geotech. Eng.,* Waterways Expt. Station, Vicksburg, MS.

Desai, C. S., Ed. 2976. *Proc. Second Int. Conf. on Num. Methods in Geomech.*, Blacksburg, Va., 1976, Vols. I, II, III, ASCE.

Desai, C. S. and Lytton, R. L. 1975. "Stability criteria of finite element schemes for parabolic equation," *Int. J. Num. Methods Eng.* 9, 721-726.

Idriss, I. M., Seed, H. B., and Seriff, N. 1974. "Seismic response by variable damping finite elements," *J. Geotech. Eng. Div. ASCE* 100(GT1), 1-13.

Kreig, R. D. and Key, S. W. 1973. "Comparison of finite element and finite difference methods," *Proc. ONR Symp. Num. Methods Struct. Mech.*, University of Illinois, Sept. 1971, Fenres, S.J., Perrone, N., Robinson, A.R., and Schnobrich, W.C., Eds., Academic Press, New York.

Newmark, N. M. 1959. "A method of computation of structural dynamics," *Proc. ASCE*, 85(EM3), 67-94.

Nickell, R. E. April 1973. "Direct integration methods in structural dynamics," *J. Eng. Mech. Div. ASCE*, 99(EM2).

10

Two- and Three-Dimensional Formulations

Introduction

So far we have discussed problems whose geometry can be idealized as the one-dimensional line. When a line is discretized we get the one-dimensional line elements. However, this simplification is not possible for many problems. In some cases the problem must be solved in two dimensions such as the plate, torsion, and two-dimensional field problems (fluid flow, heat flow, or stress wave propagation problems). The situation becomes more complicated when all the three dimensions of the problem geometry need consideration. For finite element analysis of two-dimensional or three-dimensional cases, the problem must be discretized into a number of elemental areas or volumes, respectively, giving rise to two-dimensional or three-dimensional finite elements.

In this chapter, we present basic details of two-dimensional elements that will be used in the subsequent chapters. Brief statements of typical three-dimensional elements are also given for the sake of completeness.

Two-Dimensional Formulation

A two-dimensional region can be discretized into a number of small triangular or quadrilateral elements as shown in Figure 10.1.

Based on the shape of the element the two-dimensional elements can be classified mainly under two categories — triangular elements (Figure 10.1a) and quadrilateral elements (Figure 10.1b).

Triangular Element

We shall first detail some of the properties of the triangular element.

As discussed in Chapter 3, it is advantageous to use the concept of local coordinate systems for finite element formulations. For the triangular element,

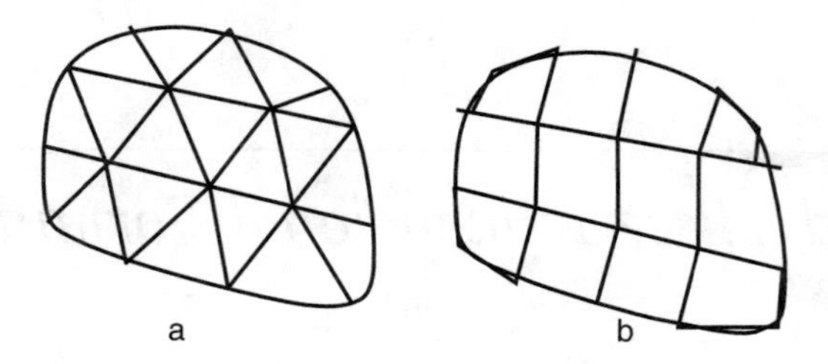

FIGURE 10.1
Discretization of a two-dimensional region.

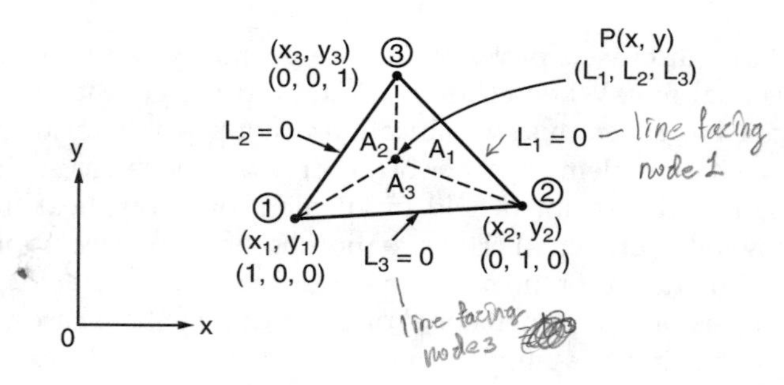

FIGURE 10.2
Triangular element.

the local or area coordinates are often defined in terms of component areas, A_1, A_2, and A_3 (Figure 10.2). Then in the nondimensional form, the local coordinates L_1, L_2, and L_3 are defined as

$$L_i = \frac{A_i}{A}, \quad i = 1, 2, 3. \tag{10.1}$$

Since $L_1 + L_2 + L_3 = 1$, there are only two independent local coordinates corresponding to the two global coordinates x and y. The relationship between the global and the local coordinates is given by

$$\begin{aligned} x &= L_1 x_1 + L_2 x_2 + L_3 x_3 = \sum_{i=1}^{3} L_i x_i, \\ y &= L_1 y_1 + L_2 y_2 + L_3 y_3 = \sum_{i=1}^{3} L_i y_i, \end{aligned} \tag{10.2a}$$

$$\begin{aligned} x &= [\mathbf{N}]\{\mathbf{x}_n\} \\ y &= [\mathbf{N}]\{\mathbf{y}_n\} \end{aligned} \tag{10.2b}$$

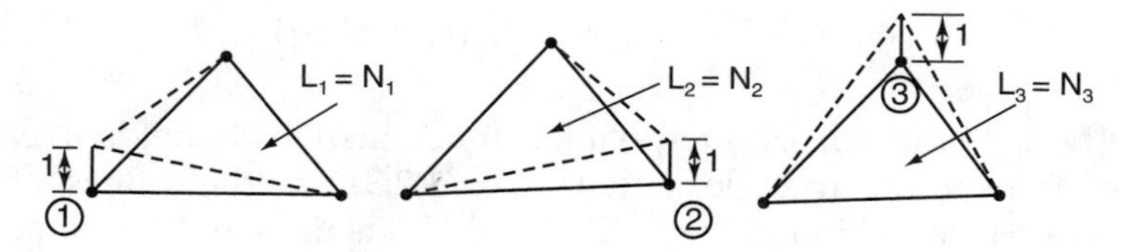

FIGURE 10.3
Distribution of interpolation functions N_i, $i = 1, 2, 3$.

where [**N**] is the matrix of interpolation functions N_i, which are the same as the local coordinates L_i; these are plotted in Figure 10.3.

$$\{\mathbf{x}_n\}^T = [x_1 \ x_2 \ x_3] \text{ and } \{\mathbf{y}_n\}^T = [y_1 \ y_2 \ y_3]$$

are the vectors of nodal coordinates. Equation 10.2 represents linear variations for coordinates at any point $P(x, y)$.

The inverse relation corresponding to Equation 10.2 is

$$\begin{aligned} N_1 &= L_1 = \frac{1}{2A}(2A_{23} + b_1 x + a_1 y), \\ N_2 &= L_2 = \frac{1}{2A}(2A_{31} + b_2 x + a_2 y), \\ N_3 &= L_3 = \frac{1}{2A}(2A_{12} + b_3 x + a_3 y), \end{aligned} \tag{10.3a}$$

or

$$\begin{Bmatrix} N_1 \\ N_2 \\ N_3 \end{Bmatrix} = \frac{1}{2A} \begin{bmatrix} 2A_{23} & b_1 & a_1 \\ 2A_{31} & b_2 & a_2 \\ 2A_{12} & b_3 & a_3 \end{bmatrix} \begin{Bmatrix} 1 \\ x \\ y \end{Bmatrix}, \tag{10.3b}$$

where $2A = a_3b_2 - a_2b_3 = a_1b_3 - a_3b_1 = a_2b_1 - a_1b_2$; A is the total area of the triangle; A_{23} is the area of the triangle whose vertices are nodes 2, 3 and the origin 0 and so on; and a_i and b_i are given as differences between various nodal coordinates:

$$\begin{aligned} a_1 &= x_3 - x_2, & b_1 &= y_2 - y_3, \\ a_2 &= x_1 - x_3, & b_2 &= y_3 - y_1, \\ a_3 &= x_2 - x_1, & b_3 &= y_1 - y_2. \end{aligned} \tag{10.3c}$$

Let us first consider only one unknown degree of freedom at a point. The unknown, represented by a general symbol ψ, can be expressed as

$$\psi = N_1\psi_1 + N_2\psi_2 + N_3\psi_3 = [\mathbf{N}]\{\mathbf{q}\} \tag{10.4a}$$

where [**N**] is the matrix of interpolation functions which yields linear variation of ψ over the triangle, and ψ_i (i = 1,2,3) are the values of nodal unknowns. Since [**N**] in Equations 10.2 and 10.4 is the same, we can call this element an isoparametric triangular element.

Requirements for the Approximation Function

As discussed in Chapter 3, the approximation function in Equation 10.4a should satisfy continuity within an element, interelement compatibility, and completeness. Equation 10.4a is the transformed version of the polynomial form of linear approximation for ψ as

$$\psi = \alpha_1 + \alpha_2 x + \alpha_3 y \tag{10.4b}$$

which is continuous within the element.

If the highest order of derivative in the functional expression of the problem is 1 then the approximation function should provide interelement compatibility up to order 1 – 1 = 0. Linear approximation indeed provides for compatibility of the primary unknown across interelement boundaries. This is illustrated in Figure 10.4. Since $\psi_2^1 = \psi_1^2$ *and* $\psi_3^1 = \psi_3^2$ and since only one straight line can pass through two points, the variations of ψ along the edge AA of element 1 and BB of element 2 must coincide. That is, at the common boundary, compatibility of ψ is fulfilled. Note that for linear approximation this does not necessarily imply interelement compatibility of higher orders, such as the first derivatives.

Since the approximation function in Equation 10.4b provides for rigid body motion (term α_1) and constant state of strains or gradients of ψ (terms $\alpha_2 x$ and $\alpha_3 y$), it is complete. Notice that in this problem, there exist two such

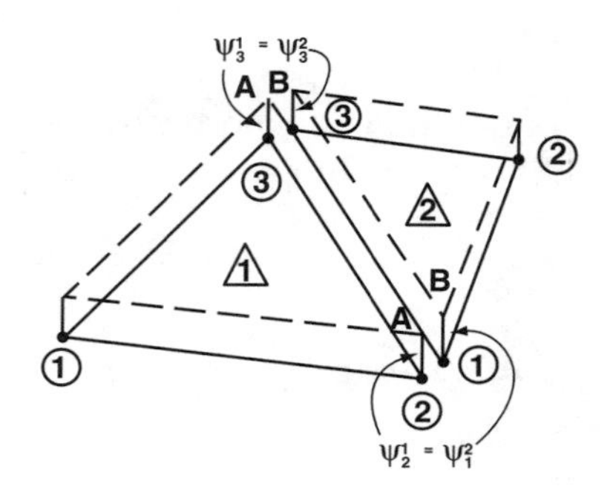

FIGURE 10.4
Interelement compatibility.

states: $\partial\psi/\partial x$ and $\partial\psi/\partial y$. Furthermore, it contains all terms up to the order $n = 1$ in the polynomial expansion for the two-dimensional problem. This idea can be explained by using the polynomial expansion represented (by Pascal's triangle) as follows:

									order or degree (n) of polynomial	
				1					$n = 0$	constant
			x		y				1	Linear
		x^2		xy		y^2			2	Quadratic
	x^3		x^2y		xy^2		y^3		3	Cubic
x^4		x^3y		x^2y^2		xy^3		y^4	4	Quartic

We can see that for the linear approximation in Equation 10.4b, all the three terms up to and including $n = 1$ are provided.

To find the values of the derivatives, we use the following chain rule of differentiation:

$$\frac{\partial\psi}{\partial x} = \frac{\partial N_1}{\partial x}\frac{\partial\psi}{\partial N_1} + \frac{\partial N_2}{\partial x}\frac{\partial\psi}{\partial N_2} + \frac{\partial N_3}{\partial x}\frac{\partial\psi}{\partial N_3} \tag{10.5a}$$

$$\frac{\partial\psi}{\partial y} = \frac{\partial N_1}{\partial y}\frac{\partial\psi}{\partial N_1} + \frac{\partial N_2}{\partial y}\frac{\partial\psi}{\partial N_2} + \frac{\partial N_3}{\partial y}\frac{\partial\psi}{\partial N_3}. \tag{10.5b}$$

For example, by differentiation of N_i ($i = 1, 2, 3$) and ψ with respect to x in Equations 10.3 and 10.4a, we obtain

$$\begin{aligned}\frac{\partial\psi}{\partial x} &= \frac{\partial}{\partial x}\left[\frac{1}{2A}(2A_{23} + b_1x + a_1y)\right]\frac{\partial}{\partial N_1}(N_1\psi_1 + N_2\psi_2 + N_3\psi_3) \\ &+ \frac{\partial}{\partial x}\left[\frac{1}{2A}(2A_{31} + b_2x + a_2y)\right]\frac{\partial}{\partial N_2}(N_1\psi_1 + N_2\psi_2 + N_3\psi_3) \\ &+ \frac{\partial}{\partial x}\left[\frac{1}{2A}(2A_{12} + b_3x + a_3y)\right]\frac{\partial}{\partial N_3}(N_1\psi_1 + N_2\psi_2 + N_3\psi_3) \\ &= \frac{b_1}{2A}\psi_1 + \frac{b_2}{2A}\psi_2 + \frac{b_3}{2A}\psi_3.\end{aligned} \tag{10.6a}$$

Since A_{23}, etc. are constants, they do not contribute to the results after differentiation. Since the ψ_i ($i = 1, 2, 3$) are values at the nodes and not functions of N_i and (x, y), they also do not contribute to the differentiation. Similarly,

$$\frac{\partial \psi}{\partial y} = \frac{a_1}{2A}\psi_1 + \frac{a_2}{2A}\psi_2 + \frac{a_3}{2A}\psi_3. \tag{10.6b}$$

Equations 10.6a and 10.6b are combined in matrix notation as

$$\begin{Bmatrix} \dfrac{\partial \psi}{\partial x} \\ \dfrac{\partial \psi}{\partial y} \end{Bmatrix} = \frac{1}{2A}\begin{bmatrix} b_1 & b_2 & b_3 \\ a_1 & a_2 & a_3 \end{bmatrix}\begin{Bmatrix} \psi_1 \\ \psi_2 \\ \psi_3 \end{Bmatrix} = [\mathbf{B}]\{\mathbf{q}\} \tag{10.6c}$$

Here **[B]** is called the strain-unknown (displacement) transformation matrix.

Integration of $[\mathbf{B}]^T[\mathbf{B}]$

Expression of the **[B]** matrix is given in Equation 10.6c. To obtain element equations we need to carry out the following integration for the element (stiffness) matrix **[k]**:

$$[k] = \iint_A [B]^T[C][B]dA \tag{10.7a}$$

or

$$[k] = C\iint_A [B]^T[B]dA \tag{10.7b}$$

For the one-dimensional 2-node linear element, as described in Chapter 3, the **[B]** matrix is 1×2, and **[C]** is a scalar quantity; thus the resulting **[k]** matrix is 2×2. When the two-dimensional field problems (Chapters 11 and 12) are analyzed using the three-node triangular element, then there is generally only one degree of freedom (the field variable) at each node. In that case the **[B]** matrix has a dimension of 2×3, and C can be simplified to a constant scalar quantity and be brought outside the integral as shown in Equation 10.7b; then the 3×3 **[k]** matrix takes the following form:

$$[k] = \frac{C}{4A^2}\iint_A \begin{bmatrix} b_1 & a_1 \\ b_2 & a_2 \\ b_3 & a_3 \end{bmatrix}\begin{bmatrix} b_1 & b_2 & b_3 \\ a_1 & a_2 & a_3 \end{bmatrix} dA \tag{10.8}$$

$$= \frac{C}{4A}\begin{bmatrix} b_1^2 + a_1^2 & b_1 b_2 + a_1 a_2 & b_1 b_3 + a_1 a_3 \\ & b_2^2 + a_2^2 & b_2 b_3 + a_2 a_3 \\ sym. & & b_3^2 + a_3^2 \end{bmatrix}$$

For the stress-deformation problem, however, there are two degrees of freedom — horizontal and vertical components of displacement at each node. In this case, dimensions of both **[B]** and **[C]** matrices increase and the **[k]** matrix becomes a 6×6 matrix. It will be discussed in more detail in Chapter 13.

Quadrilateral Element

As in the triangular element case, it is advantageous to use the concept of local coordinate system for finite element formulations using quadrilateral finite elements. A local s–t coordinate system is defined such that the quadrilateral element is mapped into a perfect square in the local coordinate system as shown in Figure 10.5(a).

Approximation Model for Unknown

Let us denote the primary unknown field variable by the symbol φ, which represents only one degree of freedom at a point. For the four-node quadrilateral element, there are thus four nodal degrees of freedom, and a bilinear model for φ at any point can be written as

$$\varphi = \alpha_1 + \alpha_2 x + \alpha_3 y + \alpha_4 xy \tag{10.9a}$$

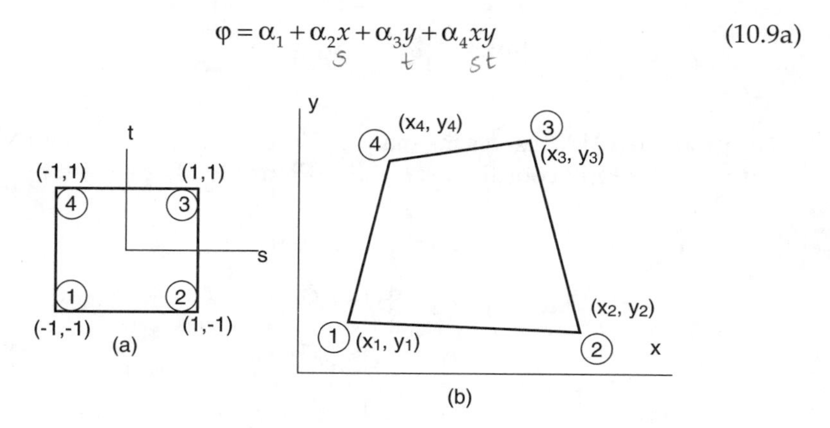

FIGURE 10.5
Quadrilateral element in: (a) Local coordinate system (s–t) and (b) Global coordinate system (x-y).

or

$$\varphi = [\Phi]\{\alpha\}, \tag{10.9b}$$

where $[\Phi] = [1\ x\ y\ xy]$ and $\{\alpha\}^T = [\alpha_1\ \alpha_2\ \alpha_3\ \alpha_4]$ is the vector of generalized coordinates. Note that the term xy yields a bilinear distribution compared to the strictly linear distribution given by the triangular element discussed earlier. This function is called bilinear because the function is linear in two directions (x = constant and y = constant) at every point. Evaluation of φ at the four nodes yields

$$\begin{aligned}
\varphi_1 &= \alpha_1 + \alpha_2 x_1 + a_3 y_1 + a_4 x_1 y_1, \\
\varphi_2 &= \alpha_1 + \alpha_2 x_2 + a_3 y_2 + a_4 x_2 y_2, \\
\varphi_3 &= \alpha_1 + \alpha_2 x_3 + a_4 y_3 + a_4 x_3 y_3, \\
\varphi_4 &= \alpha_1 + \alpha_2 x_4 + a_3 y_4 + a_4 x_4 y_4,
\end{aligned} \tag{10.10a}$$

or

$$\{\mathbf{q}_\varphi\} = [\mathbf{A}]\{\alpha\}, \tag{10.10b}$$

where $[\mathbf{A}]$ is the square matrix of coordinates of the nodes and $\{\mathbf{q}_\varphi\}^T = [\varphi_1\ \varphi_2\ \varphi_3\ \varphi_4]$. We can solve for $\{\alpha\}$ as

$$\{\alpha\} = [\mathbf{A}]^{-1}\{\mathbf{q}_\varphi\} \tag{10.10c}$$

and substitute the results into Equation 10.9b to yield

$$\varphi = [\Phi][\mathbf{A}]^{-1}\{\mathbf{q}_\varphi\} = [\mathbf{N}]\{\mathbf{q}_\varphi\} = \sum_{i=1}^{4} N_i \varphi_i. \tag{10.11}$$

The product $[\Phi]\,[A]^{-1}$ gives the matrix of interpolation functions $[\mathbf{N}]$, which in terms of the local coordinates s and t (Figure 10.5) are given by

$$\begin{aligned}
N_1 &= \tfrac{1}{4}(1-s)(1-t), \\
N_2 &= \tfrac{1}{4}(1+s)(1-t), \\
N_3 &= \tfrac{1}{4}(1+s)(1+t), \\
N_4 &= \tfrac{1}{4}(1-s)(1+t).
\end{aligned} \tag{10.12}$$

Here, s, t are nondimensionalized local coordinates. Figure 10.6 shows plots for N_i, $i = 1, 2, 3, 4$.

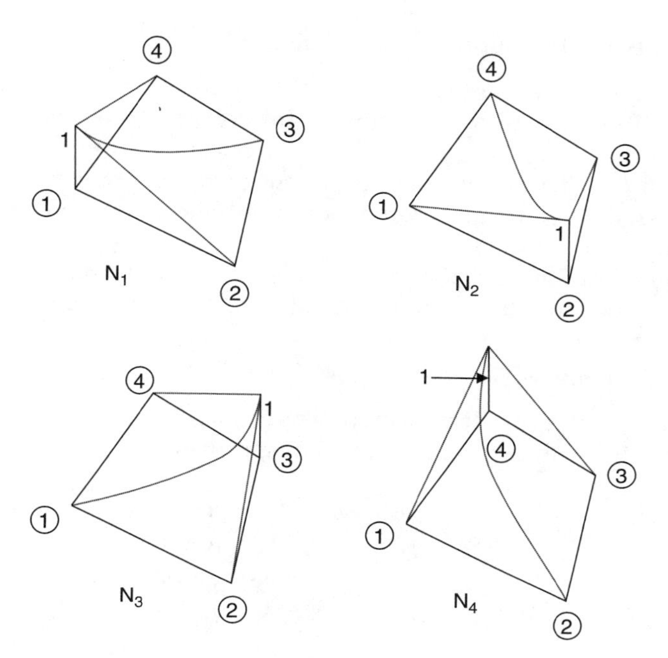

FIGURE 10.6
Distributions of interpolation functions N_i, i = 1, 2, 3,4.

The global coordinates x, y at any point in the element can also be expressed by using the same interpolation functions N_i,

$$
\begin{aligned}
x &= \sum_{i=1}^{4} N_i x_i, \\
y &= \sum_{i=1}^{4} N_i y_i,
\end{aligned}
\tag{10.13a}
$$

or

$$
\begin{Bmatrix} x \\ y \end{Bmatrix} = \begin{bmatrix} N_1 & N_2 & N_3 & N_4 & 0 & 0 & 0 & 0 \\ 0 & 0 & 0 & 0 & N_1 & N_2 & N_3 & N_4 \end{bmatrix} \begin{Bmatrix} \{x_n\} \\ \{y_n\} \end{Bmatrix}, \tag{10.13b}
$$

where $\{\mathbf{x}_n\}^T = [x_1\, x_2\, x_3\, x_4]$ and $\{\mathbf{y}_n\}^T = [y_1\, y_2\, y_3\, y_4]$.

This approach in which the geometry, that is, x, y coordinates and the unknown φ, are both expressed by using the same interpolation functions N_i is called the isoparametric concept. As stated previously, use of this concept offers a number of advantages in terms of easier differentiations and integrations.

Requirements for the Approximation Function

As discussed for the triangular element, the approximation function should satisfy continuity within an element, interelement compatibility, and completeness.

If the highest order of derivative in the functional expression of the problem is m then the approximation function should provide interelement compatibility up to order (m – 1).

The approximation function in Equations 10.9 and 10.11 satisfies the requirements of continuity, comformability, and completeness.

Secondary Unknowns

Derivatives g_x and g_y of primary unknowns can be obtained by taking partial derivatives of φ with respect to x and y, respectively,

$$g_x = \frac{\partial \varphi}{\partial x} = \frac{\partial \varphi}{\partial s}\frac{\partial s}{\partial x} + \frac{\partial \varphi}{\partial t}\frac{\partial t}{\partial x},$$
$$g_y = \frac{\partial \varphi}{\partial y} = \frac{\partial \varphi}{\partial s}\frac{\partial s}{\partial y} + \frac{\partial \varphi}{\partial t}\frac{\partial t}{\partial y}. \quad (10.14)$$

Since the N_i are expressed in terms of local coordinates s, t, we use the following mathematical results based on the chain rule of differentiation in order to find g_x and g_y:

$$\frac{\partial N_i}{\partial s} = \frac{\partial N_i}{\partial x}\frac{\partial x}{\partial s} + \frac{\partial N_i}{\partial y}\frac{\partial y}{\partial s},$$
$$\frac{\partial N_i}{\partial t} = \frac{\partial N_i}{\partial x}\frac{\partial x}{\partial t} + \frac{\partial N_i}{\partial y}\frac{\partial y}{\partial t}, \quad (10.15)$$

which as a general rule, in matrix notation, can be expressed as

$$\begin{Bmatrix} \dfrac{\partial}{\partial s} \\ \dfrac{\partial}{\partial t} \end{Bmatrix} = \begin{bmatrix} \dfrac{\partial x}{\partial s} & \dfrac{\partial y}{\partial s} \\ \dfrac{\partial x}{\partial t} & \dfrac{\partial y}{\partial t} \end{bmatrix} \begin{Bmatrix} \dfrac{\partial}{\partial x} \\ \dfrac{\partial}{\partial y} \end{Bmatrix} = [\mathbf{J}] \begin{Bmatrix} \dfrac{\partial}{\partial x} \\ \dfrac{\partial}{\partial y} \end{Bmatrix}. \quad (10.16)$$

The matrix [J] is often referred to as the Jacobian matrix. Equation 10.16 represents a set of simultaneous equations in which ∂/∂x and ∂/∂y are the unknowns. Solution by Cramer's rule gives

$$\begin{Bmatrix} \dfrac{\partial}{\partial x} \\ \dfrac{\partial}{\partial y} \end{Bmatrix} = [J]^{-1} \begin{Bmatrix} \dfrac{\partial}{\partial s} \\ \dfrac{\partial}{\partial t} \end{Bmatrix} = \frac{1}{|J|} \begin{bmatrix} \dfrac{\partial y}{\partial t} & -\dfrac{\partial y}{\partial s} \\ -\dfrac{\partial x}{\partial t} & \dfrac{\partial x}{\partial s} \end{bmatrix} \begin{Bmatrix} \dfrac{\partial}{\partial s} \\ \dfrac{\partial}{\partial t} \end{Bmatrix} \quad (10.17)$$

where the determinant $|J|$ is called the Jacobian,

$$|J| = \frac{\partial x}{\partial s}\frac{\partial y}{\partial t} - \frac{\partial x}{\partial t}\frac{\partial y}{\partial s}. \tag{10.18}$$

The terms in Equation 10.18 can be evaluated by using expressions for x and y (Equation 10.13) and N_i (Equation 10.12). For instance,

$$\begin{aligned}\frac{\partial x}{\partial s} &= \frac{\partial}{\partial s}\left[\tfrac{1}{4}(1-s)(1-t)x_1 + \tfrac{1}{4}(1+s)(1-t)x_2\right.\\ &\quad \left. + \tfrac{1}{4}(1+s)(1+t)x_3 + \tfrac{1}{4}(1-s)(1+t)x_4\right]\\ &= -\tfrac{1}{4}(1-t)x_1 + \tfrac{1}{4}(1-t)x_2 + \tfrac{1}{4}(1+t)x_3 - \tfrac{1}{4}(1+t)x_4\end{aligned} \tag{10.19a}$$

and so on, and

$$\frac{\partial N_1}{\partial s} = -\tfrac{1}{4}(1-t) \tag{10.19b}$$

and so on. Then $|J|$ is evaluated as

$$\begin{aligned}|J| &= \sum_{i=1}^{4}\sum_{j=1}^{4}\left[\left(\frac{\partial N_i}{\partial s}x_i\frac{\partial N_j}{\partial t}y_j\right) - \left(\frac{\partial N_i}{\partial t}x_i\frac{\partial N_j}{\partial s}y_j\right)\right]\\ &= \sum_{i=1}^{4}\sum_{j=1}^{4}\left[x_i\left(\frac{\partial N_i}{\partial s}\frac{\partial N_j}{\partial t} - \frac{\partial N_i}{\partial t}\frac{\partial N_j}{\partial s}\right)y_j\right].\end{aligned} \tag{10.20a}$$

By setting i = 1, 2, 3, 4 and j = 1, 2, 3, 4 for each i, the summation in Equation 10.20a leads to $|J|$ in matrix notation as

$$|J| = \tfrac{1}{8}\begin{bmatrix}x_1 & x_2 & x_3 & x_4\end{bmatrix}\begin{bmatrix}0 & 1-t & -s+t & -1+s\\ -1+t & 0 & 1+s & -s-t\\ s-t & -1-s & 0 & 1+t\\ 1-s & s+t & -1-t & 0\end{bmatrix}\begin{Bmatrix}y_1\\ y_2\\ y_3\\ y_4\end{Bmatrix}. \tag{10.20b}$$

Expansion of $|J|$ in Equation 10.20b gives

$$\begin{aligned}&= \tfrac{1}{8}\{[x_1 - x_3](y_2 - y_4) - (x_2 - x_4)(y_1 - y_3)\\ &\quad + s[(x_3 - x_4)(y_1 - y_2) - (x_1 - x_2)(y_3 - y_4)]\\ &\quad + t[(x_2 - x_3)(y_1 - y_4) - (x_1 - x_4)(y_2 - y_3)]\}\end{aligned} \tag{10.20c}$$

$$
= \tfrac{1}{8}\Big[\big(x_{13}y_{24} - x_{24}y_{13}\big) + s\big(x_{34}y_{12} - x_{12}y_{34}\big)
$$

$$
+ t\big(x_{23}y_{14} - x_{14}y_{23}\big)\Big],
$$

where $x_{ij} = x_i - x_j$, and $y_{ij} = y_i - y_j$.

Use of Equation 10.17 allows computations of $\partial s/\partial x$, $\partial t/\partial x$, $\partial s/\partial y$, and $\partial t/\partial y$ in Equation 10.14 as

$$
\frac{\partial s}{\partial x} = \frac{1}{|J|}\left(\frac{\partial y}{\partial t}\frac{\partial s}{\partial s} - \cancelto{0}{\frac{\partial y}{\partial s}\frac{\partial s}{\partial t}}\right) = \frac{1}{|J|}\frac{\partial y}{\partial t} = \frac{1}{|J|}\left(\sum_{i=1}^{4}\frac{\partial N_i}{\partial t}y_i\right). \tag{10.21a}
$$

Similarly,

$$
\frac{\partial s}{\partial y} = -\frac{1}{|J|}\left(\sum_{i=1}^{4}\frac{\partial N_i}{\partial t}x_i\right), \tag{10.21b}
$$

$$
\frac{\partial t}{\partial x} = -\frac{1}{|J|}\left(\sum_{i=1}^{4}\frac{\partial N_i}{\partial s}y_i\right), \tag{10.21c}
$$

$$
\frac{\partial t}{\partial y} = \frac{1}{|J|}\left(\sum_{i=1}^{4}\frac{\partial N_i}{\partial s}x_i\right). \tag{10.21d}
$$

Now from Equation 10.11, we have

$$
\frac{\partial \varphi}{\partial s} = \frac{\partial N_i}{\partial s}\varphi_i \quad \text{and} \quad \frac{\partial \varphi}{\partial t} = \frac{\partial N_i}{\partial t}\varphi_i. \tag{10.22}
$$

Substitution of Equations 10.21 and 10.22 into Equation 10.14 finally leads to

$$
\begin{Bmatrix} g_x \\ g_y \end{Bmatrix} = \begin{Bmatrix} \dfrac{\partial \varphi}{\partial x} \\ \dfrac{\partial \varphi}{\partial y} \end{Bmatrix} = \begin{Bmatrix} \displaystyle\sum_{i=1}^{4}\sum_{j=1}^{4}\left[\varphi_i\left(\frac{\partial N_i}{\partial s}\frac{\partial N_j}{\partial t} - \frac{\partial N_i}{\partial t}\frac{\partial N_j}{\partial s}\right)y_j\right]\frac{1}{|J|} \\ \displaystyle\sum_{i=1}^{4}\sum_{j=1}^{4}\left[\varphi_t\left(\frac{\partial N_i}{\partial s}\frac{\partial N_j}{\partial t} - \frac{\partial N_i}{\partial t}\frac{\partial N_j}{\partial s}\right)x_j\right]\frac{1}{|J|} \end{Bmatrix}. \tag{10.23a}
$$

The indicial notation in Equation 10.23a indicates double summation. For instance, the first term in the summation for the first row is obtained by setting $i = 1$ and $j = 1, 2, 3, 4$ as

$$\varphi_1\left[\left(\frac{\partial N_1}{\partial s}\frac{\partial N_1}{\partial t}-\frac{\partial N_1}{\partial t}\frac{\partial N_1}{\partial s}\right)y_1+\left(\frac{\partial N_1}{\partial s}\frac{\partial N_2}{\partial t}-\frac{\partial N_1}{\partial t}\frac{\partial N_2}{\partial s}\right)y_2\right.$$
$$\left.+\left(\frac{\partial N_1}{\partial s}\frac{\partial N_3}{\partial t}-\frac{\partial N_1}{\partial t}\frac{\partial N_3}{\partial s}\right)y_3+\left(\frac{\partial N_1}{\partial s}\frac{\partial N_4}{\partial t}-\frac{\partial N_1}{\partial t}\frac{\partial N_4}{\partial s}\right)y_4\right]. \tag{10.23b}$$

When this is added to the other three terms obtained by setting $i = 2$ and $j = 1, 2, 3, 4$; $i = 3, j = 1, 2, 3, 4$; and $i = 4, j = 1, 2, 3, 4$, we have $\partial\varphi/\partial x$. Similar evaluation gives $\partial\varphi/\partial y$.

After relevant substitutions and rearrangements,

$$\begin{Bmatrix} g_x \\ g_y \end{Bmatrix} = \begin{bmatrix} B_{11} & B_{12} & B_{13} & B_{14} \\ B_{21} & B_{22} & B_{23} & B_{24} \end{bmatrix} \begin{Bmatrix} \varphi_1 \\ \varphi_2 \\ \varphi_3 \\ \varphi_4 \end{Bmatrix} \tag{10.24a}$$

or

$$\{\mathbf{g}\} = [\mathbf{B}]\{\mathbf{q}_\varphi\}, \tag{10.24b}$$

where

$$B_{11} = \frac{1}{8|J|}(y_{24} - y_{34}s - y_{23}t), \tag{10.24c}$$

$$B_{12} = \frac{1}{8|J|}(-y_{13} + y_{34}s + y_{14}t),$$

$$B_{13} = \frac{1}{8|J|}(-y_{24} + y_{12}s - y_{14}t),$$

$$B_{14} = \frac{1}{8|J|}(y_{13} - y_{12}s + y_{23}t),$$

$$B_{21} = \frac{1}{8|J|}(-x_{24} + x_{34}s + x_{23}t),$$

$$B_{22} = \frac{1}{8|J|}(x_{13} - x_{34}s - x_{14}t),$$

$$B_{23} = \frac{1}{8|J|}(x_{24} - x_{12}s + x_{14}t),$$

$$B_{24} = \frac{1}{8|J|}(-x_{13} + x_{12}s - x_{23}t).$$

Integration of [B]ᵀ[B]

Expression of the **[B]** matrix is given in Equation 10.24. As mentioned above, to obtain the element equation and the stiffness matrix one needs to carry out the integration in the following form

$$[k] = \iint_A [B]^T [C][B] dA \tag{10.7a}$$

or

$$[k] = C\iint_A [B]^T [B] dA \tag{10.7b}$$

For the three-node triangular element the **[B]** matrix is a constant and the above integrals can be evaluated analytically in closed form; see Equation 10.8. However, elements of the **[B]** matrix for the quadrilateral element are functions of the local coordinates s and t. Variables s and t appear both in the numerator and denominator of B_{ij}. As a result, it is often difficult to obtain analytically integration of B_{ij} over the area of the element. Numerical integration schemes are then necessary for carrying out the integration.

Numerical Integration

The idea of numerical integration in finite element analysis is similar to integration by using well-known formulas such as the trapezoidal and Simpson rules. In finite element applications, the Gauss–Legendre[1] formula is often used. In general terms, numerical integration can be expressed as

$$\int_{x_1}^{x_2} F(x)dx = \sum_{i=1}^{m} F(x_i)W_i, \tag{10.25}$$

where i denotes an integration point, m is the number of such points at which the function is evaluated and summed (Figure 10.7(a)), and the W_i are the

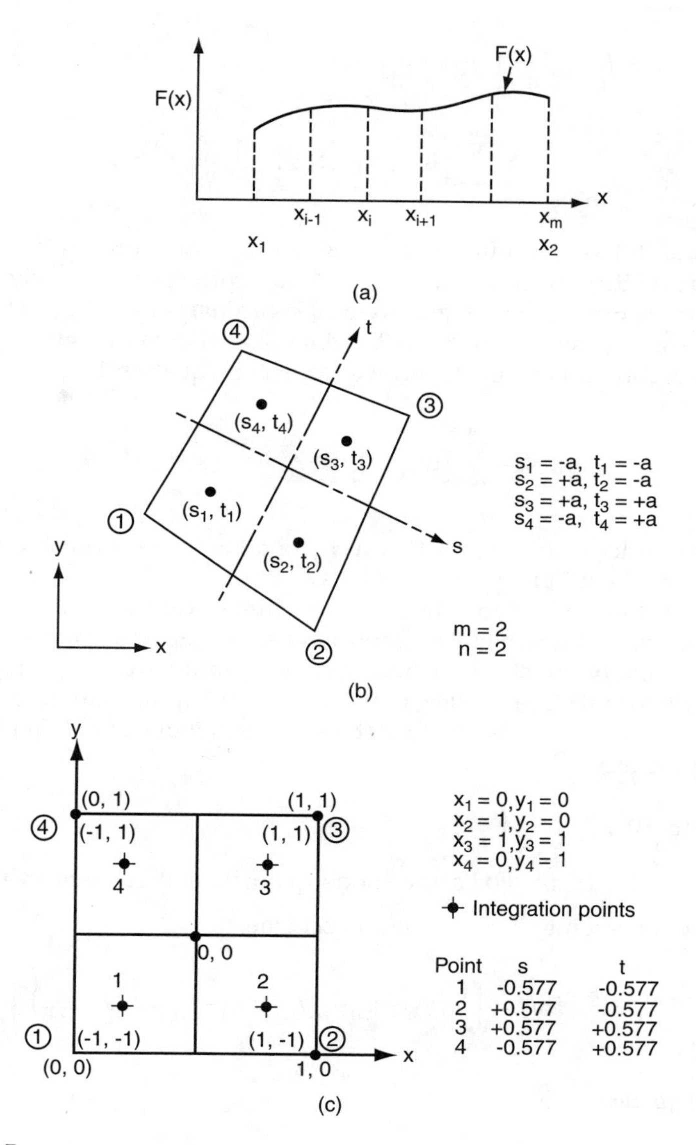

FIGURE 10.7
Numerical integration: (a) Schematic representation of numerical integration of a function, (b) Numerical integration over a quadrilateral, and (c) Example of numerical integration.

weighting functions. That is, an approximate value of the integral in Equation 10.25 is obtained by finding the summation of the values of the function at a number of points multiplied by a weighting function at each point. For the stiffness matrix in the two-dimensional space of the quadrilateral, we have

$$\left[\mathbf{k}_{\varphi}\right]=\int_{-1}^{1}\int_{-1}^{1}[\mathbf{B}]^{T}[\mathbf{B}]|J|dsdt$$
$$\simeq\sum_{i=1}^{m}\sum_{j=1}^{n}\left[\mathbf{B}(s_{i},t_{j})\right]^{T}\left[\mathbf{B}(s_{i},t_{j})\right]\left|J(s_{i},t_{j})\right|W_{i}W_{j},\tag{10.26}$$

where m and n are the number of integration points in the two coordinate directions (Figure 10.7(b)) and W_i and W_j are corresponding weights. Note that we transformed the integral to local coordinates by using $|\boldsymbol{J}|$.

As an example, we can use $m = 2$ and $n = 2$ as shown in Figure 10.7. Then total integration points are 4, and we can write Equation 10.26 as

$$[\mathbf{k}\varphi]\simeq\sum_{i=1}^{4}\left[\mathbf{B}(s_{i},t_{i})\right]^{T}\left[\mathbf{B}(s_{i},t_{i})\right]\left|J(s_{i},t_{i})\right|W_{i}.\tag{10.27}$$

The magnitudes of (s_i, t_i) and W_i can be obtained from available literature which gives them in ready-made tables.[1]

It is useful to understand the subject of numerical integration because it is used in many finite element applications. We illustrate in the following example some of the steps for numerical integration over a (square) quadrilateral (Figure 10.7(c)). In view of the fact that hand calculations of all terms can be lengthy and cumbersome, details of calculations of only one term are given in the example.

Example 10.1

Let $[k_\varphi] = \iint [B]^T[B]dA$. For a four-node quadrilateral element evaluate $k_{\varphi 11}$ using the four-point Gauss–Legendre scheme.

$$k_{\varphi 11}=\iint_{A}\left(B_{11}^{2}+B_{21}^{2}\right)dxdy=\iint_{A}\left[\left(y_{24}-y_{34}s-y_{23}t\right)^{2}+\left(-x_{24}+x_{34}s+x_{23}t\right)^{2}\right]\frac{1}{64|J|^{2}}dxdy$$

From Equation 10.20c:

$$x_{13}=x_1-x_3=0-1=-1,\qquad y_{24}=y_2-y_4=0-1=-1,$$
$$x_{24}=x_2-x_4=1-0=1,\qquad y_{13}=y_1-y_3=0-1=-1,$$
$$x_{34}=x_3-x_4=1-0=1,\qquad y_{12}=y_1-y_2=0-0=0,$$
$$x_{12}=x_1-x_2=0-1=-1,\qquad y_{34}=y_3-y_4=1-1=0,$$
$$x_{23}=x_2-x_3=1-1=0,\qquad y_{14}=y_1-y_4=0-1=-1,$$
$$x_{14}=x_1-x_4=1-1=0,\qquad y_{23}=y_2-y_3=0-1=-1,$$

and

$$|J| = \frac{1}{8}\{[(-1)(-1)-(1)(-1)]+s[(1)(0)-(-1)(0)]+t[(0)(-1)-(0)(-1)]\}$$
$$= \frac{1}{8}(1+1) = \frac{2}{8} = \frac{2A}{8} = \frac{1}{4},$$

where A is the area of the square = 1 unit^2. Note that for a (square) quadrilateral only the first term contributes to $|J|$ and its value is one fourth of the area of the square, which is constant for all points in the square. Now, the integration of Example 10.1 can be expressed by using Equation 10.27 as

$$k_{\varphi 11} = \sum_{i=1}^{4}\left[B_{11}^2(s_i, t_i) + B_{21}^2(s_i, t_i)\right]\left|J(s_i, t_i)\right| W_i, \quad i = 1, 2, 3, 4.$$

From the available numerical integration tables, such as in Reference 1, we can choose an appropriate number of points of integration and weighting function W_i. For a 2×2 or 4-point integration with Gauss–Legendre quadrature, the points of integrations are as shown in Figure 10.7(c), and the weighting functions are $W_1 = W_2 = W_3 = W_4 = 1$. Here we have used the local coordinates of the integration points only up to the third decimal; for computer implementation, values with higher decimal points are used. The computations for B_{11} and B_{21} are shown below in tabular forms:

$$B_{11} = \frac{1}{8|J|}(y_{24} - y_{34}s - y_{23}t),$$

$$B_{21} = \frac{1}{8|J|}(-x_{24} + x_{34}s + x_{23}t).$$

	Points of Integration			
	(–0.577, –0.577)	(0.577, –0.577)	(0.577, 0.577)	(–0.577, 0.577)
B_{11}	$-\frac{1.577}{2}$	$-\frac{1.577}{2}$	$-\frac{0.423}{2}$	$-\frac{0.423}{2}$
B_{21}	$-\frac{1.577}{2}$	$-\frac{0.423}{2}$	$-\frac{0.423}{2}$	$-\frac{1.577}{2}$

Therefore,

$$k_{\varphi 11} = \left[B_{11}^2(s_1, t_1) + B_{21}^2(s_1, t_1)\right]\left|J(s_1, t_1)\right| W_1$$
$$+ \left[B_{11}^2(s_2, t_2) + B_{21}^2(s_2, t_2)\right]\left|J(s_2, t_2)\right| W_2$$

$$+\left[B_{11}^2(s_3, t_3)+B_{21}^2(s_3, t_3)\right]\left|J(s_3, t_3)\right|W_3$$

$$+\left[B_{11}^2(s_4, t_4)+B_{21}^2(s_4, t_4)\right]\left|J(s_4, t_4)\right|W_4$$

$$=\frac{1}{4}\times 1\times\frac{1}{(2)^2}\left[(-1.577)^2+(-1.577)^2+(1.577)^2+(-0.423)^2\right.$$

$$\left.+(-0.423)^2+(-0.423)^2+(-0.423)^2+(-1.577)^2\right]$$

$$=\frac{1}{16}(9.9477+0.7157)=\frac{10.663}{16}=0.667.$$

Other terms of $[k_\varphi]$ and the load vector(s) can be evaluated in a similar manner.

Three-Dimensional Formulation

The main objective in this introductory text is to consider simplified one- and two-dimensional problems. However, brief statements of typical three-dimensional elements are presented below; for details the reader may consult various publications, e.g., References 2–4.

When the problem geometry is such that its dimensions in all three directions are of the same order then the problem cannot be idealized as one- or two-dimensional problems. Then the problem geometry must be discretized into three-dimensional volume elements. The two most popular volume elements are tetrahedron and brick elements, Figure 10.8.

It should be noted here that the tetrahedron has four corner nodes, six straight edges, and four planes in the bounding surface. For the brick element, on the other hand, there are 8 corner nodes, 12 straight edges, and 6 planes in the bounding surface.

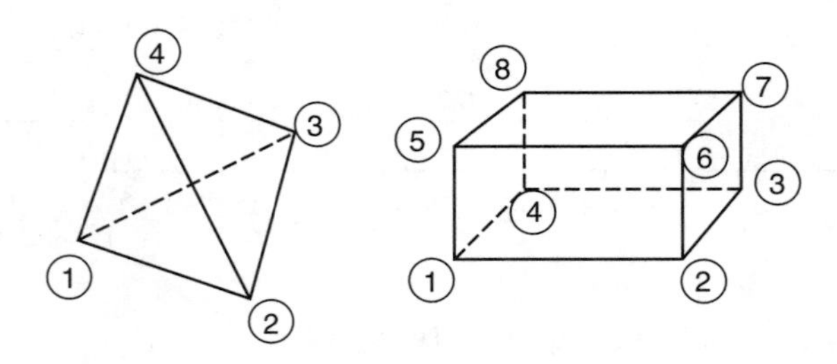

FIGURE 10.8
Tetrahedron (left figure) and brick element (right figure).

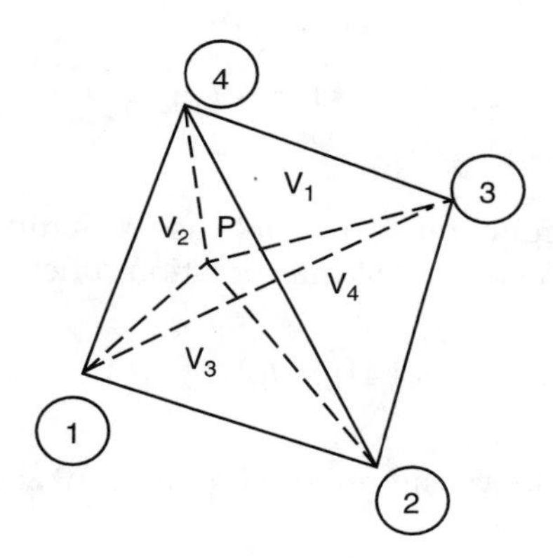

FIGURE 10.9
Point P inside a tetrahedron element.

Tetrahedron Element

Local coordinates for this element are the natural coordinates in three dimensions.

If we take a point P inside a tetrahedron element and connect it to the four vertices of the tetrahedron then the element volume is discretized into four smaller tetrahedrons of volumes V_1, V_2, V_3, and V_4 respectively as shown in Figure 10.9; e.g., V_1 is the volume of the tetrahedron formed by connecting the point P with vertices 2, 3, and 4, and V_2 is the volume of the tetrahedron whose vertices are nodes 1, 3, 4, and point P. In the same manner V_3 and V_4 can be defined. Then in the nondimensional form, the local coordinates L_1, L_2, L_3, and L_4 are defined as

$$L_i = \frac{V_i}{V} \quad i = 1, 2, 3, 4. \tag{10.28}$$

Since $L_1 + L_2 + L_3 + L_4 = 1$, there are only three independent local coordinates corresponding to the three global coordinates x, y, and z. The relationship between the global and the local coordinates is given by

$$x = L_1 x_1 + L_2 x_2 + L_3 x_3 + L_4 x_4 = \sum_{i=1}^{4} L_i x_i,$$

$$y = L_1 y_1 + L_2 y_2 + L_3 y_3 + L_4 y_4 = \sum_{i=1}^{4} L_i y_i, \tag{10.29}$$

$$z = L_1 z_1 + L_2 z_2 + L_3 z_3 + L_4 z_4 = \sum_{i=1}^{4} L_i z_i.$$

The interpolation function N_i in this case is simply equal to the local coordinate L_i. Then the matrix of interpolation functions takes the form

$$[N] = \begin{bmatrix} L_1 & L_2 & L_3 & L_4 \end{bmatrix}. \tag{10.30}$$

Using Equation 10.30 we can rewrite Equation 10.29 in the following form

$$x = [\mathbf{N}]\{\mathbf{x_n}\}, \quad y = [\mathbf{N}]\{\mathbf{y_n}\}, \quad z = [\mathbf{N}]\{\mathbf{z_n}\} \tag{10.31}$$

where $\{\mathbf{x}_n\}$, $\{\mathbf{y}_n\}$, $\{\mathbf{z}_n\}$ are 4×1 vectors of nodal coordinates.

The primary unknown φ can be expressed in terms of its values at the node points using the same interpolation functions

$$\varphi = L_1\varphi_1 + L_2\varphi_2 + L_3\varphi_3 + L_4\varphi_4 = [\mathbf{N}]\{\mathbf{q}\}. \tag{10.32}$$

Since N_1, N_2, N_3, and N_4 in Equations 10.31 and 10.32 are identical, this element is an isoparametric element.

Brick Element

Local coordinates for the brick element are introduced in such a manner that the brick is mapped into a perfect cube of dimensions $2 \times 2 \times 2$ when it is mapped in the local (r,s,t) coordinate system, Figure 10.10.

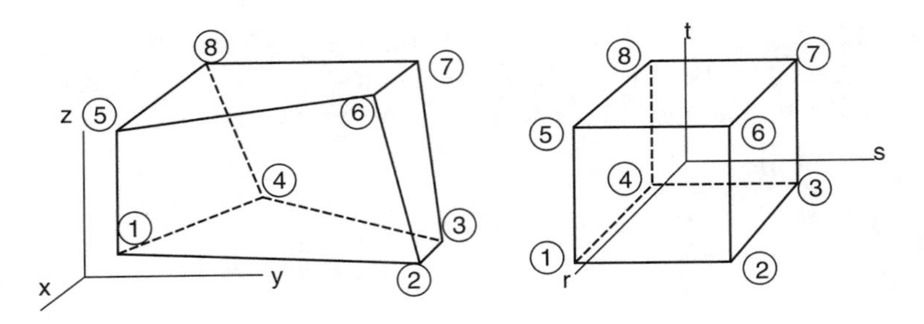

FIGURE 10.10
Brick element in global coordinate system (derived element, left figure) and local coordinate system (parent element, right figure). Local coordinates of corner nodes are (±1, ±1, ±1).

If every node has only one primary unknown or degree of freedom then the eight-node brick element should have eight interpolation functions associated with eight nodes. These interpolation functions are given by

$$
\begin{aligned}
N_1 &= \frac{1}{8}(1+r)(1-s)(1-t) \\
N_2 &= \frac{1}{8}(1+r)(1+s)(1-t) \\
N_3 &= \frac{1}{8}(1-r)(1+s)(1-t) \\
N_4 &= \frac{1}{8}(1-r)(1-s)(1-t) \\
N_5 &= \frac{1}{8}(1+r)(1-s)(1+t) \\
N_6 &= \frac{1}{8}(1+r)(1+s)(1+t) \\
N_7 &= \frac{1}{8}(1-r)(1+s)(1+t) \\
N_8 &= \frac{1}{8}(1+r)(1-s)(1+t)
\end{aligned}
\tag{10.33}
$$

Note that the value of N_i is 1 at the i-th node and 0 at all other nodes. These interpolation functions vary linearly in three mutually perpendicular directions; these are r and s constants, s and t constants, and t and r constants. In other words, it varies linearly along a line parallel to any straight edge of the element. This variation is quadratic along a line that is parallel to any of the three mutually perpendicular bounding planes (r = constant, s = constant, and t = constant) but not parallel to any straight edge. The N_i variation is cubic along a line that is neither parallel to the straight edges nor to the planes. Thus these interpolation functions are trilinear functions because these functions are linear only in three directions.

Then the matrix of interpolation functions takes the form

$$[N] = [N_1 \quad N_2 \quad N_3 \quad N_4 \quad N_5 \quad N_6 \quad N_7 \quad N_8]. \tag{10.34}$$

Global coordinates x, y, z of a point can be obtained from its local coordinates r, s, t and global coordinates of the nodes:

$$x = [\mathbf{N}]\{\mathbf{x_n}\}, \quad y = [\mathbf{N}]\{\mathbf{y_n}\}, \quad z = [\mathbf{N}]\{\mathbf{z_n}\} \tag{10.35}$$

where $\{\mathbf{x}_n\}$, $\{\mathbf{y}_n\}$, $\{\mathbf{z}_n\}$ are 8×1 vectors of nodal coordinates.

The primary unknown φ can be expressed in terms of its values at the node positions using the same interpolation functions

$$\varphi = N_1\varphi_1 + N_2\varphi_2 + N_3\varphi_3 + N_4\varphi_4 + N_5\varphi_5 + N_6\varphi_6 + N_7\varphi_7 + N_8\varphi_8 = [\mathbf{N}]\{\mathbf{q}\}. \quad (10.36)$$

Since N_i in Equations 10.35 and 10.36 are identical, this element is also an isoparametric element.

The above approximation function can be used to derive element equations using the variational or residual procedures.

Problems

10.1. Invert the matrix [A] of the following equation

$$\begin{Bmatrix} 1 \\ x \\ y \end{Bmatrix} = \begin{bmatrix} 1 & 1 & 1 \\ x_1 & x_2 & x_3 \\ y_1 & y_2 & y_3 \end{bmatrix} \begin{Bmatrix} L_1 \\ L_2 \\ L_3 \end{Bmatrix} = [\mathbf{A}] \begin{Bmatrix} L_1 \\ L_2 \\ L_3 \end{Bmatrix}$$

and obtain the matrix in Equation 10.3b.

10.2. In a heat flow problem a rectangular plate is discretized into three elements as shown in the Figure 10.11. Triangular elements (ABD and ACD) are three-node elements and the rectangular element (CDFE) is a four-node element. Let T, T_x, and T_y be the computed temperature and its gradients in x and y directions, respectively, obtained from the finite element analysis.

(a) Identify the true and false statements:

(i) T is continuous across the boundary AD.

(ii) T is continuous across the boundary CD.

(iii) T is continuous in the entire plate.

(iv) T_x is continuous across the boundary AD.

(v) T_x is continuous across the boundary CD.

(vi) T_y is continuous across the boundary AD.

(vii) T_y is continuous across the boundary CD.

(viii) Above discretization is not recommended because inter-element conformability condition is violated across the boundary CD.

(ix) If three-node elements are replaced by six-node elements then the inter-element conformability condition will be violated across the boundary CD.

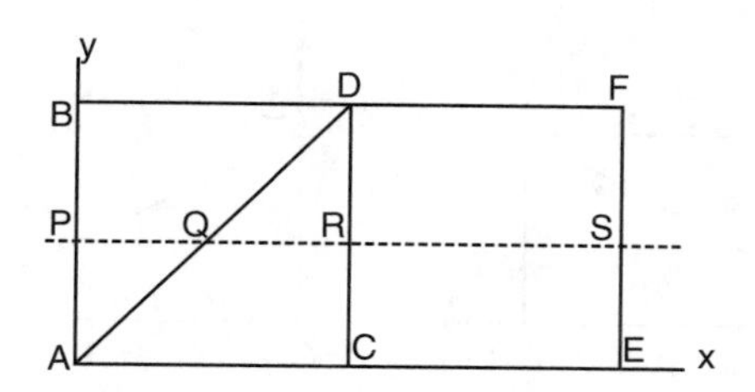

FIGURE 10.11

(b) Plot a schematic variation of T_x along the line PQRS.

(c) How should T_y vary (zero, constant, linear, quadratic etc.) along the line PQ and QR?

(d) Should T_y remain constant or vary along the line RS?

10.3. Let F denote the integral $\int_0^{t_1} f(t)dt$. F_T and F_S denote the numerical values of the same integral computed by the trapezoidal and Simpson's rule, respectively. The numerical values are obtained by dividing the integral interval into the same number of subintervals. Similar notations are used for two other functions $g(t)$ and $h(t)$. These three functions are shown in Figure 10.12. Fill the following blanks using any one of the three symbols, = (equal to), > (greater than), or < (less than).

(a) F_T _ F_S, F_T _ F,

(b) G_T _ G_S, G_T _ G,

(c) H_T _ H_S, H_T _ H

10.4. Fill the blanks using the correct words from the parenthesis (zero, constant, linear, quadratic, cubic). φ is the primary unknown; φ_x and φ_y are its derivatives with respect to x and y, respectively

(a) For a four-node quadrilateral element shown in Figure 10.13

φ_x is ________ along line 1–2, and ________ along line 1–3;

φ_y is ________ along line 1–2, and ________ along line 1–3.

(b) For a three-node triangular element shown in Figure 10.13

φ_x is ________ along line 1–2, and ________ along line 1–3;

φ_y is ________ along line 1–2, and ________ along line 1–3.

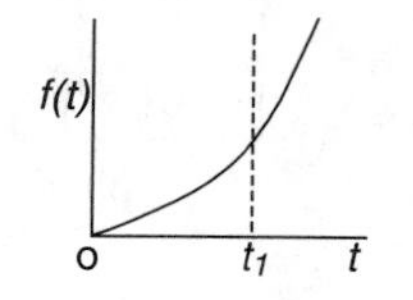

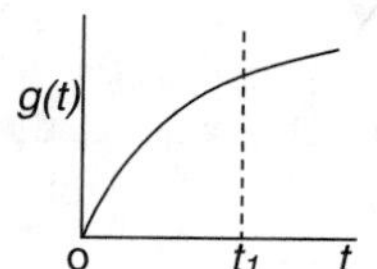

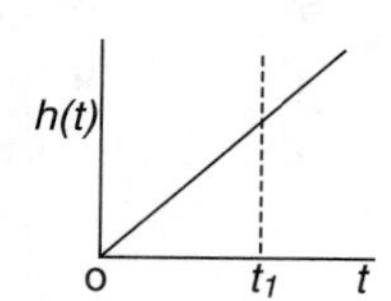

FIGURE 10.12

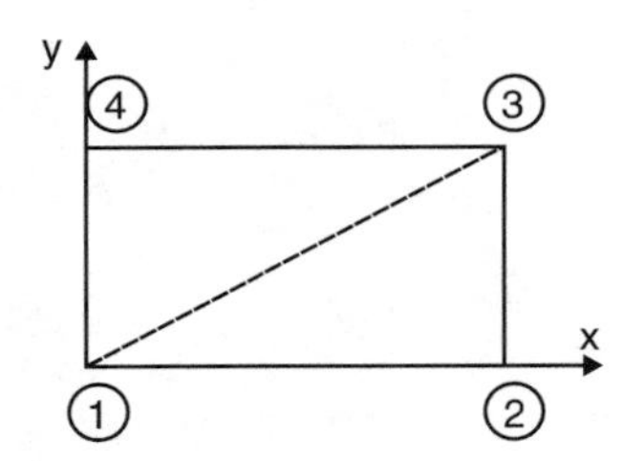

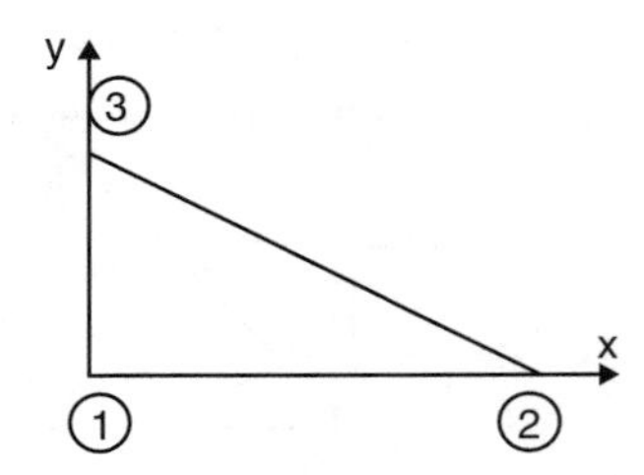

FIGURE 10.13

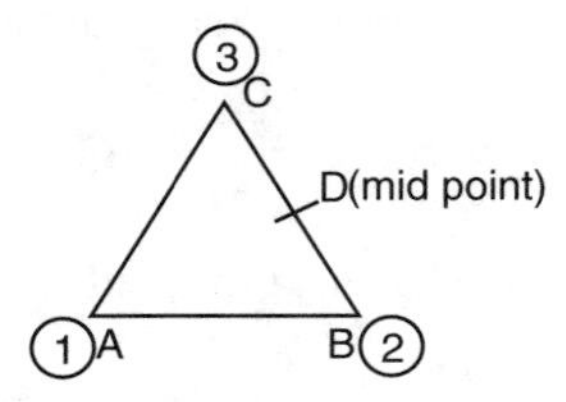

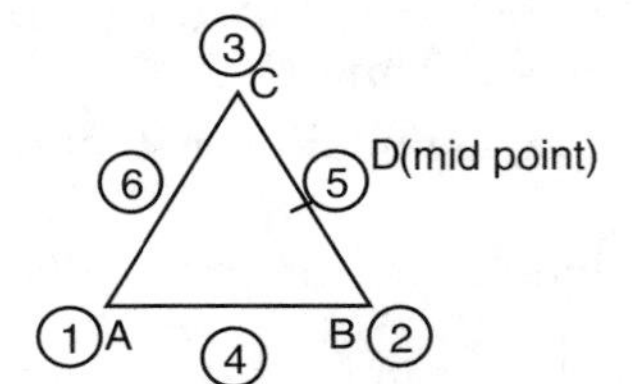

FIGURE 10.14

10.5. Plot the variations of interpolation functions N_1 and N_3 along lines AB and AD (Figure 10.14) for three-node and six-node triangular elements. In your plot give necessary values.

10.6. In a two-dimensional field problem, a plate is discretized into different elements such that the neighboring elements are not similar. State whether the interelement compatibility of the primary unknown will be satisfied across the interelement boundary in the four cases shown in Figure 10.15. Circles indicate the node positions (Figure 10.15).

10.7. (a) Express x and y in terms of s and t such that the perfect square abcd in the s–t coordinate system maps into a quadrilateral element ABCD in the x–y coordinate system as shown in Figure 10.16.

(b) Give values of x and y coordinates for the points e, f, and g. Locate these three points in the x–y coordinate system and denote those points as E, F, and G.

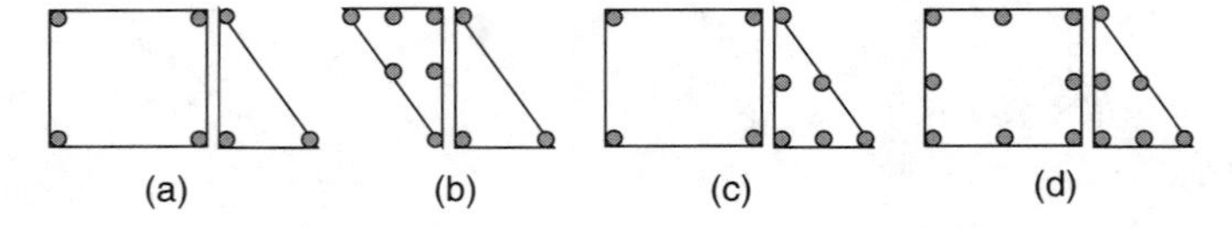

FIGURE 10.15

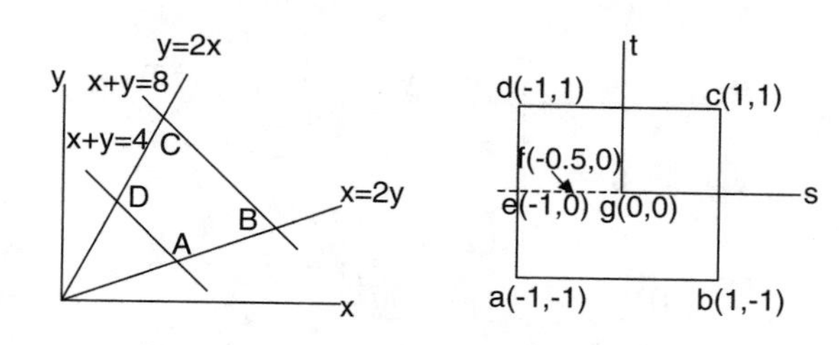

FIGURE 10.16

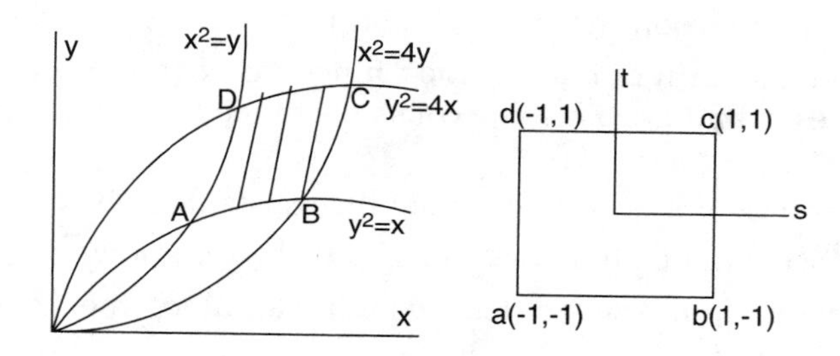

FIGURE 10.17

10.8. It is required to map the quadrilateral element shown in Figure 10.17 into a perfect square in the local s–t coordinate system.

(a) How many minimum nodal points should the element have?

(b) Write the form of the interpolation functions (that must be used for mapping the perfect square parent element into the derived element, in the global coordinate, shown in the figure) in terms of some unknown constants and coordinates s and t (i.e., $N_j = a_j + b_j s + c_j t + \ldots$).

10.9. Consider the 6-node triangular element with nodes 4, 5, and 6 located at the midpoints of the 3 sides (Figure 10.18).

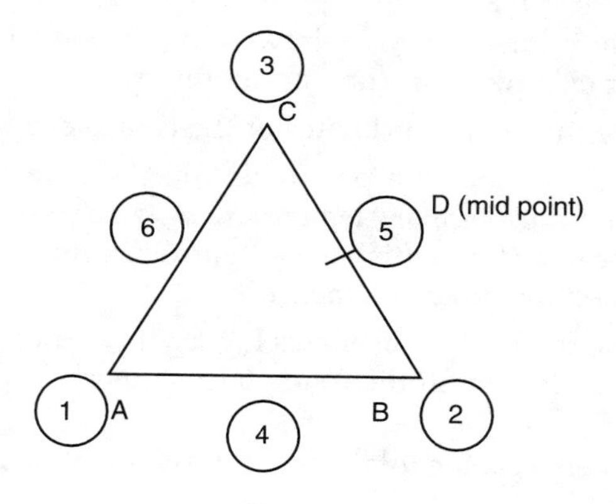

FIGURE 10.18

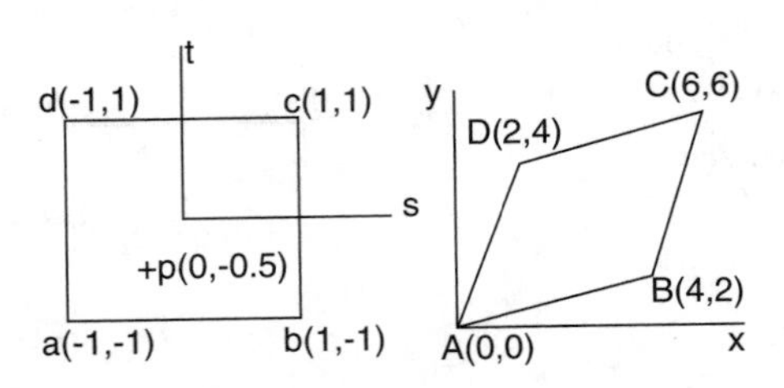

FIGURE 10.19

(a) Give local coordinates L_1, L_2, and L_3 for all six nodes.

(b) Give the form of interpolation functions N_j in terms of coordinates L_1 and L_2, and some constants in the following form, $N_j = a_j + b_jL_1 + c_jL_2 + \dots$

You do not need to evaluate the constants a_j, b_j, c_j ...

(c) Which interpolation functions should be zero along the side AB?

(d) Obtain N_1 in terms of local coordinates along the sides (i) AB and (ii) AC.

(e) How should the primary unknown vary (constant, linear, quadratic etc.) in this element along lines AB and AD?

10.10. The four-node parent element abcd in the local s–t coordinate system is mapped into the derived element ABCD in the global x–y coordinate system, as shown in the Figure 10.19. Point p(0,–0.5) in the local coordinate system is mapped into point P in the global coordinate system.

(a) Obtain the global coordinates of point P.

(b) How many lines can you draw through point P such that the primary unknown varies linearly along these lines? Draw these lines in the global coordinate system and give the coordinates of the intersection points of these lines with the element boundary.

(c) How many lines can you draw through point P along which the primary unknown can vary nonlinearly?

10.11. (a) What are the area coordinates of the centroid of a triangle?

(b) In a field problem the potentials (the field variable) at three vertices of a triangular element are ϕ_1, ϕ_2, and ϕ_3. What will be the value of the potential at the centroid if linear variation of ϕ is assumed inside the element?

10.12. Let us define the local coordinates L_{12}, L_{23}, L_{34}, and L_{41} for a 4-node quadrilateral element in the following manner

$L_{ij} = \dfrac{A_{ij}}{A}$, where A is the total area of the element and A_{ij} is the area of the triangle formed by connecting the nodes i, j, and the point P, whose coordinates are to be evaluated; see Figure 10.20.

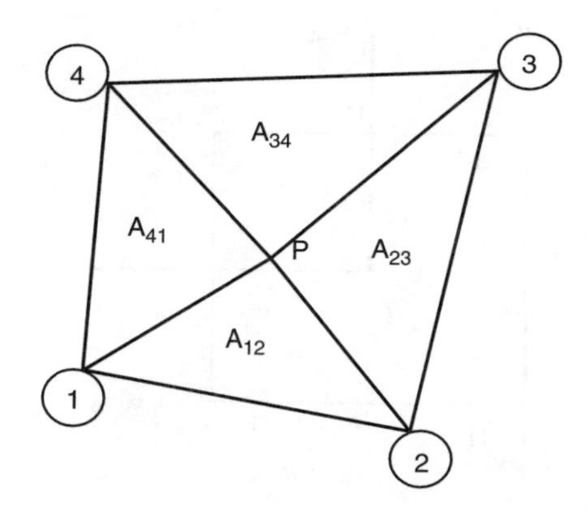

FIGURE 10.20

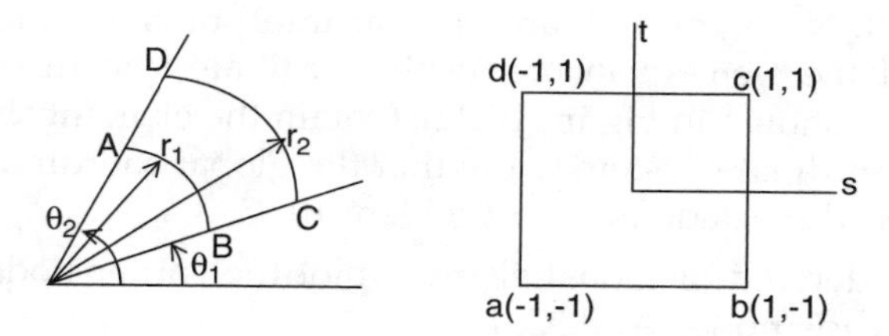

FIGURE 10.21

(a) How many of these four local coordinates should be linearly independent?

(b) Write at least one relation that must be satisfied by these coordinates.

(c) What will be the values of L_{12}, L_{14}, and $L_{23}+L_{34}$ at node 1?

(d) Which local coordinates should be 0 at nodes 2, 3, and 4?

10.13. A quadrilateral element ABCD in global coordinate system (r,θ) has the shape as shown in Figure 10.21. The element is mapped into a perfect square abcd in the local coordinate system (s,t). Obtain a relation between the global coordinates (r, θ) and the local coordinates (s,t). Give the relation in the following form:

$$s = f(r,\ \theta,\ r_1,\ r_2,\ \theta_1,\ \theta_2),\quad t = g(r,\ \theta,\ r_1,\ r_2,\ \theta_1,\ \theta_2).$$

Give explicit expressions of the functions f and g.

10.14. Relations between local (s-t) and global (r-θ) coordinates are given by,

$$r = N_1 + N_2 + 1.414N_3 + N_4$$

$$\theta = \pi(N_3 + 2N_4)/4$$

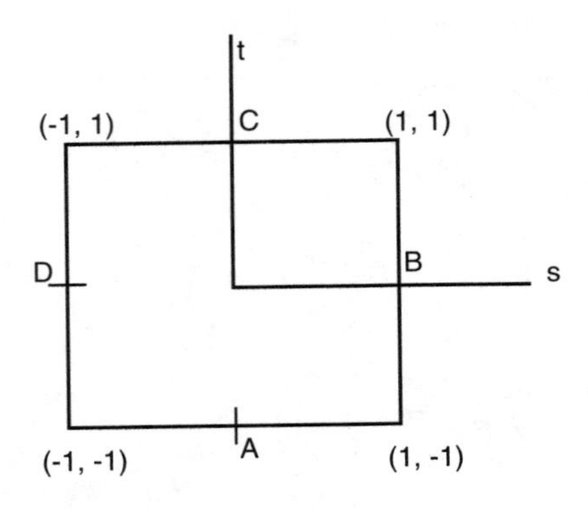

FIGURE 10.22

where N_1, N_2, N_3 and N_4 are bilinear interpolation functions. The shape of the element in the local coordinate system is a perfect square (as shown in Figure 10.22). Obtain the element shape in the global coordinate system. Note that the global coordinates are the polar coordinates defined in Problem 10.13.

In the derived or global element plot the corner nodes and the four midside nodes A, B, C, D.

References

1. Abramowitz, M. and Stegun, I. A. Eds. 1964. *Handbook Mathematical Functions with Formulas, Graphs, and Mathematical Tables*, Applied Math. Series 55, National Bureau of Standards, Washington, D.C.
2. Desai, C. S. and Abel, J. F. 1972. *Introduction to the Finite Element Method*, Van Nostrand Reinhold, New York.
3. Zienkiewics, O. C. 1977. *The Finite Element Method*, McGraw-Hill, London.
4. Bathe, K.J. 1996. *Finite Element Procedures*, Prentice-Hall, Englewood Cliffs, NJ.

11

Torsion

Introduction

Up to Chapter 9 we have considered problems that can be idealized as one-dimensional. They involved only line elements and, except for beam bending, had only one unknown at each node. In Chapter 10 we presented formulations of two- and three-dimensional problems in a general sense. Next we will consider different two-dimensional problems in this and subsequent chapters. We start with the torsion problem.

We choose torsion first because it involves only one degree of freedom or unknown at each point (node). A number of other problems, often called field problems (Chapter 12), are governed by equations similar to those for the torsion problem.

We shall illustrate here a number of formulation procedures: displacement, stress or equilibrium, and hybrid and mixed, which are based on the energy principles. For all these formulations, we approximate the behavior of a bar (Figure 11.1(a)) subjected to torsion by considering essentially the behavior of the cross section of the bar (Figure 11.1(b)). In the semi-inverse method of Saint-Venant, it is assumed that the twisting of the bar is composed of the rotations of the cross sections of the bar as in the case of a circular bar and of warping of the cross sections; the latter is constant for all cross sections.[1,2] As a consequence, no normal stress exists between the longitudinal fibers of the bar. Also there is no distortion of the planes of cross section, and hence the strain components ε_x, ε_y, and γ_{xy} vanish, and only pure shear components γ_{xz} and γ_{yz} will remain.

For the two-dimensional approximation, we shall consider discretization involving triangular elements (Figures 10.1a and 10.2); quadrilateral elements are used in Chapter 12. For the triangular case, a linear approximation for the unknown can be adopted, while for the quadrilateral isoparametric elements, a bilinear approximation can be chosen (Chapter 10). This constitutes the first two steps in the finite element formulation.

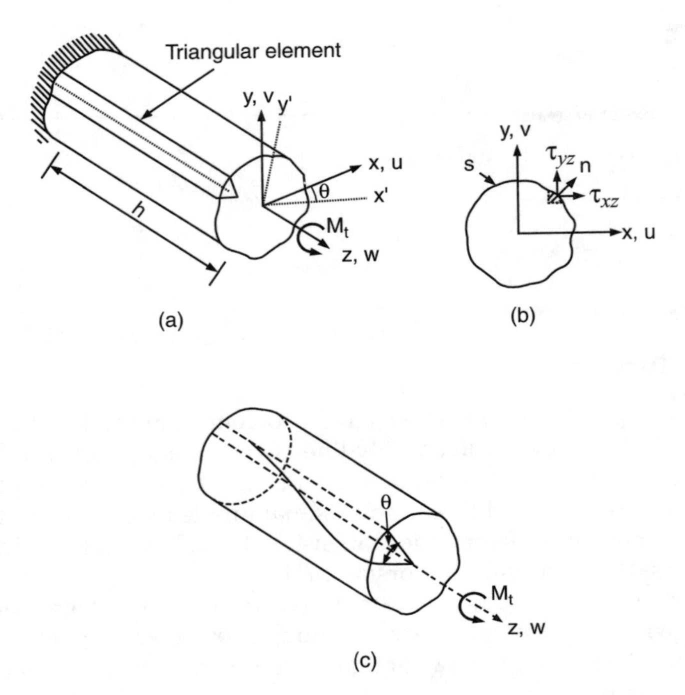

FIGURE 11.1
Torsion of bar: (a) Bar subjected to torsion, (b) Cross section of bar, and (c) Angle of twist θ.

Finite Element Formulation (Displacement Approach)

First we consider the displacement approach. According to Saint-Venant torsion,[1,2] in the absence of body forces, the differential (Laplace) equation governing torsion in a homogeneous and isotropic bar can be expressed as

$$\frac{\partial^2\psi}{\partial x^2}+\frac{\partial^2\psi}{\partial y^2}=0, \tag{11.1}$$

where x and y are the global coordinates in the plane of cross section of the bar (Figure 11.1). The warping function, ψ, is related to the displacement w in the z direction, and is assumed to be constant along the length of the bar and function only of x and y (Figure 11.1):

$$w=\theta\psi(x,y), \tag{11.2}$$

where θ is the twist of the bar per unit length under the applied twisting moment, M_t (Figure 11.1(a)). The warping function denotes a measure of warping of the cross sections of the bar (Figure 11.1(c)). The other two displacement components, u and v, are given by

$$\begin{aligned} u &= -\theta z y, \\ v &= \theta z x. \end{aligned} \tag{11.3}$$

The boundary condition along the surface of the bar is given by

$$\left(\frac{\partial \psi}{\partial x} - y\right)\frac{\partial y}{\partial s} - \left(\frac{\partial \psi}{\partial y} + x\right)\frac{dx}{ds} = 0, \tag{11.4}$$

where s is measured along the surface or boundary. Equation 11.4 indicates that the shear stress on the boundary, τ_{nz}, is 0.

Step 3. Gradient-Unknown Relation and Constitutive Law

By using Equations 11.2 and 11.3, the gradient or strain-warping function relation is[2]

$$\{\varepsilon\} = \begin{Bmatrix} \dfrac{\partial w}{\partial x} + \dfrac{\partial u}{\partial z} \\ \dfrac{\partial w}{\partial y} + \dfrac{\partial v}{\partial z} \end{Bmatrix} = \begin{Bmatrix} \gamma_{xz} \\ \gamma_{yz} \end{Bmatrix} = \begin{Bmatrix} \theta\left(\dfrac{\partial \psi}{\partial x} - y\right) \\ \theta\left(\dfrac{\partial \psi}{\partial y} + x\right) \end{Bmatrix}, \tag{11.5}$$

where $\{\varepsilon\}^T = [\gamma_{xz}\ \gamma_{yz}]$ is the vector of (shear) strain components.

In the general three-dimensional problem, there are six nonzero components of stress.[2] However, for the foregoing two-dimensional idealization, there are only two nonzero shear-stress components, $\tau_{xz} = \tau_{zx}$ and $\tau_{yz} = \tau_{zy}$ corresponding to the two shear-strain components. The constitutive relation can be expressed as

$$\{\sigma\} = \begin{Bmatrix} \tau_{xz} \\ \tau_{yz} \end{Bmatrix} = \begin{Bmatrix} G\gamma_{xz} \\ G\gamma_{yz} \end{Bmatrix} = \begin{bmatrix} G & 0 \\ 0 & G \end{bmatrix} \begin{Bmatrix} \gamma_{xz} \\ \gamma_{yz} \end{Bmatrix}, \tag{11.6}$$

where $\{\sigma\}^T = [\tau_{xz}\ \tau_{yz}]$ is the vector of (shear) stress components and G is the shear modulus.

Step 4. Derive Element Equations

For the displacement approach, the element equation can be obtained in the following manner. We start with the strain energy of the element:

$$U = \frac{1}{2}\iiint\limits_V \left(\tau_{xz}\gamma_{xz} + \tau_{yz}\gamma_{yz}\right) dV. \tag{11.7}$$

In the above expression only two terms survive since all other terms are zero. The above integral can be rewritten in the following form,

$$U = \frac{hG}{2}\iint_A \left(\gamma_{xz}^2 + \gamma_{yz}^2\right) dA \tag{11.8}$$

where h is the length of the bar and G is the shear modulus of the bar material. From Equations 11.5 and 11.8, we have

$$U = \frac{hG}{2}\iint_A \left[\theta^2\left(\frac{\partial\psi}{\partial x} - y\right)^2 + \theta^2\left(\frac{\partial\psi}{\partial y} + x\right)^2\right] dxdy. \tag{11.9}$$

In absence of any external loads in the element the strain energy is equal to the potential energy. Hence,[3,4]

$$\Pi_p = U = \frac{hG}{2}\iint_A \left[\theta^2\left(\frac{\partial\psi}{\partial x} - y\right)^2 + \theta^2\left(\frac{\partial\psi}{\partial y} + x\right)^2\right] dxdy \tag{11.10}$$

For linear variation of ψ we have

$$\begin{aligned} \psi(x,y) &= N_1\psi_1 + N_2\psi_2 + N_3\psi_3 \\ &= [\mathbf{N}_\psi]\{\mathbf{q}_\psi\}, \end{aligned} \tag{11.11}$$

where ψ (x, y) is the warping function at any point (x, y) in the triangular element and $\{\mathbf{q}_\psi\}^T = [\psi_1\ \psi_2\ \psi_3]$ is the vector of nodal values of the warping function.

Gradients of ψ or derivatives of ψ with respect to x and y can be obtained by following the steps outlined in Chapter 10 (see Equations 10.5 and 10.6):

$$\begin{Bmatrix} \dfrac{\partial\psi}{\partial x} \\ \dfrac{\partial\psi}{\partial y} \end{Bmatrix} = \frac{1}{2A}\begin{bmatrix} b_1 & b_2 & b_3 \\ a_1 & a_2 & a_3 \end{bmatrix}\begin{Bmatrix} \psi_1 \\ \psi_2 \\ \psi_3 \end{Bmatrix} = [B]\{q_\psi\} \tag{11.12}$$

Here $[\mathbf{B}]$ is the strain-warping function transformation matrix.

Hence Equation 11.5 becomes

$$\{\varepsilon\} = \begin{Bmatrix} \theta\left(\dfrac{\partial\psi}{\partial x} - y\right) \\ \theta\left(\dfrac{\partial\psi}{\partial y} + x\right) \end{Bmatrix} = \begin{Bmatrix} \theta\dfrac{\partial\psi}{\partial x} \\ \theta\dfrac{\partial\psi}{\partial y} \end{Bmatrix} + \begin{Bmatrix} -\theta y \\ \theta x \end{Bmatrix} \tag{11.13a}$$

$$= \theta[\mathbf{B}]\{\mathbf{q}_\psi\} + \begin{Bmatrix} -\theta y \\ \theta x \end{Bmatrix}. \tag{11.13b}$$

Now Equation 11.10 can be expanded as

$$\Pi_p = \frac{Gh\theta^2}{2} \iint_A \left\{ \left[\left(\frac{d\psi}{dx} \right)^2 - 2\frac{d\psi}{dx} y + y^2 \right] + \left[\left(\frac{d\psi}{dy} \right)^2 + 2\frac{d\psi}{dy} x + x^2 \right] \right\} dxdy \tag{11.14a}$$

$$= \frac{Gh\theta^2}{2} \iint_A \left(\begin{bmatrix} \frac{d\psi}{dx} & \frac{d\psi}{dy} \end{bmatrix} \begin{Bmatrix} \frac{d\psi}{dx} \\ \frac{d\psi}{dy} \end{Bmatrix} + 2\begin{bmatrix} \frac{d\psi}{dx} & \frac{d\psi}{dy} \end{bmatrix} \begin{Bmatrix} -y \\ x \end{Bmatrix} + \begin{bmatrix} y & x \end{bmatrix} \begin{Bmatrix} y \\ x \end{Bmatrix} \right) dxdy. \tag{11.14b}$$

Substitution from Equations 11.12 and 11.13 into Equation 11.14b gives

$$\Pi_p = \frac{Gh\theta^2}{2} \left(\iint_A \left[\{\mathbf{q}_\psi\}^T [\mathbf{B}]^T [\mathbf{B}] \{\mathbf{q}_\psi\} + 2\{\mathbf{q}_\psi\}^T [\mathbf{B}]^T \begin{Bmatrix} -y_m \\ x_m \end{Bmatrix} + \begin{bmatrix} y_m & x_m \end{bmatrix} \begin{Bmatrix} y_m \\ x_m \end{Bmatrix} \right] \right) dxdy, \tag{11.15}$$

where $x_m = (x_1 + x_2 + x_3)/3$ and $y_m = (y_1 + y_2 + y_3)/3$ are the assumed mean values. It is not necessary to make this assumption, however. We can substitute for x and y from Equation 10.2 and pursue the derivations, which will be somewhat more involved. Now we invoke the principle of stationary (minimum) potential energy; hence

$$\delta\Pi_p = 0 \Rightarrow \begin{cases} \frac{\partial \Pi_p}{\partial \psi_1} = 0, \\ \frac{\partial \Pi_p}{\partial \psi_2} = 0, \\ \frac{\partial \Pi_p}{\partial \psi_3} = 0, \end{cases} \tag{11.16}$$

which leads to

$$Gh\theta^2 \iint_A [\mathbf{B}]^T [\mathbf{B}] dxdy \{\mathbf{q}_\psi\} = Gh\theta^2 \iint_A [\mathbf{B}]^T \begin{Bmatrix} y_m \\ -x_m \end{Bmatrix} dxdy. \tag{11.17a}$$

Note that the last term in Π_p, being constant, does not contribute to the results. Equation 11.17a can be written in matrix form as

$$[\mathbf{k}_\psi]\{\mathbf{q}_\psi\} = \{\mathbf{Q}_\psi\}, \tag{11.17b}$$

where $[\mathbf{k}_\psi]$ is the element stiffness matrix:

$$[\mathbf{k}_\psi] = Gh\theta^2 \iint_A [\mathbf{B}]^T[\mathbf{B}]dxdy \tag{11.17c}$$

and $[\mathbf{Q}_\psi]$ is the equivalent nodal load vector

$$\{\mathbf{Q}_\psi\} = Gh\theta^2 \iint_A [\mathbf{B}]^T \begin{Bmatrix} y_m \\ -x_m \end{Bmatrix} dxdy. \tag{11.17d}$$

They can be evaluated as follows:

$$[\mathbf{k}_\psi] = \frac{Gh\theta^2}{4A^2} \iint \begin{bmatrix} b_1 & a_1 \\ b_2 & a_2 \\ b_3 & a_3 \end{bmatrix} \begin{bmatrix} b_1 & b_2 & b_3 \\ a_1 & a_2 & a_3 \end{bmatrix} dxdy. \tag{11.18a}$$

Since a_i and b_i are constants, the integral equals the area; hence

$$[\mathbf{k}_\psi] = Gh\theta^2 A[\mathbf{B}]^T[\mathbf{B}]$$

or

$$[\mathbf{k}_\psi] = \frac{Gh\theta^2}{4A} \begin{bmatrix} b_1^2 + a_1^2 & b_1b_2 + a_1a_2 & b_1b_3 + a_1a_3 \\ & b_2^2 + a_2^2 & b_2b_3 + a_2a_3 \\ & \text{sym.} & b_3^2 + a_3^2 \end{bmatrix} \tag{11.18b}$$

and

$$\{\mathbf{Q}_\psi\} = Gh\theta^2 \iint \begin{bmatrix} b_1 & a_1 \\ b_2 & a_2 \\ b_3 & a_3 \end{bmatrix} \begin{Bmatrix} y_m \\ -x_m \end{Bmatrix} dxdy \tag{11.19a}$$

$$= \frac{Gh\theta^2 A}{2A} \begin{bmatrix} b_1 & a_1 \\ b_2 & a_2 \\ b_3 & a_3 \end{bmatrix} \begin{Bmatrix} y_m \\ -x_m \end{Bmatrix}$$

$$= \frac{Gh\theta^2}{2} \begin{Bmatrix} b_1 y_m - a_1 x_m \\ b_2 y_m - a_2 x_m \\ b_3 y_m - a_3 x_m \end{Bmatrix}. \tag{11.19b}$$

We note that the term $Gh\theta^2$ is common on both sides of Equation 11.17a and can be deleted, because of the assumption that the bar is isotropic and the problem is governed by the Laplace equation (Equation 11.1).

Step 5. Assembly

Example 11.1. Torsion of Square Bar: Warping Function Approach

Equations 11.18b and 11.19b can now be used to generate stiffness matrices,and load vectors for all the elements in a discretized body. For instance, Figure 11.2(a) shows a square bar, 2 cm × 2 cm, discretized in four elements. Figure 11.2(b) shows the four elements with their local and global node numbers.

For this example assume the following properties:

$$G = 1\ \text{N/cm}^2,$$

$$h = 1\ \text{cm},$$

$$\theta = 1\ \text{rad/cm}.$$

The area of each triangle $A = 1\ \text{cm}^2$. These properties are chosen in order to illustrate the procedures; they do not necessarily refer to practical problems. The quantities a_i and b_i for the four elements are as follows:

Element 1:

$$\begin{aligned} a_1 &= x_3 - x_2 = 1 - 2 = -1, & b_1 &= y_2 - y = 0 - 1 = -1 \\ a_2 &= x_1 - x_3 = 0 - 1 = -1, & b_2 &= y_3 - y_1 = 1 - 0 = 1, \\ a_3 &= x_2 - x_1 = 2 - 0 = 2, & b_3 &= y_1 - y_2 = 0 - 0 = 0, \end{aligned} \tag{11.20a}$$

and $x_m = (0 + 2 + 1)/3 = 1$, $y_m = (0 + 0 + 1)/3 = \frac{1}{3}$.

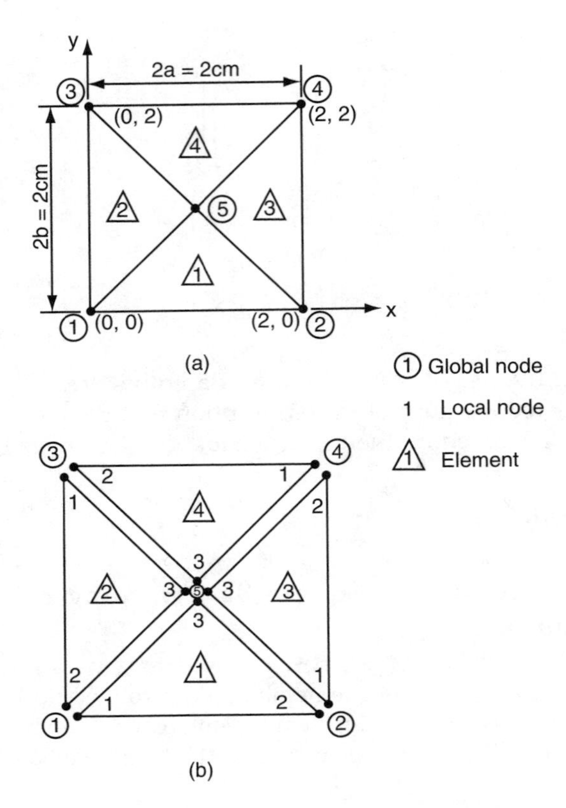

FIGURE 11.2
Torsion of square bar: (a) Square bar and mesh and (b) Elements with local and global nodes.

Element 2:

$$a_1 = 1 - 0 = 1, \qquad b_1 = 0 - 1 = -1,$$
$$a_2 = 0 - 1 = -1, \qquad b_2 = 1 - 2 = -1, \tag{11.20b}$$
$$a_3 = 0 - 0 = 0, \qquad b_3 = 2 - 0 = 2,$$

and $x_m = (1 + 0 + 0)/3 = \frac{1}{3}$, $y_m = (2 + 0 + 1)/3 = 1$.

Element 3:

$$a_1 = 1 - 2 = -1, \qquad b_1 = 2 - 1 = 1,$$
$$a_2 = 2 - 1 = 1, \qquad b_2 = 1 - 0 = 1, \tag{11.20c}$$
$$a_3 = 2 - 2 = 0, \qquad b_3 = 0 - 2 = -2,$$

and $x_m = (2 + 2 + 1)/3 = \frac{5}{3}$, $y_m = (0 + 2 + 1)/3 = 1$.

Element 4:

$$
\begin{aligned}
a_1 &= 1-0=1, & b_1 &= 2-1=1,\\
a_2 &= 2-1=1, & b_2 &= 1-2=-1,\\
a_3 &= 0-2=-2, & b_3 &= 2-2=0,
\end{aligned}
\tag{11.20d}
$$

and $x_m = (2 + 0 + 1)/3 = 1$, $y_m = (2 + 2 + 1)/3 = \frac{5}{3}$.

Substitution of these values into Equations 11.18b and 11.19b gives the following:

Element 1:

$$
\begin{array}{ll|l}
\text{Global} & \rightarrow & 1 \quad 2 \quad 5 \quad \downarrow\text{Local} \quad \downarrow\text{Global}\\
\downarrow & \text{Local} \rightarrow & 1 \quad 2 \quad 3
\end{array}
$$

$$
\begin{matrix} 1 & 1 \\ 2 & 2 \\ 5 & 3 \end{matrix} \qquad
\frac{1}{4}\begin{bmatrix} 2 & 0 & -2 \\ 0 & 2 & -2 \\ -2 & -2 & 4 \end{bmatrix}
\begin{Bmatrix} \psi_1^1 = \psi_1 \\ \psi_2^1 = \psi_2 \\ \psi_3^1 = \psi_5 \end{Bmatrix}
= \frac{1}{2}\begin{Bmatrix} \frac{2}{3} \\ \frac{4}{3} \\ -2 \end{Bmatrix}. \tag{11.21a}
$$

Element 2:

$$
\begin{array}{ll|l}
\text{Global} & \rightarrow & 3 \quad 1 \quad 5\\
\downarrow & \text{Local} \rightarrow & 1 \quad 2 \quad 3
\end{array}
$$

$$
\begin{matrix} 3 & 1 \\ 1 & 2 \\ 5 & 3 \end{matrix} \qquad
\frac{1}{4}\begin{bmatrix} 2 & 0 & -2 \\ 0 & 2 & -2 \\ -2 & -2 & 4 \end{bmatrix}
\begin{Bmatrix} \psi_1^2 = \psi_3 \\ \psi_2^2 = \psi_1 \\ \psi_3^2 = \psi_5 \end{Bmatrix}
= \frac{1}{2}\begin{Bmatrix} -\frac{4}{3} \\ -\frac{2}{3} \\ 2 \end{Bmatrix}. \tag{11.21b}
$$

Element 3:

$$
\begin{array}{ll|l}
\text{Global} & \rightarrow & 2 \quad 4 \quad 5\\
\downarrow & \text{Local} \rightarrow & 1 \quad 2 \quad 3
\end{array}
$$

$$
\begin{matrix} 2 & 1 \\ 4 & 2 \\ 5 & 3 \end{matrix} \qquad
\frac{1}{4}\begin{bmatrix} 2 & 0 & -2 \\ 0 & 2 & -2 \\ -2 & -2 & 4 \end{bmatrix}
\begin{Bmatrix} \psi_1^3 = \psi_2 \\ \psi_2^3 = \psi_4 \\ \psi_3^3 = \psi_5 \end{Bmatrix}
= \frac{1}{2}\begin{Bmatrix} \frac{8}{3} \\ -\frac{2}{3} \\ -2 \end{Bmatrix}. \tag{11.21c}
$$

Element 4:

$$\begin{array}{ccc} \text{Global} & \rightarrow & 4 \quad 3 \quad 5 \\ \downarrow\ \text{Local} & \rightarrow & 1 \quad 2 \quad 3 \end{array}$$

$$\begin{matrix} 4 & 1 \\ 3 & 2 \\ 5 & 3 \end{matrix} \quad \frac{1}{4}\begin{bmatrix} 2 & 0 & -2 \\ 0 & 2 & -2 \\ -2 & -2 & 4 \end{bmatrix}\begin{Bmatrix} \psi_1^4 = \psi_4 \\ \psi_2^4 = \psi_3 \\ \psi_3^4 = \psi_5 \end{Bmatrix} = \frac{1}{2}\begin{Bmatrix} \frac{2}{3} \\ -\frac{8}{3} \\ 2 \end{Bmatrix}. \tag{11.21d}$$

In Equation 11.21 the superscript denotes an element, and we have shown the relations between local and global node numbers. For the particular node numbering chosen in this example the coefficients of the element matrices are the same. In fact, even if a different numbering is used, if the elements are of equal dimensions, $[\mathbf{k}_\psi]$ can be generated once it is obtained for one element.

The assembly procedure is carried out by observing the fact that the values of ψ at common nodes are compatible. Thus, by adding the local coefficients in appropriate locations in the global relation, we obtain

$$\text{Global} \rightarrow \quad 1 \quad 2 \quad 3 \quad 4 \quad 5$$

$$\begin{matrix} 1 \\ 2 \\ 3 \\ 4 \\ 5 \end{matrix} \quad \frac{1}{4}\begin{bmatrix} (2+2) & 0 & 0 & 0 & (-2-2) \\ 0 & (2+2) & 0 & 0 & (-2-2) \\ 0 & 0 & (2+2) & 0 & (-2-2) \\ 0 & 0 & 0 & (2+2) & (-2-2) \\ (-2-2) & (-2-2) & (-2-2) & (-2-2) & (4+4+4+4) \end{bmatrix}\begin{Bmatrix} \psi_1 \\ \psi_2 \\ \psi_3 \\ \psi_4 \\ \psi_5 \end{Bmatrix} \tag{11.22a}$$

$$= \frac{1}{2}\begin{Bmatrix} \frac{2}{3} - \frac{2}{3} = 0 \\ \frac{4}{3} + \frac{8}{3} = 4 \\ -\frac{4}{3} - \frac{8}{3} = -4 \\ -\frac{2}{3} + \frac{2}{3} = 0 \\ -2+2-2+2=0 \end{Bmatrix}$$

or

$$\begin{bmatrix} 4 & 0 & 0 & 0 & -4 \\ 0 & 4 & 0 & 0 & -4 \\ 0 & 0 & 4 & 0 & -4 \\ 0 & 0 & 0 & 4 & -4 \\ -4 & -4 & -4 & -4 & 16 \end{bmatrix} \begin{Bmatrix} \psi_1 \\ \psi_2 \\ \psi_3 \\ \psi_4 \\ \psi_5 \end{Bmatrix} = \frac{1}{2} \begin{Bmatrix} 0 \\ 16 \\ -16 \\ 0 \\ 0 \end{Bmatrix}, \tag{11.22b}$$

or

$$[\mathbf{K}]\{\mathbf{r}\} = \{\mathbf{R}\}, \tag{11.22c}$$

where $[\mathbf{K}]$ is the assemblage stiffness matrix, $\{\mathbf{r}\}$ is the vector of global nodal warping functions, and $\{\mathbf{R}\}$ is the assemblage nodal forcing parameter vector.

Since the values of the warping function are relative, the boundary conditions can be introduced by assigning datum value of ψ to one of the nodes. Then we shall obtain relative values of nodal warping functions. For example, assume $\psi_1 = 0$. Then the modified equations are obtained by deleting the first row and column in Equation 11.22b (see Chapter 3):

$$\begin{bmatrix} 4 & 0 & 0 & -4 \\ 0 & 4 & 0 & -4 \\ 0 & 0 & 4 & -4 \\ -4 & -4 & -14 & 16 \end{bmatrix} \begin{Bmatrix} \psi_2 \\ \psi_3 \\ \psi_4 \\ \psi_5 \end{Bmatrix} = \begin{Bmatrix} 8 \\ -8 \\ 0 \\ 0 \end{Bmatrix}. \tag{11.22d}$$

Solution of these equations yields the primary unknowns as $\psi_1 = 0$ cm^2/rad (given), $\psi_2 = 2$, $\psi_3 = -2$, $\psi_4 = 0$, $\psi_5 = 0$.

Step 6. Secondary Quantities

The secondary quantities can be the shear stresses and twisting moment. Equation 11.6 can be used to find the shear stresses in the four elements:

Element 1:

$$\begin{aligned} \begin{Bmatrix} \tau_{xz} \\ \tau_{yz} \end{Bmatrix} &= G\theta \begin{Bmatrix} \dfrac{\partial \psi}{\partial x} \\ \dfrac{\partial \psi}{\partial y} \end{Bmatrix} + G\theta \begin{Bmatrix} -y_m \\ x_m \end{Bmatrix} \\ &= \frac{1 \times 1}{2} \begin{Bmatrix} b_1\psi_1 + b_2\psi_2 + b_3\psi_3 \\ a_1\psi_1 + a_2\psi_2 + a_3\psi_3 \end{Bmatrix} + 1 \times 1 \begin{Bmatrix} -y_m \\ x_m \end{Bmatrix}. \end{aligned} \tag{11.23}$$

Substitution from Equation 11.20a then gives

$$\begin{Bmatrix} \tau_{xz} \\ \tau_{yz} \end{Bmatrix} = \frac{1}{2}\begin{Bmatrix} -1\times 0+1\times 2+0\times 0 \\ -1\times 0-1\times 2+2\times 0 \end{Bmatrix} + \begin{Bmatrix} -\frac{1}{3} \\ 1 \end{Bmatrix}$$

$$= \frac{1}{2}\begin{Bmatrix} 2 \\ -2 \end{Bmatrix} + \begin{Bmatrix} -\frac{1}{3} \\ 1 \end{Bmatrix}$$

$$= \begin{Bmatrix} \frac{2}{3} \\ 0 \end{Bmatrix}.$$

Element 2:

$$\begin{Bmatrix} \tau_{xz} \\ \tau_{yz} \end{Bmatrix} = \begin{Bmatrix} 0 \\ -\frac{2}{3} \end{Bmatrix}.$$

Element 3:

$$\begin{Bmatrix} \tau_{xz} \\ \tau_{yz} \end{Bmatrix} = \begin{Bmatrix} 0 \\ \frac{2}{3} \end{Bmatrix}.$$

Element 4:

$$\begin{Bmatrix} \tau_{xz} \\ \tau_{yz} \end{Bmatrix} = \begin{Bmatrix} -\frac{2}{3} \\ 0 \end{Bmatrix}.$$

The shear stresses are plotted in Figure 11.3.

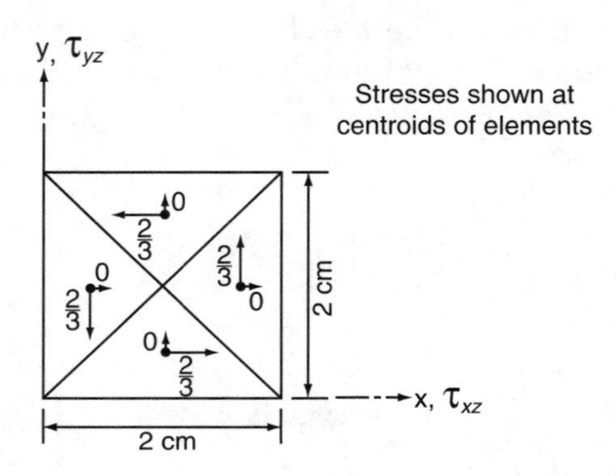

FIGURE 11.3
Plots of computed shear stresses: warping function approach.

Twisting Moment

The expression for twisting moment M_t in the bar is given by[2-4]

$$M_t = G\theta \iint_A \left(-y\frac{\partial\psi}{\partial x} + x\frac{\partial\psi}{\partial y} + x^2 + y^2 \right) dxdy. \qquad (11.24a)$$

Often $M_t/G\theta$ is called the torsional constant and is used to express solution for the twisting moment in the bar.

If we assume x_m and y_m as average constant values associated with the first two terms in the integrand and since $\partial\psi/\partial x$ and $\partial\psi/\partial y$ are constant for the linear approximation, the integral in Equation 11.24a can be approximated as

$$\frac{M_t}{G\theta} = \sum_{m=1}^{M} M_{tm} = \sum_{m=1}^{M} \left[-y_m A \left(\frac{\partial\psi}{\partial x} \right)_m + x_m A \left(\frac{\partial\psi}{\partial y} \right)_m + \overline{x}_m + \overline{y}_m \right], \qquad (11.24b)$$

where (Equation 11.34a)

$$\overline{x}_m = \frac{A}{6}\left(x_1^2 + x_2^2 + x_3^2 + x_1x_2 + x_2x_3 + x_1x_3\right)$$

and

$$\overline{y}_m = \frac{A}{6}\left(y_1^2 + y_2^2 + y_3^2 + y_1y_2 + y_2y_3 + y_1y_3\right).$$

M is the total number of elements, m denotes an element, and M_{tm} denotes the contribution to twisting moment by element m.

The total twisting moment can now be found as the sum of element twisting moments. Substitute for $\partial\psi/\partial x$ and $\partial\psi/\partial y$ from Equation 11.12, and the coordinates of the nodes of each element lead to the following:

Element 1:

$$\frac{\partial\psi}{\partial x} = \tfrac{1}{2}(-1 \times 0 + 1 \times 2 + 0 \times 0) = 1,$$

$$\frac{\partial\psi}{\partial y} = \tfrac{1}{2}(-1 \times 0 - 1 \times 2 + 2 \times 0) = -1,$$

$$\overline{x}_1 + \overline{y}_1 = \tfrac{8}{6},$$

$$M_{t_1} = -\tfrac{1}{3} \times 1 + 1 \times (-1) + \tfrac{8}{6} = -\tfrac{4}{3} + \tfrac{8}{6}.$$

Element 2:

$$\frac{\partial \psi}{\partial x} = \tfrac{1}{2}\left[-1 \times (-2) - 1 \times 0 + 2 \times 0\right] = 1,$$

$$\frac{\partial \psi}{\partial y} = \tfrac{1}{2}\left[1 \times (-2) - 1 \times 0 + 0 \times 0\right] = -1,$$

$$\bar{x}_2 + \bar{y}_2 = \tfrac{8}{6},$$

$$M_{t_2} = -1 \times 1 + \tfrac{1}{3} \times (-1) + \tfrac{8}{6} = -\tfrac{4}{3} + \tfrac{8}{6}.$$

Element 3:

$$\frac{\partial \psi}{\partial x} = \tfrac{1}{2}\left[1 \times 2 + 1 \times 0 - 2 \times 0\right] = 1,$$

$$\frac{\partial \psi}{\partial y} = \tfrac{1}{2}\left[-1 \times 2 + 1 \times 0 + 0 \times 0\right] = -1,$$

$$\bar{x}_3 + \bar{y}_3 = \tfrac{24}{6},$$

$$M_{t_3} = -1 - \tfrac{5}{3} + \tfrac{24}{6} = -\tfrac{8}{3} + \tfrac{24}{6}.$$

Element 4:

$$\frac{\partial \psi}{\partial x} = \tfrac{1}{2}\left[1 \times 0 - 1 \times (-2) + 0 \times 0\right] = 1,$$

$$\frac{\partial \psi}{\partial y} = \tfrac{1}{2}\left[1 \times 0 + 1 \times (-2) - 2 \times 0\right] = -1,$$

$$\bar{x}_4 + \bar{y}_4 = \tfrac{24}{6},$$

$$M_{t_4} = -\tfrac{5}{3} - 1 + \tfrac{24}{6} = -\tfrac{8}{3} + \tfrac{24}{6}.$$

Therefore

Element 1:

$$\begin{aligned} M_t &= M_{t_1} + M_{t_2} + M_{t_3} + M_{t_4} \\ &= -\tfrac{1}{3}(4 + 4 + 8 + 8) + \tfrac{1}{6}(8 + 8 + 24 + 24) \\ &= -\tfrac{24}{3} + \tfrac{64}{6} = -8 + 10.6666 \\ &= 2.6666 \text{ N-cm}. \end{aligned}$$

Comparisons of Numerical Predictions and Closed Form Solutions

Twisting Moment

Based on the stress function approach presented subsequently, a closed form solution for the twisting moment for a square bar is given by[2]

$$M_t \simeq 0.1406G\theta(2a)^4 \tag{11.25}$$

or

$$\frac{M_t}{G\theta} \simeq 0.1406(2a)^4,$$

where a = the half width of the bar. Hence, the closed form value of the torsional constant is

$$\begin{aligned}\frac{M_t}{G\theta} &\simeq 0.1406(2\times 1)^4\\ &\simeq 0.1406\times 16\\ &\simeq 2.250\ \text{cm}^4.\end{aligned}$$

The value computed from the finite element procedure is 2.666; hence the error is

$$\begin{aligned}\text{error} &= \frac{2.2500-2.6666}{2.2500}\times 100\\ &\simeq -18.5\%.\end{aligned}$$

Shear Stresses

From the stress function approach (see next section), the closed form values of the components of shear stresses, Equation 11.28a below, can be obtained from the following expression for closed form solution for stress function in a rectangular bar:[2]

$$\varphi^* = \frac{32G\theta a^2}{\pi^3}\sum_{n=1,3,5,\ldots}^{\infty}\frac{1}{n^3}(-1)^{\frac{n-1}{2}}\left[1-\frac{\cosh(n\pi y/2a)}{\cosh(n\pi b/2a)}\right]\cos\left(\frac{n\pi x}{2a}\right) \tag{11.26a}$$

$$\tau_{xz}^* = \frac{\partial \varphi}{\partial y} = \frac{16G\theta a}{\pi^2} \sum_{n=1,3,5,\ldots}^{\infty} \frac{1}{n^2}(-1)^{\frac{n-1}{2}} \frac{\sinh(n\pi y/2a)}{\cosh(n\pi b/2a)} \cos\left(\frac{n\pi x}{2a}\right) \tag{11.26b}$$

$$\tau_{yz}^* = -\frac{\partial \varphi}{\partial x} = \frac{16G\theta a}{\pi^2} \sum_{n=1,3,5,\ldots}^{\infty} \frac{1}{n^2}(-1)^{\frac{n-1}{2}} \left[1 - \frac{\cosh(n\pi y/2a)}{\cosh(n\pi b/2a)}\right] \sin\left(\frac{n\pi x}{2a}\right) \tag{11.26c}$$

where x and y are measured from the center point of the bar, and a and b denote half the dimensions of the cross section of the bar, respectively.

In the finite element analysis, we assumed linear approximation for the warping function; hence, the distribution of shear stresses is constant at every point in the element. This element can, therefore, be called the constant strain or stress triangle (CST). If we had used a higher-order approximation, shear stresses would vary from point to point in the element. In view of the constant values for the shear stresses, it is often customary, as shown in Figure 11.3, to attach them at the centroids of the elements. For the purpose of comparison and error analysis below, we compute the closed form or exact values of shear stresses at the centroid of an element. It must be understood that this is only an approximation, since the computed shear stresses are constant over an element as a part of the formulation, and are attached at the centroid only for convenience.

The subject of error analysis is important and wide in scope. Our purpose is to introduce the topic by presenting comparisons between computed and closed form solutions for some of the problems in this chapter. In making these comparisons we have considered only typical elements and nodes.

For Example 11.1, let us choose element 3. The centroidal coordinates of this element are

$$x = 0.667 \text{ and } y = 0.000.$$

Then use of Equations 11.26b and 11.26c gives closed form values of shear stresses as

$$\tau_{xz}^* = 0.000 \text{ N/cm}^2; \quad \tau_{yz}^* = 0.775 \text{ N/cm}^2.$$

These and the subsequent closed form solutions are obtained by using n = 25 in Equation 11.26. The errors in the two components are then given by

$$\begin{aligned} \text{error in } \tau_{xz} &= \frac{\tau_{xz}^* - \tau_{xz}}{\tau_{xz}^*} \times 100 \\ &= 0.000 - 0.000 \\ &= 0.00\%, \end{aligned}$$

$$\text{and error in } \tau_{yz} = \frac{\tau^*_{yz} - \tau_{yz}}{\tau^*_{yz}} \times 100$$

$$= \frac{0.775 - 0.667}{0.775}$$

$$= 14.00\%.$$

Comment

The foregoing comparisons show that despite linear approximation, and the rather crude mesh, the computed results yield realistic solutions of the same order of magnitude. Here we used a coarse mesh mainly to illustrate hand calculations. In general, however, one should use a finer mesh, and the solution will improve significantly. For instance, with a finer mesh, say with 200 elements, the computed value of the torsional constant for the square bar would involve an error of less than 1.0%.

Stress Approach

The torsion problem can be expressed by using the concept of stress function, φ;[1-7] this approach represents the same phenomenon as represented by the warping function, but it is an alternative or dual representation for torsion. The governing differential equation for a linear, isotropic, and homogeneous medium, in terms of φ, according to Prandtl,[1,2] is

$$\frac{1}{G}\frac{\partial^2 \varphi}{dx^2} + \frac{1}{G}\frac{\partial^2 \varphi}{dy^2} = -2\theta = -\bar{Q}(x, y) \tag{11.27}$$

where $\bar{Q}$ is the equivalent forcing parameter, which can be a prescribed known function of (x, y).

The two nonzero shear stresses in terms of φ are

$$\{\sigma\} = \begin{Bmatrix} \tau_{xz} \\ \tau_{yz} \end{Bmatrix} = \begin{Bmatrix} \dfrac{\partial \varphi}{\partial y} \\ -\dfrac{\partial \varphi}{\partial x} \end{Bmatrix}. \tag{11.28a}$$

The relation between the stress function and warping function is

$$\frac{\partial \varphi}{\partial y} = G\theta\left(\frac{\partial \psi}{\partial x} - y\right),$$
$$-\frac{\partial \varphi}{\partial x} = G\theta\left(\frac{\partial \psi}{\partial y} + x\right). \tag{11.28b}$$

The energy function associated with Equation 11.27 is the complementary energy, expressed as (see Chapter 3)

$$\Pi_c = U_c + W_c = U_c - W,$$

or

$$\Pi_c = \iint \frac{1}{2G}\left[\left(\frac{\partial \varphi}{\partial x}\right)^2 + \left(\frac{\partial \varphi}{\partial y}\right)^2\right] dxdy - \iint \bar{Q}\varphi dxdy, \tag{11.29a}$$

or

$$\Pi_c = \frac{1}{2G}\iint \begin{bmatrix} \frac{\partial \varphi}{\partial x} & \frac{\partial \varphi}{\partial y} \end{bmatrix} \begin{Bmatrix} \frac{\partial \varphi}{\partial x} \\ \frac{\partial \varphi}{\partial y} \end{Bmatrix} dxdy - \iint (2\theta)\varphi dxdy. \tag{11.29b}$$

Here U_c is the complementary strain energy, W_c is the complementary potential of external loads, and W is the work of external loads. One can derive the governing Equation 11.27 by minimizing the functional in Equation 11.29a, following the steps described in Chapter 2.

The unknown φ for a triangular element is now expressed as

$$\varphi = N_1\varphi_1 + N_2\varphi_2 + N_3\varphi_3 = [\mathbf{N}_s]\{\mathbf{q}_s\}, \tag{11.29c}$$

where $\{\mathbf{q}_s\}^T = [\varphi_1\ \varphi_2\ \varphi_3]$. The gradient–$\varphi$ relation is

$$\{\mathbf{g}\} = \begin{Bmatrix} g_x \\ g_y \end{Bmatrix} = \begin{Bmatrix} \frac{\partial \varphi}{\partial x} \\ \frac{\partial \varphi}{\partial y} \end{Bmatrix} = [\mathbf{B}]\{\mathbf{q}_s\}. \tag{11.30}$$

Substitution of Equations 11.29c and 11.30 into Equation 11.29b yields

$$\Pi_c = \tfrac{1}{2}\iint \{\mathbf{q}_s\}^T[\mathbf{B}]^T[\mathbf{D}][\mathbf{B}]\{\mathbf{q}_s\} dxdy - \iint (2\theta)[\mathbf{N}_s]\{\mathbf{q}_s\} dxdy, \tag{11.31a}$$

where

$$[\mathbf{D}] = \frac{1}{G}\begin{bmatrix} 1 & 0 \\ 0 & 1 \end{bmatrix}$$

is the strain–stress matrix. By invoking the principle of stationary (minimum) complementary energy, we have

$$\delta\Pi_c = 0 \Rightarrow \begin{cases} \dfrac{\partial \Pi_c}{\partial \varphi_1} = 0, \\ \dfrac{\partial \Pi_c}{\partial \varphi_2} = 0, \\ \dfrac{\partial \Pi_c}{\partial \varphi_3} = 0, \end{cases} \tag{11.31b}$$

or

$$\iint [\mathbf{B}]^T[\mathbf{D}][\mathbf{B}]dxdy\{\mathbf{q}_s\} = \iint (2\theta)[\mathbf{N}_s]^T dxdy, \tag{11.31c}$$

or

$$[\mathbf{k}_s]\{\mathbf{q}_s\} = \{\mathbf{Q}_s\}, \tag{11.31d}$$

where $[\mathbf{k}_s]$ is the element property or flexibility matrix and $\{\mathbf{Q}_s\}$ is the element nodal forcing parameter vector. They are evaluated as follows:

$$[\mathbf{k}_x] = \frac{1}{G}\iint [\mathbf{B}]^T[\mathbf{B}]dxdy = \frac{A}{G}[\mathbf{B}]^T[\mathbf{B}]$$

$$= \frac{1}{4GA}\begin{bmatrix} b_1^2 + a_1^2 & b_1b_2 + a_1a_2 & b_1b_3 + a_1a_3 \\ & b_2^2 + a_2^2 & b_2b_3 + a_2a_3 \\ \text{sym.} & & b_3^2 + a_3^2 \end{bmatrix} \tag{11.32}$$

and

$$\{\mathbf{Q}_s\} = 2\theta \iint \begin{Bmatrix} N_1 \\ N_2 \\ N_3 \end{Bmatrix} dxdy. \tag{11.33}$$

The integrations in Equation 11.33 can be evaluated in closed form by using the formula[4]

$$\iint_A N_1^\alpha N_2^\beta N_3^\gamma dxdy = \frac{\alpha!\beta!\gamma!}{(\alpha+\beta+\gamma+2)!}2A, \tag{11.34a}$$

where ! denotes factorial and 0! = 1. For example,

$$\iint N_1 dxdy = \iint N_1^1 N_2^0 N_3^0 dxdy$$
$$= \frac{1!0!0!}{(1+0+0+2)!} 2A = \frac{2A}{3 \times 2} = \frac{A}{3}. \tag{11.34b}$$

Hence

$$\{\mathbf{Q}_s\} = \frac{2\theta A}{3} \begin{Bmatrix} 1 \\ 1 \\ 1 \end{Bmatrix}, \tag{11.34c}$$

which implies that the applied forcing function is distributed equally at the three nodes.

Example 11.2. Torsion of Square Bar: Stress Function Approach

We now consider an example of torsion of a square bar 4 cm × 4 cm (Figure 11.4). The properties G and θ are as before. Our idea in choosing this and subsequent bars of different cross sections for different approaches is to illustrate different discretizations governed by special characteristics such as

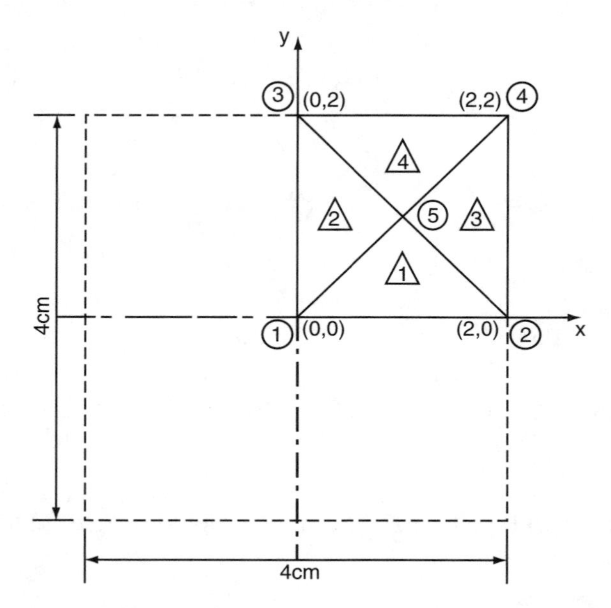

FIGURE 11.4
Torsion of square bar: stress function approach.

symmetry of the cross sections. In view of symmetry in the distribution of the stress function, we can consider and discretize only one-quarter of the bar. In fact, as is done in the next example, only an eighth of the bar can be discretized. The finite element mesh for the quarter of the bar is shown in Figure 11.4, and it is identical to the mesh used in Example 11.1. Because the coordinates and the geometry of this mesh are the same as before, the element stiffness matrices are essentially similar. The element load vector is, however, different, given by Equation 11.34c. Assembly of the four element matrices and load vectors lead to the global equations as

$$\frac{1}{4}\begin{bmatrix} 4 & 0 & 0 & 0 & -4 \\ 0 & 4 & 0 & 0 & -4 \\ 0 & 0 & 4 & 0 & -4 \\ 0 & 0 & 0 & 4 & -4 \\ -4 & -4 & -4 & -4 & 16 \end{bmatrix} \begin{Bmatrix} \varphi_1 \\ \varphi_2 \\ \varphi_3 \\ \varphi_4 \\ \varphi_5 \end{Bmatrix} = \frac{2}{3}\begin{Bmatrix} 2 \\ 2 \\ 2 \\ 2 \\ 4 \end{Bmatrix} \tag{11.35a}$$

or

$$[K]\{\mathbf{r}\} = \{\mathbf{R}\}.$$

Boundary Conditions

According to Prandtl torsion, the boundary condition is

$$\varphi = 0 \text{ or constant along the boundary.} \tag{11.36a}$$

If we choose $\varphi = 0$, we have (Figure 11.4)

$$\varphi_2 = \varphi_3 = \varphi_4 = 0. \tag{11.36b}$$

Equation 11.35a is now modified by deleting the rows and columns corresponding to φ_2, φ_3, and φ_4 to yield

$$\frac{1}{4}\begin{bmatrix} 4 & -4 \\ -4 & 16 \end{bmatrix} \begin{Bmatrix} \varphi_1 \\ \varphi_5 \end{Bmatrix} = \frac{2}{3}\begin{Bmatrix} 2 \\ 4 \end{Bmatrix}. \tag{11.35b}$$

Solution by Gaussian elimination gives

$$\varphi_1 = \tfrac{8}{3} = 2.66667\ N/\text{cm},$$

$$\varphi_5 = \tfrac{4}{3} = 1.33334,$$

and

$$\varphi_2 = \varphi_3 = \varphi_4 = 0.$$

Step 6. Secondary Quantities

Shear Stresses

The stresses can now be computed using the available values of φ. Use of the [**B**] matrix in Equation 11.12 in Equation 11.30 gives

$$\begin{Bmatrix} \dfrac{\partial\varphi}{\partial x} \\ \dfrac{\partial\varphi}{\partial y} \end{Bmatrix} = \frac{1}{2A}\begin{bmatrix} b_1 & b_2 & b_3 \\ a_1 & a_2 & a_3 \end{bmatrix}\begin{Bmatrix} \varphi_1 \\ \varphi_2 \\ \varphi_3 \end{Bmatrix}. \tag{11.37}$$

Substitution of a_i and b_i from Equations 11.20 leads to stresses in the four elements:

Element 1:

$$\begin{Bmatrix} \dfrac{\partial\varphi}{\partial x} \\ \dfrac{\partial\varphi}{\partial y} \end{Bmatrix} = \frac{1}{2}\begin{bmatrix} -1 & 1 & 0 \\ -1 & -1 & 2 \end{bmatrix}\begin{Bmatrix} \frac{8}{3} \\ 0 \\ \frac{4}{3} \end{Bmatrix} = \begin{Bmatrix} -\frac{4}{3} \\ 0 \end{Bmatrix} \text{ N/cm}^2.$$

Therefore, from Equation 11.28a, we have

$$\begin{Bmatrix} \tau_{yz} \\ \tau_{xz} \end{Bmatrix} = \begin{Bmatrix} \dfrac{\partial\varphi}{\partial y} \\ -\dfrac{\partial\varphi}{\partial x} \end{Bmatrix} = \begin{Bmatrix} 0 \\ \frac{4}{3} \end{Bmatrix} \text{ N/cm}^2.$$

Element 2:

$$\begin{Bmatrix} \tau_{xz} \\ \tau_{yz} \end{Bmatrix} = \begin{Bmatrix} -\frac{4}{3} \\ 0 \end{Bmatrix} \text{ N/cm}^2.$$

Element 3:

$$\begin{Bmatrix} \tau_{xz} \\ \tau_{yz} \end{Bmatrix} = \begin{Bmatrix} 0 \\ \frac{4}{3} \end{Bmatrix} \text{ N/cm}^2.$$

Element 4:

$$\begin{Bmatrix} \tau_{xz} \\ \tau_{yz} \end{Bmatrix} = \begin{Bmatrix} -\frac{4}{3} \\ 0 \end{Bmatrix} \text{ N/cm}^2.$$

In view of the fact that the stress function (Equation 11.29c) is linear, the shear strains and stresses in the element are constant. Hence, as before, we can call this element a constant-strain or -stress triangle (CST) element.

Twisting Moment

According to the stress approach, the twisting moment M is given by[2]

$$M_t = 2\iint_A \varphi dxdy = \sum_{m=1}^{M} 2\iint_A \varphi^m dxdy, \tag{11.38a}$$

where $m = 1, 2,\ldots, M$, M = number of elements. Equation 11.38a represents twice the volume under the stress function distribution over the cross section of the bar (Figure 11.4). For a generic element we have

$$\begin{aligned} M_{t_m} &= 2\iint \varphi^m dxdy \\ &= 2\iint \left[N_1\varphi_1^m + N_2\varphi_2^m + N_3\varphi_3^m\right] dxdy \\ &= \frac{2A}{3}\left(\varphi_1^m + \varphi_2^m + \varphi_3^m\right). \end{aligned} \tag{11.38b}$$

Summing over the four elements, we have

$$\begin{aligned} M_t &= \frac{2\times 1}{3}\left(\frac{8}{3} + 0 + \frac{4}{3}\right) \\ &+ \frac{2\times 1}{3}\left(0 + \frac{8}{3} + \frac{4}{3}\right) \\ &+ \frac{2\times 1}{3}\left(0 + 0 + \frac{4}{3}\right) \\ &+ \frac{2\times 1}{3}\left(0 + 0 + \frac{4}{3}\right) \end{aligned} \tag{11.38c}$$

$$= \frac{2}{3}\left(\frac{32}{3}\right)$$

$$= 7.1111 \text{ N/cm}.$$

Therefore, total moment for the bar is

$$M_t = 4 \times 7.1111 = 28.4444 \text{ N/cm}. \quad (11.38d)$$

Step 8. Interpretation and Plots

Figure 11.5(a) and (b) show plots of computed stress functions and shear stresses for the entire bar (4 cm × 4 cm) and the quarter bar, respectively. Note that because linear approximation is used, the shear stresses are constant within each element and are attached at the element centroids.

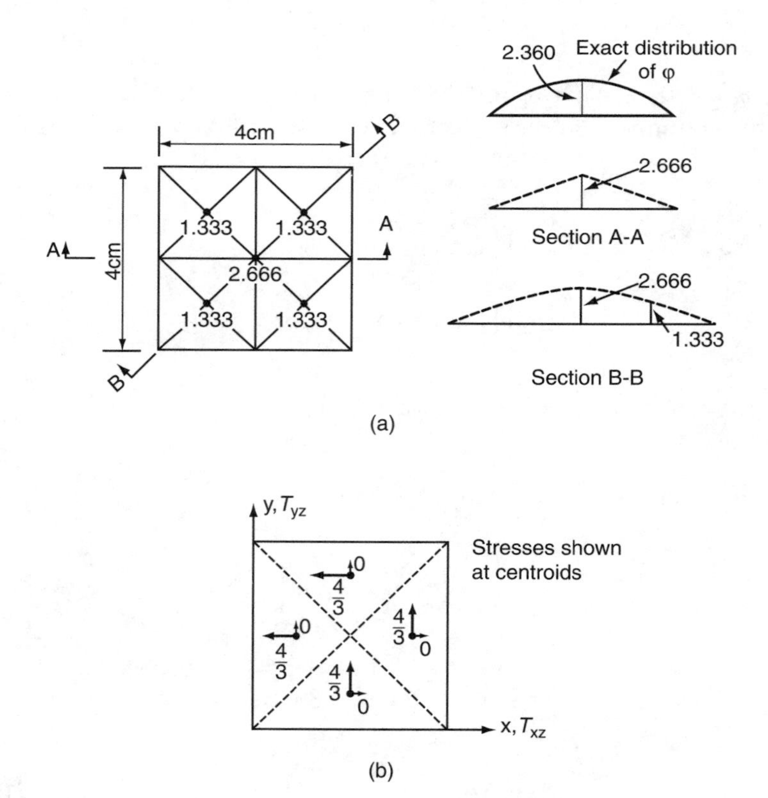

FIGURE 11.5
Results for torsion of square bar: stress function approach: (a) Distribution of computed stress functions in entire bar and (b) Shear stresses in quarter bar.

Comparisons

Stress Function. The exact value of φ at node 1 ($x = 0, y = 0$) from Equation 11.26a is

$$\varphi = 2.360$$

Therefore, the error is

$$\text{error} = \frac{2.360 - 2.666}{2.360} \times 100$$
$$= -13.00\%.$$

Shear Stresses. Use of Equations 11.26b and 11.26c gives the shear stresses in typical element number 1 ($x = 1, y = 1/3$) as

$$\tau_{xz} = -0.246 \ N/\text{cm}^2$$
$$\tau_{yz} = 1.060.$$

Therefore, the errors are

$$\text{error in } \tau_{xz} = -0.246 - 0.000$$
$$= -25.00\%$$
$$\text{error in } \tau_{yz} = \frac{1.060 - 1.333}{1.06} \times 100$$
$$= -26.00\%.$$

Torsional Constant. According to Equation 11.25, the torsional constant is

$$\frac{M_t}{G\theta} \simeq 0.1406(2 \times 2)^4$$
$$\simeq 36.0000 \text{ cm}^4.$$

From the foregoing, the error in the computed torsional constant is

$$\text{error} = \frac{36.0000 - 28.4444}{36} \times 100$$
$$\simeq 21.00\%.$$

We have used a coarse mesh so that hand calculations can be performed. Hence, the errors are rather high. The computed solutions can be improved significantly if a finer mesh is used.

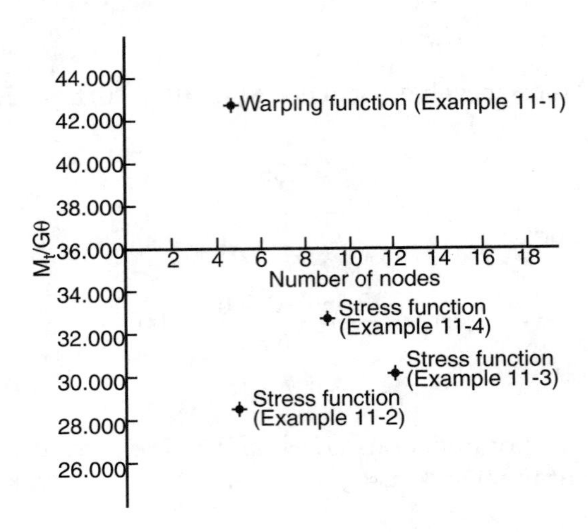

FIGURE 11.6
Bounds in torsion problem.

Bounds

Figure 11.6 shows the computed values of torsional constants from the warping (displacement) and stress function approaches in comparison with the closed form value. For this comparison, the values of $M_t/G\theta$ for the 2 cm × 2 cm bar of Example 11.1 are multiplied by 16 as per Equation 11.25. The displacement approach yields an upper bound to the true value of torsional constant, whereas the stress function approach gives a lower bound solution.

Example 11.3. Torsion of Square Bar with Finer Mesh: Stress Function Approach

Figure 11.7(a) shows a refined mesh for an eighth of the square bar in Figure 11.4. In view of the symmetry, it is possible to reduce the problem to analysis of only an eighth of the bar. Thus, now we have effectively eight elements for a quarter of the bar instead of four as in Example 11.2. Figure 11.7(b) shows local node numbers for the elements.

In the following we give brief details of computations of element equations for the four elements (with values of G and θ as before):

Element 1:

$$a_1 = 0,\ a_2 = -1,\ a_3 = 1;\quad b_1 = -1,\ b_2 = 1,\ b_3 = 0;\quad x_m = \tfrac{2}{3},\ y_m = \tfrac{1}{3};$$

$$\frac{1}{2}\begin{bmatrix} 1 & -1 & 0 \\ -1 & 2 & 0 \\ 0 & -1 & 1 \end{bmatrix}\begin{Bmatrix} \varphi_1 \\ \varphi_2 \\ \varphi_3 \end{Bmatrix} = \frac{1}{3}\begin{Bmatrix} 1 \\ 1 \\ 1 \end{Bmatrix}.$$

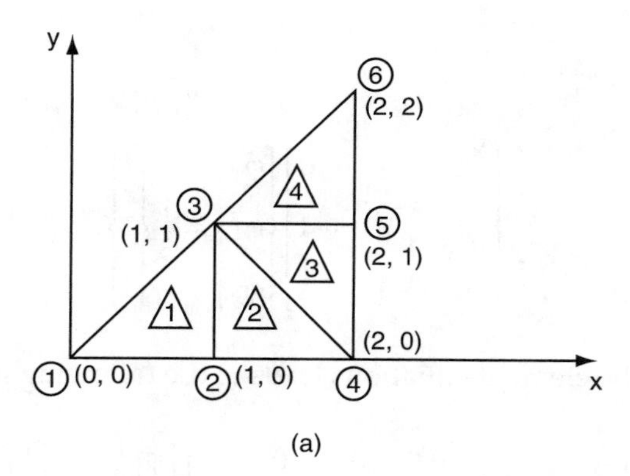

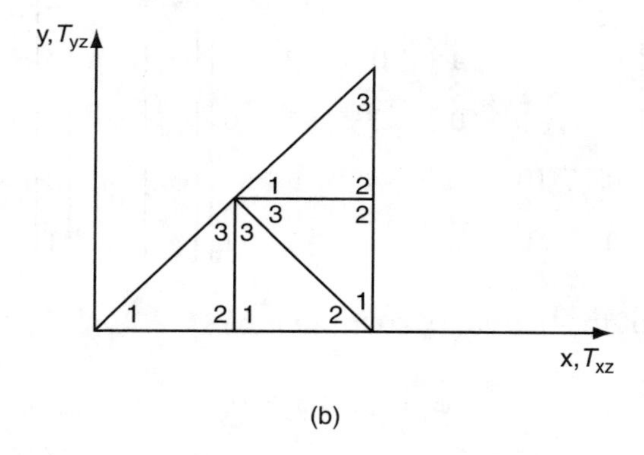

FIGURE 11.7
Torsion of square bar: stress function approach, finer mesh: (a) Mesh for eighth bar of Figure 11.4 and (b) Local node numbers.

Element 2:

$$a_1 = -1,\ a_2 = 0,\ a_3 = 1;\quad b_1 = -1,\ b_2 = 1,\ b_3 = 0;\quad x_m = \tfrac{4}{3},\ y_m = \tfrac{1}{3};$$

$$\frac{1}{2}\begin{bmatrix} 2 & -1 & -1 \\ -1 & 1 & 0 \\ -1 & 0 & 1 \end{bmatrix}\begin{Bmatrix} \varphi_2 \\ \varphi_4 \\ \varphi_3 \end{Bmatrix} = \frac{1}{3}\begin{Bmatrix} 1 \\ 1 \\ 1 \end{Bmatrix}.$$

Element 3:

$$a_1 = -1,\ a_2 = 1,\ a_3 = 0;\quad b_1 = 0,\ b_2 = 1,\ b_3 = -1;\quad x_m = \tfrac{5}{3},\ y_m = \tfrac{2}{3};$$

$$\frac{1}{2}\begin{bmatrix} 1 & -1 & 0 \\ -1 & 2 & -1 \\ 0 & -1 & 1 \end{bmatrix}\begin{Bmatrix} \varphi_4 \\ \varphi_5 \\ \varphi_3 \end{Bmatrix} = \frac{1}{3}\begin{Bmatrix} 1 \\ 1 \\ 1 \end{Bmatrix}.$$

Element 4:

$$a_1 = 0,\ a_2 = -1,\ a_3 = 1;\quad b_1 = -1,\ b_2 = 1,\ b_3 = 0;\quad x_m = \tfrac{5}{3},\ y_m = \tfrac{4}{3};$$

$$\frac{1}{2}\begin{bmatrix} 1 & -1 & 0 \\ -1 & 2 & -1 \\ 0 & -1 & 1 \end{bmatrix}\begin{Bmatrix} \varphi_3 \\ \varphi_5 \\ \varphi_6 \end{Bmatrix} = \frac{1}{3}\begin{Bmatrix} 1 \\ 1 \\ 1 \end{Bmatrix}.$$

Assembly of the element equations leads to the following global equations:

$$\begin{bmatrix} 1 & -1 & 0 & 0 & 0 & 0 \\ -1 & 4 & -2 & -1 & 0 & 0 \\ 0 & -2 & 4 & 0 & -2 & 0 \\ 0 & -1 & 0 & 2 & -1 & 0 \\ 0 & 0 & -2 & -1 & 4 & 1 \\ 0 & 0 & 0 & 0 & -1 & 1 \end{bmatrix}\begin{Bmatrix} \varphi_1 \\ \varphi_2 \\ \varphi_3 \\ \varphi_4 \\ \varphi_5 \\ \varphi_6 \end{Bmatrix} = \frac{2}{3}\begin{Bmatrix} 1 \\ 2 \\ 4 \\ 2 \\ 2 \\ 1 \end{Bmatrix}. \tag{11.39a}$$

Introduction of the boundary conditions

$$\varphi_4 = \varphi_5 = \varphi_6 = 0.0$$

leads to modified assemblage equations as

$$\begin{aligned} -\varphi_1 - \varphi_2 \qquad\qquad &= \tfrac{2}{3}, \\ -\varphi_1 + 4\varphi_2 - 2\varphi_3 &= \tfrac{4}{3}, \\ -2\varphi_2 + 4\varphi_3 &= \tfrac{8}{3}, \end{aligned} \tag{11.39b}$$

solution of which gives

$$\begin{aligned} \varphi_1 &= 2.333\ N/\text{cm}, \\ \varphi_2 &= 1.666, \\ \varphi_3 &= 1.500, \\ \left.\begin{aligned} \varphi_4 &= 0.000 \\ \varphi_5 &= 0.000 \\ \varphi_6 &= 0.000 \end{aligned}\right\} &\ \text{prescribed.} \end{aligned}$$

The shear stresses according to Equations 11.37 and 11.28a are now found as follows:

Element 1:

$$\begin{Bmatrix} \tau_{xz} \\ \tau_{yz} \end{Bmatrix} = \begin{Bmatrix} \dfrac{\partial \varphi}{\partial y} \\ -\dfrac{\partial \varphi}{\partial x} \end{Bmatrix} = \begin{Bmatrix} -0.166 \\ 1.666 \end{Bmatrix} \text{ N/cm}^2.$$

Element 2:

$$\begin{Bmatrix} \tau_{xz} \\ \tau_{yz} \end{Bmatrix} = \begin{Bmatrix} -0.166 \\ 1.666 \end{Bmatrix} \text{ N/cm}^2.$$

Element 3:

$$\begin{Bmatrix} \tau_{xz} \\ \tau_{yz} \end{Bmatrix} = \begin{Bmatrix} 0.000 \\ 1.5000 \end{Bmatrix} \text{ N/cm}^2.$$

Element 4:

$$\begin{Bmatrix} \tau_{xz} \\ \tau_{yz} \end{Bmatrix} = \begin{Bmatrix} 0.000 \\ 1.5000 \end{Bmatrix} \text{ N/cm}^2.$$

Figure 11.8 shows plots of computed values of φ and shear stresses for the refined mesh.

Comparisons

Torsional Constant. The torsional constant for the eighth bar according to Equation 11.38 is

$$M_t = \frac{2}{3} \times \frac{1}{2}\left[\left(\frac{7}{3} + \frac{5}{3} + \frac{4.5}{3}\right) + \left(\frac{5}{3} + \frac{4.5}{3}\right) + \frac{4.5}{3} + \frac{4.5}{3}\right]$$
$$= 3.8888 \text{ N-cm}.$$

Then the total value of torsional constant is

$$\sum M_t = 8 \times 3.8888 = 31.1100 \text{ N/cm}.$$

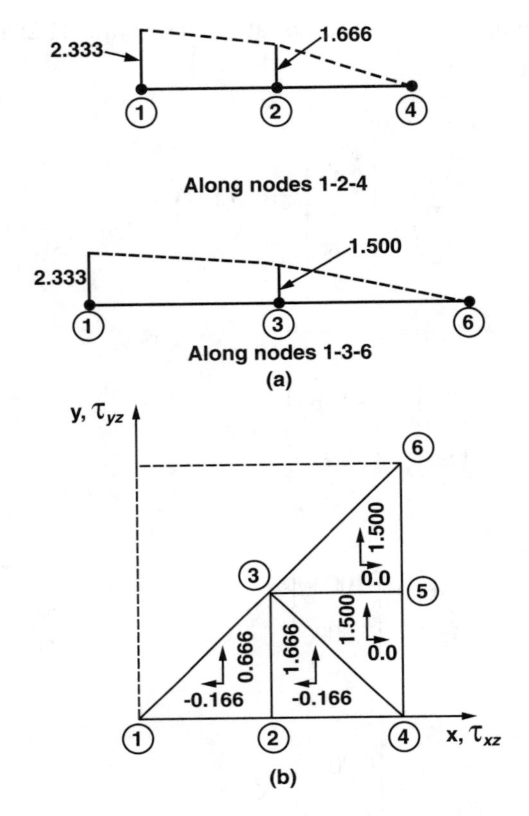

FIGURE 11.8
Results for torsion of square bar: stress function approach, refined mesh: (a) Distribution of computed stress functions in quarter bar and (b) Computed shear stresses in eighth bar.

Therefore, the error in the numerical solution is

$$\text{error} = \frac{36.0000 - 31.1100}{36.0000} \times 100$$

$$\simeq 14.00\%.$$

The computed value of the torsional constant from the finer mesh is shown in Figure 11.6 and yields an improved lower bound to the true value of $M_t/G\theta$.

Stress Function. From Equation 11.26a, the closed form value of φ at node 1 ($x = 0$, $y = 0$) is found to be

$$\varphi = 2.360;$$

hence, the error is

$$\text{error} = \frac{2.360 - 2.333}{2.360} \times 100$$
$$= 1.00\%.$$

Shear Stresses. For element 3, ($x = 5/3$, $y = 2/3$), use of Equations 11.26b and 11.26c yields errors as

$$\text{error in } \tau_{xz} = -0.194 - 0.000$$
$$= -19.00\%$$
$$\text{error in } \tau_{yz} = \frac{1.900 - 1.500}{1.900} \times 100$$
$$= 21.00\%.$$

For the foregoing comparisons, it can be seen that the numerical solutions with the finer mesh are closer to the closed form values. The error in the torsional constant reduced from 21.00 to 14.00%, and that in the shear stresses from 25.00 to 19.00 and from 26.00 to 21.00% in τ_{xz} and τ_{yz}, respectively. Note that the comparisons of errors in the shear stresses are not rigorous, since they are not necessarily at the same point; however, the general trend shows reduction in the error.

The numerical solution can be improved significantly with progressively refined meshes. Such refinement should follow certain criteria for consistent comparisons; these are briefly discussed in Chapter 13 (Example 13.5) and in Reference 2.

Example 11.4. Computer Solution for Torsion of Square Bar

To examine the influence of a different (higher-) order element on the numerical solution, we now consider a square bar divided into four (square) quadrilateral elements (Figure 11.9(a)). Note that the size of the bar here is 2 cm × 2 cm, while in the previous examples with the stress function approach the size of the bar was 4 cm × 4 cm.

The computer code FIELD-2D detailed in Chapter 12 and Appendix 3 was used to solve this problem. This code is based on a four-node isoparametric quadrilateral element, which is covered in Chapter 12. This element has a bilinear distribution of the stress function as compared to the linear distribution within the triangular element used in this chapter. Thus, there is an improvement in the assumed approximation within the element.

Figure 11.9(b) shows the computed values of stress functions at the node points. Computed values of the shear stresses τ_{xz} and τ_{yz} are plotted in Figure 11.9(c). The computed value of the torsional constant is 2.046.

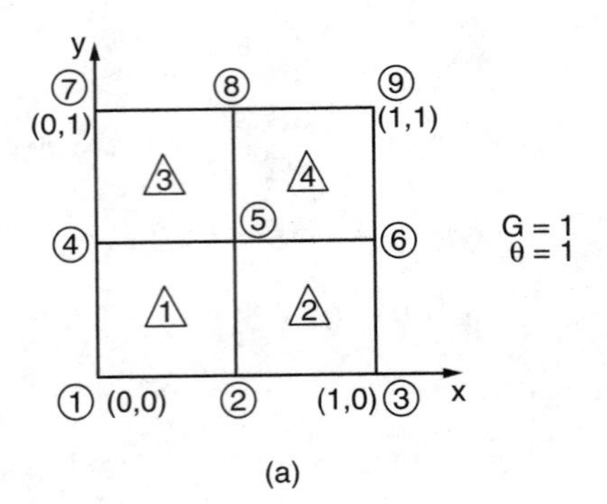

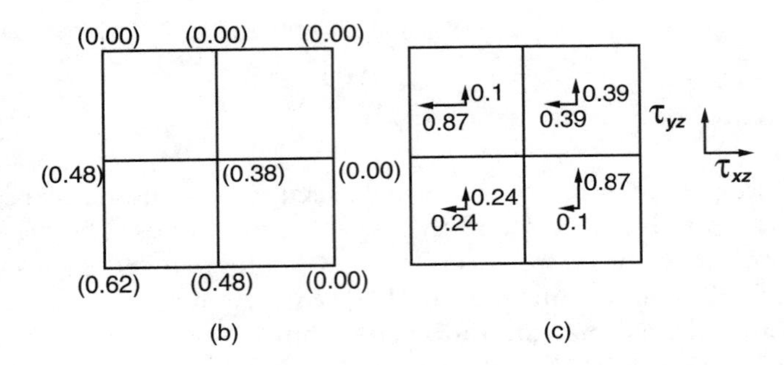

FIGURE 11.9
Torsion of square bar with quadrilateral element: (a) Finite element mesh for quarter bar, (b) Computed nodal stress functions in parentheses, and (c) Computed shear stresses.

Comparisons

Torsional Constant. The error in the computed and exact values of the torsional constant is

$$\text{error} = \frac{2.250 - 2.046}{2.250} \times 100$$

$$\simeq 9.00\%.$$

The value of $M_t/G\theta = 2.046 \times 16 = 32.74$ is shown in Figure 11.6 and yields a lower bound to the exact solution.

Stress Function. The closed form value of φ (Equation 11.26a) at node 1 ($x = 0$, $y = 0$) is found to be

$$\varphi = 0.590;$$

hence, the error is

$$\text{error} = \frac{0.590 - 0.620}{0.59} \times 100$$

$$= -5.00\%$$

Shear Stresses. The shear stresses are compared for two typical elements, 1 and 2:

Element 1 ($x = 0.25$, $y = 0.25$)

$$\text{error in } \tau_{xz} = \frac{-0.238 - (-0.24)}{0.238} \times 100$$

$$= 1.00\%,$$

$$\text{error in } \tau_{yz} = \frac{0.238 - 0.240}{0.238} \times 100$$

$$= -1.00\%.$$

Element 2 ($x = 0.75$, $y = 0.25$)

$$\text{error in } \tau_{xz} = \frac{-0.104 - (-0.100)}{0.104} \times 100$$

$$= 4.00\%,$$

$$\text{error in } \tau_{yz} = \frac{0.853 - 0.870}{0.853} \times 100$$

$$= -2.00\%.$$

Use of the four-node isoparametric quadrilateral provides an overall improvement in the solution for the values of φ, shear stresses, and torsional constant.

For better accuracy one should use finer meshes with a computer code. With about 100 elements one can expect results of acceptable accuracy for all practical purposes. The cost of such computations with a computer code is not high.

Review and Comments

In the displacement approach, we assume approximate displacement or warping functions. These functions are chosen such that they satisfy physical continuity of the body or continuum up to a certain degree or level of approximation. The potential energy Π_p is then expressed in terms of the assumed functions, and its stationary value yields approximate equilibrium equations. In other words, for the assumed compatible function, equilibrium of forces is satisfied only in an approximate sense.

In the case of the stress approach, the assumed stress functions that satisfy equilibrium are substituted into the complementary energy Π_c. Its stationary value leads to approximate compatibility equations.

Thus in both cases, for the assumed unknowns, either the equilibrium or compatibility is fulfilled only approximately. For example, in the case of the displacement formulation for torsion, the stress boundary conditions over the surface are not necessarily satisfied. That is, as required by Saint-Venant's torsion, the shearing stress τ_n normal to the boundary computed from the displacement procedure may not be zero. Figure 11.10(b) shows the variation of τ_n along the boundary AB of the discretized region of a triangle (Figure 11.10(a)) subjected to torsion[3-5] computed from a displacement and a hybrid formulation. It can be seen that τ_n is not zero as required by the theoretical assumptions.

We can overcome some of these deficiencies by using hybrid and mixed procedures.

Hybrid Approach

A wide variety of procedures is available for formulating the hybrid approach.[4,8-11] One can assume displacement inside the element and stresses on the boundaries of the element. Conversely, it is possible to assume stresses inside the element and displacement along the boundaries. Here we shall consider a special case of the former.[9]

Since it is found that the major error in the results from the displacement formulation occur in the stresses at the boundaries, we shall choose the elements on the boundary, shown shaded in Figure 11.11, for the hybrid formulation. For the elements in the interior the displacement formulation will be used. Since the displacement approach has already been formulated, we give now details of the hybrid approach for the elements on the boundary.

Consider a generic boundary triangular element (Figure 11.11). The stress function is assumed within the element as

$$\begin{aligned}\varphi &= N_1\varphi_1 + N_2\varphi_2 + N_3\varphi_3 \\ &= [\mathbf{N}_s]\{\mathbf{q}_s\},\end{aligned} \tag{11.29c}$$

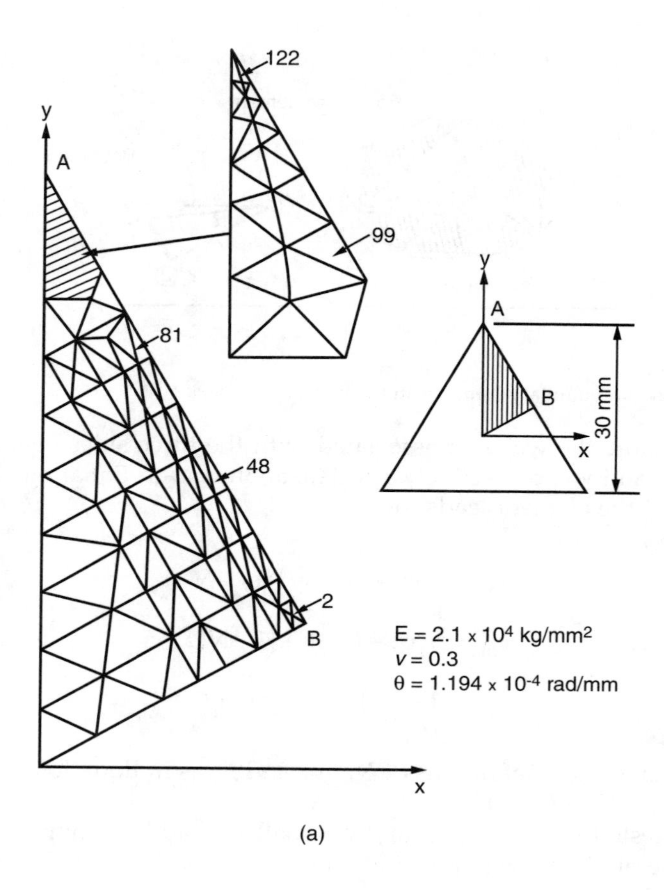

(a)

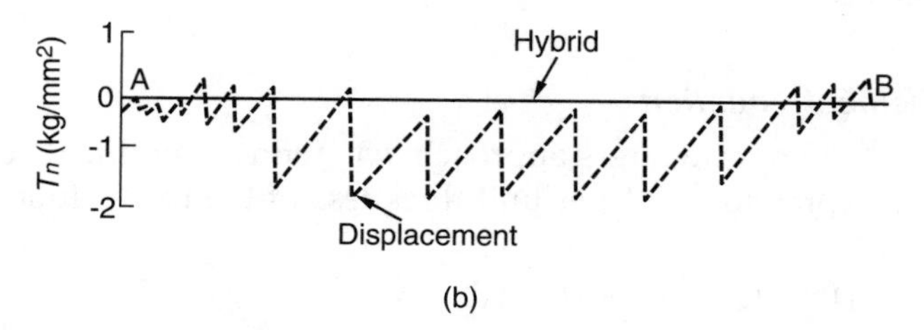

(b)

FIGURE 11.10
Torsion of triangular bar with hybrid approach:[4] (a) Finite element mesh and (b) Distribution of normal shear stress on edge *AB*.

where the subscript *s* denotes stress function. The general expression for the warping function in the element is defined as

$$\begin{aligned} \psi(x,y) &= N_1\psi_1 + N_2\psi_2 + N_3\psi_3 \\ &= [\mathbf{N}_\psi]\{\mathbf{q}_\psi\}, \end{aligned} \tag{11.11}$$

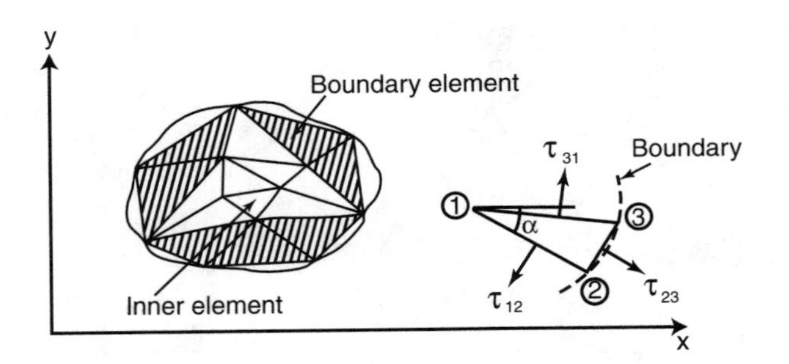

FIGURE 11.11
Discretization and boundary elements in the hybrid approach.

If we denote the warping associated with the three sides of the element as ψ_{23}, ψ_{31}, and ψ_{12}, respectively, specialization of ψ in Equation 11.11 along the sides of the element leads to

$$\begin{aligned} \psi_{23} &= 0 \cdot \psi_1 + (1 - N_3)\psi_2 + N_3\psi_3, \\ \psi_{31} &= N_1\psi_1 + 0 \cdot \psi_2 + (1 - N_3)\psi_3, \\ \psi_{12} &= (1 - N_2)\psi_1 + N_2\psi_2 + 0 \cdot \psi_3. \end{aligned} \tag{11.40}$$

Here we have used definitions of N_1, N_2, and N_3 as in Equation 10.3 and the relation $N_1 + N_2 + N_3 = 1$.

As suggested by Yamada et al.,[9] we shall also add the twisting moment M_t and the angle of twist per unit length θ to the formulation and follow their procedure.

Step 4. Element Equations

In the case of the hybrid stress approach, we define a modified complementary energy expression, Π_{ch}, per unit thickness of the bar as follows:[9,11]

$$\begin{aligned} \Pi_{ch} &= U_c + W_{ph} = U_c - W_h, \\ \Pi_{ch} &= \frac{1}{2G}\iint_A \left[\left(\frac{\partial \varphi}{\partial x}\right)^2 + \left(\frac{\partial \varphi}{\partial y}\right)^2\right] dxdy - \int_s \{\overline{\mathbf{T}}\}^T \{\psi_b\} dS \\ &= \frac{1}{2}\iint_A \{\sigma\}^T [\mathbf{D}]\{\sigma\} dxdy - \int_s \{\overline{\mathbf{T}}\}^T \{\psi_b\} dS \end{aligned} \tag{11.41}$$

Here W_h denotes the work of boundary forces, and we have used the subscript h to denote the hybrid formulation. In very simple words, the procedure is called hybrid, because in Π_{ch} it combines complementary strain energy U_c as in the case of stress approach and potential of external loads W_{ph} as in the case of the displacement approach.

In Equation 11.41, $\{\bar{\mathbf{T}}\}^T = [M_t\ \tau_{23}\ \tau_{31}\ \tau_{12}]$ is the vector of twisting moment and normal components of shear stresses on the boundary, $\{\psi_b\}^T = [\theta\ \psi_{23}\ \psi_{32}\ \psi_{12}]$ vector of corresponding angle of twist and warping associated with the sides of the element, $\{\sigma\}$ is defined in Equation 11.28a, and

$$[\mathbf{D}] = \frac{1}{G}\begin{bmatrix} 1 & 0 \\ 0 & 1 \end{bmatrix},$$

which is the inverse of stress–strain matrix $[\mathbf{C}]$. We assume linear elastic behavior.

To evaluate U_c and W_{ph} in Π_{ch}, we need to derive the following results for the boundary element. For the triangular element, the stress vector $\{\sigma\}$ is given by

$$\{\sigma\} = \begin{Bmatrix} \tau_{xz} \\ \tau_{yz} \end{Bmatrix} = \begin{Bmatrix} \dfrac{\partial \varphi}{\partial y} \\ -\dfrac{\partial \varphi}{\partial x} \end{Bmatrix} = \frac{1}{2A}\begin{bmatrix} a_1 & a_2 & a_3 \\ -b_1 & -b_2 & -b_3 \end{bmatrix}\begin{Bmatrix} \varphi_1 \\ \varphi_2 \\ \varphi_3 \end{Bmatrix} \tag{11.42a}$$

Because the element lies on the boundary such that the side 2–3 is along the boundary, $\varphi_2 = \varphi_3 =$ constant, which we assume as zero; then Equation 11.42a specializes to

$$\begin{Bmatrix} \tau_{xz} \\ \tau_{yz} \end{Bmatrix} = \frac{1}{2A}\begin{Bmatrix} a_1 \\ -b_1 \end{Bmatrix}\varphi_1 \tag{11.42b}$$

or

$$\{\sigma\} = [\mathbf{P}]\{\beta\}, \tag{11.42c}$$

where

$$[\mathbf{P}] = \frac{1}{2A}\begin{Bmatrix} a_1 \\ b_1 \end{Bmatrix}$$

and $\{\beta\} = \{\varphi_1\} = \varphi_1$ is just a scalar. Substitution of Equation 11.42c into U_c (Equation 11.41) leads to

$$U_c = \tfrac{1}{2}\{\beta\}^T \iint_A [\mathbf{P}]^T[\mathbf{D}][\mathbf{P}]dxdy\{\beta\} \tag{11.43a}$$

$$= \tfrac{1}{2}\{\beta\}^T[\mathbf{H}]\{\beta\} \tag{11.43b}$$

$$= \frac{\varphi_1}{2 \times 2A \times 2A}\iint_A [a_1 \quad -b_1]\frac{1}{G}\begin{bmatrix}1 & 0\\ 0 & 1\end{bmatrix}\begin{Bmatrix}a_1\\ -b_1\end{Bmatrix}dxdy\varphi_1 \tag{11.43c}$$

$$= \frac{A\varphi_1^2}{8A^2}\frac{(a_1^2+b_1^2)}{G} = \frac{l_{23}^2\varphi_1^2}{8GA}, \tag{11.43d}$$

where $l_{23}^2 = (a_1^2 + b_1^2) = (x_3 - x_2)^2 + (y_2 - y_3)^2$ denotes the length of the side 2–3 and

$$\begin{aligned}[\mathbf{H}] &= \iint_A [\mathbf{P}]^T[\mathbf{D}][\mathbf{P}]dxdy\\ &= \frac{l_{23}^2}{4GA}.\end{aligned} \tag{11.44}$$

The second part in Equation 11.41 denotes the potential of the surface tractions $\{\bar{\mathbf{T}}\}$ over the entire surface of the element per unit thickness and can be expressed as

$$W_{ph} = M_t\theta + \int_{2-3}\tau_{23}\psi_{23}dS + \int_{3-1}\tau_{31}\psi_{31}dS + \int_{1-2}\tau_{12}\psi_{12}dS \tag{11.45a}$$

$$= \{\mathbf{T}\}^T\int\{\psi_b\}dS. \tag{11.45b}$$

Here τ_{23} and so on are constants because of the linear interpolation function. The components of $\{\bar{\mathbf{T}}\}^T = [M_t\ \psi_{23}\ \psi_{31}\ \psi_{12}]$ are evaluated as follows:

$$\begin{aligned}M_t &= \iint_A (-y_m\tau_{xz} + x_m\tau_{yz})dxdy\\ &= \frac{A}{2A}(-y_m a_1 - x_m b_1)\varphi_1\\ &= \frac{1}{2}(y_m a_1 + x_m b_1)\varphi_1,\end{aligned} \tag{11.46}$$

and by referring to Figure 11.11

$$\tau_{12} = \tau_{xz}\sin\alpha - \tau_{yz}\cos\alpha$$
$$= \frac{-(a_1b_3 - b_1a_3)\varphi_1}{2Al_{12}},$$
$$\tau_{23} = \frac{-(a_1b_1 - b_1a_1)\varphi_1}{2Al_{23}} = 0, \tag{11.47}$$
$$\tau_{31} = \frac{-(a_1b_2 - b_1a_2)\varphi_1}{2Al_{31}}.$$

It is important to note that the stress τ_{23} normal to the boundary is identically zero, as required by the torsion problem. In matrix notation, we have

$$\begin{Bmatrix} M_t \\ \tau_{23} \\ \tau_{31} \\ \tau_{12} \end{Bmatrix} = \frac{1}{2A}\begin{bmatrix} y_mA & -x_mA \\ b_1/l_{23} & a_1/l_{23} \\ b_2/l_{31} & a_2/l_{31} \\ b_3/l_{12} & a_3/l_{12} \end{bmatrix}\begin{Bmatrix} -a_1 \\ b_1 \end{Bmatrix}\varphi_1 \tag{11.48a}$$

or

$$\{\bar{\mathbf{T}}\} = [\mathbf{R}]\{\beta\}. \tag{11.48b}$$

The surface integrations relevant to $\{\psi_b\}$ of Equation 11.45b in W_{ph} can be performed easily. For instance, substitution for ψ_{23} from Equation 11.40 into Equation 11.45 gives

$$\int_{2-3}\psi_{23}dS = \int_0^1\left[(1-N_3)\psi_2 + N_3\psi_2\right]l_{23}dN_3 \tag{11.49a}$$

because $dS = l_{23}dN_3$ along side 2–3. Finally, integration of the expression in Equation 11.49a leads to

$$\int_{2-3}\psi_{23}dS = \frac{l_{23}}{2}(\psi_2 + \psi_3). \tag{11.49b}$$

The other two integrals are evaluated similarly. Then Equation 11.45 becomes

$$W_{ph} = \{\overline{\mathbf{T}}\}^T \int \{\psi_b\} dS \tag{11.50a}$$

$$= \{\beta\}^T [\mathbf{R}]^T \frac{1}{2} \begin{Bmatrix} \theta \\ l_{23}(\psi_2 + \psi_3) \\ l_{31}(\psi_3 + \psi_1) \\ l_{12}(\psi_1 + \psi_2) \end{Bmatrix} \tag{11.50b}$$

$$= \frac{1}{2A} \varphi_1 \begin{bmatrix} -a_1 & b_1 \end{bmatrix} \begin{bmatrix} y_m A & (b_2 + b_3)/2 & (b_3 + b_1)/2 & (b_1 + b_2)/2 \\ -x_m A & (a_2 + a_3)/2 & (a_3 + a_1)/2 & (a_1 + a_2)/2 \end{bmatrix} \begin{Bmatrix} \theta \\ \psi_1 \\ \psi_2 \\ \psi_3 \end{Bmatrix} \tag{11.50c}$$

$$= \{\beta\}^T [\mathbf{G}] \{\mathbf{q}_\psi\} = \{\beta\}^T [\mathbf{P}]^T [\mathbf{L}] \{\mathbf{q}_w\}. \tag{11.50d}$$

Here

$$\begin{aligned} \{\mathbf{q}_\psi\}^T &= [\theta \quad \psi_1 \quad \psi_2 \quad \psi_3], \\ [\mathbf{G}] &= [\mathbf{P}]^T [\mathbf{L}], \end{aligned} \tag{11.50e}$$

and

$$\{\mathbf{L}\} = \frac{1}{2} \begin{bmatrix} -2y_m A & b_1 & b_2 & b_3 \\ 2x_m A & a_1 & a_2 & a_3 \end{bmatrix}. \tag{11.50f}$$

Here we used the relations $a_1 + a_2 + a_3 = 0$ and $b_1 + b_2 + b_3 = 0$.

Now, we substitute for U_c and W_{ph} in Equation 11.41 to obtain

$$\Pi_{ch} = \frac{1}{2} \{\beta\}^T [\mathbf{H}] \{\beta\} - \{\beta\}^T [\mathbf{G}] \{\mathbf{q}_\psi\}. \tag{11.51}$$

By differentiating Π_{ch} with respect to $\{\beta\} = \varphi_1$ and equating the results to zero, we obtain

$$\frac{\partial \Pi_{ch}}{\partial \varphi_1} = [\mathbf{H}] \{\beta\} - [\mathbf{G}] \{\mathbf{q}_\psi\} = 0, \tag{11.52a}$$

or

$$[\mathbf{H}]\{\beta\} = [\mathbf{G}]\{\mathbf{q}_\psi\}, \tag{11.52b}$$

or

$$\{\beta\} = [\mathbf{H}]^{-1}[\mathbf{G}]\{\mathbf{q}_\psi\}, \tag{11.52c}$$

which expresses the relation between nodal stress functions and nodal warping functions for the boundary elements.

Element Stiffness Matrix

We now derive element stiffness matrix $[\mathbf{k}_h]$ relevant to the hybrid approach. For this, we express the strain energy as

$$U = \tfrac{1}{2}\{\mathbf{q}_\psi\}^T[\mathbf{k}_h]\{\mathbf{q}_\psi\}, \tag{11.53a}$$

where $\{\mathbf{q}_\psi\}^T = [\theta\ \psi_1\ \psi_2\ \psi_3]$ is the vector of generalized nodal displacements. The complementary strain energy expression U_c, after substitution of Equation 11.52c, leads to

$$\begin{aligned} U_c &= \tfrac{1}{2}\{\beta\}^T[\mathbf{H}]\{\beta\} \\ &= \tfrac{1}{2}\{\mathbf{q}_\psi\}^T[\mathbf{G}]^T\left([\mathbf{H}]^{-1}\right)^T[\mathbf{H}][\mathbf{H}]^{-1}[\mathbf{G}]\{\mathbf{q}_\psi\} \\ &= \tfrac{1}{2}\{\mathbf{q}_\psi\}^T[\mathbf{G}]^T[\mathbf{H}]^{-1}[\mathbf{G}]\{\mathbf{q}_\psi\}. \end{aligned} \tag{11.53b}$$

Here, since $[\mathbf{H}]$ defined by Equation 11.44 is usually symmetric, we have $([\mathbf{H}]^{-1})^T = [\mathbf{H}]^{-1}$.

Equating the above two expressions for U and U_c we obtain the element stiffness matrix $[\mathbf{k}_h]$ as

$$\begin{aligned} [\mathbf{k}_h] &= [\mathbf{G}]^T[\mathbf{H}]^{-1}[\mathbf{G}] \\ &= [\mathbf{G}]^T\frac{4GA}{l_{23}^2}[\mathbf{G}] \end{aligned} \tag{11.54a}$$

$$= \frac{G}{Al_{23}^2}\begin{bmatrix} A^2(y_m a_1 + x_m b_1)^2 & 0 & -A(y_m a_1 + x_m b_1)\left(\frac{-2A}{2}\right) & -A(y_m a_1 + x_m b_1)(2A) \\ 0 & 0 & 0 & 0 \\ -A(y_m a_1 + x_m b_1)\left(\frac{-2A}{2}\right) & 0 & \frac{1}{4}(-2A)(-2A) & \frac{1}{4}(-2A)(2A) \\ -A(y_m a_1 + x_m b_1)\left(\frac{-2A}{2}\right) & 0 & \frac{1}{4}(2A)(-2A) & \frac{1}{4}(2A)(2A) \end{bmatrix}$$

$$=\frac{G}{Al_{23}^2}\begin{bmatrix} (y_m a_1 + x_m b_1)^2 & 0 & (y_m a_1 + x_m b_1) & -(y_m a_1 + x_m b_1) \\ 0 & 0 & 0 & 0 \\ y_m a_1 + x_m b_1 & 0 & 1 & -1 \\ -(y_m a_1 + x_m b_1) & 0 & -1 & 1 \end{bmatrix}. \quad (11.54b)$$

Here we used again the following relations:

$$a_1 + a_2 + a_3 = 0, \quad b_1 + b_2 + b_3 = 0,$$

and

$$a_3 b_2 - a_2 b_3 = a_1 b_3 - a_3 b_1 = a_2 b_1 - a_1 b_2 = 2A.$$

Inner Elements

We choose to use the displacement approach to define the stiffness matrix for the elements inside the bar. By following the procedure outlined previously in this chapter, the stiffness matrix with warping functions can be derived as

$$[\mathbf{k}_\psi] = \frac{G}{4A}\begin{bmatrix} 4A^2(x_m^2 + y_m^2) & 2A(a_1 x_m - b_1 y_m) & 2A(a_2 x_m - b_2 y_m) & 2A(a_3 x_m - b_3 y_m) \\ & b_1^2 + a_1^2 & b_1 b_2 + a_1 a_2 & b_1 b_3 + a_1 a_3 \\ & \text{sym.} & b_2^2 + a_2^2 & b_2 b_3 + a_2 a_3 \\ & & & b_3^2 + a_3^2 \end{bmatrix}. \quad (11.55)$$

Both $[\mathbf{k}_h]$ and $[\mathbf{k}_\psi]$ express relations between generalized forces and displacements $\{\mathbf{q}_\psi\}$ and have identical forms as

$$[\mathbf{k}_h]\{\mathbf{q}_\psi\} = \{\mathbf{Q}\} \quad (11.56a)$$

and

$$[\mathbf{k}_\psi]\{\mathbf{q}_\psi\} = \{\mathbf{Q}\} \quad (11.56b)$$

where

$$\{\mathbf{Q}\} = -\begin{bmatrix} Ay_m & -Ax_m \\ \frac{b_1}{2} & \frac{a_1}{2} \\ \frac{b_2}{2} & \frac{a_2}{2} \\ \frac{b_3}{2} & \frac{a_3}{2} \end{bmatrix}\begin{Bmatrix} \tau_{xz} \\ \tau_{yz} \end{Bmatrix} = \begin{Bmatrix} M_t \\ F_1 \\ F_2 \\ F_3 \end{Bmatrix} \quad (11.56c)$$

and F_1, F_2, and F_3 represent one half of the resultant shearing forces on the sides 2–3, 3–1, and 1–2, respectively.

Computation of Boundary Shear Stresses

For an element with side 2-3 along the boundary, we have, from Equations 11.42) and (11-52c),

$$\begin{Bmatrix}\tau_{xz}\\ \tau_{yz}\end{Bmatrix}=\frac{1}{2A}\begin{Bmatrix}a_1\\ -b_1\end{Bmatrix}\{\beta\}=\frac{1}{2A}\begin{Bmatrix}a_1\\ -b_1\end{Bmatrix}[\mathbf{H}]^{-1}[\mathbf{G}]\{\mathbf{q}_\psi\}$$

$$=\frac{G}{l_{23}^2}\begin{bmatrix}-a_1(a_1y_m+b_1x_m) & 0 & -a_1 & a_1\\ b_1(a_1y_m-b_1x_m) & 0 & b_1 & -b_1\end{bmatrix}\begin{Bmatrix}\theta\\ \psi_1\\ \psi_2\\ \psi_3\end{Bmatrix}. \tag{11.57}$$

Assembly

For elements on the boundary and inside the bar, we use element equations 11.56a and 11.56b, respectively. Since both are expressed in terms of nodal displacements, the assembly follows the same rule that the displacements at common nodes are compatible.

In view of the fact that the hybrid approach satisfies the stress boundary conditions, the results from the proposed procedure can yield improved accuracy in the stresses at and around the boundary. In the following, first we present an example using hand calculations and then results from computer solutions by Yamada et al.[9]

Example 11.5. Torsion of Square Bar: Hybrid Approach

We consider torsion of the square bar 2 cm × 2 cm shown in Figure 11.2. In this example the local node numbering is changed as shown in Figure 11.12 to comply with the numbering for the boundary element in the foregoing hybrid formulation.

All the four elements lie on the boundary and hence are treated as boundary elements in the sense of the hybrid approach. It is incidental that all elements are boundary elements; in general, both boundary and inner elements will occur. Hence it is necessary to use Equations 11.54b and 11.55, respectively, for deriving the element matrices for the boundary and inner elements. In the following are given salient details of the element and assemblage equations:

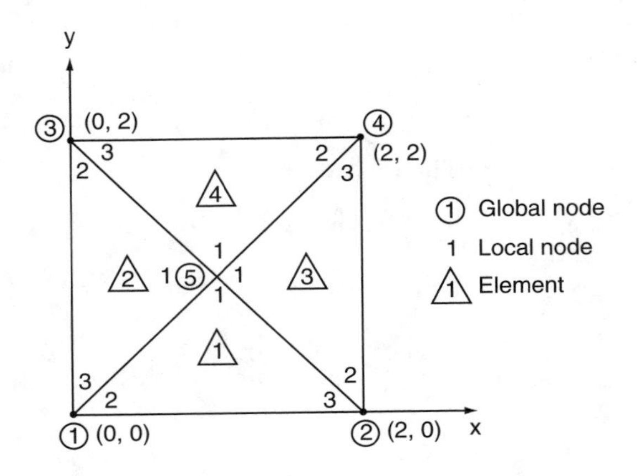

FIGURE 11.12
Torsion of square bar: hybrid approach.

Element 1:

$$a_1 = x_3 - x_1 = 2, \qquad b_1 = y_2 - y_3 = 0,$$

$$a_2 = x_1 - x_3 = -1, \qquad b_2 = y_3 - y_1 = -1,$$

$$a_3 = x_2 - x_1 = -1, \qquad b_3 = y_1 - y_2 = 1,$$

$$x_m = 1, \qquad y_m = \tfrac{1}{3},$$

$$l_{23} = 2, \qquad \frac{GA}{l_{23}^2} = \frac{1}{4}.$$

Here $A = 1\ \text{cm}^2$, and we have assumed that $G = 1\ \text{N/cm}^2$.
Use of Equation 11.54b leads to element equations:

I 5 1 2 ← Global

I 1 2 3 ← Local

$$[\mathbf{k}_h]_1 = \frac{1}{4}\begin{bmatrix} \frac{4}{9} & 0 & \frac{2}{3} & -\frac{2}{3} \\ 0 & 0 & 0 & 0 \\ \frac{2}{3} & 0 & 1 & -1 \\ -\frac{2}{3} & 0 & -1 & 1 \end{bmatrix}\begin{Bmatrix} \theta \\ \psi_1 \\ \psi_2 \\ \psi_3 \end{Bmatrix} = \begin{Bmatrix} M_{t_1} \\ 0 \\ 0 \\ 0 \end{Bmatrix}. \qquad \begin{matrix} \text{I} & \text{I} \\ 1 & 5 \\ 2 & 1 \\ 3 & 2 \end{matrix}$$

We assume that θ is uniform over the bar, and hence a special number I common to both local and global numbering is assigned to this degree of freedom. M_{t_1} denotes the twisting moment for element 1 and so on.

Element 2:

$$
\begin{aligned}
a_1 &= 0, & b_1 &= 2, \\
a_2 &= 1, & b_2 &= -1, \\
a_3 &= -1, & b_3 &= -1, \\
x_m &= \tfrac{1}{3}, & y_m &= 1 \\
l_{23} &= 2, & \frac{GA}{l_{23}^2} &= \frac{1}{4},
\end{aligned}
$$

I 5 3 1 ⟵ Global

I 1 2 3 ⟵ Local

$$
[\mathbf{k}_h]_2 = \frac{1}{4}\begin{bmatrix} \frac{4}{9} & 0 & \frac{2}{3} & -\frac{2}{3} \\ 0 & 0 & 0 & 0 \\ \frac{2}{3} & 0 & 1 & -1 \\ \frac{2}{3} & 0 & -1 & 1 \end{bmatrix} \begin{Bmatrix} \theta \\ \psi_1 \\ \psi_2 \\ \psi_3 \end{Bmatrix} = \begin{Bmatrix} M_{t_2} \\ 0 \\ 0 \\ 0 \end{Bmatrix}. \quad \begin{matrix} \text{I} & \text{I} \\ 1 & 5 \\ 2 & 3 \\ 3 & 1 \end{matrix}
$$

Element 3:

$$
\begin{aligned}
a_1 &= 0, & b_1 &= -2, \\
a_2 &= -1, & b_2 &= 1, \\
a_3 &= 1, & b_3 &= 1, \\
x_m &= \tfrac{5}{3}, & y_m &= 1, \\
l_{23} &= 2, & \frac{GA}{l_{23}^2} &= \frac{1}{4},
\end{aligned}
$$

I 5 2 4 ⟵ Global

I 1 2 3 ⟵ Local

$$
[\mathbf{k}_h]_3 = \frac{1}{4}\begin{bmatrix} \frac{100}{9} & 0 & -\frac{10}{3} & \frac{10}{3} \\ 0 & 0 & 0 & 0 \\ -\frac{10}{3} & 0 & 1 & -1 \\ \frac{10}{3} & 0 & -1 & 1 \end{bmatrix} \begin{Bmatrix} \theta \\ \psi_1 \\ \psi_2 \\ \psi_3 \end{Bmatrix} = \begin{Bmatrix} M_{t_3} \\ 0 \\ 0 \\ 0 \end{Bmatrix}. \quad \begin{matrix} \text{I} & \text{I} \\ 1 & 5 \\ 2 & 2 \\ 3 & 4 \end{matrix}
$$

Element 4:

$$
\begin{aligned}
a_1 &= -2, & b_1 &= 0, \\
a_2 &= 1, & b_2 &= 1, \\
a_3 &= 1, & b_3 &= -1, \\
x_m &= 1, & y_m &= \tfrac{5}{3}, \\
l_{23} &= 2, & \frac{GA}{l_{23}^2} &= \frac{1}{4},
\end{aligned}
$$

I 5 4 3 ← Global

I 1 2 3 ← Local

$$
[\mathbf{k}_h]_4 = \frac{1}{4}\begin{bmatrix} \frac{100}{9} & 0 & -\frac{10}{3} & \frac{10}{3} \\ 0 & 0 & 0 & 0 \\ -\frac{10}{3} & 0 & 1 & -1 \\ \frac{10}{3} & 0 & -1 & 1 \end{bmatrix} \begin{Bmatrix} \theta \\ \psi_1 \\ \psi_2 \\ \psi_3 \end{Bmatrix} = \begin{Bmatrix} M_{t_4} \\ 0 \\ 0 \\ 0 \end{Bmatrix}. \quad \begin{matrix} \text{I} & \text{I} \\ 1 & 5 \\ 2 & 4 \\ 3 & 3 \end{matrix}
$$

Combination of the foregoing four element equations by observing compatibility of nodal ψ's leads to assemblage equations:

$$
\frac{1}{4}\begin{bmatrix} \frac{208}{9} & 0 & -\frac{12}{3} & \frac{12}{3} & 0 & 0 \\ 0 & 2 & -1 & -1 & 0 & 0 \\ -\frac{12}{3} & -1 & 2 & 0 & -1 & 0 \\ \frac{12}{3} & -1 & 0 & 2 & -1 & 0 \\ 0 & 0 & -1 & -1 & 2 & 0 \\ 0 & 0 & 0 & 0 & 0 & 0 \end{bmatrix} \begin{Bmatrix} \theta \\ \psi_1 \\ \psi_2 \\ \psi_3 \\ \psi_4 \\ \psi_5 \end{Bmatrix} = \begin{Bmatrix} \sum_{m=1} M_{t_m} \\ 0 \\ 0 \\ 0 \\ 0 \\ 0 \end{Bmatrix}.
$$

These equations can now be modified for the boundary condition $\psi_1 = 0$. Then for $\theta = 1$ the solution is

$$
\begin{aligned}
&\theta = 1, \\
&\psi_1 = 0 \text{ (prescribed)}, \\
&\psi_2 = 2,
\end{aligned}
$$

$$\psi_3 = -2,$$

$$\psi_4 = 0,$$

$$\psi_5 = 0.$$

These results for ψ's are essentially the same as in the case of Example 11.1 with the warping function approach. We have considered this rather simple example mainly to illustrate hand calculations. For acceptable accuracy and in order to exploit the advantage of improved boundary stresses in the hybrid procedure, one should use a finer mesh. In Example 11.6 we shall consider a computer solution for the mesh in Figure 11.10 where both boundary and inside elements occur. The results in Example 11.6 will illustrate the advantage of the hybrid approach.

Shear Stresses

For the four elements at the boundary, we use Equation 11.57 to evaluate the shear stresses:

Element 1:

$$\begin{Bmatrix} \tau_{xz} \\ \tau_{yz} \end{Bmatrix} = \frac{1}{4} \begin{bmatrix} -2(2 \times \frac{1}{3}) & 0 & -2 & 2 \\ 0 & 0 & 0 & 0 \end{bmatrix} \begin{Bmatrix} 1 \\ 0 \\ 0 \\ 2 \end{Bmatrix}$$

$$= \begin{Bmatrix} 0.666 \\ 0.000 \end{Bmatrix} \text{N/cm}^2.$$

Element 2:

$$\begin{Bmatrix} \tau_{xz} \\ \tau_{yz} \end{Bmatrix} = \frac{1}{4} \begin{bmatrix} 0 & 0 & 0 & 0 \\ 2 \times \frac{2}{3} & 0 & 2 & -2 \end{bmatrix} \begin{Bmatrix} 1 \\ 0 \\ -2 \\ 0 \end{Bmatrix}$$

$$= \begin{Bmatrix} 0.000 \\ -0.666 \end{Bmatrix} \text{N/cm}^2.$$

Element 3:

$$\begin{Bmatrix} \tau_{xz} \\ \tau_{yz} \end{Bmatrix} = \frac{1}{4}\begin{bmatrix} 0 & 0 & 0 & 0 \\ -\frac{20}{3} & 0 & -2 & 2 \end{bmatrix}\begin{Bmatrix} 1 \\ 0 \\ 2 \\ 0 \end{Bmatrix}$$

$$= \begin{Bmatrix} 0.000 \\ 0.666 \end{Bmatrix} \text{N/cm}^2.$$

Element 4:

$$\begin{Bmatrix} \tau_{xz} \\ \tau_{yz} \end{Bmatrix} = \frac{1}{4}\begin{bmatrix} -\frac{20}{3} & 0 & 2 & -2 \\ 0 & 0 & 0 & 0 \end{bmatrix}\begin{Bmatrix} 1 \\ 0 \\ 0 \\ -2 \end{Bmatrix}$$

$$= \begin{Bmatrix} -0.666 \\ 0.000 \end{Bmatrix} \text{N/cm}^2.$$

These results are the same as those in Example 11.1 (Figure 11.3); they show that the shear stress normal to the boundary is zero. It was only incidental that with the four-element mesh chosen the warping function approach also satisfied the zero stress boundary condition. With a different and arbitrary mesh, the warping function approach, in general, would not satisfy this condition.

Twisting Moment

The torsional constant can be found by substituting the computed results for ψ into the first of the above assemblage equations:

$$\frac{M_t}{G\theta} = \frac{1}{4}\left[\frac{208}{9} \times 1 - 4 \times 2 + 4 \times (-2)\right]$$

$$= 1.7777 \ \text{N/cm}.$$

This value is the same as that from the stress function approach and has an error of about 21.00%.

Example 11.6. Torsion of Triangular Bar: Hybrid Approach, Computer Solution

Figure 11.10(a) shows a bar in the shape of an equilateral triangle. Analysis for torsion for this bar using the hybrid method is presented by Yamada et al.[9] Due to symmetry, only one sixth of the triangle is discretized. The properties of the system are

$$\text{Young's modulus,}\quad E = 2.1 \times 10^4 \ \text{kg/mm}^2,$$

$$\text{Poisson's ratio,}\quad = 0.3,$$

$$\text{Shear modulus,}\quad G = \frac{E}{2(1+\nu)} = \frac{2.1 \times 10^4}{2(1+0.3)} = 0.81 \times 10^4 \ \text{kg/mm}^2$$

$$\text{Angle of twist,}\quad \theta = 1.194 \times 10^{-4}\ \text{rad/mm}.$$

Table 11.1 compares the values of warping along the side AB (Figure 11.10(a)) obtained from the closed form solution, the displacement method, and the hybrid method. The two numerical results are close to the exact solution, without significant difference between them.

As noted before, Figure 11.10(b) shows comparisons between computed values of the shear stress normal to the boundary τ_n from the displacement and the hybrid methods. In contrast to the solution by the displacement method, the hybrid approach yields zero values of τ_n as required by the theory.

TABLE 11.1

Warping along the Contour AB[9]

Node		Exact Solution	Hybrid Method	Displacement Method
B	11	0.370	0.368	0.375
↑	13	0.957	0.942	0.975
	21	1.924	1.884	1.921
	29	2.772	2.721	2.766
	36	3.471	3.419	3.467
	43	3.887	3.821	3.868
	49	3.973	3.913	3.960
	55	3.962	3.632	3.659
	60	2.978	2.918	2.960
	63	2.398	2.352	2.385
	66	1.865	1.834	1.854
	69	1.354	1.330	1.347
	73	0.898	0.885	0.896
↓	76	0.530	0.524	0.529
A	79	0.270	0.268	0.270

Note: $w \times 10^3$ mm.

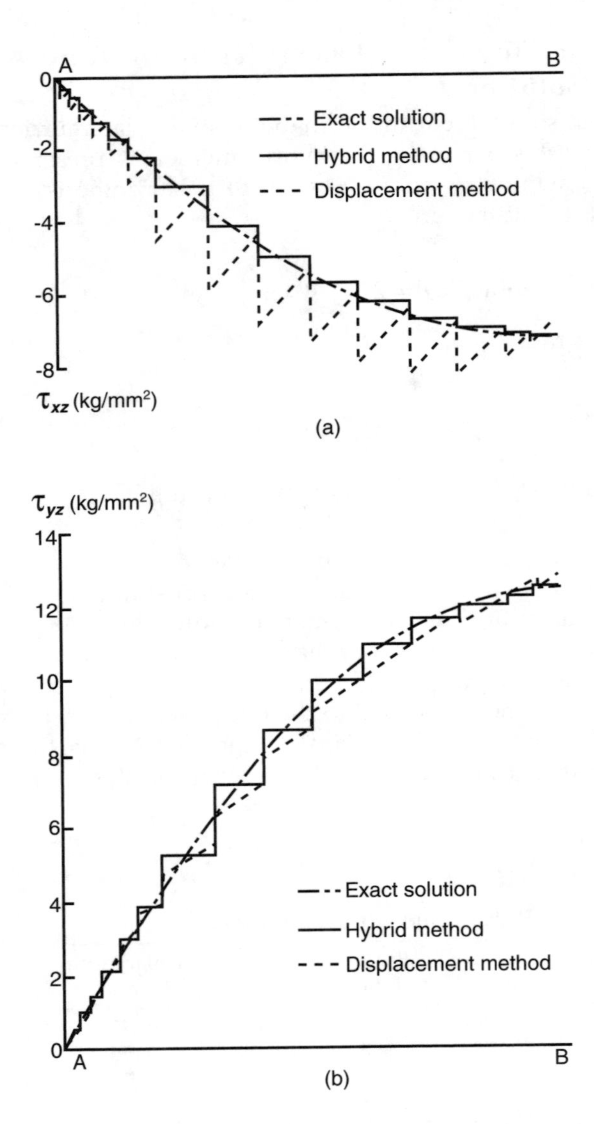

FIGURE 11.13
Comparisons for shear stresses:[9] (a) Shear stress τ_{xz} along contour AB and (b) Shear stress τ_{yz} along contour AB.

Figures 11.13(a) and (b) show comparisons between shear stresses τ_{xz} and τ_{yz} along *AB* as computed from exact solution and the displacement and hybrid methods. The latter improves computations of the shear stresses not only along the boundary but in the interior also; this aspect is indicated in Table 11.2, which lists the shearing stresses at the centroids of typical elements adjacent to side *AB* (Figure 11.10). These results also show the improvement in accuracy provided by the hybrid approach as compared to the results from the displacement approach.

TABLE 11.2

Comparison of the Shearing Stresses at the Centroid of Triangular Elements Adjacent to AB[9]

	Element				
	2	48	81	99	122
τ_{xz}					
Exact solution	–7.136	–6.354	–4.351	–2.259	–0.128
Hybrid method	–7.168	–6.264	–4.117	–2.251	–0.125
Displacement method	–7.145	–6.884	–4.759	–2.563	–0.225
τ_{yz}					
Exact solution	12.354	10.803	7.209	3.348	0.168
Hybrid method	12.416	10.853	7.131	3.900	0.216
Displacement method	12.541	10.511	6.653	3.628	0.071
τ_n					
Exact solution	–0.003	–0.101	–0.164	–0.282	–0.027
Hybrid method	0.000	0.000	0.000	0.000	0.000
Displacement method	0.083	–0.706	–0.795	–0.406	–0.159

Note: Unit: kg/mm^2.

Mixed Approach

In the mixed formulation, both the displacements (warping) and stresses (stress function) within (including the boundaries) the element are assumed to be unknown. Hence, for a triangular element,

$$\psi = \left[\mathbf{N}_\psi\right]\left\{\mathbf{q}_\psi\right\} = \left[N_1 \quad N_2 \quad N_3\right]\begin{Bmatrix}\psi_1\\ \psi_2\\ \psi_3\end{Bmatrix}, \tag{11.58a}$$

$$\{\sigma\} = \left[\mathbf{N}_\sigma\right]\left\{\mathbf{q}_\sigma\right\} = \begin{bmatrix} N_1 & N_2 & N_3 & 0 & 0 & 0\\ 0 & 0 & 0 & N_1 & N_2 & N_3\end{bmatrix}\left\{\mathbf{q}_\sigma\right\}, \tag{11.58b}$$

where $\{\sigma\}^T = [\tau_{xz}\ \tau_{yz}]$ is the vector of shear stresses at any point, $[\mathbf{N}_\sigma]$ and $[\mathbf{N}_\psi]$ are the interpolation functions, $\{\mathbf{q}_\psi\}$ is the vector of nodal warping, and $\{\mathbf{q}_\sigma\}^T = [\tau_{xz_1}\ \tau_{xz_2}\ \tau_{xz_3}\ \tau_{xy_1}\ \tau_{yz_2}\ \tau_{yz_3}]$ is the vector of nodal shear stresses.

The strain displacement relation is given in Equation 11.5, and the constitutive relation for linearly elastic and homogeneous material is

$$\{\sigma\} = \begin{Bmatrix}\tau_{xz}\\ \tau_{yz}\end{Bmatrix} = G\begin{bmatrix}1 & 0\\ 0 & 1\end{bmatrix}\begin{Bmatrix}\gamma_{xz}\\ \gamma_{yz}\end{Bmatrix}. \tag{11.59}$$

Step 4. Derive Element Equations

We shall follow essentially the procedure presented by Noor and Andersen.[12,13] A variational function for the mixed approach can be expressed as[4,12]

$$\Pi_R = \iiint_V \left(\{\sigma\}^T\{\varepsilon\} - \theta\{\sigma\}^T\{x\}\right)dV - \iiint_V dU_c - \iint_{S_2} \psi\{\bar{\sigma}\}\{\mathbf{n}\}dS, \quad (11.60)$$

where dU_c is the complementary energy density (Equation 11.29), given by

$$dU_c = \frac{1}{2}\begin{bmatrix}\tau_{xz} & \tau_{yz}\end{bmatrix}\frac{1}{G}\begin{bmatrix}1 & 0\\ 0 & 1\end{bmatrix}\begin{Bmatrix}\tau_{xz}\\ \tau_{yz}\end{Bmatrix}dV$$

$$= \tfrac{1}{2}\{\sigma\}^T[\mathbf{D}]\{\sigma\}dV;$$

$\{\bar{\sigma}\}^T = [\bar{\tau}_{xz}\ \ \bar{\tau}_{yz}]$ is the shear stresses at the boundaries; $\{n\}$ is the vector of outward normal to the boundary; V is the volume of the element; $\{x\}^T = [x\ y]$; and S_2 is the part of boundary on which tractions are prescribed. Substitution for $\{\sigma\}$ in Equation 11.58a, $\{\varepsilon\}$ in Equation 11.5, and dU_c in Π_R leads to

$$\begin{aligned}\Pi_R = &\iiint_V \{\mathbf{q}_\sigma\}^T[\mathbf{N}_\sigma]^T[\mathbf{B}]\{\mathbf{q}_\psi\}dV\\ &- \iiint_V \{\mathbf{q}_\sigma\}^T[\mathbf{N}_\sigma]^T[\mathbf{D}][\mathbf{N}_\sigma]\{\mathbf{q}_\sigma\}dV\\ &- \theta\iiint_V \{\mathbf{q}_\sigma\}^T[\mathbf{N}_\sigma]^T[\mathbf{N}_\sigma]\{\mathbf{x}_n\}dV\\ &- \iint_{S_2} \{\mathbf{q}_\psi\}^T[\mathbf{N}_\psi]^T\{\mathbf{n}\}[\mathbf{N}_\sigma]\{\bar{\mathbf{q}}_\sigma\}dS.\end{aligned} \quad (11.61)$$

Here $\{\mathbf{x}_n\} = [x_1\, x_2\, x_3\, y_1\, y_2\, y_3]$ is the vector of nodal coordinates, given by

$$\begin{Bmatrix}x\\ y\end{Bmatrix} = \begin{bmatrix}N_1 & N_2 & N_3 & 0 & 0 & 0\\ 0 & 0 & 0 & N_1 & N_2 & N_3\end{bmatrix}\{\mathbf{x}_n\}, \quad (11.62)$$

and $[\mathbf{B}]$ is defined in Equation 11.12.

Application of the variational principle to Π_R leads to its stationary value. The variations or differentiations are performed independently and simultaneously with respect to nodal stresses and nodal warping functions; thus

$$\delta\Pi_R = 0 \Rightarrow \begin{cases} \dfrac{\partial \Pi_R}{\partial\{\mathbf{q}_\sigma\}} = 0, \\ \dfrac{\partial \Pi_R}{\partial\{\mathbf{q}_\psi\}} = 0. \end{cases} \tag{11.63}$$

This leads to two sets of equations,

$$\begin{bmatrix} [\mathbf{k}_\sigma] & [\mathbf{k}_{\sigma\psi}] \\ [\mathbf{k}_{\sigma\psi}]^T & [0] \end{bmatrix} \begin{Bmatrix} \{\mathbf{q}_\sigma\} \\ \{\mathbf{q}_\psi\} \end{Bmatrix} = \begin{Bmatrix} \{\mathbf{Q}_\sigma\} \\ \{0\} \end{Bmatrix}, \tag{11.64}$$

where, for the triangular element (Figure 10.2)

$$\underset{(6\times 6)}{[\mathbf{k}_\sigma]} = -\iiint \underset{(6\times 2)}{[\mathbf{N}_\sigma]^T} \underset{(2\times 2)}{[\mathbf{D}]} \underset{(2\times 6)}{[\mathbf{N}_\sigma]} dxdydz, \tag{11.65a}$$

$$\underset{(6\times 3)}{[\mathbf{k}_{\sigma\psi}]} = \iiint_V \underset{(6\times 2)}{[\mathbf{N}_\sigma]^T} \underset{(2\times 3)}{[\mathbf{B}]} \, dxdydz, \tag{11.65b}$$

and

$$\underset{(6\times 1)}{\{\mathbf{Q}_\sigma\}} = \theta \iint \underset{(6\times 2)}{\{\mathbf{N}_\sigma\}^T} \underset{(2\times 6)}{[\mathbf{N}]} \, dxdy \underset{(6\times 1)}{\{\mathbf{x}_n\}}. \tag{11.65c}$$

The last term in Equation 11.61 represents boundary conditions. Here we have assumed that the boundary conditions occur as prescribed shear stresses and that their components normal to the boundary vanish. The tangential components do not contribute when the variation, Equation 11.63, is performed, hence the last term in Equation 11.61 does not appear in the calculation of forcing parameter vector $\{\mathbf{Q}_\sigma\}$.

Evaluation of Element Matrices and Load Vector

We now illustrate computations of the element properties for the triangular element. For unit length,

$$[\mathbf{k}_\sigma] = -\frac{1}{G}\iint \begin{Bmatrix} N_1 & 0 \\ N_2 & 0 \\ N_3 & 0 \\ 0 & N_1 \\ 0 & N_2 \\ 0 & N_3 \end{Bmatrix} \begin{bmatrix} N_1 & N_2 & N_3 & 0 & 0 & 0 \\ 0 & 0 & 0 & N_1 & N_2 & N_3 \end{bmatrix} dxdy$$

$$= -\frac{1}{G}\begin{bmatrix} N_1^2 & N_1N_2 & N_1N_3 & 0 & 0 & 0 \\ & N_2^2 & N_2N_3 & 0 & 0 & 0 \\ & & N_3^2 & 0 & 0 & 0 \\ & \text{sym.} & & N_1^2 & N_1N_2 & N_1N_3 \\ & & & & N_2^2 & N_2N_3 \\ & & & & & N_3^2 \end{bmatrix} dxdy.$$

The integrations can be performed in closed form; for instance,

$$\iint_A N_1^2 dA = 2A\frac{2!0!0!}{(2+2)!} = \frac{A}{6}$$

and

$$\iint_A N_1N_2 dA = 2A\frac{1!1!0!}{(4)!} = \frac{A}{12}.$$

Therefore,

$$\left[\mathbf{k}_\sigma\right] = -\frac{A}{12G}\begin{bmatrix} 2 & 1 & 1 & 0 & 0 & 0 \\ 1 & 2 & 1 & 0 & 0 & 0 \\ 1 & 1 & 2 & 0 & 0 & 0 \\ 0 & 0 & 0 & 2 & 1 & 1 \\ 0 & 0 & 0 & 1 & 2 & 1 \\ 0 & 0 & 0 & 1 & 1 & 2 \end{bmatrix}.$$

Now

$$\left[\mathbf{k}_{\sigma\psi}\right] = \iint \begin{bmatrix} N_1 & 0 \\ N_2 & 0 \\ N_3 & 0 \\ 0 & N_1 \\ 0 & N_2 \\ 0 & N_3 \end{bmatrix} \frac{1}{2A}\begin{bmatrix} b_1 & b_2 & b_3 \\ a_1 & a_2 & a_3 \end{bmatrix} dA$$

$$= \frac{1}{2A}\iint \begin{bmatrix} N_1 b_1 & N_1 b_2 & N_1 b_3 \\ N_2 b_1 & N_2 b_2 & N_2 b_3 \\ N_3 b_1 & N_3 b_2 & N_3 b_3 \\ N_1 a_1 & N_1 a_2 & N_1 a_3 \\ N_2 a_1 & N_2 a_2 & N_2 a_3 \\ N_3 a_1 & N_3 a_2 & N_3 a_3 \end{bmatrix} dA$$

$$= \frac{1}{6}\begin{bmatrix} b_1 & b_2 & b_3 \\ b_1 & b_2 & b_3 \\ b_1 & b_2 & b_3 \\ a_1 & a_2 & a_3 \\ a_1 & a_2 & a_3 \\ a_1 & a_2 & a_3 \end{bmatrix}$$

and

$$\{\mathbf{Q}_\sigma\} = \theta \iint \begin{bmatrix} N_1 & 0 \\ N_2 & 0 \\ N_3 & 0 \\ 0 & N_1 \\ 0 & N_2 \\ 0 & N_3 \end{bmatrix} \begin{bmatrix} N_1 & N_2 & N_3 & 0 & 0 & 0 \\ 0 & 0 & 0 & N_1 & N_2 & N_3 \end{bmatrix} \begin{Bmatrix} x_1 \\ x_2 \\ x_3 \\ y_1 \\ y_2 \\ y_3 \end{Bmatrix}$$

$$= \frac{\theta A}{12}\begin{Bmatrix} 2x_1 + x_2 + x_3 \\ x_1 + 2x_2 + x_3 \\ x_1 + x_2 + 2x_3 \\ 2y_1 + y_2 + y_3 \\ y_1 + 2y_2 + y_3 \\ y_1 + y_2 + 2y_3 \end{Bmatrix}.$$

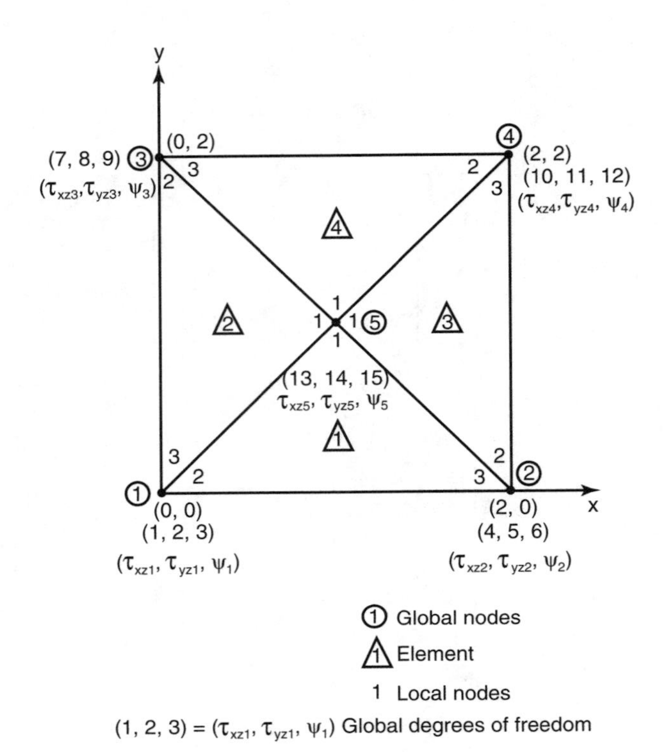

FIGURE 11.14
Torsion of square bar: mixed approach.

Example 11.7. Torsion of Square Bar: Mixed Approach

The node numbers and details for the problem of the square bar (2×2 cross section) in Figure 11.2 are shown in Figure 11.14. In the following are given essential details of the computations of element and assemblage matrices based on Equations 11.64 and 11.65.

Equation 11.64a can be written in the expanded form as

$$\frac{A}{12G}\begin{bmatrix} -2 & -1 & -1 & 0 & 0 & 0 & 2b_1 & 2b_2 & 2b_3 \\ -1 & -2 & -1 & 0 & 0 & 0 & 2b_1 & 2b_2 & 2b_3 \\ -1 & -1 & -2 & 0 & 0 & 0 & 2b_1 & 2b_2 & 2b_3 \\ 0 & 0 & 0 & -2 & -1 & -1 & 2a_1 & 2a_2 & 2a_3 \\ 0 & 0 & 0 & -1 & -2 & -1 & 2a_1 & 2a_2 & 2a_3 \\ 0 & 0 & 0 & -1 & -1 & -2 & 2a_1 & 2a_2 & 2a_3 \\ 2b_1 & 2b_1 & 2b_1 & 2a_1 & 2a_1 & 2a_1 & 0 & 0 & 0 \\ 2b_2 & 2b_2 & 2b_2 & 2a_2 & 2a_2 & 2a_2 & 0 & 0 & 0 \\ 2b_3 & 2b_3 & 2b_3 & 2a_3 & 2a_3 & 2a_3 & 0 & 0 & 0 \end{bmatrix}\begin{Bmatrix} \tau_{xz_1} \\ \tau_{xz_2} \\ \tau_{xz_3} \\ \tau_{yz_1} \\ \tau_{yz_2} \\ \tau_{yz_3} \\ \psi_1 \\ \psi_2 \\ \psi_3 \end{Bmatrix} = \frac{QA}{12}\begin{Bmatrix} 2x_1 + x_2 + x_3 \\ x_1 + 2x_2 + x_3 \\ x_1 + x_2 + 2x_3 \\ 2y_1 + y_2 + y_3 \\ y_1 + 2y_2 + y_3 \\ y_1 + y_2 + 2y_3 \\ 0 \\ 0 \\ 0 \end{Bmatrix} \tag{11.64b}$$

Element 1:

$$a_1 = 2, \qquad b_1 = 0,$$
$$a_2 = -1, \qquad b_2 = -1,$$
$$a_3 = -1, \qquad b_3 = 1,$$
$$x_m = 1, \qquad y_m = \tfrac{1}{3},$$

$$\begin{array}{ccccccccc} 13 & 1 & 4 & 14 & 2 & 5 & 15 & 3 & 6 \leftarrow \text{Global} \end{array}$$

$$\begin{bmatrix} -2 & -1 & -1 & 0 & 0 & 0 & 0 & -2 & 2 \\ -1 & -2 & -1 & 0 & 0 & 0 & 0 & -2 & 2 \\ -1 & -1 & -2 & 0 & 0 & 0 & 0 & -2 & 2 \\ 0 & 0 & 0 & -2 & -1 & -1 & 4 & -2 & -2 \\ 0 & 0 & 0 & -1 & -2 & -1 & 4 & -2 & -2 \\ 0 & 0 & 0 & -1 & -1 & -2 & 4 & -2 & -2 \\ 0 & 0 & 0 & 4 & 4 & 4 & 0 & 0 & 0 \\ -2 & -2 & -2 & -2 & -2 & -2 & 0 & 0 & 0 \\ 2 & 2 & 2 & -2 & -2 & -2 & 0 & 0 & 0 \end{bmatrix} \begin{Bmatrix} \tau_{xz_1} \\ \tau_{xz_2} \\ \tau_{xz_3} \\ \tau_{yz_1} \\ \tau_{yz_2} \\ \tau_{yz_3} \\ \psi_1 \\ \psi_2 \\ \psi_3 \end{Bmatrix} = \begin{Bmatrix} 4 \\ 3 \\ 5 \\ 2 \\ 1 \\ 1 \\ 0 \\ 0 \\ 0 \end{Bmatrix}. \quad \begin{matrix} 13 \\ 1 \\ 4 \\ 14 \\ 2 \\ 5 \\ 15 \\ 3 \\ 6 \end{matrix}$$

We have assigned global numbers by rearranging the unknowns as shown in Figure 11.14. That is, in the subsequent assemblage equations, the global numbers are assigned such that the three unknowns at a given node appear consecutively. This is done simply for convenience.

Element 2:

$$a_1 = 0, \qquad b_1 = 2,$$
$$a_2 = 1, \qquad b_2 = -1,$$
$$a_3 = -1, \qquad b_3 = -1,$$
$$x_m = \tfrac{1}{3}, \qquad y_m = 1,$$

$$
\begin{array}{ccccccccc}
13 & 7 & 1 & 14 & 8 & 2 & 15 & 9 & 3
\end{array} \leftarrow \text{Global}
$$

$$
\begin{bmatrix}
-2 & -1 & -1 & 0 & 0 & 0 & 4 & -2 & -2 \\
-1 & -2 & -1 & 0 & 0 & 0 & 4 & -2 & -2 \\
-1 & -1 & -2 & 0 & 0 & 0 & 4 & -2 & -2 \\
0 & 0 & 0 & -2 & -1 & -1 & 0 & 2 & -2 \\
0 & 0 & 0 & -1 & -2 & -1 & 0 & 2 & -2 \\
0 & 0 & 0 & -1 & -1 & -2 & 0 & 2 & -2 \\
4 & 4 & 4 & 0 & 0 & 0 & 0 & 0 & 0 \\
-2 & -2 & -2 & 2 & 2 & 2 & 0 & 0 & 0 \\
-2 & -2 & -2 & -2 & -2 & -2 & 0 & 0 & 0
\end{bmatrix}
\begin{Bmatrix}
\tau_{xz_1} \\ \tau_{xz_2} \\ \tau_{xz_3} \\ \tau_{yz_1} \\ \tau_{yz_2} \\ \tau_{yz_3} \\ \psi_1 \\ \psi_2 \\ \psi_3
\end{Bmatrix}
=
\begin{Bmatrix}
2 \\ 1 \\ 1 \\ 4 \\ 5 \\ 3 \\ 0 \\ 0 \\ 0
\end{Bmatrix}.
\quad
\begin{array}{c}
13 \\ 7 \\ 1 \\ 14 \\ 8 \\ 2 \\ 15 \\ 9 \\ 3
\end{array}
$$

Element 3:

$$
\begin{aligned}
a_1 &= 0, & b_1 &= -2, \\
a_2 &= -1, & b_2 &= 1, \\
a_3 &= 1, & b_3 &= 1, \\
x_m &= \tfrac{5}{3}, & y_m &= 1,
\end{aligned}
$$

$$
\begin{array}{ccccccccc}
13 & 4 & 10 & 14 & 5 & 11 & 15 & 6 & 12
\end{array} \leftarrow \text{Global}
$$

$$
\begin{bmatrix}
-2 & -1 & -1 & 0 & 0 & 0 & -4 & 2 & 2 \\
-1 & -2 & -1 & 0 & 0 & 0 & -4 & 2 & 2 \\
-1 & -1 & -2 & 0 & 0 & 0 & -4 & 2 & 2 \\
0 & 0 & 0 & -2 & -1 & -1 & 0 & -2 & 2 \\
0 & 0 & 0 & -1 & -2 & -1 & 0 & -2 & 2 \\
0 & 0 & 0 & -1 & -1 & -2 & 0 & -2 & 2 \\
-4 & -4 & -4 & 0 & 0 & 0 & 0 & 0 & 0 \\
2 & 2 & 2 & -2 & -2 & -2 & 0 & 0 & 0 \\
2 & 2 & 2 & 2 & 2 & 2 & 0 & 0 & 0
\end{bmatrix}
\begin{Bmatrix}
\tau_{xz_1} \\ \tau_{xz_2} \\ \tau_{xz_3} \\ \tau_{yz_1} \\ \tau_{yz_2} \\ \tau_{yz_3} \\ \psi_1 \\ \psi_2 \\ \psi_3
\end{Bmatrix}
=
\begin{Bmatrix}
6 \\ 7 \\ 7 \\ 4 \\ 3 \\ 5 \\ 0 \\ 0 \\ 0
\end{Bmatrix}.
\quad
\begin{array}{c}
13 \\ 4 \\ 10 \\ 14 \\ 5 \\ 11 \\ 15 \\ 6 \\ 12
\end{array}
$$

Element 4:

$$a_1 = -2, \qquad b_1 = 0,$$
$$a_2 = 1, \qquad b_2 = 1,$$
$$a_3 = 1, \qquad b_3 = -1,$$
$$x_m = 1, \qquad y_m = \tfrac{5}{3},$$

$$\begin{array}{ccccccccc} 13 & 10 & 7 & 14 & 11 & 8 & 15 & 12 & 9 \leftarrow \text{Global} \end{array}$$

$$\begin{bmatrix} -2 & -1 & -1 & 0 & 0 & 0 & 0 & 2 & -2 \\ -1 & -2 & -1 & 0 & 0 & 0 & 0 & 2 & -2 \\ -1 & -1 & -2 & 0 & 0 & 0 & 0 & 2 & -2 \\ 0 & 0 & 0 & -2 & -1 & -1 & -4 & 2 & 2 \\ 0 & 0 & 0 & -1 & -2 & -1 & -4 & 2 & 2 \\ 0 & 0 & 0 & -1 & -1 & -2 & -4 & 2 & 2 \\ 0 & 0 & 0 & -4 & -4 & -4 & 0 & 0 & 0 \\ 2 & 2 & 2 & 2 & 2 & 2 & 0 & 0 & 0 \\ -2 & -2 & -2 & 2 & 2 & 2 & 0 & 0 & 0 \end{bmatrix} \begin{Bmatrix} \tau_{xz_1} \\ \tau_{xz_2} \\ \tau_{xz_3} \\ \tau_{yz_1} \\ \tau_{yz_2} \\ \tau_{yz_3} \\ \psi_1 \\ \psi_2 \\ \psi_3 \end{Bmatrix} = \begin{Bmatrix} 4 \\ 5 \\ 3 \\ 6 \\ 7 \\ 7 \\ 0 \\ 0 \\ 0 \end{Bmatrix}. \quad \begin{array}{c} \downarrow \\ 13 \\ 10 \\ 7 \\ 14 \\ 11 \\ 8 \\ 15 \\ 12 \\ 9 \end{array}$$

The foregoing four element equations can now be assembled by observing interelement compatibility.

It is difficult to solve by hand calculations the (modified) assemblage equations (involving 14 unknowns), and it becomes necessary to use an (available) equation solver for large sets of equations (see Appendix 2). In this elementary treatment, it will suffice only to illustrate the mixed approach and the foregoing steps, because detailed consideration of the topic is considered beyond our scope. Solutions for this and the other relevant exercises in the Problems are left to the inquisitive and advanced reader.

The foregoing completes our treatment of the torsion problem. Before we proceed to the next chapter in which a four-node quadrilateral element is described, it is appropriate to present a brief description of the topic of (static) condensation, which is often used in finite element applications.

Static Condensation

Often it may be convenient and necessary to use nodes within an element, particularly when it is required to use higher-order elements. Sometimes a single element is divided into subcomponent elements by using the inner nodes. For instance, Figure 11.15 shows a line and a quadrilateral element each with one inner node. The line element has two end or external or primary nodes and one inner node, and the quadrilateral has four corners or primary nodes and one inside node.

The finite element equations

$$[\mathbf{k}]\{\mathbf{q}\} = \{\mathbf{Q}\} \tag{11.66a}$$

are derived on the basis of unknown degrees of freedom at all the nodes. Thus the formulation includes improved (higher-order) approximation yielded by the use of the inner node. In the case of the quadrilateral (Figure 11.15(b)), the element equations are often obtained by adding individual element equations of the four component triangles. Thus the total number of element unknowns in $\{\mathbf{q}\}$ are 3 and 10 for the line and the quadrilateral, respectively. These element equations can now be assembled to obtain global equations, in which the degrees of freedom at the inner node(s) will appear as additional unknowns, and to that extent the number of equations to be solved will be increased.

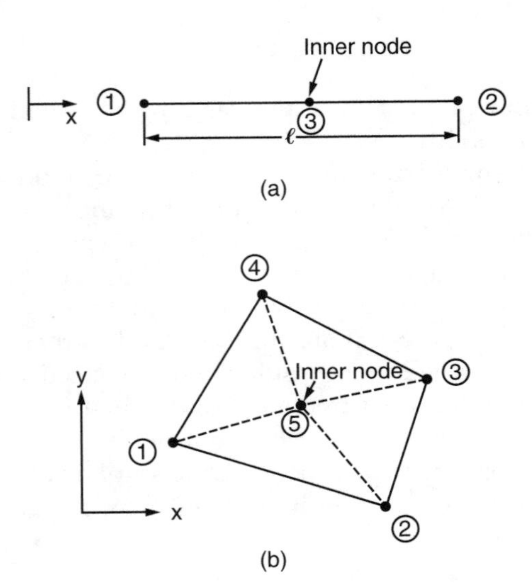

FIGURE 11.15
Static condensation: (a) Line element and (b) Quadrilateral.

It is possible to temporarily eliminate the unknowns at the inner node by creating an equivalent element relation in place of Equation 11.66a. The procedure of doing this is called static condensation. It is possible to do this because the unknowns at the inner node do not participate in the interelement compatibility at the element sides; that is, the unknowns at the inner node are not needed for the direct stiffness assembly procedure.

The procedure of static condensation involves solution of the unknowns at the inner node in terms of those at the primary nodes. To understand it, we write Equation 11.66a in a partitioned form as

$$\begin{bmatrix} [\mathbf{k}_{pp}] & [\mathbf{k}_{ip}] \\ [\mathbf{k}_{ip}]^T & [\mathbf{k}_{ii}] \end{bmatrix} \begin{Bmatrix} \{\mathbf{q}_p\} \\ \{\mathbf{q}_i\} \end{Bmatrix} = \begin{Bmatrix} \{\mathbf{Q}_p\} \\ \{\mathbf{Q}_i\} \end{Bmatrix}, \tag{11.66b}$$

where $\{\mathbf{q}_p\}$, $\{\mathbf{Q}_p\}$ and $\{\mathbf{q}_i\}$, $\{\mathbf{Q}_i\}$ denote vectors of nodal unknowns and loads at the primary and inner nodes, respectively. The second equation in Equation 11.66b can be used to solve for $\{\mathbf{q}_i\}$as

$$\{\mathbf{q}_i\} = [\mathbf{k}_{ii}]^{-1}\left(\{\mathbf{Q}_i\} - [\mathbf{k}_{ip}]^T\{\mathbf{q}_p\}\right). \tag{11.67a}$$

Substitution of $\{\mathbf{q}_i\}$into the first equation of Equation 11.66b leads to

$$\left([\mathbf{k}_{pp}] - [\mathbf{k}_{ip}][\mathbf{k}_{ii}]^{-1}[\mathbf{k}_{ip}]^T\right)\{\mathbf{q}_p\} = \{\mathbf{Q}_p\} - [\mathbf{k}_{ip}][\mathbf{k}_{ii}]^{-1}\{\mathbf{Q}_t\} \tag{11.67b}$$

or

$$[\bar{\mathbf{k}}]\{\mathbf{q}_p\} = \{\bar{\mathbf{Q}}\},$$

where $[\bar{\mathbf{k}}]$ and $\{\bar{\mathbf{Q}}\}$ are the condensed element (stiffness) and nodal load vectors, respectively. Notice that the vector of unknown $\{\mathbf{q}_p\}$ now contains the degrees of freedom only at the external primary nodes; thus when $[\bar{\mathbf{k}}]$ and $\{\bar{\mathbf{Q}}\}$ are used to assemble element equations, we have reduced from the total equations to be solved equations corresponding to the unknowns at the inner node. After the assembled equations are solved, if desired, it is possible to retrieve the unknowns at the inner node by using Equation 11.67a. Further details of static condensation can be found in many publications, including Reference 4.

Problems

11.1. Obtain the governing equation and natural boundary conditions associated with the functional of Equation 11.29a.

11.2. Use the mesh in Figure 11.7 and treat the problem as torsion of a triangular bar. By employing the warping function approach and setting the boundary condition as $\psi_3 = 0$, compute nodal warping functions, shear stresses, and twisting moments.

11.3. With the mesh in Problem 11.2, solve for torsion using the hybrid approach.

11.4. With the mesh in Problem 11.2, derive element equations using the mixed approach.*

11.5. Solve for torsion of the square bar in Figure 11.16 by using the warping function approach.

11.6. Solve Problem 11.5 by using the stress function approach.

11.7. Solve Problem 11.5 by using the hybrid approach. Consider elements 4, 6, and 8 as boundary elements and the remaining as inner elements.

11.8. Solve Problem 11.5 by using the mixed approach.

11.9. Consider the mesh in Figure 11.17 and solve for torsion by using the stress function approach. Here, it may be necessary to use a computer. The code FIELD-2D can be used for solving the entire problem. Alternatively, the assemblage matrix can be computed by hand calculations, and its solution can be obtained by using an available code for solution of simultaneous equations.

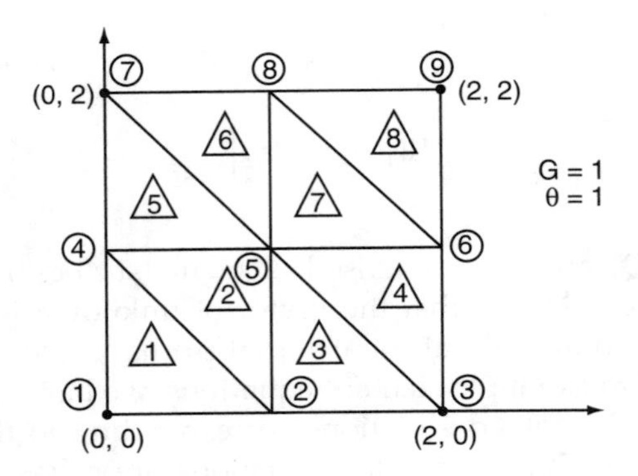

FIGURE 11.16

* This and some of the other problems will require use of an available equation solver once the global equations are obtained.

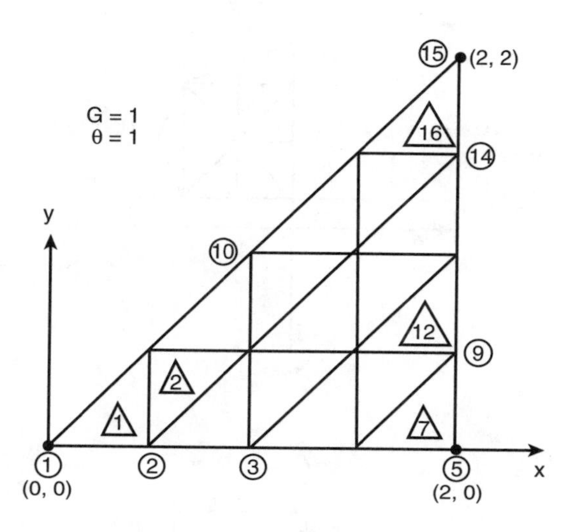

FIGURE 11.17

11.10. For the mesh in Figure 11.18, use the mixed procedure and derive the element equations (Equation 11.64a).

11.11. For the torsion problem in Figure 11.4, treat the bar as having a cross section of 2 cm × 2 cm. With boundary conditions $\varphi_1 = \varphi_2 = \varphi_4 = \varphi_5 = 0$, compute φ_5 using the stress function approach. Evaluate torsional constant and discuss its accuracy.

Answer: $\varphi_5 = 0.6667$; $M_t/G\theta = 1.777$, error = 21.00%.

11.12. For the circular bar of 2-cm diameter, compute nodal stress functions for the mesh shown in Figure 11.19 Find shear stresses and the twisting moment by using a computer code. Assume $G = 10^7$ N/cm^2 and $\theta = 0.005$ deg/cm.

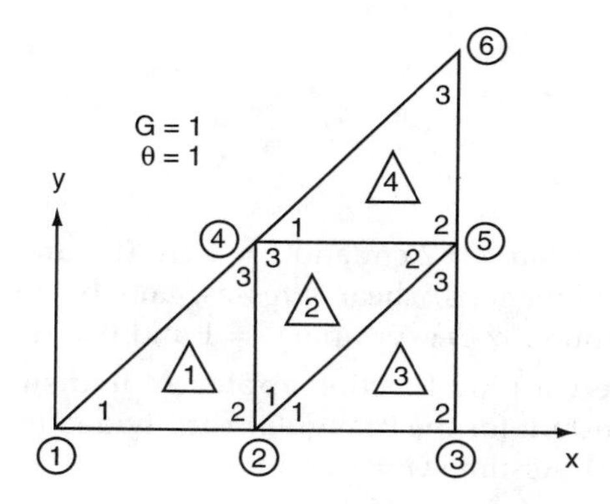

FIGURE 11.18

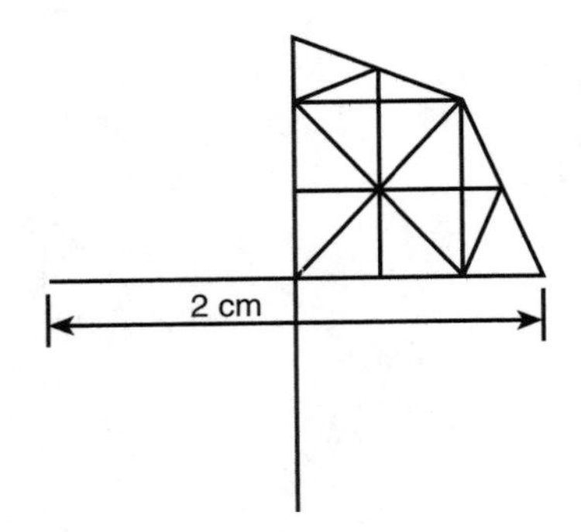

FIGURE 11.19

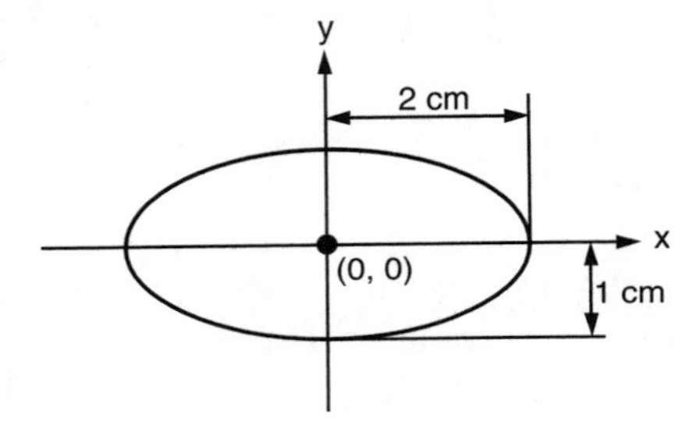

FIGURE 11.20

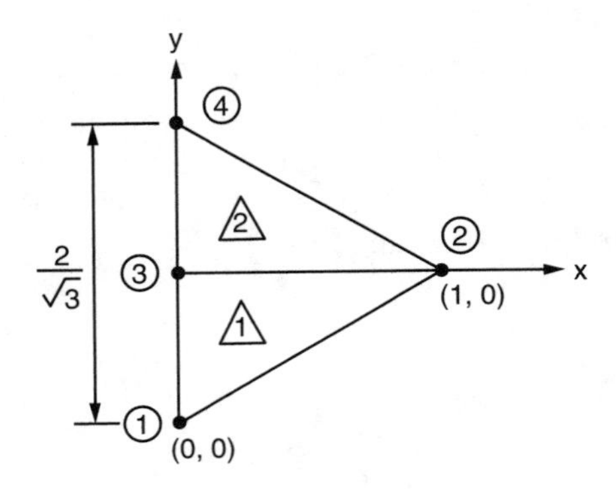

FIGURE 11.21

11.13. For the elliptic bar $a = 2$ cm and $b = 1$ cm (Figure 11.20), compute nodal stress functions, shear stresses, and twisting moment by using a computer code. Assume $G = 1$ and $\theta = 1$.

11.14. By using the warping function approach, find shear stresses and torsional constant for the triangular bar divided into two elements (Figure 11.21). Assume $G = 1$ and $\theta = 1$.

11.15. Use the stress function approach to solve for torsion of the triangular bar in Problem 11.14.

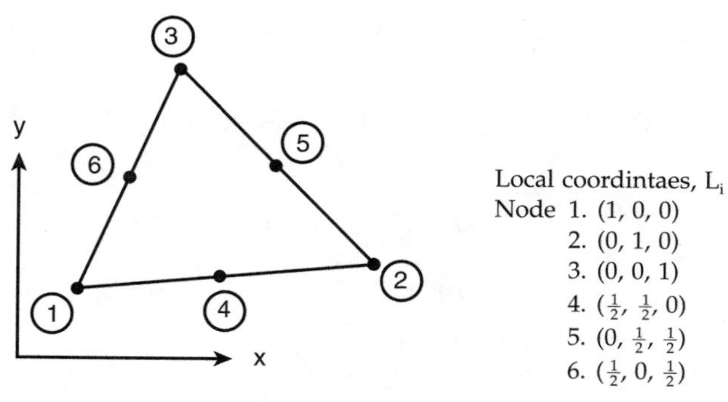

FIGURE 11.22

11.16. By using the code FIELD-2D or other available code, refine the mesh progressively (Example 11.4), say 1,4, 16, 64, etc. elements, and obtain results for stress functions φ and shear stresses. Plot the twisting moment M_t vs. number of elements or nodes and study the convergence behavior of the numerical solution.

11.17. Higher-order approximation. Use the quadratic interpolation function for φ with a triangular element (6 nodes) and derive the element matrix and load vector. See Figure 11.22.

Partial results: Assume

$$\varphi = [\mathbf{N}]\{\mathbf{q}_0\},$$

$$\{\mathbf{q}_0\}^T = [\varphi_1 \quad \varphi_2 \quad \varphi_3 \quad \varphi_4 \quad \varphi_5 \quad \varphi_6],$$

Note that this function includes all terms up to $n = 2$ in the polynomial expansion discussed under "Requirements for the Approximation Function."

$$[\mathbf{N}]^T = \begin{Bmatrix} L_1(2L_1 - 1) \\ L_2(2L_2 - 1) \\ L_3(2L_3 - 1) \\ 4L_1L_2 \\ 4L_2L_3 \\ 4L_3L_1 \end{Bmatrix},$$

$$\underset{(2\times 1)}{\begin{Bmatrix} g_x \\ g_y \end{Bmatrix}} = \begin{bmatrix} [\mathbf{B}_x] & [0] \\ [0] & [\mathbf{B}_y] \end{bmatrix} \{\mathbf{q}_\varphi\} = [\mathbf{B}]\{\mathbf{q}_\varphi\},$$

where

$$[B_x] = [(4L_1 - 1)b_1 \ (4L_2 - 1)b_2 \ (4L_3 - 1)b_3 \ 4(L_1b_2 + L_2b_1) \ 4(L_2b_3 + L_3b_2) \ 4(L_3b_1 + L_1b_3)]$$

and

$$[B_y] = [(4L_1 - 1)a_1 \ (4L_2 - 1)a_2 \ (4L_3 - 1)a_3 \ 4(L_1a_2 + L_2a_1) \ 4(L_2a_3 + L_3a_2) \ 4(L_3a_1 + L_1a_3)]$$

Computation of {**Q**s} will require integrations over element area and relevant boundary surfaces.

References

1. Love, A. E. H. 1994. *The Mathematical Theory of Elasticity,* Dover, New York.
2. Timoshenko, S. and Goodier, J. N. 1951. *Theory of Elasticity,* McGraw-Hill, New York.
3. Herrmann, L. R. 1965. "Elastic torsional analysis of irregular shapes," *J. Eng. Mech. Div. ASCE,* 91(EM6), 11-19.
4. Desai, C. S. and Abel, J. F. 1972. *Introduction to the Finite Element Method,* Van Nostrand Reinhold, New York.
5. Bach, C. and Baumann, R. 1924. *Elastizitat und Festigheit,* Springer-Verlag, Berlin.
6. Murphy, G. 1946. *Advanced Mechanics of Materials,* McGraw-Hill, New York.
7. Valliappan, S. and Pulmano, V. A. Jan. 1974. "Torsion of nonhomogeneous anisotropic bars," *J. Struct. Div. ASCE,* 100(ST1), 286-295.
8. Yamada, Y., Kawai, T., Yoshimura, N., and Sakurai, T. 1968. "Analysis of the elastic-plastic problems by matrix displacement method," Proc. 2nd Conf. Matrix Methods in Structural Mechanics, Wright Patterson Air Force Base, Dayton, OH, 1271-1289.
9. Yamada, Y., Nakagiri, S., and Takatsuka, K. 1969. "Analysis of Saint-Venant torsion problem by a hybrid stress model," in Proc. Japan-U.S. Seminar on Matrix Methods of Structural Analysis and Design, Tokyo.
10. Yamada, Y., Nakagiri, S., and Takatsuka, K. 1972. "Elastic-plastic analysis of Saint-Venant problem by a hybrid stress model," *Int. J. Num. Methods in Engineering,* 5, 193-207.
11. Pian, T. H. H. and Tong, P. 1972. "Finite element methods in continuum mechanics," in *Advances in Applied Mechanics,* Vol. 12, Academic Press, New York.
12. Noor, A. K. and Andersen, C. M. 1975. "Mixed isoparametric elements for Saint-Venant torsion," *Comput. Methods Appl. Mech. Eng.,* 6, 195-218.
13. Noor, A. K. Private communication.

12

Other Field Problems: Potential, Thermal, Fluid, and Electrical Flow

Introduction

The problem of torsion (Chapter 11) and a number of other problems that we shall consider in this chapter are often known as field problems. They are governed essentially by similar differential equations, which are special cases of the following general equation:[1]

$$\frac{\partial}{\partial x}\left(k_x \frac{\partial \varphi}{\partial x}\right)+\frac{\partial}{\partial y}\left(k_y \frac{\partial \varphi}{\partial y}\right)+\frac{\partial}{\partial z}\left(k_z \frac{\partial \varphi}{\partial z}\right)+\overline{Q}=c\frac{\partial \varphi}{\partial t}. \tag{12.1a}$$

The associated boundary conditions are

$$\varphi=\overline{\varphi}(t) \quad \text{on } S_1 \tag{12.2a}$$

$$k_x\frac{\partial \varphi}{\partial x}\ell_x+k_y\frac{\partial \varphi}{\partial y}\ell_y+k_z\frac{\partial \varphi}{\partial z}\ell_z+\alpha(\varphi-\varphi_0)+\overline{q}(t)=0, \qquad \text{on } S_2 \text{ and } S_3. \tag{12.2b}$$

Here φ is the unknown (warping, stress function, velocity potential, stream function, temperature, electrical potential, fluid head, or potential); k_x, k_y, and k_z, are material properties in the x, y, and z directions, respectively; $\overline{Q}$ is the applied (heat, fluid, and so on) flux; c is specific heat or effective porosity, and so on; S_1 is the part of the boundary on which φ is prescribed; and S_2 is the part of the boundary on which the intensity of flux $\overline{q}$ is prescribed. In the case of heat flow α $(\varphi - \varphi_o)$ is prescribed on S_3, in which α is the transfer coefficient and φ_o is the surrounding temperature, l_x, l_y, and l_z are the direction cosines of the outward normal to the boundary, t denotes time, and the overbar denotes a prescribed quantity.

In this book, we shall consider only two-dimensional steady-state problems; that is, the problem is independent of time and the right-hand side of

Equation 12.la vanishes. Also, for simplicity, only homogeneous materials are considered; then Equation 12.la reduces to

$$k_x \frac{\partial^2 \varphi}{\partial x^2} + k_y \frac{\partial^2 \varphi}{\partial y^2} + \overline{Q} = 0 \tag{12.1b}$$

and the boundary conditions to

$$\varphi = \overline{\varphi} \qquad \text{on } S_1 \tag{12.2c}$$

and

$$k_x \frac{\partial \varphi}{\partial x} \ell_x + k_y \frac{\partial \varphi}{\partial y} \ell_y + \overline{q} = 0 \qquad \text{on } S_2. \tag{12.2d}$$

We note that Equations 11.1 and 11.27 have the same form as Equation 12.1b.

Potential Flow

The potential flow of fluids in absence of fluid sink (or source) and fluid influx across the boundary is governed by a special form of Equation 12.1b.

$$\frac{\partial^2 \varphi}{\partial x^2} + \frac{\partial^2 \varphi}{\partial y^2} = 0$$

or

$$\nabla^2 \varphi = 0. \tag{12.3}$$

which is called the Laplace equation. Here ∇^2 is a differential operator. The assumptions commonly made are that the flow is irrotational; that is, fluid particles do not experience net rotation during flow, the friction between the fluid and surfaces is ignored, and the fluid is incompressible.[2] Some practical problems where this kind of flow can be assumed are flow over weirs and through pipes (with obstructions).

Equation 12.3 is based on the basic requirement that the flow is continuous; that is,

$$\frac{\partial v_x}{\partial x} + \frac{\partial v_y}{\partial y} = 0, \tag{12.4}$$

where v_x and v_y are components of velocity in the x and y direction, respectively. The flow problem can be represented in terms of either the velocity potential or the fluid head φ or the stream function ψ. This dual representation is similar to that for the stress-deformation problem in the sense that φ and ψ are used in the same manner as the displacement (warping) and stress functions, respectively.

The relations between the velocity components and φ and ψ are given by[2]

$$
\begin{aligned}
-v_x &= \frac{\partial \varphi}{dx} \\
-v_y &= \frac{\partial \varphi}{dy}
\end{aligned}
\tag{12.5a}
$$

and

$$
\begin{aligned}
v_x &= \frac{\partial \psi}{\partial y}, \\
v_y &= -\frac{\partial \psi}{\partial x}.
\end{aligned}
\tag{12.5b}
$$

Substitution of v_x and v_y into Equation 12.4 leads to the Laplace equation as in Equation 12.3.

Boundary Conditions

For flow through a cylindrical pipe (Figure 12.1) a boundary condition is that the fluid and the wall have the same normal velocity. Hence, if the wall is stationary, we have

$$
\mathbf{V} \cdot n = \mathbf{V}_w \cdot n = 0, \tag{12.6}
$$

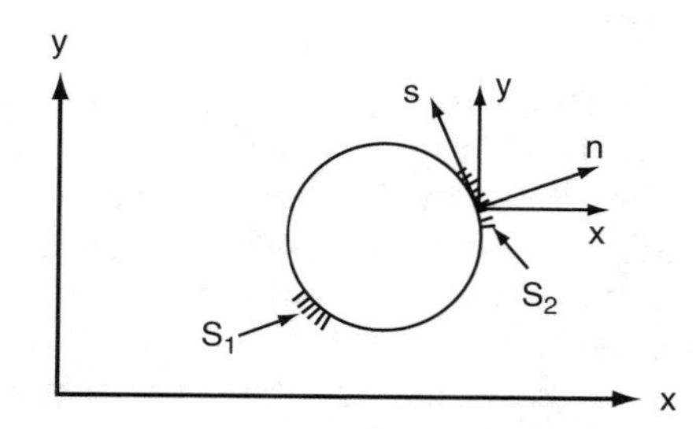

FIGURE 12.1
Flow in pipe.

where $\mathbf{V}$ and $\mathbf{V}_w$ are the velocities of the fluid and the wall, respectively, and n is the unit normal vector. Substitution of Equation 12.5a into Equation 12.6 yields

$$\frac{\partial \varphi}{\partial x} n_x + \frac{\partial \varphi}{\partial y} n_y = 0. \tag{12.7a}$$

From Figure 12.1,

$$n_x = \frac{\partial x}{\partial n} = \frac{\partial y}{\partial s}, \tag{12.8a}$$

$$n_y = \frac{\partial y}{\partial n} = -\frac{\partial x}{\partial s}. \tag{12.8b}$$

Hence Equation 12.7a transforms to

$$\frac{\partial \varphi}{\partial x}\frac{\partial x}{\partial n} + \frac{\partial \varphi}{\partial y}\frac{\partial y}{\partial n} = \frac{\partial \varphi}{dn} = 0, \tag{12.7b}$$

which is similar to Equation 12.2b. This is the flow or Neumann-type boundary condition (Chapters 2 and 3). The potential or Dirichlet boundary condition is

$$\varphi = \overline{\varphi} \qquad \text{on } S_1. \tag{12.8c}$$

Often both the flow and potential boundary conditions occur together, which is called the mixed condition.

Finite Element Formulation

For two-dimensional idealization, we can use a triangular or quadrilateral element. Formulation with a triangular element will be essentially identical to the one described for torsion in Chapter 11. Hence, we present a formulation by using a four-node quadrilateral element (Figure 12.2). It is possible to use either φ or ψ (or both) for the finite element formulation. We first consider a formulation with φ.

The velocity potential (φ) is a scalar quantity and has one value at any point. For the four-node quadrilateral element there are thus four nodal degrees of freedom. Following the steps described in Chapter 10 (Equations 10.9 to 10.24

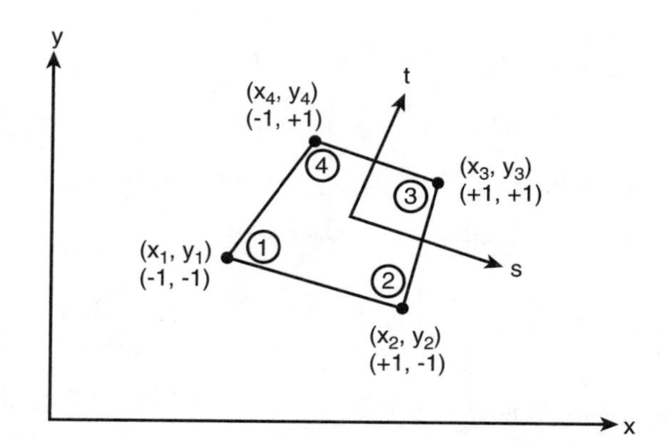

FIGURE 12.2
Quadrilateral isoparametric element.

one can obtain the relation between the gradient vector {**g**} and the vector of nodal degrees of freedom $\{\mathbf{q}_\varphi\}$ for this problem,

$$\begin{Bmatrix} g_x \\ g_y \end{Bmatrix} = \begin{Bmatrix} \dfrac{\partial \varphi}{\partial x} \\ \dfrac{\partial \varphi}{\partial y} \end{Bmatrix} = \begin{bmatrix} B_{11} & B_{12} & B_{13} & B_{14} \\ B_{21} & B_{22} & B_{23} & B_{24} \end{bmatrix} \begin{Bmatrix} \varphi_1 \\ \varphi_2 \\ \varphi_3 \\ \varphi_4 \end{Bmatrix} \tag{10.24a}$$

or

$$\{g\} = [B]\{q_\varphi\}. \tag{10.24b}$$

Step 4. Derive Element Equations

Use of either a variational or a residual procedure for the problem governed by Equation 12.3 will yield essentially the same results. The appropriate functional for Equation 12.3 is

$$\Omega_p(\varphi) = \iint_A \frac{1}{2}\left[\left(\frac{\partial \varphi}{\partial x}\right)^2 + \left(\frac{\partial \varphi}{\partial x}\right)^2\right] dxdy. \tag{12.9}$$

Governing Equation 12.3 is only valid for a flow problem without any source term and when fluid influx across the boundary is zero. If the control volume has a fluid source (or sink), and there is a fluid influx or outflux across the control volume boundary, then the governing equation takes the following form:

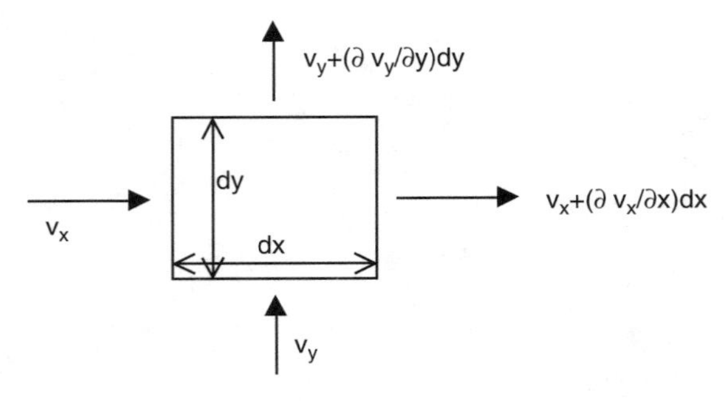

FIGURE 12.3
A two-dimensional control volume.

$$k\left(\frac{\partial^2\varphi}{\partial x^2}+\frac{\partial^2\varphi}{\partial y^2}\right)+\overline{Q}=0 \tag{12.10}$$

and the natural boundary condition is

$$k\frac{\partial\varphi}{\partial n}=k\frac{\partial\varphi}{\partial x}n_x+k\frac{\partial\varphi}{\partial y}n_y=\bar{q} \tag{12.11}$$

where k is the permeability coefficient; for an isotropic medium it is the same in both x and y directions. $\overline{Q}$ is the fluid source per unit volume and $\bar{q}$ is the fluid influx per unit length of the boundary.

Derivation of the Governing Equation

The governing equation for the two-dimensional fluid flow can be derived in the following manner.

Let the fluid velocity components v_x and v_y be functions of x and y and the fluid source function $\overline{Q}$ be a function of x and y. Then the fluid enters into the control volume (CV), Figure 12.3, with a velocity v_x across the left boundary and v_y across the bottom boundary. The fluid leaves the control volume with a velocity v_x + $(\partial v_x/\partial x)dx$ across the right boundary and v_y + $(\partial v_y/\partial y)dy$ across the top boundary. Then the total volume of the fluid entering into the CV is equal to $h(v_x dy + v_y dx + \overline{Q}dxdy)$ and the volume of fluid leaving the CV is equal to $h[\{v_x + (\partial vv_x/\partial x)dx\}\mathrm{d}y + \{v_y + (\partial v_y/\partial y)dy\}dx]$, where h is the thickness of the CV. For incompressible flow,

$$h\left(v_x dy+v_y dx+\overline{Q}dxdy\right)=h\left[\left\{v_x\left(\partial v_x/\partial x\right)dx\right\}dy+\left\{v_y+\left(\partial v_y/\partial y\right)dy\right\}dx\right] \tag{12.12}$$

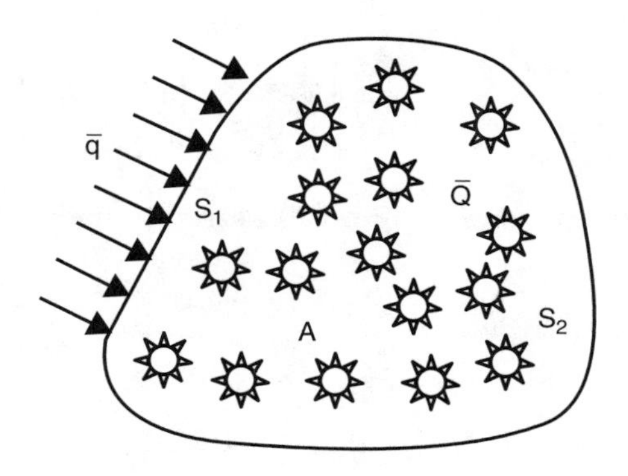

FIGURE 12.4
Control volume showing the fluid source $\bar{Q}$ per unit volume (shown by stars) and fluid entering across the surface area S_1 by an amount $\bar{q}$ per unit area.

or

$$\frac{\partial v_x}{\partial x} + \frac{\partial v_y}{\partial y} - \bar{Q} = 0. \tag{12.13}$$

After substitution of v_x and v_y in terms of the fluid head φ (Equation 12.5a), one obtains

$$\frac{\partial^2 \varphi}{\partial x^2} + \frac{\partial^2 \varphi}{\partial y^2} + \bar{Q} = 0 \tag{12.14}$$

In addition to the specified fluid source $\bar{Q}$ per unit volume, say, it is also specified that the fluid enters into the control volume across the boundary S_1 by an amount $\bar{q}$ per unit surface area, as shown in Figure 12.4. Then the boundary condition on S_1 will be specified by the derivative of the fluid potential. Thus this boundary condition will be a natural boundary condition (see Chapter 2).

The functional corresponding to the governing of Equation 12.14 is given by

$$\Omega_p(\varphi) = \iint_A \frac{1}{2}\left[\left(\frac{\partial \varphi}{\partial x}\right)^2 + \left(\frac{\partial \varphi}{\partial y}\right)^2\right] dxdy - \iint_A \bar{Q}\varphi dxdy \tag{12.15}$$

If the fluid influx $\bar{q}$ per unit surface area is specified across the boundary S_1 (as in Figure 12.4) then to accommodate this natural boundary condition the functional should be the following (see Chapter 2, section on the variational calculus),

$$\Omega_p(\varphi) = \iint_A \frac{1}{2}\left[\left(\frac{\partial \varphi}{\partial x}\right)^2 + \left(\frac{\partial \varphi}{\partial y}\right)^2\right] dxdy - \iint_A \overline{Q}\varphi dxdy - \int_{S_1} \overline{q}\varphi dS \quad (12.16)$$

Now we substitute {**g**} from Equation 10.24 into Equation 12.16 to yield

$$\Omega_p = \frac{1}{2}\{q_\varphi\}^T \iint_A [B]^T[B]dxdy\{q_\varphi\} - \{q_\varphi\}^T \iint_A [N]^T \overline{Q}dxdy - \{q_\varphi\}^T \int_{S_1} [N]^T \overline{q}dS \quad (12.17)$$

Clearly for the functional expressions of Equations 12.15 and 12.9 the right-hand side of Equation 12.17 will be reduced to two terms and one term, respectively, from its current expression involving three terms.

Taking derivatives of Ω_p with respect to $\{\mathbf{q}_\varphi\}$, we obtain

$$\frac{\partial \Omega_p}{\partial \{q_\varphi\}} = 0, \quad (12.18)$$

that is,

$$\left.\begin{matrix} \dfrac{\partial \Omega_p}{\partial \varphi_1} \\ \dfrac{\partial \Omega_p}{\partial \varphi_2} \\ \dfrac{\partial \Omega_p}{\partial \varphi_3} \\ \dfrac{\partial \Omega_p}{\partial \varphi_3} \end{matrix}\right\} = 0 \quad (12.19)$$

which leads to

$$\iint_A [B]^T[B]dxdy\{q_\varphi\} - \iint_A [N]^T \overline{Q}dxdy - \int_{S_1} [N]^T \overline{q}dS = 0 \quad (12.20)$$

or

$$[k_\phi]\{q_\varphi\} = \{Q\} \quad (12.21a)$$

where $[\mathbf{k}_\varphi]$ is the element property matrix similar to the element stiffness matrix for the stress-deformation problem and {**Q**} is the element level forcing parameter vector or load vector,

$$\left[k_\varphi\right] = \iint_A [B]^T [B] dxdy$$

$$\{Q\} = \iint_A [N]^T \bar{Q} dxdy + \int_{S_1} [N]^T \bar{q} dS = \{Q_1\} + \{Q_2\}. \tag{12.22}$$

Integrals of Equation 12.22 are carried out numerically as discussed in Chapter 10 (Equation 10.26).

Step 5. Assembly

Equation 12.21a is assembled such that the potentials at common nodes are compatible. The final assemblage equation is

$$\left[K_\varphi\right]\left\{r_\varphi\right\} = \{R\}. \tag{12.21b}$$

Evaluation of {Q}

The load vector {**Q**} (Equation 12.22) can be evaluated by using numerical integration. The expanded forms of the two parts of {**Q**} are given below. The first part is

$$\begin{aligned}\{\mathbf{Q}_1\} &= \iint_A [\mathbf{N}]^T \{\bar{\mathbf{Q}}\} dxdy = \int_{-1}^{1}\int_{-1}^{1} [\mathbf{N}]^T \{\bar{\mathbf{Q}}\} |J| dsdt \\ &= \sum_{i=1}^{4} \left[\mathbf{N}(s_i, t_i)\right]^T \{\bar{\mathbf{Q}}\} |J(s_i, t_i)| W_i \\ &= \sum_{i=1}^{4} \begin{Bmatrix} N_1(s_i, t_i) \\ N_2(s_i, t_i) \\ N_3(s_i, t_i) \\ N_4(s_i, t_i) \end{Bmatrix} \bar{\mathbf{Q}} |J(s_i, t_i)| W_i .\end{aligned} \tag{12.23}$$

Here we assumed $\bar{Q}$ to be a uniform source.

The second part is relevant to the boundary only. As an illustration, consider side 1–2 of an element (Figure 12.5) subjected to components $\bar{q}_x$ and $\bar{q}_y$. Then the second part for $\bar{q}_x$ specializes to

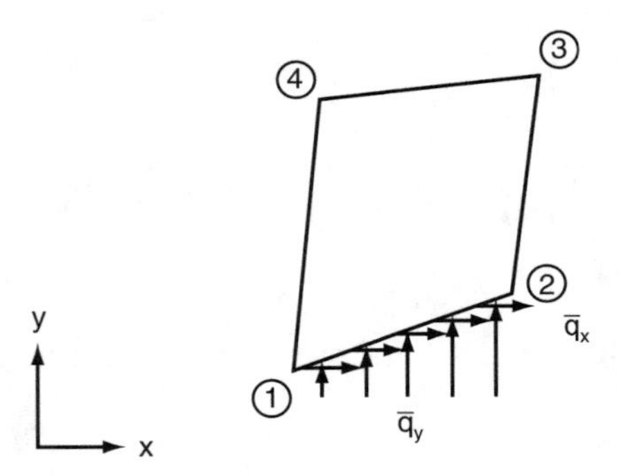

FIGURE 12.5
Boundary fluxes.

$$\{\mathbf{Q}_2\} = \int_{s_2} \begin{Bmatrix} N_1 \\ N_2 \\ N_3 \\ N_4 \end{Bmatrix} \bar{q}_x dS. \tag{12.24}$$

Example 12.1

Compute $\{\mathbf{Q}_2\}$ for the element shown in Figure 12.5.

We note that along side 1–2, coordinate t has a constant value equal to –1, and s varies from –1 to +1. Therefore, the N_i specialize to $N_1 = \frac{1}{2}(1 - s)$, $N_2 = \frac{1}{2}(1 + s)$, $N_3 = 0$, $N_4 = 0$. Hence, we have

$$\{\mathbf{Q}_2\} = \int_{s_2} \begin{Bmatrix} \frac{1}{2}(1-s) \\ \frac{1}{2}(1+s) \\ 0 \\ 0 \end{Bmatrix} \bar{q}_x dS = \int_{-1}^{1} \begin{Bmatrix} \frac{1}{2}(1-s) \\ \frac{1}{2}(1+s) \\ 0 \\ 0 \end{Bmatrix} \bar{q}_x |J| dS. \tag{12.25}$$

Since $s = 2S/l_1$, we have

$$dS = \frac{l_1}{2} ds = |J| ds, \tag{12.26}$$

where l_1 is the length of side 1–2. Therefore,

$$\{\mathbf{Q}_2\} = \frac{l_1}{4}\begin{Bmatrix} 1-s \\ 1+s \\ 0 \\ 0 \end{Bmatrix}\bar{q}_x ds, \tag{12.27}$$

which reduces to

$$\{\mathbf{Q}_2\} = \frac{\bar{q}_x l_1}{2}\begin{Bmatrix} 1 \\ 1 \\ 0 \\ 0 \end{Bmatrix}. \tag{12.28}$$

This indicates that the total flux on side 1–2 in the x direction is divided equally between nodes 1 and 2. Contributions of other applied fluxes in both the x and y directions can be evaluated in an identical manner.

Stream Function Formulation

Formulation by using stream function ψ as an unknown is similar to that with velocity potential φ. In absence of any fluid source or sink term $\bar{Q}$, the differential equation for flow with ψ is

$$\frac{\partial^2\psi}{\partial x^2} + \frac{\partial^2\psi}{\partial y^2} = 0 \quad \text{or} \quad \nabla^2\psi = 0, \tag{12.29}$$

and the corresponding variational functional is

$$\Omega_c(\psi) = \iint_A \left[\left(\frac{\partial\psi}{\partial x}\right)^2 + \left(\frac{\partial\psi}{\partial y}\right)^2\right] dxdy. \tag{12.30}$$

Stream function ψ can now be expressed as

$$\psi = [\mathbf{N}_\psi]\{\mathbf{q}_\psi\} = \sum_{i=1}^{4} \mathbf{N}_{\psi i}\psi_i. \tag{12.31}$$

By following a procedure similar to the one above, the element equations are

$$[\mathbf{k}_{\psi}]\{\mathbf{q}_{\psi}\} = \{0\}, \tag{12.32}$$

where $[\mathbf{k}_{\psi}]$ is the element property matrix and $\{\mathbf{q}_{\psi}\}^T = [\psi_1\ \psi_2\ \psi_3\ \psi_4]$ is the vector of nodal stream functions.

The boundary conditions are expressed as

$$\psi = \overline{\psi} \quad \text{on } S_1,$$

where S_1 is the part on which ψ is prescribed.

Secondary Quantities

The velocities can be computed by using Equation 12.5. For the velocity potential approach,

$$-\begin{Bmatrix} v_x \\ v_y \end{Bmatrix} = \begin{bmatrix} B_{11} & B_{12} & B_{13} & B_{14} \\ B_{21} & B_{22} & B_{23} & B_{24} \end{bmatrix} \begin{Bmatrix} \varphi_1 \\ \varphi_2 \\ \varphi_3 \\ \varphi_4 \end{Bmatrix}, \tag{12.33}$$

and for the stream function approach,

$$\begin{Bmatrix} v_x \\ v_y \end{Bmatrix} = \begin{bmatrix} B_{21} & B_{22} & B_{23} & B_{24} \\ -B_{11} & -B_{12} & -B_{13} & -B_{14} \end{bmatrix} \begin{Bmatrix} \psi_1 \\ \psi_2 \\ \psi_3 \\ \psi_4 \end{Bmatrix}. \tag{12.34}$$

The quantity of flow Q_f across a section A–A in an element (Figure 12.6) can be now found as

$$\dot{Q}_{f(A-A)} = AV_n, \tag{12.35}$$

where A is the cross-sectional area at section A–A and V_n is the velocity normal to the section. V_n can be computed from the two components v_x and v_y for an element as

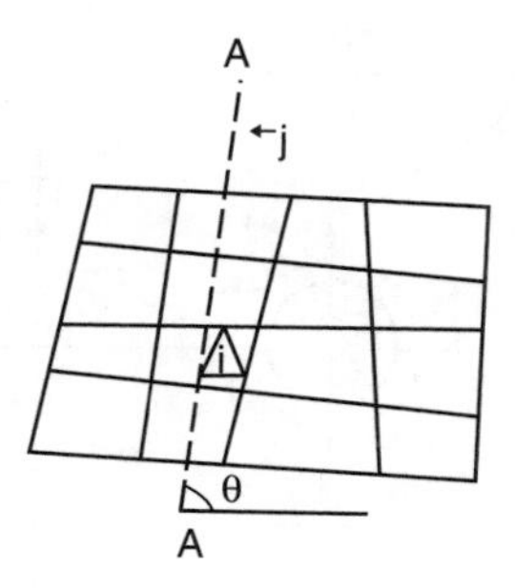

FIGURE 12.6
Computation of quantity of flow.

$$V_{n_i} = v_{x_i} \sin\theta - v_{y_i} \cos\theta, \tag{12.36a}$$

where i denotes an element and θ is the angle between the section A–A and the horizontal. The total flow across a section can be found as

$$Q_{fj} = \sum_{i=1}^{M} A_i V_{ni}, \tag{12.36b}$$

where j denotes the cross section and M is the total number of elements across the cross section.

Example 12.2. Potential Flow Around Cylinder

(a) *Velocity Potential Solution.* Figure 12.7(a) shows the problem of uniform fluid flow around a cylinder of unit radius. The flow domain is 8 × 8 units. Due to symmetry, only one half of the domain is discretized, as shown in Figure 12.7(b).

The boundary conditions for the velocity potential are

$$\begin{aligned} &\varphi = 1 \text{ unit along the upstream boundary,} \quad &&\text{nodes 1–5} \\ &\varphi = 0 \text{ unit along the downstream boundary,} \quad &&\text{nodes 51–55.} \end{aligned} \tag{12.37}$$

The computer code FIELD-2D (for further details, see Appendix 3) was used to obtain numerical solutions. Solutions for velocity potential are obtained by setting NTYPE = 2 in the code.

Figure 12.8 shows computed distribution values of velocity potentials at nodes in half of the flow domain. Table 12.1 gives comparisons between closed form solutions from Streeter[2] and numerical results for φ at the nodes along the line $y = 0$. The formula for the closed form solution is given by[2]

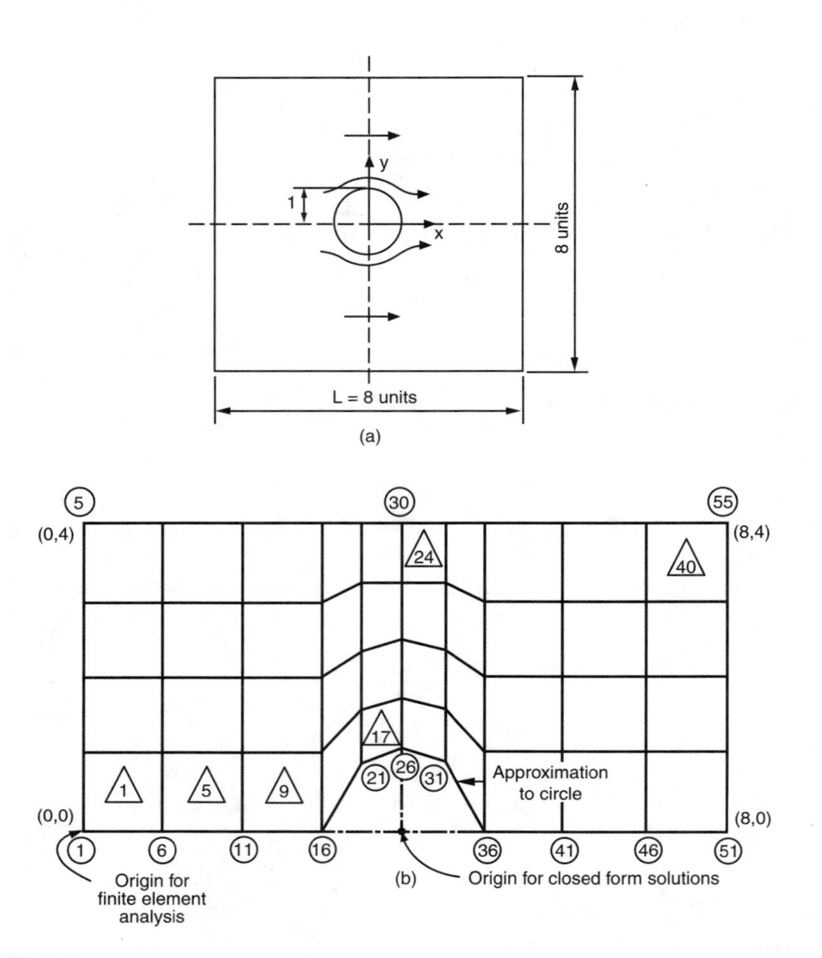

FIGURE 12.7
Analysis for potential flow around cylinder: (a) Flow around cylinder and (b) Finite element mesh for half flow domain. (0, 0) etc. denote coordinate.

$$\varphi = U\left(r + \frac{a^2}{r}\right)\cos\theta + 0.50, \tag{12.38}$$

where U is the uniform undisturbed velocity in the negative x direction = $[\varphi\ (x = 0) - \varphi\ (x = L)]/\mathrm{L} = (1 - 0)/8 = 0.125$, L = length of the flow domain = 8 (Figure 12.7[a]), $r = (x^2 + y^2)^{1/2}$, a is the radius of cylinder, and θ is the angle measured from x axis.

(b) *Stream Function Solution*. Specification of NTYPE = 3 in the code FIELD-2D permits solution for the stream function approach. For this solution, the following boundary conditions were used:

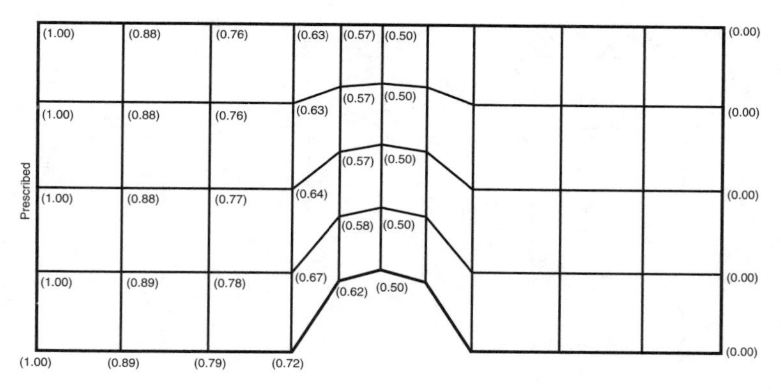

FIGURE 12.8
Computed nodal potentials. (*Notes:* (1) Potentials shown in parentheses. (2) Potentials on the right-hand half can be obtained by identical potential drop.)

TABLE 12.1
Comparison of Computed and Exact Solutions for φ

Node	Exact	Computed
26	0.5000	0.5000
31	0.3743	0.3780
36	0.2500	0.2765
41	0.1875	0.2132
51	−0.03125	0.0000[a]

[a] Prescribed.

$$\begin{aligned} &\psi = 0.5 \text{ along the top nodes } 5\text{–}55, \\ &\psi = 0.0 \text{ along the bottom nodes, } 1\text{–}16\text{–}26\text{–}36\text{–}51. \end{aligned} \tag{12.39}$$

Figure 12.9 shows the distribution of computed values of ψ. Table 12.2 gives comparisons between closed form solutions and numerical predictions for ψ along the line of nodes 26–30. The closed form solution is given by[2]

$$\psi = U\left(r - \frac{a^2}{r}\right)\sin\theta. \tag{12.40}$$

For the line considered, $\theta = 90$ deg, and $r = (x^2 + y^2)^{1/2} = y$. The value of $U = (0.5 - 0.0)/4 = 0.125$.

Figure 12.10 shows comparisons between nondimensionalized values of the x component of velocity v_x along the section of nodes 26–30. The closed form value is obtained from[2]

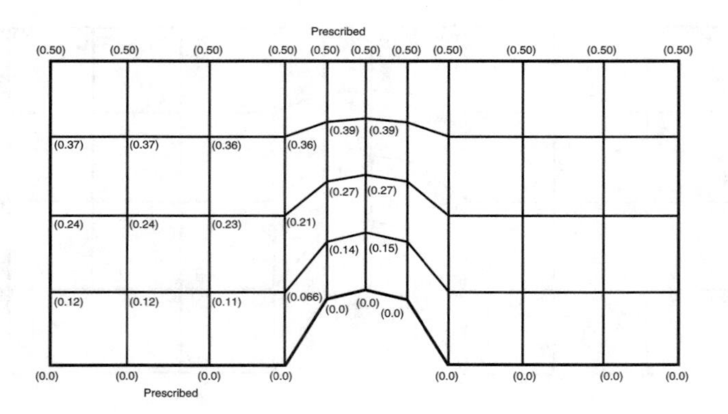

FIGURE 12.9
Computed nodal stream functions. (*Notes:* (1) Stream functions shown in parentheses. (2) Stream functions on the right-hand half are symmetrical.)

TABLE 12.2
Comparisons between Computed and Exact Solutions for ψ

Node	Exact	Computed
26	0.0000	0.0000[a]
27	0.1473	0.1529
28	0.2625	0.2746
29	0.3678	0.3892
30	0.4688	0.5000[a]

[a] Prescribed.

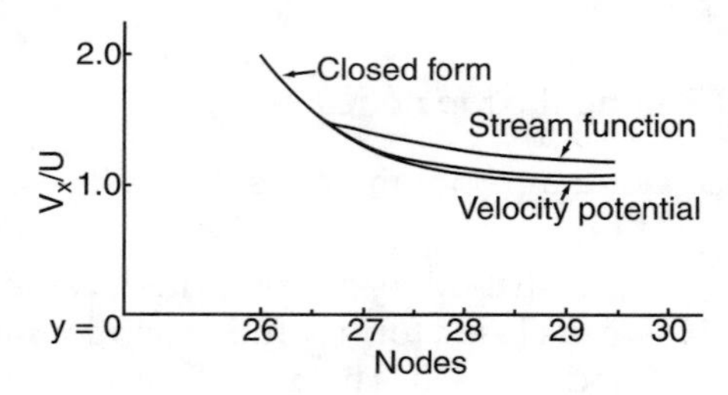

FIGURE 12.10
Comparisons of velocities at midsection.

$$v_x = \frac{\partial \psi}{\partial y} = U\left(1 + \frac{a^2}{r^2}\right)\sin\theta = U\left(1 + \frac{1}{y^2}\right)$$

or

$$\frac{v_x}{U} = 1 + \frac{1}{y^2}. \tag{12.41}$$

The numerical results plotted in Figure 12.10 are the values of v_x in element(s) adjoining the section.

The comparisons in Tables 12.1 and 12.2 indicate that the finite element predictions yield satisfactory solutions for the flow problem. The plots for v_x/U in Figure 12.10 show that both the velocity potential and stream function approach give satisfactory comparisons. The velocity potential approach yields the lower bound and the stream function approach the upper bound of the exact velocity. If necessary, the numerical solutions can be improved by using finer meshes.

Thermal or Heat Flow Problem

In the case of the heat flow problem, the general governing equation is essentially the same as Equation 12.la except that the meanings of various terms are different. The unknown can now be the temperature T at a point; k_x, k_y, and k_z are thermal conductivites; $\bar{Q}$ is the (internal) applied heat flux; and $\bar{q}$ is the (surface) intensity of heat input. In addition, there is the possibility of heat transfer due to the difference between the temperature of the medium, T, and the surroundings, T_0. The differential equation for two-dimensional steady-state heat flow then becomes

$$k_x \frac{\partial^2 T}{\partial x^2} + k_y \frac{\partial^2 T}{\partial y^2} + \bar{Q} = 0, \tag{12.42}$$

and the boundary conditions are

$$T = \bar{T} \quad \text{on } S_1 \tag{12.43a}$$

and

$$k_x \frac{\partial T}{\partial x} \ell_x + k_y \frac{\partial T}{\partial y} \ell_y + \alpha (T - T_0) - \bar{q} = 0 \quad \text{on } S_2 \text{ and } S_3, \tag{12.43b}$$

where α is the heat transfer coefficient.

The finite element formulation with the quadrilateral element will result in equations similar to Equations 12.21 and 12.22 with the addition of the following term in Ω_p (Equation 12.17),

$$\int_{S_2} \tfrac{1}{2} \alpha (T - T_0)^2 \, dS, \tag{12.44}$$

which will lead to additional contributions $[\mathbf{k}_\alpha]$ and $\{\mathbf{Q}_3\}$ to the element matrix and the forcing parameter vector, respectively:

$$[\mathbf{k}_\alpha] = \int_{S_2} \alpha[\mathbf{N}]^T[\mathbf{N}]dS, \tag{12.45a}$$

$$[Q_3] = \int_{S_3} \alpha[\mathbf{N}]^T[\mathbf{T}_0]dS. \tag{12.45b}$$

For evaluation of $[\mathbf{k}_\alpha]$ we need to integrate along the sides (boundaries) of the element. As an illustration, consider side 1–2 (Figure 12.5) as before:

$$\begin{aligned}
[\mathbf{k}_\alpha] &= \frac{\alpha l_1}{2}\int_{-1}^{1}[\mathbf{N}]^T[\mathbf{N}]dS \\
&= \frac{\alpha l_1}{2}\int_{-1}^{1}\begin{Bmatrix} \frac{1}{2}(1-s) \\ \frac{1}{2}(1+s) \\ 0 \\ 0 \end{Bmatrix}\left[\tfrac{1}{2}(1-s)\ \ \tfrac{1}{2}(1+s)\ \ 0\ \ 0\right]dS \\
&= \frac{\alpha l_1}{2}\int_{-1}^{1}\begin{Bmatrix} \frac{1}{4}(1-s)^2 & \frac{1}{4}(1-s^2) & 0 & 0 \\ \frac{1}{4}(1-s^2) & \frac{1}{4}(1+s)^2 & 0 & 0 \\ 0 & 0 & 0 & 0 \\ 0 & 0 & 0 & 0 \end{Bmatrix} dS \\
&= \frac{a l_1}{2}\begin{bmatrix} \frac{1}{3} & \frac{1}{6} & 0 & 0 \\ \frac{1}{6} & \frac{1}{3} & 0 & 0 \\ 0 & 0 & 0 & 0 \\ 0 & 0 & 0 & 0 \end{bmatrix}.
\end{aligned} \tag{12.46}$$

Computation of $\{\mathbf{Q}_2\}$ is similar to that for $\bar{q}_x$ in Equation 12.24.

Finally the element equations for heat flow will be

$$\begin{aligned}
&([\mathbf{k}]+[\mathbf{k}_\alpha])\{\mathbf{q}_T\} = \{Q\} = \{Q_1\}+\{Q_2\}+\{Q_3\}, \\
&\text{where } \{\mathbf{q}_T\}^T = [T_1\ T_2\ T_3\ T_4].
\end{aligned} \tag{12.47}$$

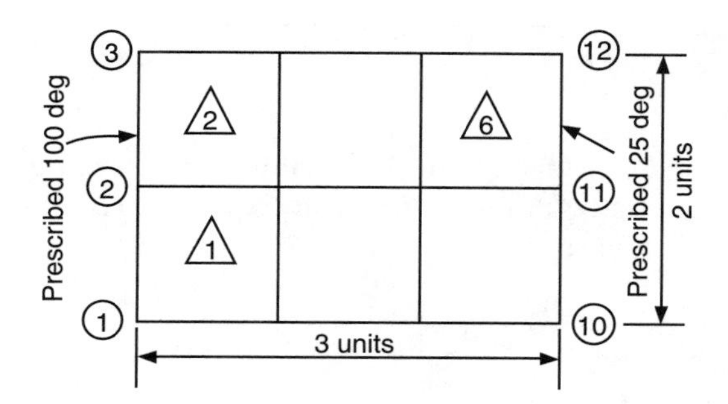

FIGURE 12.11
Heat flow in two-dimensional body.

Example 12.3. Two-Dimensional Heat Flow

As a simple illustration, consider steady-state heat flow in a rectangular block of unit thickness (Figure 12.11). The heat flow domain is divided into 9 elements with 12 nodes. The boundary conditions are assumed as follows:

$$\begin{aligned} T(0,y) &= 100 \text{ deg}, \\ T(3,y) &= 25 \text{ deg}. \end{aligned} \tag{12.48}$$

The thermal conductivities are assumed to be equal as $k_x = k_y = 1$.

Table 12.3 shows the computed values of temperatures at various nodes obtained by using the code FIELD-2D.

TABLE 12.3
Numerical Results for Temperature

Node	Temperature (deg)
1	100.0000[a]
2	100.0000[a]
3	100.0000[a]
4	74.9998
5	74.9998
6	74.9998
7	49.9999
8	49.9999
9	49.9999
10	25.0000[a]
11	25.0000[a]
12	25.0000[a]

[a] Prescribed.

Seepage

Seepage is defined as flow of fluid, usually water, through porous (soil) media. The governing equation is similar to Equation 12.la except the meanings of various terms are

$$
\begin{aligned}
\text{Unknown, } \varphi &= \frac{p}{\gamma} + z \text{ is the total fluid head or potential,} \\
k_x, k_y, k_z &= \text{coefficients of permeability in the } x, y, z \\
&\quad \text{directions, respectively,} \\
\overline{Q} &= \text{applied (internal) fluid flux,} \\
\overline{q} &= \text{applied (surface) intensity of fluid,}
\end{aligned}
\qquad (12.49)
$$

where p is the fluid pressure, γ = density of water, and z is the elevation head (Figure 12.12(a)).

For two-dimensional steady-state confined flow, the element equations are identical to Equations 12.21 and 12.22 except for the meaning of the terms. Often seepage is defined as confined or unconfined.[3] In the confined category, seepage occurs through a saturated medium subjected to prescribed boundary conditions and does not involve the so-called free or phreatic surface. In the case of unconfined seepage, on the other hand, we encounter the free or phreatic surface. Both categories can involve either steady, or unsteady or transient conditions.

When the seepage occurs with a free surface and surface of seepage (Figure 12.12(b)), we have the mixed boundary conditions on these surfaces:

$$\varphi = z \qquad (12.50a)$$

and

$$\frac{\partial \varphi}{\partial n} = 0. \qquad (12.50b)$$

At the free surface, the pressure is atmospheric; hence, $p = 0$ and $\varphi = z$. The second condition implies that the velocity of flow normal to the free surface is zero. The two conditions (Equation 12.50) must be satisfied simultaneously. The mixed conditions render the problem nonlinear. Consequently, in contrast to the steady confined flow, which requires only one finite element solution, it is often necessary to perform iterative analysis in the case of free surface flow. This requires a number of finite element solutions. This topic is beyond the scope of this text; the interested reader can consult References 3–5.

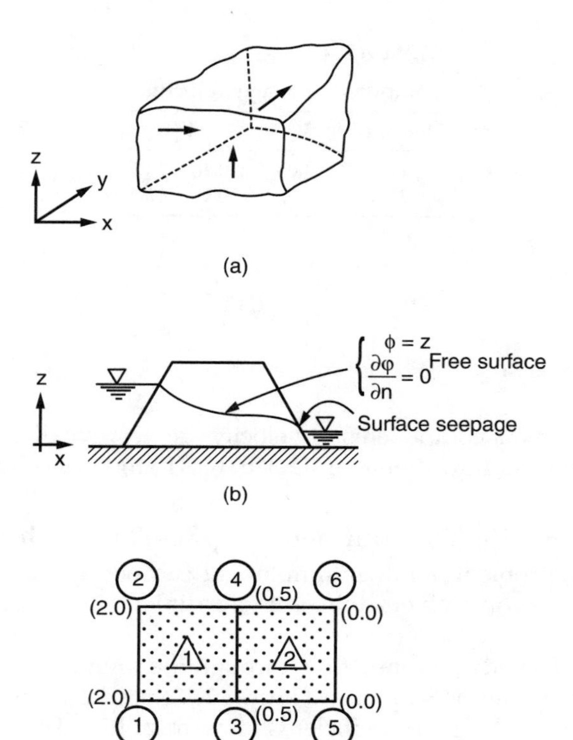

FIGURE 12.12
Seepage through porous media: (a) Three-dimensional seepage, (b) Free surface and mixed boundary conditions, and (c) Steady seepage in porous medium.

Example 12.4. Steady Confined Seepage

Figure 12.12(c) shows a rather simple problem of seepage through porous (soil) media with coefficients of permeability $k_x = k_y = 0.005(L/T)$ and thickness = 1 unit. The boundary conditions are

$$\varphi = 2 \quad \text{at } x = 0, \tag{12.51a}$$

$$\varphi = 0 \quad \text{at } x = 2 \text{ units}. \tag{12.51b}$$

The computed values of φ at various nodes are shown in Figure 12.12(c) in parentheses.

The computed values of velocities are shown in Table 12.4. These values compare closely with the velocities computed from Darcy's law:

TABLE 12.4

Computed Seepage Velocities

Element	V_x	V_y
1	0.005	-0.48×10^{-8}
2	0.005	-0.18×10^{-8}

$$v_x = -k\frac{\partial \varphi}{\partial x} = 0.005 \times \frac{2}{2}$$
$$= 0.005.$$

As expected, the y components of velocity are very small, almost equal to zero. The quantity of flow (Equation 12.36b) was found to be equal to 0.01 unit.

Example 12.5. Steady Confined Seepage Through Foundation
The foregoing problem is rather simple and can permit hand calculations. Now we consider a problem that is more difficult and needs use of the computer.

A problem of steady confined flow through the foundation soil of a structure (dam, sheet pile) is shown in Figure 12.13(a). The foundation consists of two layers with different coefficients of permeability. The soil is assumed to be isotropic, that is, $k_x = k_z$; this assumption is not necessary because anisotropic properties can be included in the finite element procedure. The structure itself is assumed to be impervious. The two-layered foundation rests on a material which has a very low permeability; hence, at the depth of 10 m, impervious base is assumed. This implies a zero flow (natural or Neumann) boundary condition.

Steady fluid heads $\bar{\varphi}_u$ and $\bar{\varphi}_d$ act on the upstream and downstream sides, respectively. For this problem, we have assumed

$$\bar{\varphi}_u = 1\text{ m},$$
$$\bar{\varphi}_d = 0, \tag{12.52}$$

which constitutes the geometric (or forced or Dirichlet) boundary condition. Because the problem is linear, results for this boundary condition can be extended to any multiple of $\bar{\varphi}_u$ and $\bar{\varphi}_d$ with proportional difference in the head. For instance, the subsequent results can be used to derive solutions for $\bar{\varphi}_u = 100$ m and $\bar{\varphi}_d = 50$ m by multiplying the computed nodal heads by 50.

The foundation medium extends toward infinity in the lateral direction. Because we can include only a finite part of this extent in the analysis, an approximation must be made to fix the discretized end boundaries. For instance, in Figure 12.13(b) we have chosen the end boundaries at a distance of 20 m (equal to the width of the structure) from the edges of the structure;

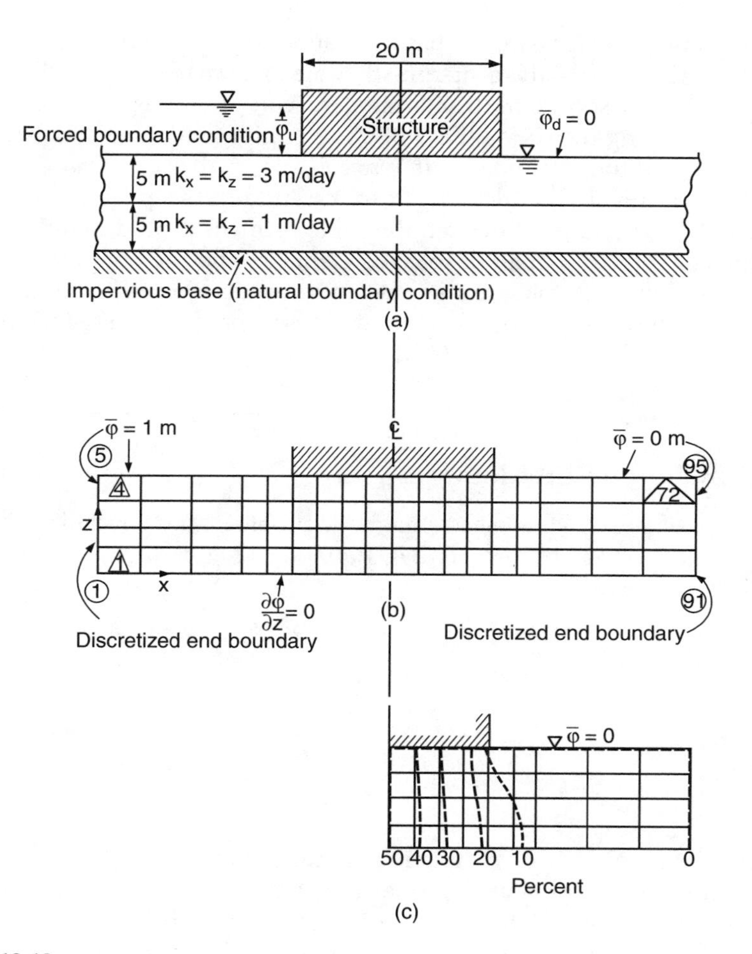

FIGURE 12.13
Steady confined seepage through layered foundation: (a) Details of foundation, (b) Finite element mesh, and (c) Solution for nodal potentials: equipotential lines.

this involves an assumption that at that distance the fluid potentials will be approximately equal to the applied potentials on the upstream and downstream sides. Often, it may be necessary to perform parametric studies to find the distances at which such conditions can be assumed. Further discussion on this topic appears in Chapter 13.

Figure 12.13(b) shows a finite mesh with 95 modes and 72 elements. A nodal line is needed at the interface between the two layers. The geometric boundary conditions at the upstream and downstream nodes are $\bar{\varphi}_u = 1$ m and $\bar{\varphi}_d = 0$ m, respectively. At the bottom boundary we have $\partial\varphi/\partial z = 0$; being the natural boundary condition, it is satisfied automatically in an integrated sense. Consequently, in the finite element analysis, we computed fluid heads at nodes at this boundary. Since the problem is symmetric, we could have solved it by considering mesh only for one half of the medium (Figure 12.13(c)). Then at the centerline we can assume a natural boundary condition; that is, $\partial \varphi / \partial x = 0$. This is valid since at the centerline the rate of change of φ vanishes.

Figure 12.13(c) shows the finite element computations for nodal heads obtained by using the code FIELD-2D. The equipotential lines are shown as a percentage of the total head difference $\bar{\varphi}_u - \bar{\varphi}_d$.

From the results in Figure 12.13(c), we can find the seepage forces causing uplift on the structure. Moreover, the finite element procedure allows computations of velocities in all or selected elements and the quantity of flow at selected sections. All this information can be used for analysis and design of structures founded on porous soil foundations through which seepage occurs.

Electromagnetic Problems

In the case of steady-state electromagnetic problems the governing differential equation, in the absence of $\bar{Q}$ in Equation 12.la, reduces to the Laplace equation

$$\frac{\partial}{\partial x}\left(k_x \frac{\partial V}{\partial x}\right) + \frac{\partial}{\partial y}\left(k_y \frac{\partial V}{\partial y}\right) = 0, \tag{12.53a}$$

or if $k_x = k_y$,

$$\nabla^2 V = 0, \tag{12.53b}$$

where V is the electric or magnetic potential, ∇^2 is the Laplacian operator, and k_x and k_y are the material properties; in the case of electrical flow they are the permittivities in the x and y directions, respectively.

The finite element formulations and details will follow essentially the same procedures as in the case of the field problems covered in this chapter and Chapter 11. For further details on this topic the reader can consult various publications.[6,7]

Thus, we once again observe the similarities of various phenomena in different disciplines in engineering and physics and how the finite element method provides a common ground for their solutions.

Computer Code

FIELD-2D is a general code for field problems and permits solution of torsion, potential flow, steady-state seepage, and heat flow. The user needs to supply appropriate properties relevant to the specific problem and specify codes for the type of problem as follows:

Problem	Code, NTYPE
Torsion	1
Potential flow	
Velocity potential	2
Stream function	3
Seepage	4
Heat flow	5

In fact, the code can be used easily for other steady-state problems such as electrostatic and magnetic flow. Further details of the code are given in Chapter 6 and Appendix 3.

Problems

12-1. Derive in detail [**k**] and {**Q**} for potential flow by using a triangular element (see Chapter 10).

12.2. Find {$\mathbf{Q}_2$} (Equations 12.24 and 12.25) if $\bar{q}_y$ and $\bar{q}_x$ were applied on sides 1–2 and 2–3, respectively.

12.3. Starting from the functional expression given in Equation 12.9 obtain the associated governing equation and natural boundary conditions.

12.4. By hand calculation, find the element matrix [**k**] for steady-state heat flow in a square plate of unit thickness (Figure 12.14) discretized in four equal square elements. The boundary conditions are

$$T(0, y) = 100 \text{ deg},$$

$$T(2, y) = 0 \text{ deg},$$

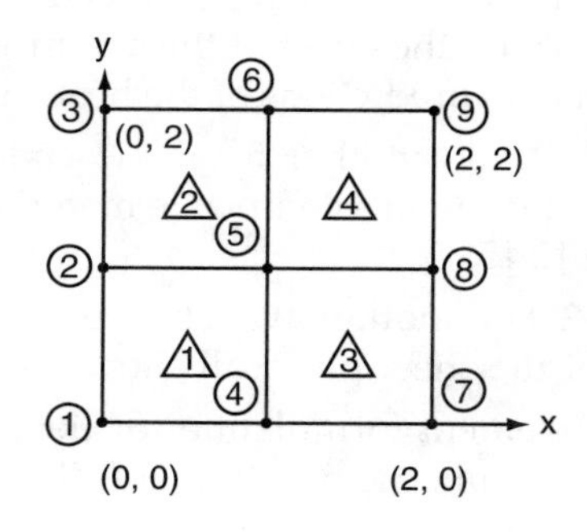

FIGURE 12.14

and $k_x = k_y = 1$ unit. Assemble the equations for the 4 elements and solve for temperatures at nodes 4, 5, and 6. Hint: You may use the results indicated in Equation 10.26 and the numerical integration shown in Example 10.1. Solution: T at nodes 4, 5, and 6 = 50 deg.

12.5. In Problem 12.4, consider heat flux $\bar{q} = 0.1$ unit per unit length along the side 3–6–9 and compute temperatures at nodes 4, 5, and 6. Hint: You may use the results in Equation 12.24 for finding the forcing function vector $\{\mathbf{Q}\}$.

12.6. In Problem 12.4, consider a source (or sink) $Q = 0.1$ units at node 6 and find the temperatures at nodes 4, 5, and 6. Hint: You may use Equation 12.23 to evaluate $\{\mathbf{Q}\}$, which will need numerical integration similar to that used in Example 10.1.

12.7. Use the same domain as in Problem 12.4 but composed of a porous medium such as a soil with

$$k_x = k_y = 0.1 \text{ cm/s}$$

and applied fluid potentials equal to

$$\bar{\varphi}(0, y) = 10 \text{ cm},$$

$$\bar{\varphi}(2, y) = 5 \text{ cm}.$$

Compute (a) fluid potentials at nodes 4, 5, and 6; (b) velocities v_x and v_y in all the elements; and (c) the quantity of flow across a cross section.

12.8. Consider Problem 12.7 but include $\bar{q}$ = 0.1 cm^3/sec-cm along the side 3–6–9 and find fluid heads at nodes 4, 5, and 6.

12.9. Consider Problem 12.7 but include $\bar{Q}$ = 0.1 cm^3 at node 6 and compute fluid heads at nodes 4, 5, and 6.

12.10. Consider Example 12.5. Prepare three meshes with discretized boundaries at 10, 20, and 40 m from the edges of the structure. By using a computer code, compare the results from the three analyses and offer comments on the effect of the distance of the end boundaries on the numerical predictions of the heads under the structure.

12.11. By using FIELD-2D or another available code, solve for steady temperature distribution in a composite material for the conditions shown in Figure 12.15.

12.12. By using FIELD-2D or another available code, solve for seepage in the foundation of the sheet pile wall shown in Figure 12.16.

12.13. Derive the finite element formulation for field problems by using an eight-node isoparametric quadrilateral (Figure 12.17).

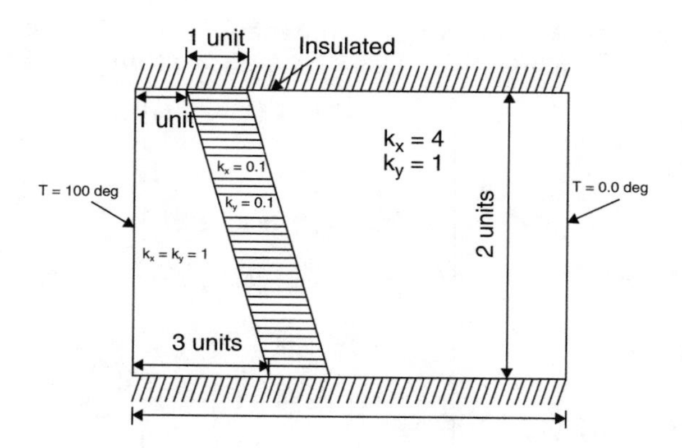

FIGURE 12.15

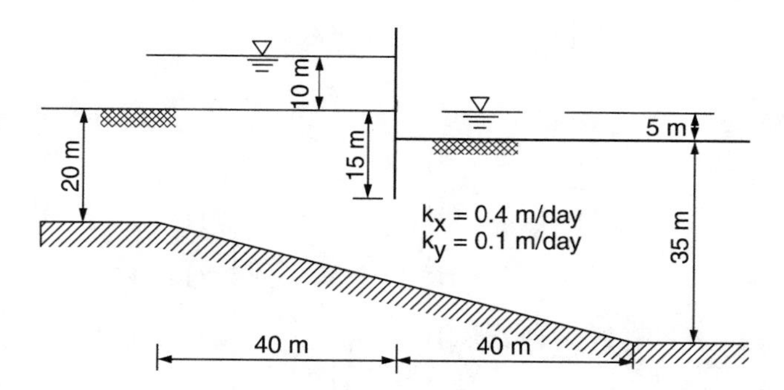

FIGURE 12.16

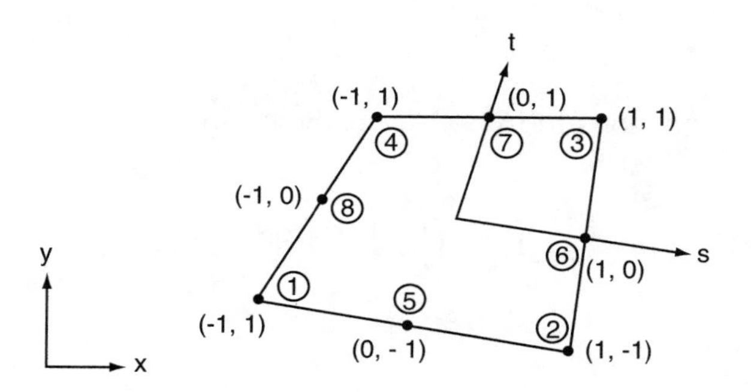

FIGURE 12.17

Partial results: Assume u = [**N**]{**q**}, where

$$[N]^T = \begin{Bmatrix} -\frac{1}{4}(1-s)(1-t)(1+s+t) \\ -\frac{1}{4}(1+s)(1-t)(1-s+t) \\ -\frac{1}{4}(1+s)(1+t)(1-s-t) \\ -\frac{1}{4}(1-s)(1+t)(1+s-t) \\ \frac{1}{2}\left(1-s^2\right)(1-t) \\ \frac{1}{2}(1+s)\left(1-t^2\right) \\ \frac{1}{2}\left(1-s^2\right)(1+t) \\ \frac{1}{2}(1-s)\left(1-t^2\right) \end{Bmatrix}$$

$$\{\mathbf{q}\}^T = \begin{bmatrix} u_1 & u_2 & u_3 & u_4 & u_5 & u_6 & u_7 & u_8 \end{bmatrix},$$

and

$$\frac{\partial}{\partial s}[\mathbf{N}]^T = \begin{Bmatrix} \frac{1}{4}(1-t)(2s+t) \\ \frac{1}{4}(1-t)(2s-t) \\ \frac{1}{4}(1+t)(2s+t) \\ \frac{1}{4}(1+t)(2s-t) \\ -s(1-t) \\ \frac{1}{2}\left(1-t^2\right) \\ -s(1+t) \\ -\frac{1}{2}\left(1+t^2\right) \end{Bmatrix}, \quad \frac{\partial}{\partial t}[\mathbf{N}]^T = \begin{Bmatrix} \frac{1}{4}(1-s)(2t+s) \\ \frac{1}{4}(1+s)(2t-s) \\ \frac{1}{4}(1+s)(2t+s) \\ \frac{1}{4}(1-s)(2t-s) \\ -\frac{1}{2}\left(1-s^2\right) \\ -t(1+s) \\ -\frac{1}{2}\left(1-s^2\right) \\ -t(1-s) \end{Bmatrix}$$

Use of Equations 10.16, 10.17, etc. then leads to

$$[\mathbf{k}] = \iint [\mathbf{B}]^T [\mathbf{C}][\mathbf{B}] dA$$

and

$$\{\mathbf{Q}\} = \iint_A [\mathbf{N}]^T \{\overline{\mathbf{Q}}\} dxdy + \int_{S_2} [\mathbf{N}]^T \{\overline{\mathbf{q}}\} dS.$$

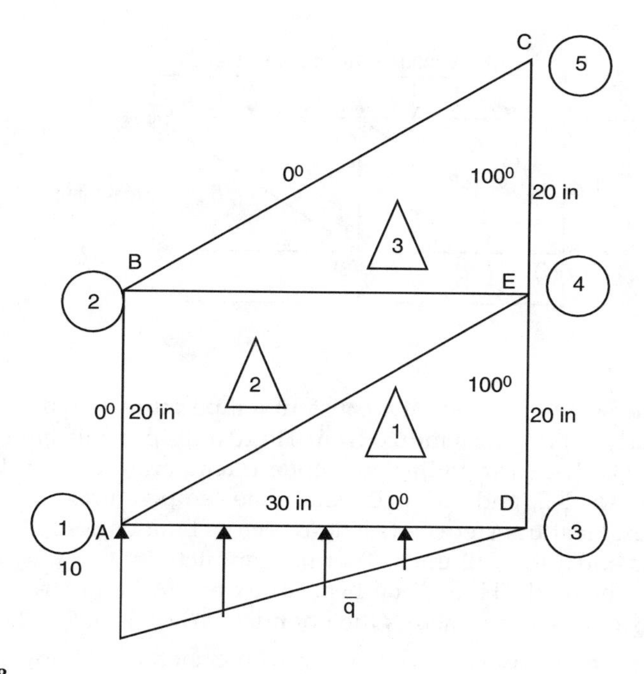

FIGURE 12.18

12.14. Starting from the functional expressions given in Equations 12.15 and 12.16 obtain the associated governing equation and natural boundary conditions for each case.

12.15. Starting from the following functional expression obtain the associated governing equation and natural boundary conditions by using the variational calculus principles described in Chapter 2.

$$\Omega_p(T) = \iint_A \frac{1}{2}\left[k_x\left(\frac{\partial T}{\partial x}\right)^2 + k_y\left(\frac{\partial T}{\partial y}\right)^2\right] dxdy - \iint_A \overline{Q}Tdxdy$$
$$- \int_{S_1} \overline{q}TdS + \int_{S_3} \frac{\alpha}{2}(T - T_0)^2 dS$$

12.16. A plate ABCD is discretized into three triangular elements as shown in Figure 12.18.

The global node numbering is shown in the figure. Global nodes 1, 2, 3, 4, and 5 correspond to the points A, B, D, E, and C respectively. For all 3 elements the local node number 1 is associated with the corner node corresponding to the smallest angle, local number 2 is for the node associated with the right angle, and local number 3 is for the remaining node.

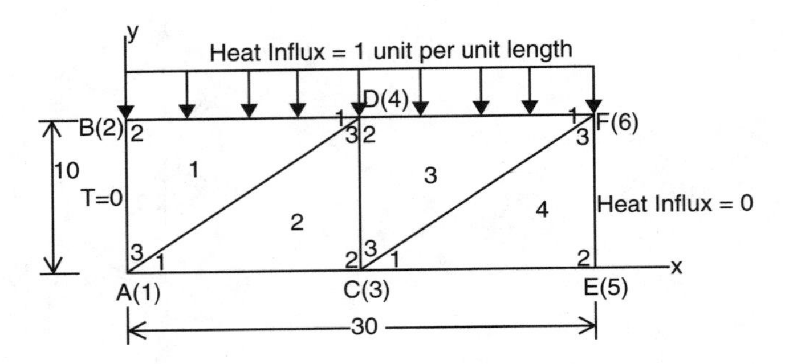

FIGURE 12.19

The side CD is kept at a constant temperature of 100°; there is a linearly varying heat influx (from 0 to 10 units per unit length) along side AD. The surrounding atmospheric temperature around boundaries AB, BC, and AD is 0°. Calculate temperatures of all global nodes. Heat transfer coefficient (α) across boundaries AB, AD, and BC is equal to 1/12 units. Thermal conductivity $k_x = k_y = 1$ for the plate material. (Hint: Total heat influx across boundaries AB and BC is $\alpha(T_0 - T)$ and across the boundary AD it is $\alpha(T_0 - T) + \bar{q}$.)

12.17. The problem geometry for the two-dimensional heat flow problem is shown in Figure 12.19. Thickness (h) of the rectangular plate is 1. The plate material is isotropic, $k_x = k_y = 1$. Boundary AB is at 0°C. Heat influx across AE and EF is zero. Heat influx across the boundary BF is 1 unit per unit length.

(a) How many unknown degrees of freedom does this problem have?

(b) After the finite element solution of the problem will there be any discontinuity in temperature across line CD?

(c) In the finite element solution will there be any discontinuity in $\frac{\partial T}{\partial x}$ across line CD?

(d) In a few sentences justify your answers to parts (b) and (c).

(e) Compute element stiffness matrices and element load vectors for all four elements.

(f) Obtain the global equation in the form [K]{r} = {R}.

(g) Apply proper boundary conditions to reduce the global equation in the following form, $[\bar{K}]\{\bar{r}\} = \{\bar{R}\}$.

12.18. Solve the above problem (12.17) using the computer code.

(a) Discretize the problem geometry into 4 triangular elements as shown in the figure of 12.17 and compute temperatures at points A, B, C, D, E, and F.

(b) Discretize the problem geometry into two quadrilateral elements (each element should have four nodes) and compute temperatures at points A, B, C, D, E, and F.

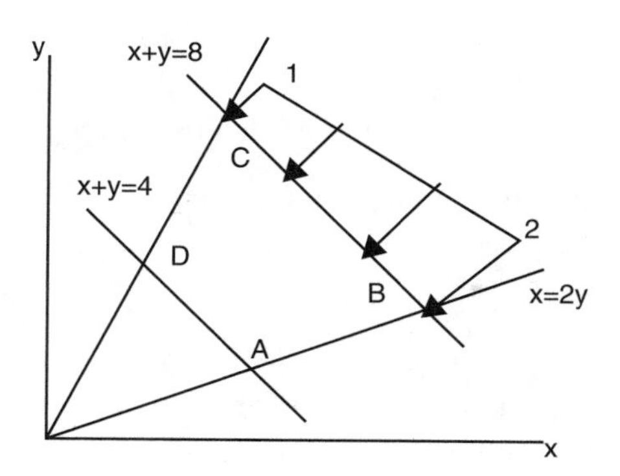

FIGURE 12.20

12.19. Consider a 3-node triangular element in which $\bar{Q}$, the fluid source per unit area, varies linearly from 0 at node 1 to Q_0 at nodes 2 and 3. Compute the element load vector for this element.

12.20. Consider a three-node triangular element. There is a constant fluid influx q_0 per unit length across the boundary line 1–2. No fluid passes across the other two boundary lines. There is no other fluid source or sink ($\bar{Q} = 0$) in the element. Compute the element load vector for this element.

12.21. Consider the quadrilateral element ABCD in Figure 12.20. Fluid is flowing into the element across the boundary BC as shown. The fluid influx is varying linearly from 1 to 2 unit per unit length. Obtain the element load vector.

12.22. In a two-dimensional heat flow problem the functional is given by

$$\Omega = \frac{1}{2}\iint_A \left[\left(\frac{\partial T}{\partial x}\right)^2 + \left(\frac{\partial T}{\partial y}\right)^2\right] dxdy + \frac{1}{2}\int_S \alpha\left(T - T_0\right)^2 ds$$

where A is the area of the element and S is its boundary.

(a) Derive the element equation in the form **[k]{q}** = **{Q}**; use the following notation:

$$\begin{Bmatrix} \dfrac{\partial T}{\partial x} \\ \dfrac{\partial T}{\partial x} \end{Bmatrix} = [B]\{q\}, \quad T = [N]\{q\}.$$

(b) Let us assume that the problem geometry has only one 4-node quadrilateral element with four sides of the element forming the boundary S of the problem. T_0 is 5.0 on the side connecting nodes 2 and 3, and it is 0 on the other 3 sides. Calculate {**Q**} for this problem.

12.23. Consider the functional Ω defined in a two-dimensional space A that is bounded by the boundary S, $\Omega = \iint_A \left(T_x^2 + TT_x + TT_y + T_y^2 + T^2\right) dA - \int_S TT_0 dS$, where $T_x = \frac{\partial T}{\partial x}$, $T_y = \frac{\partial T}{\partial y}$, and T_0 is a given function of x and y. For the three-node triangular element the primary unknown T is expressed in terms of the interpolation functions N_1, N_2, and N_3 in the following manner.

$$T = N_1T_1 + N_2T_2 + N_3T_3 = [N]\{q\}$$

and

$$\begin{Bmatrix} T_x \\ T_y \end{Bmatrix} = \begin{bmatrix} b_1 & b_2 & b_3 \\ a_1 & a_2 & a_3 \end{bmatrix} \begin{Bmatrix} T_1 \\ T_2 \\ T_3 \end{Bmatrix} = \begin{bmatrix} [B] \\ [A] \end{bmatrix} \begin{Bmatrix} T_1 \\ T_2 \\ T_3 \end{Bmatrix} = [C]\{q\}.$$

Starting from the above functional obtain the stiffness matrix [**k**]. Give your results in the following form

$$[k] = \iint_A ([W] + [X] + [Y] + [Z]) dA$$

where [W], [X], [Y], and [Z] are 2 × 2 matrices expressed in terms of the matrices [A], [B], [C], and [N].

References

1. Desai, C. S. and Abel, J. F. 1972. *Introduction to the Finite Element Method,* Van Nostrand Reinhold, New York.
2. Streeter, V. L. 1948. *Fluid Dynamics,* McGraw-Hill, New York.
3. Desai, C. S. and Christian, J. T., Eds. 1977. *Numerical Methods in Geotechnical Engineering,* McGraw-Hill, New York.
4. Desai, C. S. 1972. "Finite element procedures for seepage analysis using an isoparametric element," in *Proc. Symp. on Appl. of FEM in Geotech. Eng.,* Waterways Expt. Station, Vicksburg, MS.

5. Desai, C. S. 1972. "Seepage analysis of earth banks under drawdown," *J. Soil Mech. Found. Eng., ASCE*, 98, SMII, 1143-1162.
6. Silvester, P. and Chari, M. V. K. 1970. "Finite element solution of saturable magnetic field problems," *IEEE Trans.*, PAS 89, 1642-1651.
7. Chari, M. V. K. and Silvester, P. 1971. "Analysis of turbo-alternator magnetic fields by finite elements," *IEEE Trans. Power Apparatus and Systems*, PAS 90, 454-464.

Bibliography

Desai, C. S. 1975. "Finite element methods for flow in porous media," in *Finite Elements in Fluids*, Vol. 1, Gallagher, R. H., Oden, J. T., Taylor, C., and Zienkiewicz, O. C., Eds., John Wiley & Sons, New York, Chap. 8.

Doherty, W. P., Wilson, E. L., and Taylor, R. L. Jan. 1969. "Stress analysis of axisymmetric solids utilizing higher order quadrilateral finite elements," Report 69-3, Struct. Eng. Lab., Univ. of Calif., Berkeley.

Ergatoudis, I., Irons, B. M., and Zienkiewicz, O. C. 1968. "Curved, isoparametric quadrilateral elements for finite element analysis," *Int. J. Solids Struct.*, 4(1), 31-42.

Zienkiewicz, O. C., Mayer, P., and Cheung, Y. K. 1966. "Solution of anisotropic seepage by finite elements," *J. Eng. Mech. Div. ASCE*, 92(EM1), 111-120.

13

Two-Dimensional Stress-Deformation Analysis

Introduction

After studying a number of one-dimensional problems and two-dimensional field problems with only one unknown or degree of freedom at a point, we are now ready to consider a different class of two-dimensional problems. This class involves analysis of stress and deformations with more than one degree of freedom at a point.

Most real problems are three-dimensional. Under certain assumptions, which can depend on the geometrical and loading characteristics, it is possible to approximate many of them as two-dimensional. Such two-dimensional approximations generally involve two categories: plane deformations and bending deformations. In the case of plane deformation, we encounter subcategories such as plane stress, plane strain, and axisymmetric; in the case of bending, we deal with problems such as bending of plates, slabs, and pavements.

Plane Deformations

Plane Stress Idealization

Figure 13.1 shows a (thin) beam and a plate subjected to loads that are applied in the plane of the structure, that is, in the x–y plane. The thickness is small compared to the x–y dimensions of the body. Such loadings are often referred to as in-plane or membrane (stretching). Under these conditions, and the assumption that the variation of stresses with respect to z, that is, across the body, is constant, it is reasonable to assume that out of the six components of stresses in a three-dimensional body[1-3] three of them, σ_z, τ_{zx}, and τ_{yz}, can be ignored in comparison to the remaining three, σ_z, σ_y, and τ_{xy}.

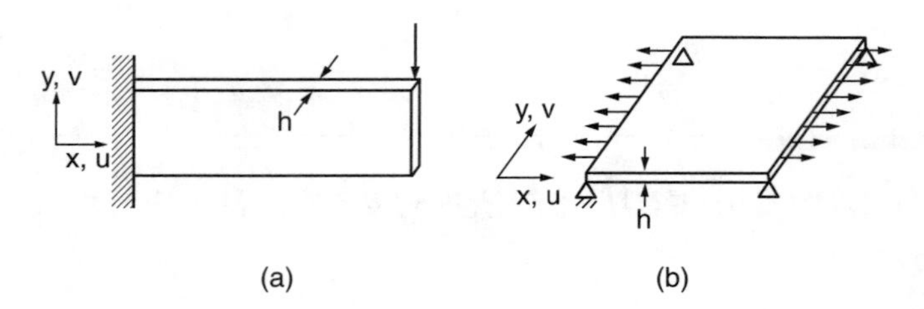

FIGURE 13.1
Plane stress approximation: (a) Beam and (b) Plate.

This idealization is called plane stress and involves only the following non-zero stress components,

$$\{\sigma\} = \begin{Bmatrix} \sigma_x \\ \sigma_y \\ \tau_{xy} \end{Bmatrix}, \tag{13.1}$$

which are functions of the coordinates x, y only. The corresponding components of strains are

$$\{\varepsilon\} = \begin{Bmatrix} \varepsilon_x \\ \varepsilon_y \\ \gamma_{xy} \end{Bmatrix}. \tag{13.2}$$

In Equations 13.1 and 13.2, σ and ε denote components of normal stress and strain and τ and γ denote components of shear stress and strain, respectively, and $\{\sigma\}^T = [\sigma_x \; \sigma_y \; \tau_{xy}]$ and $\{\varepsilon\}^T = [\varepsilon_x \; \varepsilon_y \; \gamma_{xy}]$ are the vectors of stress and strain components. In view of the plane stress assumption, we need to consider only two components of displacements at a point, u and v, in the x and y directions, respectively.

If we restrict ourselves to linear, elastic, and isotropic materials, the material behavior can be expressed by using the generalized Hooke's law for the three components of stress and strain.[1,2] Thus

$$\begin{aligned} \sigma_x &= \frac{E}{1-\nu^2}\varepsilon_x + \frac{\nu E}{1-\nu^2}\varepsilon_y, \\ \sigma_y &= \frac{\nu E}{1-\nu^2}\varepsilon_x + \frac{E}{1-\nu^2}\varepsilon_y, \\ \tau_{xy} &= \frac{E}{2(1+\nu)}\gamma_{xy} \end{aligned} \tag{13.3a}$$

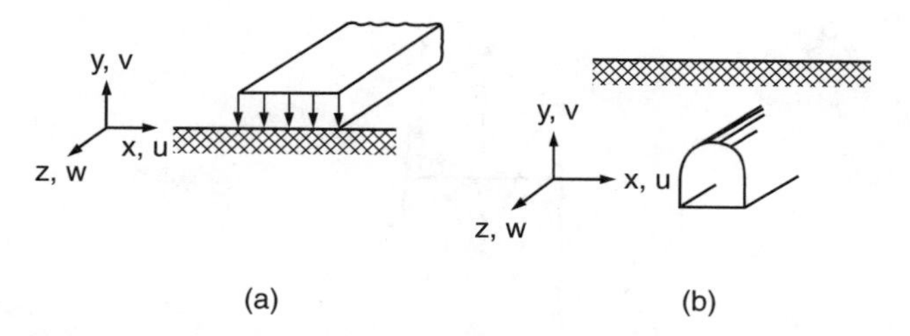

FIGURE 13.2
Plane strain approximation: (a) Strip load and (b) Long underground tunnel.

or in matrix notation,

$$\{\sigma\} = [\mathbf{C}]\{\varepsilon\} = \frac{E}{1-\nu^2}\begin{bmatrix} 1 & v & 0 \\ v & 1 & 0 \\ 0 & 0 & \frac{1-\nu}{2} \end{bmatrix}\{\varepsilon\}, \tag{13.3b}$$

where [**C**] is stress–strain or constitutive matrix and E and v are Young's modulus and Poisson's ratio, respectively.

Plane Strain Idealization

In cases where the thickness is large compared to the x–y dimensions (Figure 13.2) and where the loads are acting only in the plane of the structure, that is, the x–y plane, it can be assumed that the displacement component w in the z direction is negligible and that the in-plane displacements u, v are independent of z. This approximation is called plane strain, in which case the nonzero stress components are given by

$$\{\sigma\} = \begin{Bmatrix} \sigma_x \\ \sigma_y \\ \tau_{xy} \end{Bmatrix} \tag{13.4}$$

and $\sigma_z = \nu(\sigma_x + \sigma_y)$. The stress–strain relationship for this idealization is expressed as

$$\{\sigma\} = [\mathbf{C}]\{\varepsilon\} = \frac{E}{(1+\nu)(1-2\nu)}\begin{bmatrix} 1-\nu & \nu & 0 \\ & 1-\nu & 0 \\ \text{sym.} & & \frac{1-2\nu}{2} \end{bmatrix}\{\varepsilon\}. \tag{13.5}$$

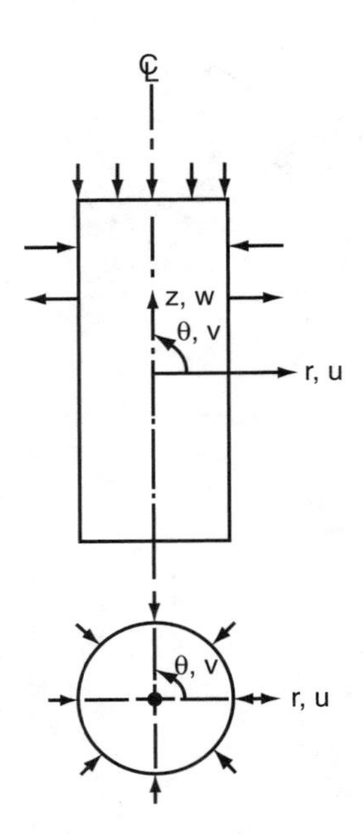

FIGURE 13.3
Axisymmetric approximation.

Axisymmetric Idealization

Figure 13.3 shows a body symmetrical about its centerline axis and subjected to a load symmetrical about the axis. In view of the symmetry, the components of stress are independent of the circumferential coordinate θ. As a consequence, we have the following nonzero stress and strain components:[2]

$$\{\sigma\} = \begin{Bmatrix} \sigma_r \\ \sigma_\theta \\ \sigma_z \\ \tau_{rz} \end{Bmatrix}, \qquad \{\varepsilon\} = \begin{Bmatrix} \varepsilon_r \\ \varepsilon_\theta \\ \varepsilon_z \\ \gamma_{rz} \end{Bmatrix}. \tag{13.6}$$

The constitutive or stress–strain relation for this case is

$$\{\sigma\} = [\mathbf{C}]\{\varepsilon\} = \frac{E}{(1+\nu)(1-2\nu)} \begin{bmatrix} 1-\nu & \nu & \nu & 0 \\ & 1-\nu & \nu & 0 \\ & & 1-\nu & 0 \\ & \text{sym.} & & \dfrac{1-2\nu}{2} \end{bmatrix} \{\varepsilon\}. \tag{13.7}$$

Strain-Displacement Relations

From the theory of elasticity, assuming small strains and deformations, we can define the following strain-displacement relations for the three idealizations:[2]

Plane Stress and Strain:

$$\begin{Bmatrix} \varepsilon_x \\ \varepsilon_y \\ \gamma_{xy} \end{Bmatrix} = \begin{Bmatrix} \dfrac{\partial u}{\partial x} \\ \dfrac{\partial v}{\partial y} \\ \dfrac{\partial u}{\partial y} + \dfrac{\partial v}{\partial x} \end{Bmatrix} \tag{13.8a}$$

Axisymmetric:

$$\begin{Bmatrix} \varepsilon_r \\ \varepsilon_\theta \\ \varepsilon_z \\ \gamma_{rz} \end{Bmatrix} = \begin{Bmatrix} \dfrac{\partial u}{\partial r} \\ \dfrac{u}{r} \\ \dfrac{\partial w}{\partial z} \\ \dfrac{\partial u}{\partial z} + \dfrac{\partial w}{\partial r} \end{Bmatrix} \tag{13.8b}$$

Initial Stress and Strain: As explained in Chapter 5, it is possible to include initial or residual strains or stresses existing in the structure before the load is applied.

The initial strain (or stress) state may be caused by factors such as known temperature, (fluid) pressure, creep effects, and geostatic stresses. For instance, in the case of temperature,

$$\varepsilon_0 = \int_{T_0}^{T} \alpha dT, \tag{13.9}$$

where $dT = T - T_0$ is the change in temperature and α is the coefficient of thermal expansion.

We define total strain e as the sum of the effective elastic strain, ε^e, and the initial strain:

$$\{\varepsilon\} = [\mathbf{C}]^{-1}\{\sigma\} + \{\varepsilon_0\}, \tag{13.10}$$

where $[\varepsilon_0]$ is the vector of initial strains and

$$\{\sigma\} = [\mathbf{C}](\{\varepsilon\} - \{\varepsilon_0\}) = [\mathbf{C}]\{\varepsilon^e\}. \tag{13.11}$$

The matrix $[\mathbf{C}]^{-1} = [\mathbf{D}]$, in which $[\mathbf{D}]$, the strain–stress matrix, for the plane strain case is

$$[\mathbf{D}] = \frac{1-\nu^2}{E}\begin{bmatrix} 1 & \dfrac{-\nu}{1-\nu} & 0 \\ \dfrac{-\nu}{1-\nu} & 1 & 0 \\ 0 & 0 & \dfrac{2}{1-\nu} \end{bmatrix}. \tag{13.12}$$

Finite Element Formulation

As shown in Figure 13.4, the finite element discretization will involve two-dimensional elements such as triangles and quadrilaterals (squares, rectangles, trapezoids) in the x–y plane. The third dimension (z) is generally included by specifying unit thickness for plane strain or a thickness h in the case of plane stress.

Detailed properties of the triangular and quadrilateral elements have been covered in Chapter 10. Here we shall discuss in detail the use of the quadrilateral element and then state briefly the use of the triangular element.

As stated previously, there are two unknown displacements u, v at a point $P(x, y)$ (Figure 13.4). We can write approximation models for u, v at a point in the element as

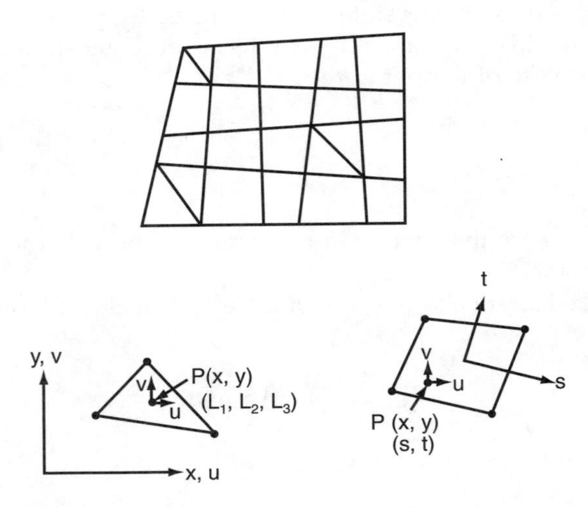

FIGURE 13.4
Discretization with triangular and quadrilateral elements.

$$u(x,y) = \alpha_1 + \alpha_2 x + \alpha_3 y + \alpha_4 xy,$$
$$v(x,y) = \beta_1 + \beta_2 x + \beta_3 y + \beta_4 xy, \qquad (13.13a)$$

or

$$\{\mathbf{u}\} = [\Phi]\{\alpha\}, \qquad (13.13b)$$

where $\{\mathbf{u}\}^T = [u \; v]$, $\{\alpha\}^T = [\alpha_1 \; \alpha_2 \; \alpha_3 \; \alpha_4 \; \beta_1 \; \beta_2 \; \beta_3 \; \beta_4]$ and $[\Phi]$ is the matrix of coordinates. Evaluation of u and v at the four nodes gives

$$u_i = \alpha_1 + \alpha_2 x_i + \alpha_3 y_i + \alpha_4 x_i y_i,$$
$$v_i = \beta_1 + \beta_2 x_i + \beta_3 y_i + \beta_4 x_i y_i, \qquad i = 1, 2, 3, 4, \qquad (13.14a)$$

or

$$\underset{(8\times 1)}{\{\mathbf{q}\}} = \underset{(8\times 8)}{[\mathbf{A}]} \underset{(8\times 1)}{\{\alpha\}}, \qquad (13.14b)$$

where $\{\mathbf{q}\}^T = [u_1 \; u_2 \; u_3 \; u_4 \; v_1 \; v_2 \; v_3 \; v_4]$ is the vector of nodal displacements and $[\mathbf{A}]$ is the square (8×8) matrix of nodal coordinates. Solution of Equation 13.14b for $\{\alpha\}$ gives

$$\{\alpha\} = [\mathbf{A}]^{-1}\{\mathbf{q}\}, \qquad (13.14c)$$

which when substituted into Equation 13.13b yields u, v at any point in terms of the nodal displacements:

$$\underset{(2\times 1)}{\{\mathbf{u}\}} = [\Phi][\mathbf{A}]^{-1}\{\mathbf{q}\} = \underset{(2\times 8)}{[\mathbf{N}]} \underset{(8\times 1)}{\{\mathbf{q}\}}. \qquad (13.15)$$

The product $[\Phi][\mathbf{A}]^{-1}$ results in the matrix of interpolation functions $[\mathbf{N}]$, where the N_i, as defined in Equation 12.12, are

$$N_1 = \tfrac{1}{4}(1-s)(1-t),$$
$$N_2 = \tfrac{1}{4}(1+s)(1-t),$$
$$N_3 = \tfrac{1}{4}(1+s)(1+t), \qquad (13.16)$$
$$N_4 = \tfrac{1}{4}(1-s)(1+t).$$

Here s, t are local coordinates (Figures 12.2 and 13.4).

The geometry, that is, the x, y coordinates at any point in the element, can be expressed by using the same interpolation functions N_i:

$$x = \sum_{i=1}^{4} N_i x_i, \quad i = 1, 2, 3, 4,$$
$$y = \sum_{i=1}^{4} N_i y_i, \quad i = 1, 2, 3, 4, \tag{13.17a}$$

or

$$\underset{(2\times 1)}{\begin{Bmatrix} x \\ y \end{Bmatrix}} = \underset{(2\times 8)}{\begin{bmatrix} [\mathbf{N}] & [0] \\ [0] & [\mathbf{N}] \end{bmatrix}} \underset{(8\times 1)}{\begin{Bmatrix} \{x_n\} \\ \{y_n\} \end{Bmatrix}}, \tag{13.17b}$$

where $\{\mathbf{x}_n\}^T = [x_1\ x_2\ x_3\ x_4]$ and $\{\mathbf{y}_n\}^T = [y_1\ y_2\ y_2\ y_2]$.

This leads to the definition of the element as the four-node isoparametric element.

Requirements for Approximation Function

The approximation function must yield continuous values of u and v within the element. This is satisfied since we have chosen the functions in the polynomial form as in Equation 13.13.

For plane problems, the approximation models must satisfy the interelement compatibility at least up to derivative of order zero; that is, the displacement between adjoining elements must be compatible. This is tied in with the highest order of derivative n in the energy function, Equation 13.21 below. Since $n = 1$ in Equation 13.21, the minimum order for interelement compatibility is equal to $n - 1 = 0$.

The approximation of Equations 13.13 and 13.15 yields continuous bilinear distributions of u and v within an element and linear distributions along the element boundaries. It can be seen (Figure 13.5) that the displacements along common sides of two elements are compatible since only the straight line can pass through two nodal displacements common to both elements. For the two-dimensional plane deformation problems, the approximation function provides for rigid body displacements and constant states of strains: ε_x, ε_y, γ_{xy}. However, the function does not include all terms in the polynomial expansion represented by the Pascal's triangle, Chapter 10.

Plane Stress Idealization

First we consider the plane stress idealization. By following the procedure in Chapter 10 the strain components can be evaluated as

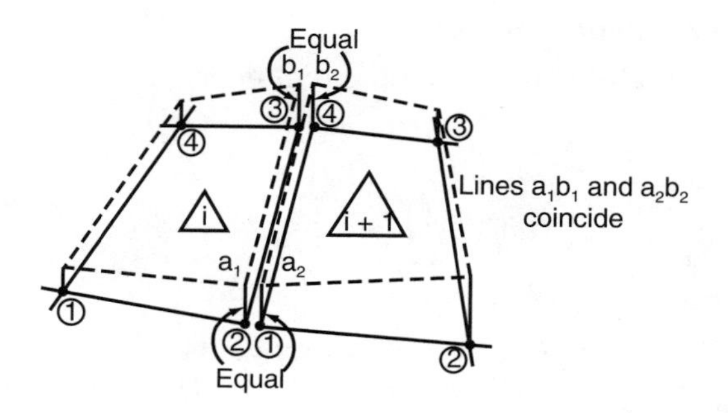

FIGURE 13.5
Interelement compatibility.

$$\varepsilon_x = \frac{\partial u}{\partial x} = \frac{\partial u}{\partial s}\frac{\partial s}{\partial x} + \frac{\partial u}{\partial t}\frac{\partial t}{\partial x}$$
$$= \frac{1}{|J|}\sum_{i=1}^{4}\sum_{j=1}^{4}\left[u_i\left(\frac{\partial N_i}{\partial s}\frac{\partial N_j}{\partial t} - \frac{\partial N_i}{\partial t}\frac{\partial N_j}{\partial s}\right)y_j\right]. \tag{13.18a}$$

Similarly,

$$\varepsilon_y = \frac{\partial v}{\partial y} = \frac{\partial v}{\partial s}\frac{\partial s}{\partial y} + \frac{\partial v}{\partial t}\frac{\partial t}{\partial y} \tag{13.18b}$$

and

$$\gamma_{xy} = \frac{\partial u}{\partial y} + \frac{\partial v}{\partial x} = \left(\frac{\partial u}{\partial s}\frac{\partial s}{\partial y} + \frac{\partial y}{\partial t}\frac{\partial t}{\partial y}\right) + \left(\frac{\partial v}{\partial s}\frac{\partial s}{\partial x} + \frac{\partial v}{\partial t}\frac{\partial t}{\partial x}\right). \tag{13.18c}$$

Finally, we have

$$\begin{Bmatrix} \varepsilon_x \\ \varepsilon_y \\ \gamma_{xy} \end{Bmatrix} = \begin{bmatrix} B_{11} & B_{12} & B_{13} & B_{14} & 0 & 0 & 0 & 0 \\ 0 & 0 & 0 & 0 & B_{21} & B_{22} & B_{23} & B_{24} \\ B_{21} & B_{22} & B_{23} & B_{24} & B_{11} & B_{12} & B_{13} & B_{14} \end{bmatrix} \{\mathbf{q}\} \tag{13.19}$$

or

$$\{\varepsilon\} = [\mathbf{B}]\{\mathbf{q}\}, \tag{13.20}$$

where the B_{ij} are defined in Equation 10.24c and $\{\mathbf{q}\}$ is defined in Equation 13.14b.

Step 4. Derive Element Equations

We use the principle of stationary potential energy; the potential energy is given by[3]

$$\Pi_p = \frac{h}{2}\iint_A \{\varepsilon\}^T[\mathbf{C}]\{\varepsilon\}\,dxdy - h\iint_A \{\mathbf{u}\}^T\{\bar{\mathbf{X}}\}\,dxdy - h\int_{S_1}\{\mathbf{u}\}^T\{\bar{\mathbf{T}}\}\,dS, \quad (13.21)$$

where $\{\bar{\mathbf{X}}\}^T = [\bar{X}\ \bar{Y}]$ is the vector of components of body forces per unit volume, $\{\bar{\mathbf{T}}\}^T = [\bar{T}_x\ \ \bar{T}_y]$ is the vector of components of surface tractions per unit surface area in the x and y directions, respectively; and h is the (uniform) thickness of the element.

Substitution for $\{\mathbf{u}\}$ and $\{\varepsilon\}$ from Equations 13.15 and 13.20 in Equation 13.21 yields

$$\begin{aligned}\Pi_p = {} & \frac{h}{2}\underset{(1\times 8)}{\{\mathbf{q}\}^T}\iint_A \underset{(8\times 3)}{[\mathbf{B}]^T}\ \underset{(3\times 3)}{[\mathbf{C}]}\ \underset{(3\times 8)}{[\mathbf{B}]}\ dxdy\ \underset{(8\times 1)}{\{\mathbf{q}\}} \\ & - h\underset{(1\times 8)}{\{\mathbf{q}\}^T}\iint_A \underset{(8\times 2)}{[\mathbf{N}]^T}\ \underset{(2\times 1)}{\{\bar{\mathbf{X}}\}}\ dxdy \\ & - h\underset{(1\times 8)}{\{\mathbf{q}\}^T}\int_{S_1} \underset{(8\times 2)}{[\mathbf{N}]^T}\ \underset{(2\times 1)}{\{\bar{\mathbf{T}}\}}\ dS.\end{aligned} \quad (13.22)$$

By taking partial derivatives of Π_p with respect to u_1, v_1, etc., and equating to zero,

$$\frac{\partial \Pi_p}{\partial\{\mathbf{q}\}} = 0, \quad (13.23)$$

which leads to the eight element equilibrium equations as

$$\underset{(8\times 8)}{[\mathbf{k}]}\ \underset{(8\times 1)}{\{\mathbf{q}\}} = \underset{(8\times 1)}{\{\mathbf{Q}\}} = \underset{(8\times 1)}{\{\mathbf{Q}_1\}} + \underset{(8\times 1)}{\{\mathbf{Q}_2\}}, \quad (13.24)$$

where $[\mathbf{k}]$ is the element stiffness matrix,

$$[\mathbf{k}] = h\iint_A [\mathbf{B}]^T[\mathbf{C}][\mathbf{B}]\,dxdy, \quad (13.25)$$

and $\{\mathbf{Q}\}$ is the element nodal load vector,

$$\{\mathbf{Q}\} = \{\mathbf{Q}_1\} + \{\mathbf{Q}_2\} = h\iint_A [\mathbf{N}]^T\{\bar{\mathbf{X}}\}dxdy + h\int_{S_1}[\mathbf{N}]^T\{\bar{\mathbf{T}}\}dS \quad (13.26)$$

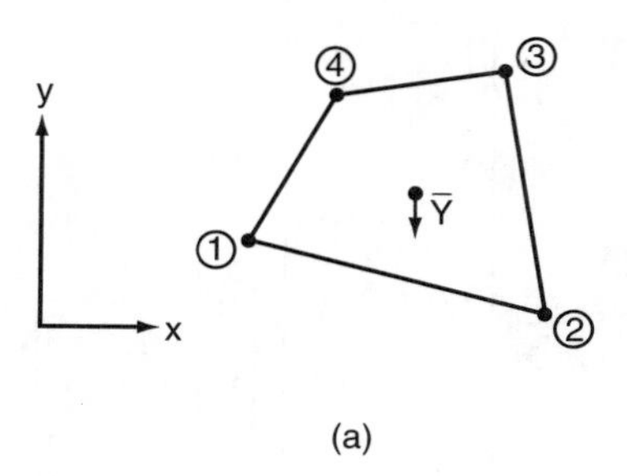

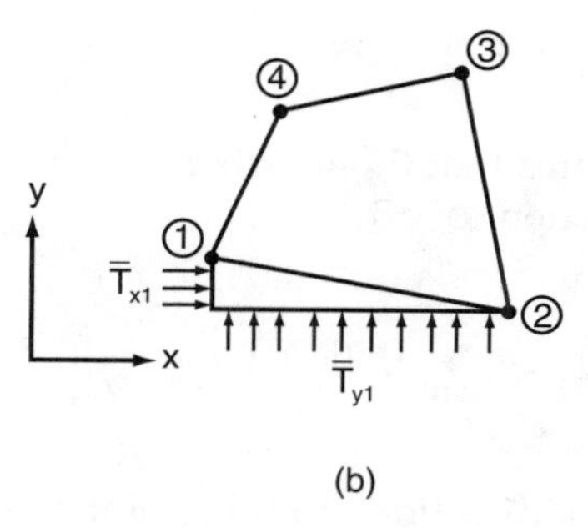

FIGURE 13.6
Loading on quadrilateral element: (a) Body force and (b) Surface tractions.

Evaluation of [k] and {Q}

The coefficients of [**k**] are functions of local coordinates s and t, and it is more convenient to perform numerical integration as follows:

$$[\mathbf{k}] \simeq h\sum_{i=1}^{N}\left[\mathbf{B}(s_i,\, t_i)\right]^T[\mathbf{C}]\left[\mathbf{B}(s_i,t_i)\right]\left|J(s_i,t_i)\right|W_i, \tag{13.27}$$

where $(s_i,\, t_i)$ denote the local coordinates of the integration point; and so on. Often, for quadrilateral elements, 2 × 2 or 4-point (N = 4) integration (Figure 10.7) is used.

The first part $\{\mathbf{Q}_1\}$ of the load vector can be computed as follows:

$$\{\mathbf{Q}_1\} = h\sum_{i=1}^{N}\left[\mathbf{N}(s_i,t_i)\right]^T\{\overline{\mathbf{X}}\}W_i. \tag{13.28a}$$

If we assume a uniform body force intensity $\overline{Y}$ (per unit volume) (Figure 13.6(a)), $\overline{X} = 0$, the expanded form of $\{\mathbf{Q}_1\}$ is

$$\{\mathbf{Q}_1\} = h\sum_{i=1}^{N} \begin{bmatrix} N_1 & 0 \\ N_2 & 0 \\ N_3 & 0 \\ N_4 & 0 \\ 0 & N_1 \\ 0 & N_2 \\ 0 & N_3 \\ 0 & N_4 \end{bmatrix}_{(s_i,t_i)} \begin{Bmatrix} 0 \\ \overline{Y} \end{Bmatrix} \left| J(s_i,t_i) \right| W_i. \tag{13.28b}$$

The subscript (s_i, t_i) denotes that the matrix is evaluated at points (s_i, t_i). For instance, the fifth component of $\{\mathbf{Q}_1\}$,

$$Q_{1(5)} = h\sum_{i=1}^{N} N_1(s_i,t_i) \left| J(s_i,t_i) \right| W_i \overline{Y}, \tag{13.28c}$$

gives nodal force in the y direction at node point 1, and so on.

The second part $\{\mathbf{Q}_2\}$ arises due to surface tractions applied on the boundary of an element. Often it is possible to evaluate this part by using closed form integration. For instance, consider $\overline{T}_{x1}$ and $\overline{T}_{y1}$ as applied tractions on side 1–2 (Figure 13.6(b)). Then

$$\{\mathbf{Q}_2\} = h\int_{-1}^{+1} \begin{bmatrix} N_1 & 0 \\ N_2 & 0 \\ N_3 & 0 \\ N_4 & 0 \\ 0 & N_1 \\ 0 & N_2 \\ 0 & N_3 \\ 0 & N_4 \end{bmatrix} \begin{Bmatrix} \overline{T}_{x1} \\ \overline{T}_{y1} \end{Bmatrix} dS. \tag{13.29a}$$

Now, along side 1, s = -1 to 1 and t = -1; therefore,

$$\begin{aligned} N_1 &= \tfrac{1}{4}(1-s)(1-t) = \tfrac{1}{2}(1-s), \\ N_2 &= \tfrac{1}{4}(1+s)(1-t) = \tfrac{1}{2}(1+s), \end{aligned} \tag{13.29b}$$

$$N_3 = \tfrac{1}{4}(1+s)(1+t) = 0,$$

$$N_4 = \tfrac{1}{4}(1-s)(1+t) = 0,$$

substitution of which leads to a line integral as

$$\{\mathbf{Q}_2\} = \frac{hl_1}{2}\int_{-1}^{1} \begin{bmatrix} (1-s)/2 & 0 \\ (1+s)/2 & 0 \\ 0 & 0 \\ 0 & 0 \\ 0 & (1-s)/2 \\ 0 & (1+s)/2 \\ 0 & 0 \\ 0 & 0 \end{bmatrix} \begin{Bmatrix} \overline{T}_{x1} \\ \overline{T}_{y1} \end{Bmatrix} dS. \tag{13.29c}$$

We used here the transformation relation in Equation 12.38. Upon required integrations, we have

$$\{\mathbf{Q}_2\} = \frac{hl_1}{2} \begin{Bmatrix} \overline{T}_{x1} \\ \overline{T}_{x1} \\ 0 \\ 0 \\ \overline{T}_{y1} \\ \overline{T}_{y1} \\ 0 \\ 0 \end{Bmatrix}, \tag{13.29d}$$

where l_1 = length of side 1–2. This implies that the applied load is distributed equally at the two nodes pertaining to the side 1–2. This is a consequence of the fact that the interpolation function (Equation 13.15) is linear along the sides of the quadrilateral. If we use a different (higher-order) approximation, the results may not be similar. Furthermore, in the case of the higher order approximation, it may be easier to perform numerical integration, as was done for $\{\mathbf{Q}_1\}$.

The assembly of element equations can be achieved by following the principle that the displacements at the common nodes are compatible. The procedure is essentially the same as illustrated in Chapters 3, 4, 5, 7, 10, 11, and 12 and as illustrated subsequently in Examples 13.1 and 13.2.

The assemblage equations are modified for the boundary conditions in terms of prescribed values of u and v on parts of the boundary. Solution of the resulting equations gives nodal displacements. Then strains and stresses are computed by using Equations 13.20; and 13.3, 13.5, and 13.7, respectively.

Triangular Element

After the derivations in Chapters 10 and 11, it is relatively straightforward to derive element equations for the plane stress idealization using the triangular element (Figure 10.2). Use of the linear function, Equation 10.4a for both u and v satisfies various requirements (for two-dimensional plane deformations).

Various terms for the triangular element are stated below:

$$\{\mathbf{u}\}^T = [u \quad v], \tag{13.30a}$$

$$[\mathbf{N}] = \begin{bmatrix} N_1 & N_2 & N_3 & 0 & 0 & 0 \\ 0 & 0 & 0 & N_1 & N_2 & N_3 \end{bmatrix}, \tag{13.30b}$$

$$\{\mathbf{q}\}^T = [u_1 \quad u_2 \quad u_3 \quad v_1 \quad v_2 \quad v_3], \tag{13.30c}$$

$$[\mathbf{B}] = \frac{1}{2A}\begin{bmatrix} b_1 & b_2 & b_3 & 0 & 0 & 0 \\ 0 & 0 & 0 & a_1 & a_2 & a_3 \\ a_1 & a_2 & a_3 & b_1 & b_2 & b_3 \end{bmatrix}, \tag{13.30d}$$

where the N_i are defined in Equation 10.3b and the b_i and a_i are given in Equation 10.3c.

The general form of the element equations is identical to that for the quadrilateral element (Equation 13.24); only the orders of various matrices are different. The stiffness matrix has the order of 6×6 and is given by

$$\begin{aligned}[\mathbf{k}] &= h[\mathbf{B}]^T[\mathbf{C}][\mathbf{B}]\iint_A dA \\ &= hA[\mathbf{B}]^T[\mathbf{C}][\mathbf{B}].\end{aligned} \tag{13.31}$$

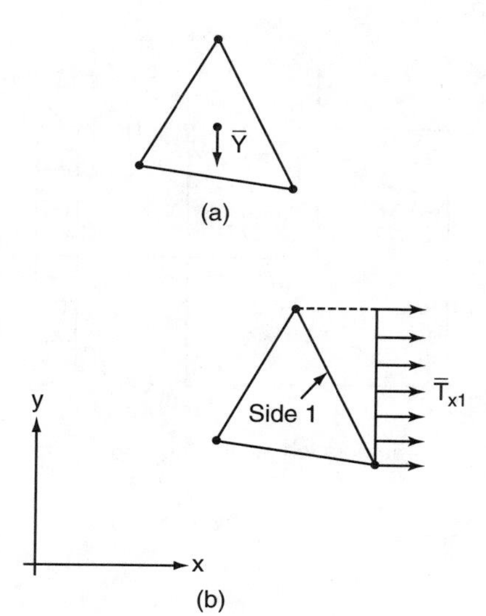

FIGURE 13.7
Loading on triangular element: (a) Body force and (b) Surface traction.

The load vector has the order 6 × 1. For a uniform body weight $\bar{Y}$ ($\bar{X} = 0$) (Figure 13.7(a)),

$$\{\mathbf{Q}_1\}^T = \frac{hA}{3}\begin{bmatrix}0 & 0 & 0 & \bar{Y} & \bar{Y} & \bar{Y}\end{bmatrix}. \tag{13.32a}$$

The uniform surface traction is divided equally among the two nodes belonging to the side on which traction is applied. For instance, for T_{x1} acting on side 1 (Figure 13.7(b)) we have

$$\{\mathbf{Q}_2\}^T = \frac{h\bar{T}_{x1}l_1}{2}\begin{bmatrix}0 & 1 & 1 & 0 & 0 & 0\end{bmatrix}, \tag{13.32b}$$

where l_1 is the length of side 1. Contributions of tractions in the y directions and on other sides can be similarly evaluated.

Example 13.1 Details of Quadrilateral Element

Figure 13.8 shows a square isoparametric element (see also Example 10.1, Figure 10.6) subjected to a surface traction $\bar{T}_x$ equal to 1 kg/cm on side 2–3. Material properties are E = 10,000 kg/cm² and ν = 0.30.

First we show brief details for the computation of matrix [**B**]. Referring to Equation 10.20c, we have

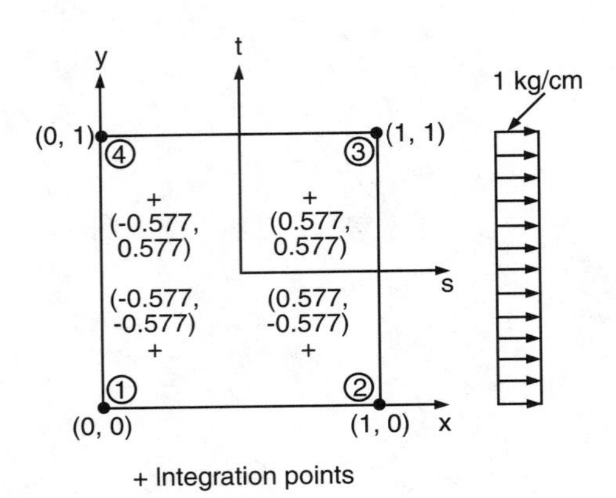

FIGURE 13.8
Integration over square element.

$$
\begin{aligned}
x_{13} &= 0-1=-1, & y_{24} &= 0-1=-1,\\
x_{24} &= 1-0=1, & y_{13} &= 0-1=-1,\\
x_{34} &= 1-0=1, & y_{12} &= 0-0=0,\\
x_{12} &= 0-1=-1, & y_{34} &= 1-1=0,\\
x_{23} &= 1-1=0, & y_{14} &= 0-1=-1,\\
x_{14} &= 0-0=0, & y_{23} &= 0-1=-1.
\end{aligned}
$$

Then

$$
\begin{aligned}
|J| &= \tfrac{1}{8}(-1)(-1)-(-1)(-1)+s\left[1\times 0-(-1)(0)\right]\\
&\quad +t\left[(0)(-1)-(0)(-1)\right]\\
&= \tfrac{1}{8}(1+1)\\
&= 0.25,
\end{aligned}
$$

which is one-fourth of the area of the square. Therefore, use of Equation 10.24c gives

$$
\begin{aligned}
B_{11} &= \frac{1}{8\times 0.25}\left[-1-0\times s-(-1)t\right]=\frac{1}{2}(-1+t),\\
B_{12} &= \frac{1}{2}\left[+1+0\times s+(-1)t\right]=\frac{1}{2}(1-t),\\
B_{13} &= \frac{1}{2}\left[1+0\times s+(1)t\right]=\frac{1}{2}(1+t),
\end{aligned}
$$

$$
\begin{aligned}
B_{21} &= \frac{1}{2}\left[-1+1\times s+0\times t\right]=\frac{1}{2}(-1+s),\\
B_{22} &= \frac{1}{2}\left[-1-1\times s-0\times t\right]=\frac{1}{2}(-1-s),\\
B_{23} &= \frac{1}{2}\left[1-(-1)s+0\times t\right]=\frac{1}{2}(1+s),\\
B_{24} &= \frac{1}{2}\left[1+(-1)s-0\times t\right]=\frac{1}{2}(1-s).
\end{aligned}
$$

Now the product $[\mathbf{B}]^T[\mathbf{C}][\mathbf{B}]$ in $[\mathbf{k}]$ (Equation 13.27) is given by

$$
[\mathbf{B}]^T[\mathbf{C}][\mathbf{B}] = \begin{bmatrix}
C_{11}B_{11}^2 + C_{33}B_{21}^2 & C_{12}B_{21}B_{11} + C_{33}B_{21}B_{11} & C_{11}B_{11}B_{12} + C_{33}B_{21}B_{22} & C_{12}B_{21}B_{22} + C_{33}B_{21}B_{12} & C_{11}B_{11}B_{13} + C_{33}B_{21}B_{23} & C_{12}B_{11}B_{23} + C_{33}B_{21}B_{13} & C_{11}B_{11}B_{14} + C_{33}B_{21}B_{24} & C_{12}B_{11}B_{24} + C_{33}B_{21}B_{14} \\
 & C_{22}B_{21}^2 + C_{33}B_{11}^2 & C_{12}B_{21}B_{12} + C_{33}B_{11}B_{22} & C_{22}B_{21}B_{22} + C_{33}B_{11}B_{12} & C_{12}B_{21}B_{13} + C_{33}B_{11}B_{23} & C_{22}B_{21}B_{23} + C_{33}B_{11}B_{13} & C_{12}B_{14}B_{21} + C_{33}B_{11}B_{24} & C_{22}B_{21}B_{24} + C_{33}B_{11}B_{14} \\
 & & C_{11}B_{12}^2 + C_{33}B_{22}^2 & C_{12}B_{12}B_{22} + C_{33}B_{22}B_{12} & C_{11}B_{12}B_{13} + C_{33}B_{22}B_{23} & C_{11}B_{12}B_{23} + C_{33}B_{22}B_{13} & C_{11}B_{12}B_{14} + C_{33}B_{22}B_{24} & C_{12}B_{12}B_{24} + C_{33}B_{22}B_{14} \\
 & & & C_{22}B_{22}^2 + C_{33}B_{12}^2 & C_{12}B_{22}B_{13} + C_{33}B_{12}B_{23} & C_{22}B_{22}B_{23} + C_{33}B_{12}B_{13} & C_{12}B_{22}B_{14} + C_{33}B_{12}B_{24} & C_{22}B_{22}B_{24} + C_{33}B_{12}B_{14} \\
 & & & & C_{11}B_{13}^2 + C_{33}B_{23}^2 & C_{12}B_{13}B_{23} + C_{33}B_{23}B_{13} & C_{11}B_{13}B_{14} + C_{33}B_{23}B_{24} & C_{12}B_{13}B_{24} + C_{33}B_{23}B_{14} \\
 & & & & & C_{22}B_{23}^2 + C_{33}B_{13}^2 & C_{12}B_{23}B_{14} + C_{33}B_{13}B_{24} & C_{22}B_{23}B_{24} + C_{33}B_{13}B_{14} \\
 & & & & & & C_{11}B_{14}^2 + C_{33}B_{24}^2 & C_{12}B_{14}B_{24} + C_{33}B_{24}B_{14} \\
 & & & & & & & C_{22}B_{24}^2 + C_{33}B_{14}^2
\end{bmatrix} \tag{13.33}
$$

Here we have arranged the vector $\{\mathbf{q}\}$ as

$$\{\mathbf{q}\}^T = \begin{bmatrix} u_1 & v_1 & u_2 & v_2 & u_3 & v_3 & u_4 & v_4 \end{bmatrix}.$$

As indicated in Equation 13.27, we need to compute the terms one by one at each of the four integration points (Figure 13.8) and add them together to obtain the matrix $[\mathbf{k}]$. As an illustration, the values of $B_{11}(s_i, t_i)$ at the four points are found as

$$B_{11}(s_1, t_1) = \tfrac{1}{2}(-1 - 0.577) = -0.789,$$

$$B_{11}(s_2, t_2) = \tfrac{1}{2}(1 - 0.577) = 0.212,$$

$$B_{11}(s_3, t_3) = \tfrac{1}{2}(1 + 0.577) = 0.789,$$

$$B_{11}(s_4, t_4) = \tfrac{1}{2}(-1 + 0.577) = -0.212,$$

and so on. The values of the C_{ij} are obtained by using E and ν.

The final result for the [**k**] matrix is

$$[\mathbf{k}] = 10^2 \begin{bmatrix} 49.45 & 17.86 & -30.22 & -13.74 & -24.73 & -17.86 & 5.49 & 1.38 \\ & 49.45 & 1.38 & 5.49 & -17.86 & -24.73 & -1.38 & -30.2 \\ & & 49.45 & -17.86 & 5.49 & -1.38 & -24.73 & 17.86 \\ & & & 49.45 & 1.38 & -30.22 & 17.86 & -24.73 \\ & & & & 49.45 & 17.86 & -30.22 & -1.38 \\ & & \text{sym.} & & & 49.45 & 1.38 & 5.49 \\ & & & & & & 49.45 & -17.86 \\ & & & & & & & 49.45 \end{bmatrix}. \quad (13.34a)$$

Use of Equation 13.29d leads to the load vector {Q} as

$$\{\mathbf{Q}\}^T = [0.0 \quad 0.0 \quad 0.50 \quad 0.0 \quad 0.50 \quad 0.0 \quad 0.0 \quad 0.0]. \quad (13.34b)$$

Introduction of the boundary conditions

$$u_1 = v_1 = v_2 = v_3 = u_4 = v_4 = 0$$

leads to two equations

$$4945u_2 + 549u_3 = 0.50,$$

$$549u_1 + 4945u_3 = 0.50.$$

Solution by Gaussian elimination yields

$$u_2 = u_3 = 0.91 \times 10^{-4} \text{ cm}.$$

Use of Equation [σ] = [**C**][**B**]{**q**} leads to element stresses as

$$\sigma_x = 1.000 \text{ kg/cm}^2,$$
$$\sigma_y = 0.300,$$
$$\tau_{xy} = 0.000,$$

and the principal stresses

$$\sigma_1 = 1.000 \text{ kg/cm}^2,$$
$$\sigma_2 = 0.300$$

are computed from

$$\sigma_{1,2} = \frac{\sigma_x + \sigma_y}{2} \pm \sqrt{(\sigma_x - \sigma_y)^2 + \tau_{xy}^2}. \tag{13.34c}$$

Thus we obtain the quantities — displacements, stresses, and strains — required for analysis and design of a structure idealized as a two-dimensional plane problem.

Example 13.2. Triangular Element

Evaluate element stiffness matrices for the problem in Example 13.1 subdivided into two triangular elements (Figure 13.9). Assemble and solve for displacement and stresses with the following data.

The surface traction $\bar{T}_y$ on side 2–4 of element 2 = 1 kg/cm per unit thickness.

Boundary conditions:

$$\begin{aligned} u(0,0) = v(0,0) &= 0.0, \\ v(1,0) = \qquad &= 0.0, \\ u(0,1) = v(0,1) &= 0.0, \\ v(1,1) = \qquad &= 0.0. \end{aligned} \tag{13.35}$$

We first consider element 1 (Figure 13.9). Here we have numbered local degrees of freedom as 1, 2, 3, 4, 5, and 6, corresponding to local displacements u_1, u_2, u_3, v_1, v_2, and v_3, respectively. The global degrees of freedom are numbered by assigning consecutively two indices to each node corresponding to displacement components u, v, respectively. Thus in Figure 13.9 there are a total of eight degrees of freedom. The terms required for finding the matrix **[B]** are evaluated as follows:

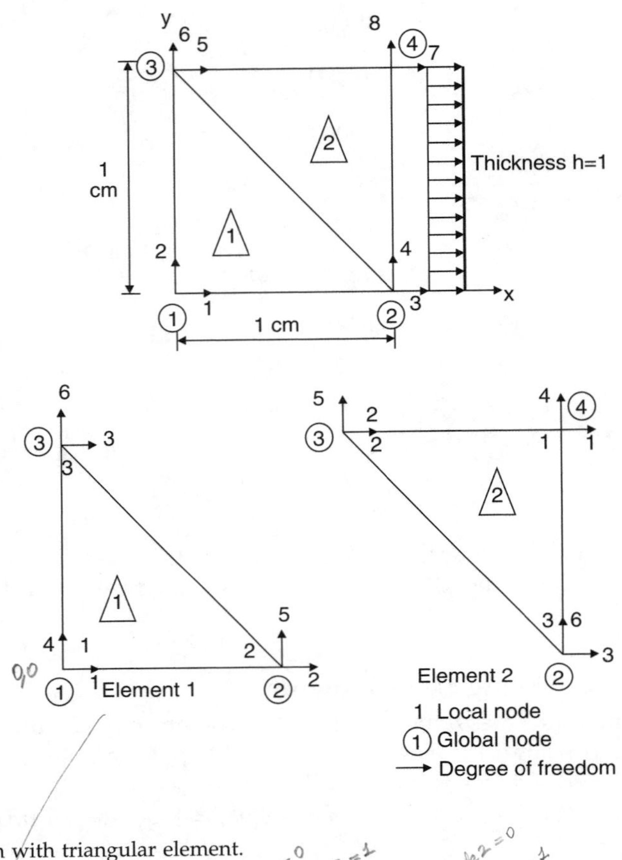

FIGURE 13.9
Discretization with triangular element.

$$a_1 = x_3 - x_2 = -1, \quad b_1 = y_2 - y_3 = -1,$$

$$a_2 = x_1 - x_3 = 0, \quad b_2 = y_3 - y_1 = 1,$$

$$a_3 = x_2 - x_1 = 1, \quad b_3 = y_1 - y_2 = 0,$$

and

$$2A = a_3 b_2 - a_2 b_3 = 1 \times 1 - 0 \times 0 = 1 \text{ cm}^2.$$

Therefore, from Equation 13.30d,

$$B = \frac{1}{1}\begin{bmatrix} -1 & 1 & 0 & 0 & 0 & 0 \\ 0 & 0 & 0 & -1 & 0 & 1 \\ -1 & 0 & 1 & -1 & 1 & 0 \end{bmatrix}. \tag{13.36}$$

For plane stress conditions, we use Equation 13.3(b)) and 13.31 to yield

$$[\mathbf{k}] = hA[\mathbf{B}]^T[\mathbf{C}][\mathbf{B}]$$

$$= hA\begin{bmatrix} -1 & 0 & -1 \\ 1 & 0 & 0 \\ 0 & 0 & 1 \\ 0 & -1 & -1 \\ 0 & 0 & 1 \\ 0 & 1 & 0 \end{bmatrix} \frac{E}{1-\nu^2}\begin{bmatrix} 1 & \nu & 0 \\ \nu & 1 & 0 \\ 0 & 0 & \dfrac{1-\nu}{2} \end{bmatrix}\begin{bmatrix} -1 & 1 & 0 & 0 & 0 & 0 \\ 0 & 0 & 0 & -1 & 0 & 1 \\ -1 & 0 & 1 & -1 & 1 & 0 \end{bmatrix} \tag{13.37a}$$

and hence

$$\begin{array}{c} \textit{Element 1} \left\{ \begin{array}{l} \text{Global} \rightarrow 1 \\ \text{Local} \downarrow \end{array} \right. \quad 3 \quad 5 \quad 2 \quad 4 \quad 6 \end{array} \tag{13.37b}$$

$$[\mathbf{k}] = \frac{E}{2(1-\nu^2)} \begin{array}{c} 1 \\ 2 \\ 3 \\ 4 \\ 5 \\ 6 \end{array} \begin{bmatrix} \dfrac{3-\nu}{2} & -1 & -\dfrac{(1-\nu)}{2} & \dfrac{1+\nu}{2} & -\dfrac{(1-\nu)}{2} & -\nu \\ & 1 & 0 & -\nu & 0 & \nu \\ & & \dfrac{1-\nu}{2} & -\dfrac{(1-\nu)}{2} & \dfrac{1-\nu}{2} & 0 \\ & & & \dfrac{3-\nu}{2} & -\dfrac{(1-\nu)}{2} & 0 \\ & & \text{sym.} & & \dfrac{1-\nu}{2} & 0 \\ & & & & & 1 \end{bmatrix} \begin{array}{c} 1 \\ 2 \\ 3 \\ 4 \\ 5 \\ 6 \end{array}$$

$$7 \quad 5 \quad 3 \quad 8 \quad 6 \quad 4 \leftarrow \begin{array}{c} \uparrow \\ \text{Local} \\ \text{Global} \end{array} \Big\} \textit{ Element 2}$$

The stiffness matrix for element 2, which has the same area and dimensions as element 1, can be deduced from that of element 1 simply by properly exchanging the node numbers. For instance, in Figure 13.9 we have marked local numbers for element 2, and the corresponding global numbers are shown at the bottom of Equation 13.37b.

With each node having two degrees of freedom, we assign global and local numbers to them. Thus the global numbers are (1, 2), (3, 4), (5, 6), and (7, 8) for the 4 nodes, 1, 2, 3, and 4, respectively. The local numbers for the nodes are (1, 4), (2, 5), and (3, 6) for the 3 local nodes, 1, 2, and 3, respectively.

Since there is no load on any of the sides of element 1, there is no contribution to the load vector. In the case of element 2, the surface traction of 1 kg/cm on side 2–4 yields

$$\{\mathbf{Q}\} = \frac{1 \times 1}{2}\begin{Bmatrix} 1 \\ 0 \\ 1 \\ 0 \\ 0 \\ 0 \end{Bmatrix}. \tag{13.37c}$$

Local	*Element 1* Global	*Element 2* Global
1	1	7
2	3	5
3	5	3
4	2	8
5	4	6
6	6	4

If we assign E = 10,000 and $v = 0.3$, Equation 13.37b reduces to

Element 1 {Global → 1 3 5 2 4 6; Local ↓}

$$[\mathbf{k}] = \frac{10{,}000}{1.82}\begin{array}{c} 1 \\ 2 \\ 3 \\ 4 \\ 5 \\ 6 \end{array}\begin{bmatrix} 1.35 & -1.00 & -0.35 & 0.65 & -0.35 & -0.30 \\ -1.00 & 1.00 & 0.00 & -0.30 & 0.00 & 0.30 \\ -0.35 & 0.00 & 0.35 & -0.35 & 0.35 & 0.00 \\ 0.65 & -0.30 & -0.35 & 1.35 & -0.35 & -1.00 \\ -0.35 & 0.00 & 0.35 & -0.35 & 0.35 & 0.00 \\ -0.30 & 0.30 & 0.00 & -1.00 & 0.00 & 1.00 \end{bmatrix}\begin{array}{c} 1 \\ 2 \\ 3 \\ 4 \\ 5 \\ 6 \end{array} \tag{13.38}$$

7 5 3 8 6 4 ← Global; Local ↑ } *Element 2*

Assembly of the stiffness matrices and load vectors yield global relations as follows:

Global → 1 3 5 7 2 4 6 8 (columns); Global ↓ (rows)

$$\frac{10{,}000}{1.82}\begin{array}{c} 1 \\ 3 \\ 5 \\ 7 \\ 2 \\ 4 \\ 6 \\ 8 \end{array}\begin{bmatrix} 1.35 & -1.00 & -0.35 & 0.00 & 0.65 & -0.35 & -0.30 & 0.00 \\ -1.00 & 1.00+0.35 & 0.00 & -0.35 & -0.30 & 0.00 & 0.30+0.35 & -0.35 \\ -0.35 & 0.00 & 0.35+1.00 & -1.00 & -0.35 & 0.35+0.30 & 0.00 & -0.30 \\ 0.00 & -0.35 & -1.00 & 1.35 & 0.00 & -0.30 & -0.35 & 0.65 \\ 0.65 & -0.30 & -0.35 & 0.00 & 1.35 & -0.35 & -1.00 & 0.00 \\ -0.35 & 0.00 & 0.35+0.30 & -0.30 & -0.35 & 0.35+1.00 & 0.00 & -1.00 \\ -0.30 & 0.30+0.35 & 0.00 & -0.35 & -1.00 & 0.00 & 1.00+0.35 & -0.35 \\ 0.00 & -0.35 & -0.30 & 0.65 & 0.00 & -1.00 & -0.35 & 1.35 \end{bmatrix}\begin{Bmatrix} u_1 \\ u_2 \\ u_3 \\ u_4 \\ v_1 \\ v_2 \\ v_3 \\ v_4 \end{Bmatrix} = \frac{1}{2}\begin{Bmatrix} 0 \\ 0+1.0 \\ 0 \\ 0+1.0 \\ 0 \\ 0 \\ 0 \\ 0 \end{Bmatrix},$$

$$\frac{10{,}000}{1.82}\begin{bmatrix} 1.35 & -1.00 & -0.35 & 0.00 & 0.65 & -0.35 & 0.30 & 0.00 \\ & 1.35 & 0.00 & -0.35 & -0.30 & 0.00 & 0.65 & -0.35 \\ & & 1.35 & -1.00 & -0.35 & 0.65 & 0.00 & -0.30 \\ & & & 1.35 & 0.00 & -0.30 & -0.35 & 0.65 \\ & & & & 1.35 & -0.35 & -1.00 & 0.00 \\ & & & & & 1.35 & 0.00 & -1.00 \\ & & & & & & 1.35 & -0.35 \\ & & & & & & & 1.35 \end{bmatrix}\begin{Bmatrix} u_1 \\ u_2 \\ u_3 \\ u_4 \\ v_1 \\ v_2 \\ v_3 \\ v_4 \end{Bmatrix} = \frac{1}{2}\begin{Bmatrix} 0.0 \\ 1.0 \\ 0.0 \\ 1.0 \\ 0.0 \\ 0.0 \\ 0.0 \\ 0.0 \end{Bmatrix}$$

Introduction of the four boundary conditions in Equation 13.35 is achieved by deleting the rows and columns corresponding to u_1, v_1, v_2, u_3, v_3, and v_4, that is, global degrees of freedom 1, 2, 4, 5, 6, and 8, respectively. This leaves only two equations:

$$1.35u_2 - 0.35u_4 = \frac{1 \times 1.82}{2 \times 10{,}000} = 0.000091$$

and

$$-0.35u_2 + 1.35u_4 = 0.000091.$$

Solution by Gaussian elimination gives

$$u_2 = u_4 = 0.91 \times 10^{-4} \text{ cm}.$$

These results agree with the solution by using the quadrilateral element.

Comment on Convergence

In the case of the displacement formulation, algebraic value of the potential energy in the system is higher than the exact energy at equilibrium. As the quality of the formulation is improved by using higher-order approximation and/or by using refined mesh, the energy converges to the exact value. Hence, an element which has a lesser value of approximated stiffness can be considered to be superior (see Figure 3.15).

Often, the trace of the stiffness matrix is used as a measure of the stiffness. The trace is defined as $\sum K_{ii}$, that is, the sum of the diagonal elements. The trace for the quadrilateral element is $4945 \times 8 = 39{,}560$, whereas the trace for the triangular element is $7418 \times 8 = 59{,}344$. This shows that use of the quadrilateral element which contains an extra term (xy) in the approximation model can yield better solutions. Such an inprovement may not be evident with the use of only one or two elements. However, for larger problems the improvement can be significant.

For general mathematical analysis we need to use the concept of the eigen system of the matrix. This topic is discussed in advanced texts.

Computer Code

The computer code PLANE-2D (see Chapter 6 and Appendix 3) permits linear elastic analysis of bodies idealized as plane stress, plane strain, or axisymmetric. The code incorporates the four-node isoparametric element (Figure 13.4). The quadrilateral can be degenerated to be a triangle by repeating the last node. In the following are presented typical problems solved by using this code.

Example 13.3. Analysis of Shear Wall

Figure 13.10(a) shows a shear wall, which may constitute part of a building frame. The wall has an opening 2 m wide × 5 m deep in the lower floor level. The properties of the wall are

E (columns, beams, and wall) = 2.1×10^9 kg/m^2,

v (columns, beams, and wall) = 0.3,

Thickness of columns and beams = 0.3 m,

Thickenss of wall = 0.15 m,

Loading is as shown in Figure 13.10(a).

By using the code PLANE-2DFE or other available codes, analyze the load-deformation behavior of the wall assuming plane stress conditions. Obtain two sets of results with two values of E for the shear wall with lateral loads only.

1. $E_{\text{wall}} = 0.21 \times 10^9$ kg/m^2,
2. $E_{\text{wall}} = 0.0021 \times 10^9$ kg/m^2.

Plot the deflected shape of the wall and the distribution of vertical stress σ_y for the first value of $E_{wall} = 0.21 \times 10^9$ kg/m^2 at selected sections. Compare the results, qualitatively, with the results expected from the conventional beam bending theory and offer comments concerning the influence of the stiffness of the shear wall on the load-deformation behavior.

Figure 13.10(b) shows a finite element mesh for the shear wall; relatively finer mesh is used near the two horizontal beams. Figure 13.11(a) shows the computed deflected shape of the left end vertical side of the structure for the values of stiffnesses for the shear wall. Figure 13.11(b) shows the deflected shapes of the entire shear wall for the two conditions.

Figure 13.12 shows vertical (bending) stress along three horizontal sections A–A, B–B, and C–C (Figure 13.10(a)). We have assigned the stresses at the center of the element for sketching the variation shown in Figure 13.12.

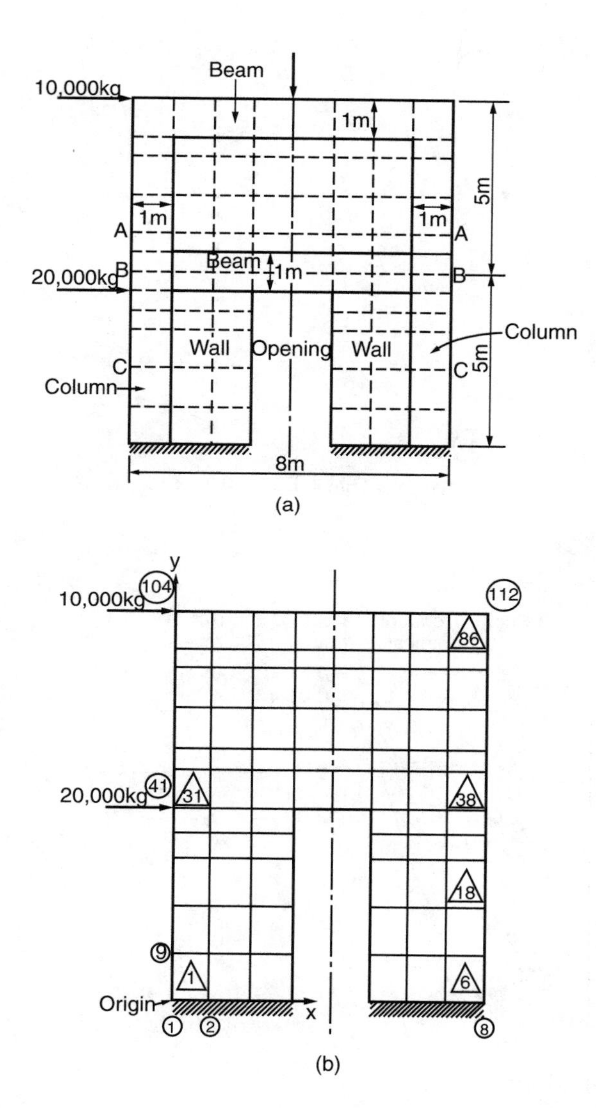

FIGURE 13.10
Analysis of shear wall: (a) Elevation of shear wall and (b) Finite element mesh. (*Note:* Scales in (a) and (b) are different.)

Comment

Conventionally, shear walls are often designed by assuming there to be thick beams idealized as one-dimensional. Here the usual assumptions of beam bending are considered to be valid (see Chapter 7). In view of the irregular geometry and the large area of the structure, such an assumption may not be valid and could give results that are different from reality.

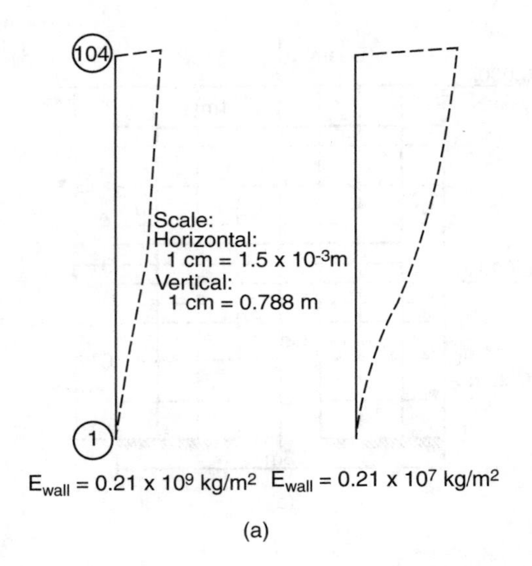

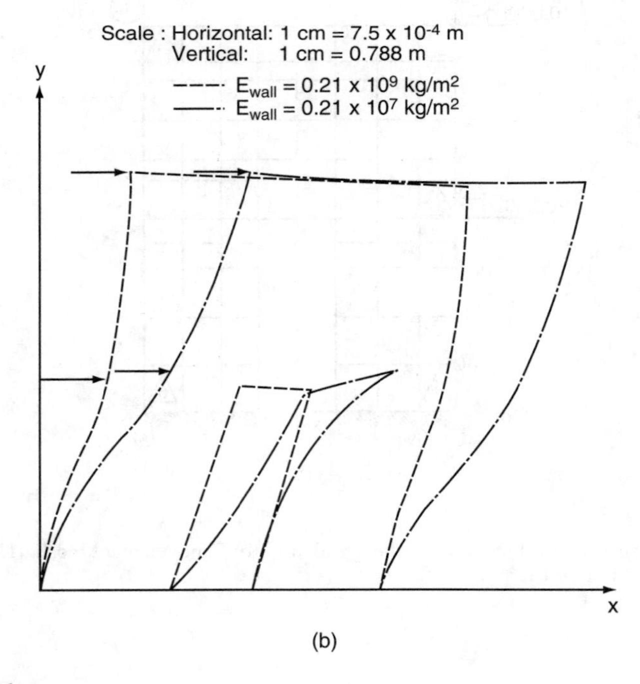

FIGURE 13.11
Computed deflections for shear wall: (a) Deflected shapes for left vertical side and (b) Deflected shapes of shear wall.

A two-dimensional finite element analysis permits inclusion of irregular geometry and the multidimensional aspects of the problem. It allows computations of displacements, stresses, and then bending moments in all elements;

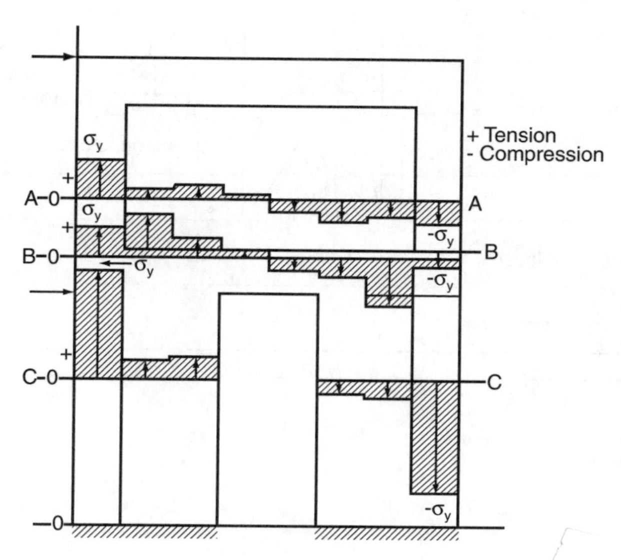

FIGURE 13.12
Computed distributions of σ_y at *A–A, B–B, C–C* (Figure 13.10(a)) for $E_{wall} = 0.21 \times 10^9$ kg/cm².

thus one can compute concentrations of stresses at critical locations and perform improved analysis and design as compared with the conventional procedures. For instance, Figure 13.12 shows that the distributions of bending stress σ_y at various sections are not linear as assumed in the conventional beam bending theory. The bending stresses show irregular and nonlinear behavior depending on the geometry, material properties, and loading conditions. Moreover, it is possible to perform parametric studies to include the influence of the stiffness of the wall on the behavior of the frame consisting of beams and columns. With a finer mesh, it is possible to identify the zones of stress concentrations, particularly near the corners.

Example 13.4. Analysis of Dam on Layered Foundation

Figure 13.13 shows a concrete dam resting on a foundation consisting of two layers; the bottom layer is underlain by a rock mass extending to considerable depth. There exists a crack at the crest of the gallery. By using PLANE-2D and assuming plane strain conditions, solve for displacements and stresses under the self-weight of the dam and foundation. The properties are given as follows:

Concrete in dam:

$$E = 432 \times 10^6 \text{ psf} \left(= 2.1 \times 10^9 \text{ kg/m}^2\right),$$

$$\nu = 0.3,$$

$$\gamma = 150 \text{ pcf} \left(= 2400 \text{ kg/m}^3\right).$$

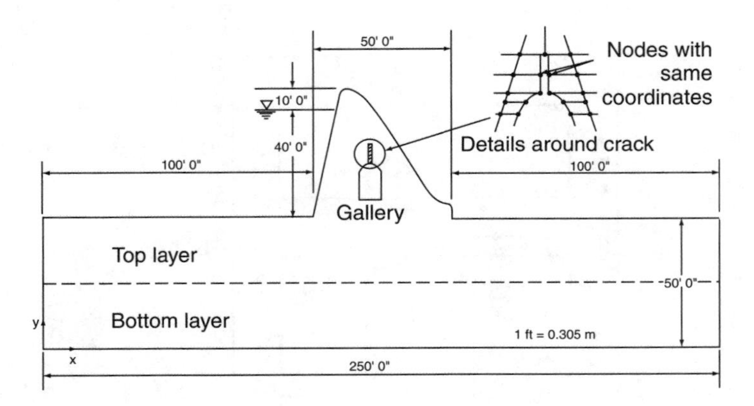

FIGURE 13.13
Dam on layered foundation.

Layer 1:

$$E = 144 \times 10^4 \text{ psf } \left(= 7.0 \times 10^6 \text{ kg/m}^2\right),$$

$$\nu = 0.4,$$

$$\gamma = 100 \text{ pcf } \left(= 1600 \text{ kg/m}^3\right).$$

Layer 2:

$$E = 144 \times 10^6 \text{ psf } \left(= 7.00 \times 10^8 \text{ kg/m}^2\right),$$

$$\nu = 0.3,$$

$$\gamma = 100 \text{ pcf } \left(= 1760 \text{ kg/m}^3\right).$$

Partial Results

Figure 13.14 shows a finite element discretization for the dam. In this figure, the nodes and elements are numbered in the x direction. It may be computationally economical to number them in the y direction, because according to Equation 13.39 below, this numbering will reduce the bandwidth B. At the same time, it will need a greater number of data cards. The mode of numbering will depend on the given problem and capabilities of the code and the computer. In general, however, selection of a numbering system that minimizes B may be a better strategy.

Discretization of infinite masses such as the foundation of the dam requires special treatment. Since only a finite zone can be included in the finite element mesh, we must include adequate extents in the mesh in the vertical and lateral

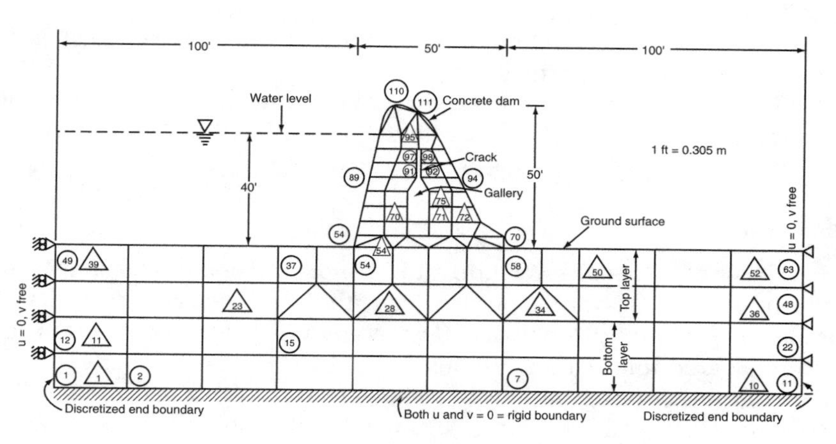

FIGURE 13.14
Finite element mesh for dam.

directions so that approximate boundary conditions can be defined. For instance, if in the mesh we include sufficiently large distances in the two directions, it can be assumed that the displacements at such large distances are negligible. Thus in Figure 13.14 we have included a lateral distance of about twice the width of the dam from the edge of the dam and have assumed that the horizontal displacement u at that distance is approximately zero. For the problem in Figure 13.13, a rigid rock boundary is available at a distance of 50 ft (15.25 m) below the base of the dam; hence, we have assumed both displacements to be zero. If such a rigid boundary is not available, we can go to a sufficiently large depth, say about three to five times the width of the dam, and assume either $v = 0$ and u free or both $u = v = 0$ at such a discretized boundary. For new problems, the analyst may need to perform a parametric study to decide the extents of discretized boundaries.

The crack at the crest of the gallery is introduced by providing node numbers on both sides of the crack. In Figure 13.15 is shown the distribution

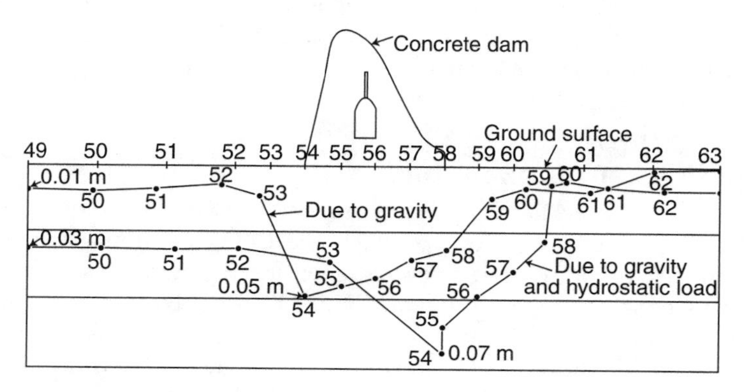

FIGURE 13.15
Vertical deflections at ground surface.

of vertical displacements along the ground level for gravity loads and for both gravity and hydrostatic loads.

Detailed analyses of stress concentrations around the crack and of such other aspects as the effect of hydrostatic forces are left to the reader.

Comment

On the basis of the results in Figures 13.14 and 13.15 and the results on stress distributions which can be plotted from the same computer results, it is possible to perform (preliminary) design analysis for the behavior of dams on multilayered foundation systems.

The computed concentrations of stresses near the crack can show us if the tensile (shear) stress in the concrete is within allowable limits. Similarly, we can check whether the compressive stress is within allowable compressive strength of the concrete. Similar conclusions can be drawn regarding the stresses in the foundation soils. Moreover, if the design limits allowable deformations, it may be possible to check if the loading causes deformations within allowable limits.

In this example, we have considered essentially only loading caused by gravity and have assumed linear material behavior. Hence, the results can be treated as preliminary. For a final design one needs to consider other loadings such as hydrostatic load due to water in the reservior, earthquake loading, uplift due to seepage through foundation, and loading caused by variations in temperature difference between the dam and the surrounding atmosphere. Nonlinear behavior of the soil and rock foundations will influence the behavior of the dam. Finite element procedures that can allow inclusion of these factors are available and will be stated in advanced study of the method and its applications.

For the present, we note that it is possible to perform a detailed stress-deformation analysis of dams on nonlinear foundations; the latter can include both soils and jointed rock masses. In addition to computations of stresses and deformations for various loading conditions, the method can also be used for performing parametric studies. For instance, it is possible to find optimum locations for instrumentation and for (underground) structures such as caverns and galleries.

Example 13.5. Beam Bending

As discussed in the earlier chapters, a numerical procedure should generally yield results that converge or approach the exact solution as the mesh is refined consistently.[3] To study convergence for the two-dimensional problems, let us consider a (deep) beam (Figure 13.16) subjected to a uniformly distributed load of 1000 kg/m. The beam is divided into three progressively refined meshes as shown in Figure 13.16. Note that a refined mesh includes the previous coarse mesh; this is required for mathematical convergence. This and other requirements for mesh refinement are discussed in Reference 3.

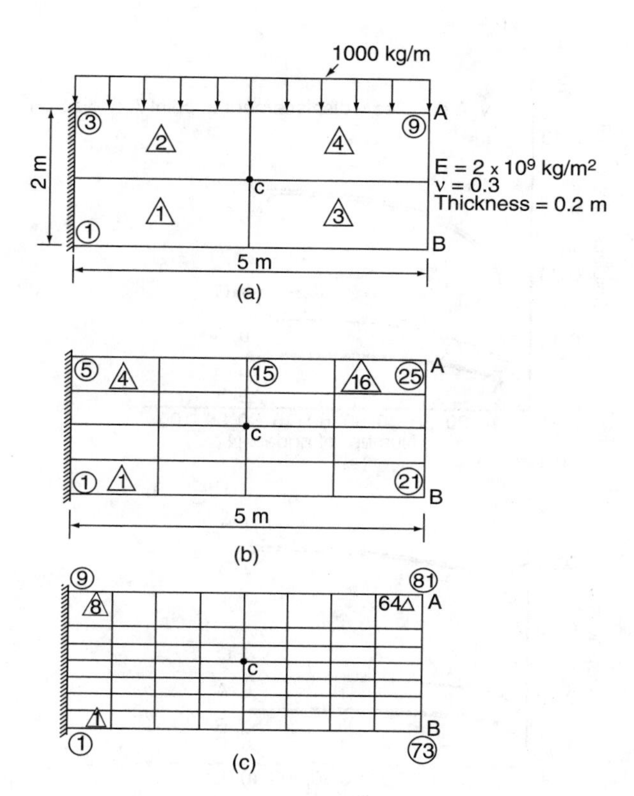

FIGURE 13.16
Beam bending: (a) 4-element mesh, (b) 16-element mesh, and (c) 81-element mesh.

TABLE 13.1
Results for the Uniform Loading

				Displacement × 10⁻³ (m)					
				Point A		Point B		Point C	
N	D	$B = (D + 1) f$	NB^2	u	v	u	v	u	v
9	4	10	900	0.0478	–0.1948	–0.0448	–0.1924	–0.00045	–0.0766
25	6	14	4,900	0.0678	–0.2821	–0.0645	–0.2796	–0.00073	–0.1119
81	10	22	39,204	0.0761	–0.3198	–0.0726	–0.3173	0.00079	–0.1278

Table 13.1 shows computed values of vertical and horizontal displacements at points A, B, and C (Figure 13.16). The number of nodes N, the maximum difference between any two node numbers D, the semibandwidth B, and the quantity NB^2 are also shown in the table. The semibandwidth B is computed from the formula

$$B = (D + 1) f, \tag{13.39}$$

where f is the number of degrees of freedom at a node; here $f = 2$.

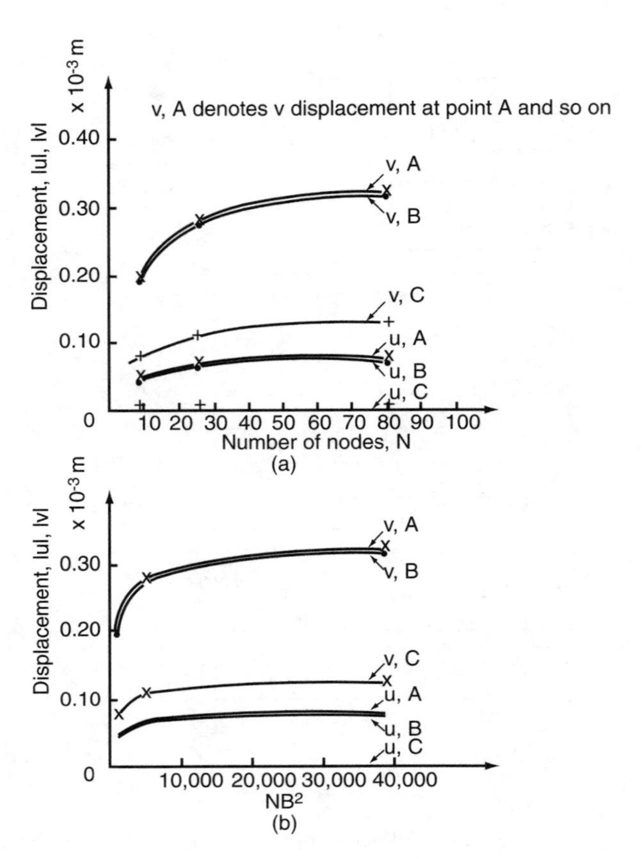

FIGURE 13.17
Behavior of numerical solutions for displacements: (a) Displacement vs. number of nodes: points *A*, *B*, and *C* (Figure 13.16) and (b) Displacements vs. NB^2: points *A*, *B*, and *C* (Figure 13.16).

Figure 13.17(a) shows convergence behavior of computed displacements u, v at points *A*, *B*, and *C* with the number of nodes, and Figure 13.17(b) shows their relationship with NB^2. In both cases the computed displacements tend to converge as N and NB^2 increase.

The quantity NB^2 is proportional to the time required for solution of the global equations, which constitutes a major portion of the time for the finite element solution. Figure 13.17(b) shows that refinement of mesh results in a faster increase in NB^2, that is, computational effort or cost. In other words, for a given desired accuracy, if the number of nodes are increased, the quantity NB^2 increases at a higher rate than the gain in accuracy. Thus there exists a trade-off between accuracy and computer effort or cost of a finite element solution.[4]

Figures 13.18(a), (b), and (c) show distributions of the bending stress σ_x for the three meshes. We can see that the distribution of bending stress improves with mesh refinement.

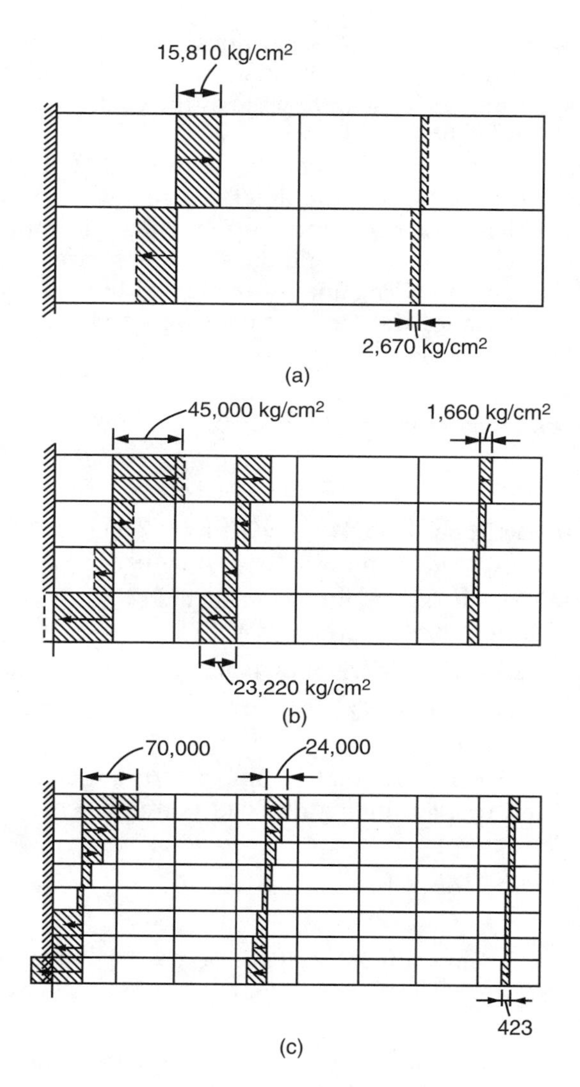

FIGURE 13.18
Distributions of bending stress, σ_x, at typical sections.

Comment

The results of this example illustrate that, in general, we can improve a finite element solution by progressive mesh refinement. Obviously, as the mesh is refined, the number of elements increase with corresponding increases in the effort for data preparation and computer time.

As noted in previous chapters, a finite element solution can also be improved by using higher-order elements. For instance, for the plane problems, we can use quadratic or second-order approximation for displacement within the element. This will require additional nodes; for the triangle there will be six nodes, and for the quadrilateral there will be eight nodes. There will be a corresponding increase in the size of the element matrices and the number of equations to be solved.

Which of the two approaches, refinement of mesh with a lower-order approximation or higher-order approximation, should be used will depend on the type of problem. There are trade-offs in using one approach over the other. This subject is wide in scope and usually needs parametric studies for a specific problem on hand. The analyst may need to use both approaches and then derive criteria for their use for the given problem.

Problems

13.1. Compute the initial load vector $\{\mathbf{Q}_0\}$ for a given initial strain $\{\varepsilon_0\}$ for the triangular element number 1 in Figure 13.9.

13.2. Compute $[\mathbf{A}]^{-1}$ in Equation 13.14c and then find the interpolation functions N_i in matrix $[\mathbf{N}]$ (Equation 13.15).

13.3. Fill in the details of the derivation of ε_x, ε_y, and γ_{xy} in Equation 13.18.

13.4. Obtain the load vector $\{\mathbf{Q}_2\}$ for a traction load $\overline{T}_y$ applied on side 4–1 (Figure 13.6(b)).

13.5. Evaluate the stiffness matrix for the triangular element in Example 13.2 by rearranging the load vector as

$$\{\mathbf{q}\}^T = [u_1 \quad v_1 \quad u_2 \quad v_2 \quad u_3 \quad v_3].$$

13.6. Find the element stiffness matrix for the triangular element in Figure 13.9 for plane strain conditions. Hint: Use $[\mathbf{C}]$ from Equation 13.5.

13.7. Evaluate the element stiffness matrix for the triangular element in Figure 13.9 for axisymmetric idealization.

13.8. By using PLANE-2D or another available program, solve the shear wall problem (Figure 13.10) with an additional vertical load of 10,000 kg at the centerline of the wall. Plot the results in terms of displacements and stresses and compare with those without the vertical load.

13.9. In Example 13.4, consider hydrostatic force caused by a water level of 40 ft (12 m) above the ground surface, in addition to the load due to the weight of the dam. Use PLANE-2D and obtain distributions of displacement along the ground surface and plot σ_x, σ_y, and τ_{xy} within the dam. Compare results with those from the anal-

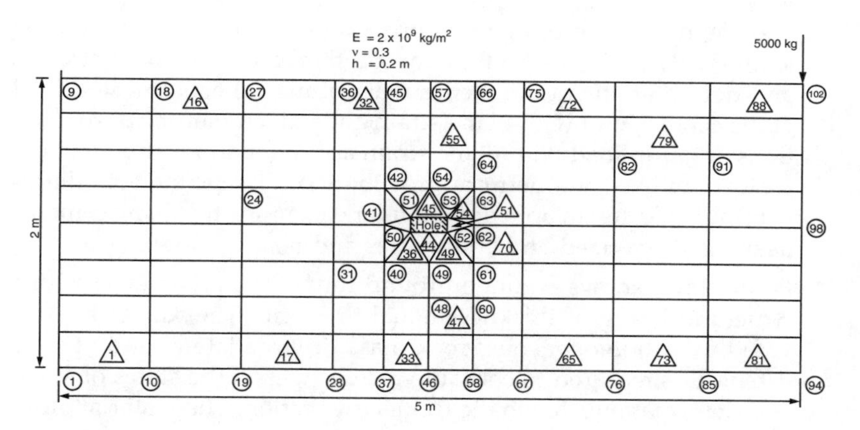

FIGURE 13.19
Problem of stress concentration around hole in beam.

yses without the hydrostatic force. Figure 13.15 shows the deflected shape of the ground surface with hydrostatic loading.

13.10. Vary the relative values of E in the two foundation layers in Figure 13.13 and study their influence on displacements and stresses.

13.11. For the beam in Figure 13.16, consider a concentrated vertical load of 5000 v at point A, and study convergence of displacements vs. N and NB^2.

13.12. Figure 13.19 shows a finite element mesh for the beam in Example 13.5, with a hole around the centerline. For the concentrated load shown, obtain the finite element solution and study the concentration of stresses around (the corners of) the hole.

13.13. Figure 13.20 shows a medium subjected to external loading. The medium is of infinite extent. Assuming discretized boundaries ($u = v = 0$) at distances of four times the width of the loading in the lateral and vertical directions, decide on a suitable mesh. Choose

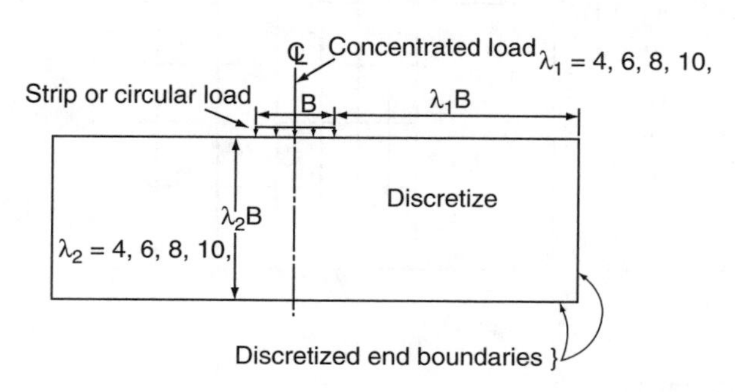

FIGURE 13.20
Load-deformation analysis for infinite medium.

your own value of B, material properties of the medium, and the loading. By using PLANE-2D or another code, evaluate stresses and deformations in the medium assuming (a) concentrated load at the center point with plane strain and axisymmetric approximations, (b) strip load with plane strain approximation, and (c) circular load with axisymmetric approximation. Change the boundaries to 6B, 8B, and 10B in both directions and examine the displacements near the discretized end boundaries and near the loading.

13.14. Figure 13.21 shows an underground tunnel (3-m diameter) with a structural lining of thickness equal to 30 cm. The tunnel is excavated in a (homogeneous) rock mass of large lateral and vertical extent. It is required to compute the changes in the states of stress and deformations due to the tunnel excavation. The material properties are shown in the figure. By using the mesh shown in the figure (or any other suitable mesh) and a computer code, compute stresses and deformations (a) before the tunnel is excavated, that is, under the gravity loading. Here use the entire mesh including that in the zone of the tunnel and lining. (b) Consider the elements in the tunnel zone to be removed, and include the elements for the

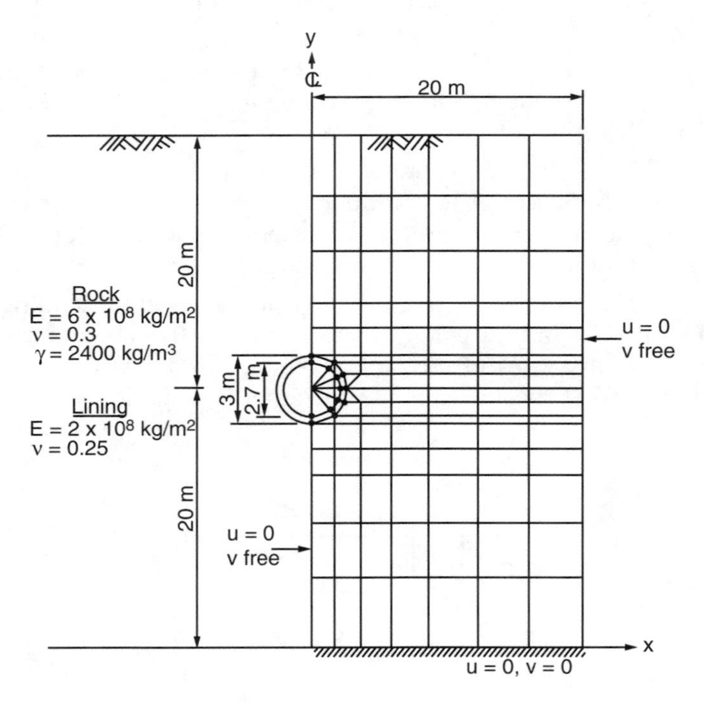

FIGURE 13.21
Analysis of underground tunnel.

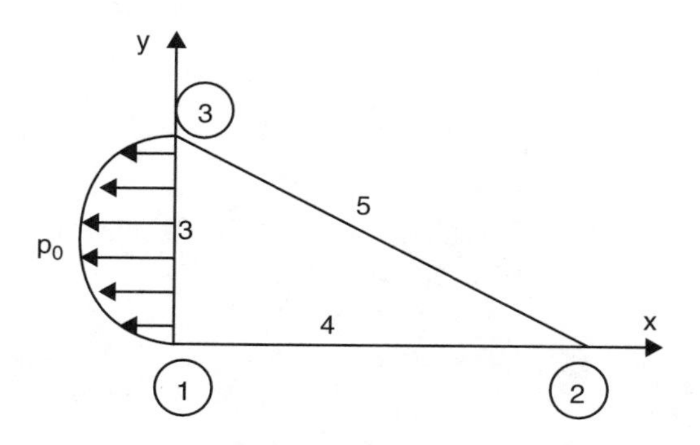

FIGURE 13.22

lining. It may be necessary to renumber the nodes and elements. Compare the two analyses and draw conclusions regarding the influence of tunnel excavation on the states of stresses and deformations, particularly in the vicinity of the tunnel.

13.15. A triangular element is subjected to a parabolic surface traction (maximum value p_0 per unit length) along the side 1–3 as shown in Figure 13.22. Lengths of the 3 sides are 3, 4, and 5, respectively.

The load vector for this element can be given by $\{Q\} = A_1 \int_{b_1}^{c_1} [N]_1^T \{T\}_1 \, dL_1$ or $\{Q\} = A_1 \int_{b_3}^{c_3} [N]_3^T \{T\}_3 dL_3$.

(a) Give values of A_1, b_1, c_1, A_3, b_3, and c_3.

(b) Give $[N]_1$ and $\{T\}_1$ in terms of the integration variable L_1.

(c) Give $[N]_3$ and $\{T\}_3$ in terms of the integration variable L_3.

(d) Compute {Q} using both equations, shown above, and see if you obtain identical results.

13.16. Consider the six-noded triangular element shown in Figure 13.23. Nodes 4, 5, and 6 are at the center points of the 3 sides.

(a) Give local area coordinates L_1, L_2, L_3 for all six nodes.

(b) Write down the general expression of the interpolation function N_i in terms of some unknown constants a_i, b_i, c_i etc. and the local coordinates L_1 and L_2, in the following form,

$$N_i = a_i + b_i L_1 + c_i L_2 + \ldots\ldots$$

Do not try to evaluate the constants a_i, b_i, c_i, etc.

(c) Which interpolation functions should be zero along the side AB?

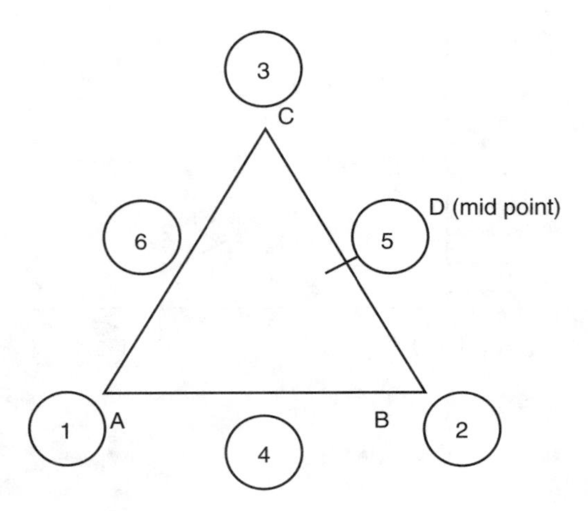

FIGURE 13.23

(d) Obtain N_1 in terms of local coordinates along the sides (i) AB and (ii) AC.

(e) How should the displacements (u,v) and strains (ε_x, ε_y, γ_{xy}) vary (constant, linear, quadratic, cubic, etc.) along lines (i) AB and (ii) AD?

13.17. A 4-node quadrilateral element is subjected to a linearly varying surface traction along the boundary 2–3 as shown in Figure 13.24. The surface traction varies from 0 to 10 units per unit length. Calculate the load vector for this quadrilateral element when it is used for the two-dimensional stress-deformation analysis.

13.18. A quadrilateral element bounded by 4 straight lines y = 2x, 2y = x, x + y = 4 and x + y = 8. It is subjected to surface tractions along its two sides 1–2 and 2–3, as shown in Figure 13.25. Surface traction along 1–2 is constant at 1 unit per unit length. The surface traction along side 2–3 varies linearly from 1 to 2 units per unit length. There is no body force acting in the element. Calculate the load vector {Q} for this element.

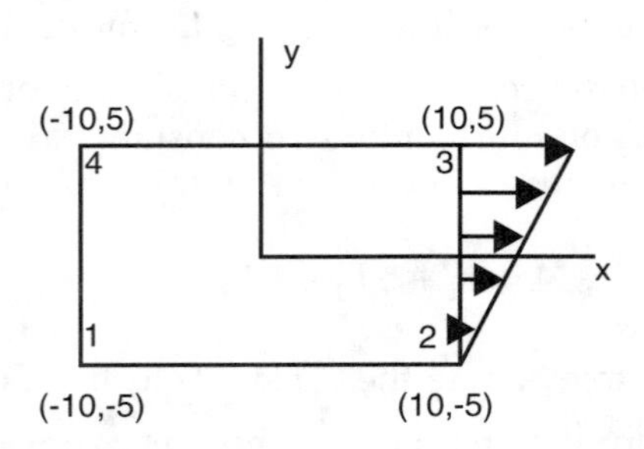

FIGURE 13.24

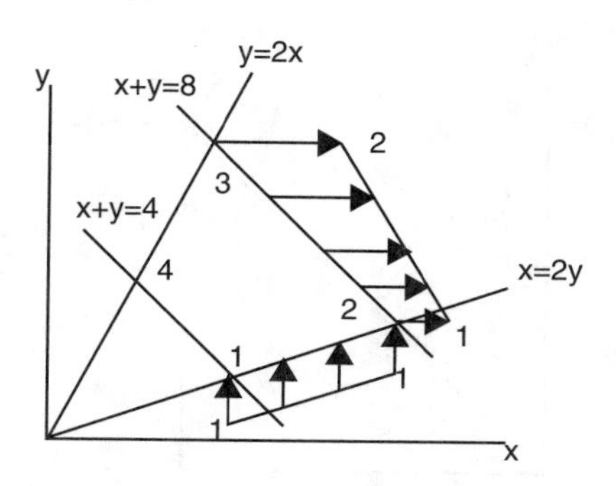

FIGURE 13.25

13.19. Consider the above problem geometry as described in Problem 13.18. However, now discretize the rectangular region into 2 triangular elements Δ123 and Δ134. For both these elements compute the load vectors $\{Q\}_1$ and $\{Q\}_2$ for the above loading. For both triangular elements consider point A as node 1. Assemble the two element load vectors to obtain the global load vector for the rectangular region.

13.20. Consider the problem geometry described in Problem 13.18. However, now discretize the rectangular region into 2 triangular elements Δ124 and Δ234. For both these elements compute the load vectors $\{Q\}_1$ and $\{Q\}_2$ for the above loading. Assemble the two element load vectors to obtain the global load vector for the rectangular region.

13.21. Discretize the 30 in. × 20 in. rectangular plate into 2 triangular elements as shown in Figure 13.26. Corner points A and B have pin supports (these points cannot move horizontally or vertically), point C has a roller support (C cannot move vertically but it is free to move horizontally), and point D is free. Point loads 20 kips and 10 kips act at points C and D as shown in the figure. Calculate the

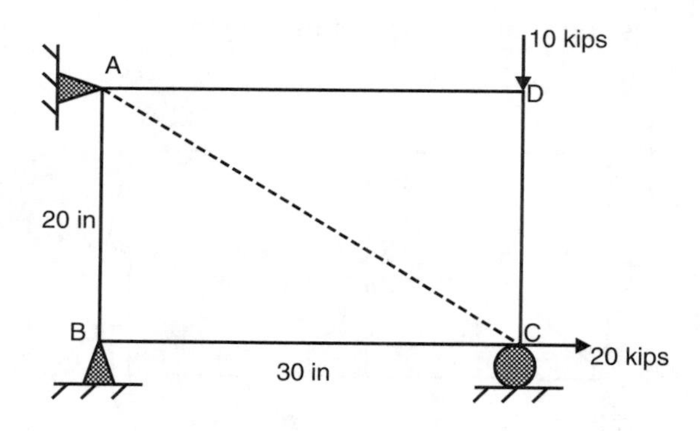

FIGURE 13.26

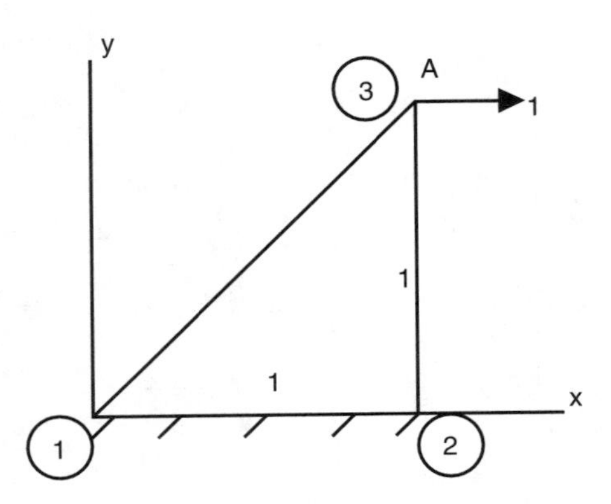

FIGURE 13.27

displacements at four corner points of the plate. Draw schematically the deformed shape of the plate. Calculate the stress variation in the plate. Young's modulus $E = 30 \times 10^6$ psi, Poisson's ratio $\nu = 0.3$, and thickness $t = 1$ in.

13.22. A triangular plate is fixed along its bottom face and is loaded by the unit load as shown in Figure 13.27. Considering the plate as a single CST element find the displacement components at point A. Adopt Young's modulus $E = 1000$, Poisson's ratio $\nu = 0$, and thickness $h = 0.04$.

13.23. During a stress analysis of a rectangular plate the plate is discretized into two triangular elements and one quadrilateral element as shown in Figure 13.28. The plate is pin supported at corners A and B.

(a) Identify true and false statements (u, ε_{xx} and ε_{yy} are computed displacement and strain components).

(i) u is continuous across line AD.

(ii) u is continuous across line CD.

(iii) u is continuous in the entire domain of the plate.

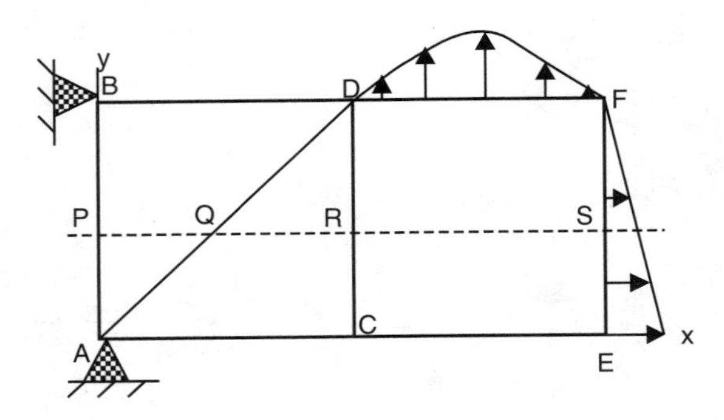

FIGURE 13.28

(iv) ε_{xx} is continuous across line AD.

(v) ε_{yy} is continuous across line CD.

(vi) Above discretization is not recommended because the interelement conformability condition is violated across CD.

(vii) If three-node triangular elements are replaced by six-noded triangular elements (with one middle node on every side of the triangle) keeping the four-node quadrilateral element unchanged then the interelement conformability condition will be violated across line CD.

(b) Plot schematically the variation of ε_{xx} along the line PQRS in the following figure. In your plot indicate in which regions ε_{xx} remains constant and where it varies linearly or quadratically, where it is continuous and, where you expect a jump.

(c) How would ε_{yy} vary (zero, constant, linear, quadratic, etc.) along line PQ and QR?

(d) Should ε_{yy} remain constant or vary along line RS?

(e) What will be the size of the global stiffness matrix for this problem after you incorporate all boundary conditions?

(f) In the final global load vector (after boundary conditions have been properly incorporated) how many entries will be zero and how many should be nonzero?

13.24. A triangular plate is analyzed using one triangular element as shown in Figure 13.29. For this plate Young's modulus E = 1000, Poisson's ratio ν = 0.3, and thickness h = 0.1.

(a) Obtain the [B] matrix.

(b) Obtain the load vector $\{\bar{Q}\}$ after boundary conditions have been properly incorporated.

(c) After you solve the finite element equation what nodal degrees of freedom should have nonzero values? (Note that you do not need to solve the global equation set to answer part [c].)

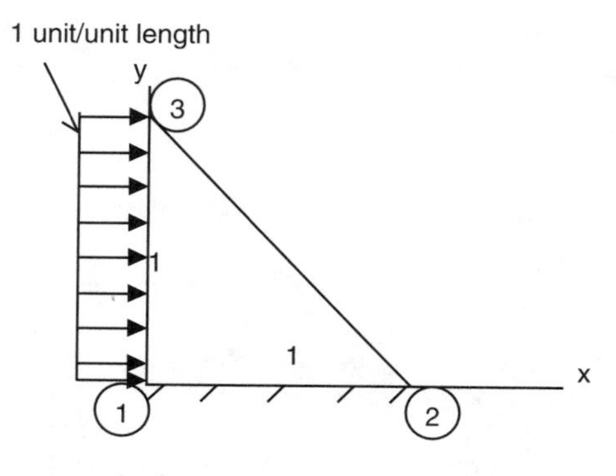

FIGURE 13.29

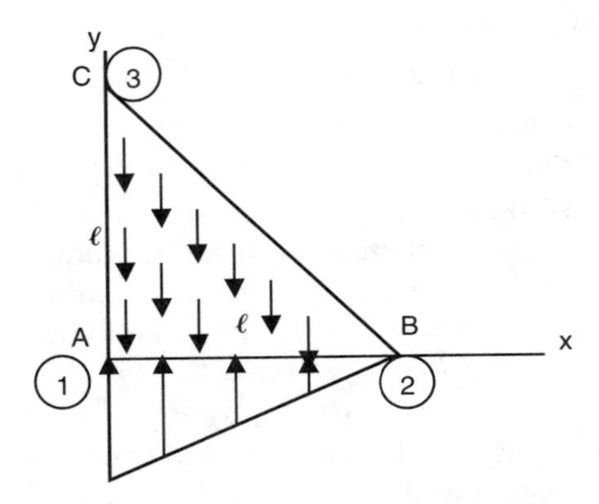

FIGURE 13.30

13.25. A triangular plate ABC of unit thickness and boundary lengths l, l and $1.414l$ ($l \gg 1$) is shown in the x–y coordinate system (Figure 13.30). The plate is resting on the smooth surface AB and is in equilibrium. It is analyzed by using one element. Points D and E are midpoints of BC and AD, respectively.

(a) Give local area coordinates of D and E.

(b) The plate is subjected to a constant body force that is the gravitational load due to its own weight W, and a linearly varying surface traction load on its boundary 1–2. The surface traction is generated at the smooth contact surface 1–2 to keep the plate in equilibrium. Compute the load vector for the element considering the effects of both the body force and the surface traction.

(c) If you now incorporate the boundary conditions that points A and B are fixed would you obtain a downward or upward displacement of E?

(d) In addition to the boundary AB if you also fix the boundary AC would you obtain a downward or upward displacement of E when you analyze the problem by one element?

References

1. Popov, E. P. 1968. *Introduction to Mechanics of Solids*, Prentice-Hall, Englewood Cliffs, NJ.
2. Timoshenko, S. and Goodier, J. N. 1951. *Theory of Elasticity*, McGraw-Hill, New York.
3. Desai, C. S. and Abel, J. F. 1972. *Introduction to the Finite Element Method*, Van Nostrand Reinhold, New York.
4. Abel, J. F. and Desai, C. S. Sept. 1972. "Comparison of finite elements for plate bending," *Proc. ASCE, J. Struct. Div.* 98(ST9), 2143-2148.

14

Multicomponent Systems: Building Frame and Foundation

Introduction

Very often, in practice, configuration of a system or structure is such that its approximate simulation may require the use of or it may be beneficial to use more than one type of idealization. For instance, if we need to idealize a three-dimensional building frame and its foundation (Figure 14.1), it is convenient and economical to treat the building frame as composed of one-dimensional beams and columns and two-dimensional slabs and plates. As a rather crude approximation, the foundation can be included by assuming the structure resting on a bed of (individual) springs representing the foundation (Figure 14.1); this approach is often referred to as a Winkler foundation. Thus, the system contains three components: beam-columns, slabs or plates, and foundation.[1]

Many other situations in stress-deformation analyses and field problems require such multicomponent idealizations. We shall illustrate here the problem of a building frame (Figure 14.1). For simplicity we shall consider an orthogonal frame with only horizontal and vertical members. Extension to the general case of inclined members can be achieved by appropriate transformations.[1-3]

Various Components

As stated earlier, we need to consider structural components that act as beam-columns, slabs and plates, and foundation (springs).

Beam-Column

In Chapter 7 we have already derived the element equations for a beam-column. However, now we need to consider the possibility of loads causing

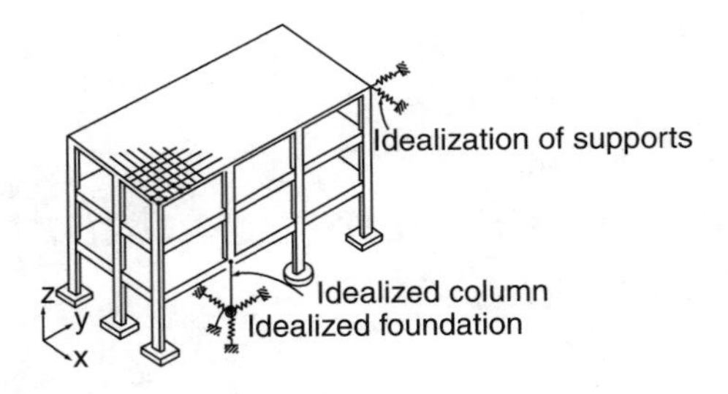

FIGURE 14.1
Building frame: multicomponent system.

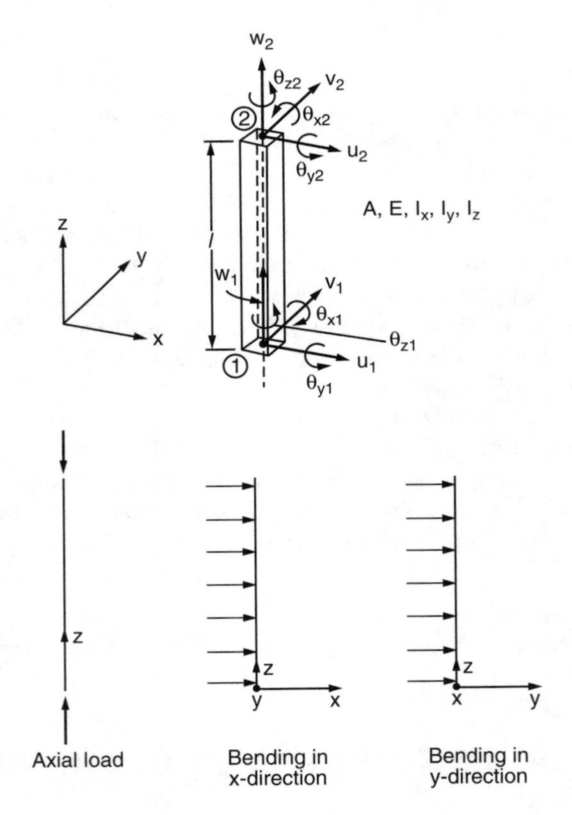

FIGURE 14.2
General beam-column element.

bending in both x and y directions. Figure 14.2 shows the beam-column idealized as one-dimensional.

The degrees of freedom for the element consist of axial displacement w, lateral displacements u and v, and rotations θ_x, θ_y, and θ_z, about the three axes, respectively. Here the subscript denotes derivative; for instance, $\theta_x = \partial w/\partial x$. Hence, the beam-column element has 6 degrees of freedom at each node and a total of 12 degrees of freedom. For the time being, if we do not consider rotation or twist around the z axis, we have a total of 10 degrees of freedom. The nodal displacement vector is then given by

$$\{\mathbf{q}\}^T = \left[u_1 \quad \theta_{x1} \quad u_2 \quad \theta_{x2} \quad v_1 \quad \theta_{y1} \quad v_2 \quad \theta_{y2} \mid w_1 \quad w_2 \right], \tag{14.1}$$

where $\theta_{x1} = (\partial w/\partial x)_1$, $\theta_{y1} = (\partial w/\partial y)_1$, and so on.

As in Chapter 7, we assume the following approximation functions:

$$u(z) = \alpha_1' + \alpha_2 z + \alpha_3 z^2 + \alpha_4 z^3, \tag{14.2a}$$

$$v(z) = \alpha_5 + \alpha_6 z + \alpha_7 z^2 + \alpha_8 z^3, \tag{14.2b}$$

$$w(z) = \alpha_9 + \alpha_{10} z. \tag{14.2c}$$

Here u(z) and v(z) correspond to bending displacements in the x and y directions, respectively, and w(z) denotes shortening or extension due to axial (end) loads. As discussed in Chapter 7, Equation 14.2 can be transformed to express u(z), v (z), and w(z) in terms of interpolation functions and their nodal values as (Equation 7.37)

$$u(z) = N_{x1} u_1 + N_{x2} \theta_{x1} + N_{x3} u_2 + N_{x4} \theta_{x2}, \tag{14.3a}$$

$$v(z) = N_{y1} v_1 + N_{y2} \theta_{y1} + N_{y3} v_2 + N_{y4} \theta_{y2}, \tag{14.3b}$$

$$w(z) = N_1 w_1 + N_2 w_2, \tag{14.3c}$$

where

$$\begin{aligned} N_{x1} &= N_{y1} = 1 - 3s^2 + 2s^3, \\ N_{x2} &= N_{y2} = ls(1 - 2s + s^2), \\ N_{x3} &= N_{y3} - s^2(3 - 2s), \\ N_{x4} &= N_{y4} = ls^2(s - 1), \end{aligned} \tag{14.3d}$$

and

$$N_1 = 1 - s,$$
$$N_2 = s, \tag{14.3e}$$

where s is the local coordinate given by

$$s = \frac{\bar{z}}{l}, \tag{14.4}$$

where $\bar{z} = z - z_1$ and z_1 is the coordinate of node 1.

By following the procedure outlined in Chapter 7, we can derive the following element equations:

$$[\mathbf{k}]\{\mathbf{q}\} = \{\mathbf{Q}\} = \begin{Bmatrix} \{\mathbf{Q}_b\} \\ \{\mathbf{Q}_\alpha\} \end{Bmatrix}, \tag{14.5a}$$

where

$$[\mathbf{k}] = \begin{bmatrix} \alpha_x[\mathbf{k}_x] & [0] & [0] \\ [0] & \alpha_y[\mathbf{k}_y] & [0] \\ [0] & [0] & [\mathbf{k}_w] \end{bmatrix}, \tag{14.5b}$$

$$[\mathbf{k}_x] = [\mathbf{k}_y] = \begin{bmatrix} 12 & 6l & -12 & 6l \\ & 4l^2 & -6l & 2l^2 \\ & \text{sym.} & 12 & -6l \\ & & & 4l^2 \end{bmatrix}, \tag{14.5c}$$

$$\alpha_x = \frac{EI_y}{l^3}, \quad \alpha_y = \frac{EI_x}{l^3}, \tag{14.5d}$$

$$[\mathbf{k}_w] = \frac{AE}{l}\begin{bmatrix} 1 & -1 \\ -1 & 1 \end{bmatrix}, \tag{14.5e}$$

$$\{\mathbf{Q}\}^T = A\int_0^l [\mathbf{N}]^T\{\overline{\mathbf{X}}\}dz + \int_0^l [\mathbf{N}]^T\{\overline{\mathbf{T}}\}dz, \tag{14.6}$$

and the subscripts a and b denote axial and bending modes, respectively.

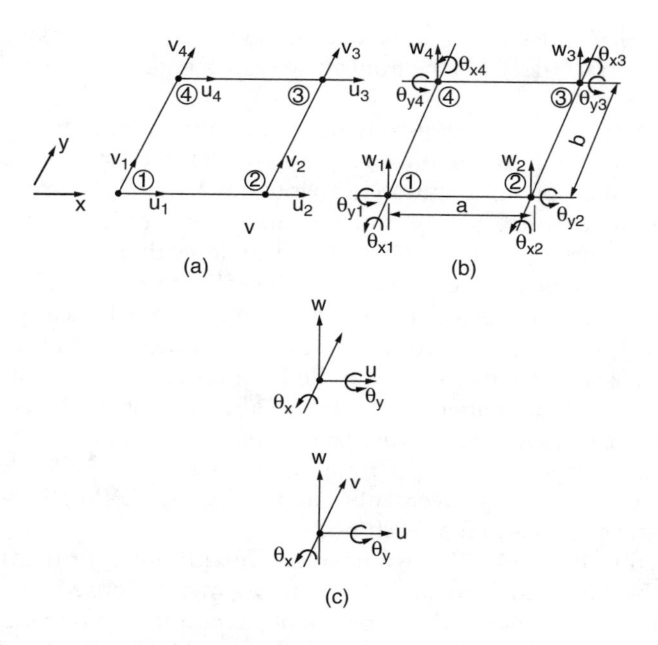

FIGURE 14.3
Plate elements and interelement compatibility: (a) Membrane or in-plane behavior, (b) Bending behavior, and (c) Compatibility between plate and beam-column.

Plate or Slab

The general loading conditions on the frame can cause plane deformations, bending, and twisting; for convenience we consider only the first two. Figure 14.3 shows the two effects, which can be superimposed if we assume small strains and deformations.

Membrane Effects

For the in-plane or membrane loading we can assume the plane stress idealization and use the element equations derived in Chapter 13, Equations 13.24 and 13.31 for the quadrilateral and triangular elements, respectively.

Bending

For isotropic flat plates, the governing differential equation is written as:

$$D\left(\frac{\partial^4 w^*}{\partial x^4} + \frac{2\partial^4 w^*}{\partial x^2 \partial y^2} + \frac{\partial^4 w^*}{\partial y^4}\right) = p \tag{14.7}$$

where $D = Eh^3/12(1 - \nu^2)$, E is the elastic modulus, h is the plate thickness, ν is the Poisson's ratio, w^* is the transverse displacement, and p is the applied surface traction.

Our aim herein is to show essentially the treatment of a multicomponent system for a building frame in which plates or slabs occur as one of the main components. It is not our intention to go into much detail of the plate bending problem. It will suffice to say that the finite element procedure for plate bending is essentially similar to the other problems that we have previously considered. Here we shall describe only the salient features of the problem.

For convenience, we consider only orthogonal frames; hence, we have only orthogonal (horizontal or vertical) plates. For a general formulation, plates with other configurations can be handled without much difficulty.

For the rectangular element in Figure 14.3(b) we consider three degrees of freedom at each node, namely the transverse displacement w; the rotation about y axis, $\theta_x = \partial w/\partial x$; and the rotation about x axis, $\theta_y = \partial w/\partial y$. Together with the two in-plane displacements u and v (Figure 14.3(a)), there are a total of five degrees of freedom at each node.

As shown in Figure 14.3(c), we have an equal number of degrees of freedom for the beam-column and the plate element. Consequently, we can observe the requirement of interelement compatibility between adjacent plate and beam-column elements.

The transverse displacement w can be approximated by using hermite interpolation functions as in the case of the beam bending problem (Chapter 7 and Equation 14.3d). In fact, the interpolation functions for the two-dimensional problem can be generated by proper multiplication of the hermitian functions defined for the x and y directions. For example, we can adopt an approximation for w as[4,5]

$$\begin{aligned} w(x,y) &= N_{x1}N_{y1}w_1 + N_{x2}N_{y1}\theta_{x1} + N_{x1}N_{y2}\theta_{y1} \\ &\quad + N_{x3}N_{y1}w_2 + N_{x4}N_{y1}\theta_{x2} + N_{x3}N_{y2}\theta_{y2} \\ &\quad + N_{x3}N_{y3}w_3 + N_{x4}N_{y3}\theta_{x3} + N_{x3}N_{y4}\theta_{y3} \\ &\quad + N_{x1}N_{y3}w_4 + N_{x2}N_{y3}\theta_{x4} + N_{x1}N_{y4}\theta_{y4} \\ &= [N_1 \;\; N_2 \;\; N_3 \;\; \cdots \;\; N_{12}]\{\mathbf{q}_b\} \\ &= [\mathbf{N}_b]\{\mathbf{q}_b\}, \end{aligned} \tag{14.8a}$$

where $[\mathbf{N}_b]$ is the matrix of interpolation functions, $N_1 = N_{x1}\,N_{y1}$, $N_2 = N_{x3}\,N_{y1}$, etc., where

$$\begin{aligned} N_{x1} &= 1 - 3s^2 + 2s^3, & N_{y1} &= 1 - 3t^2 + 2t^3, \\ N_{x2} &= as(s-1)^2, & N_{y2} &= bt(t-1)^2, \end{aligned} \tag{14.8b}$$

$$N_{x3} = s^2(3-2s), \qquad N_{y3} = t^2(3-2t),$$
$$N_{x4} = as^2(s-1), \qquad N_{y4} = bt^2(t-1).$$

Here $(a \times b)$ denotes the size of the element (Figure 14.3(b)), $s = x/a$, $0 \leq s \leq 1$, and $t = y/b$, $0 \leq t \leq 1$, and

$$\{\mathbf{q}_b\}^T = [w_1 \ \ \theta_{x1} \ \ \theta_{y1} \ \ w_2 \ \ \theta_{x2} \ \ \theta_{y2} \ \ w_3 \ \ \theta_{x3} \ \ \theta_{y3} \ \ w_4 \ \ \theta_{x4} \ \ \theta_{y4}]$$

is the vector of nodal unknowns.

The interpolation functions N_1, N_2, etc. satisfy the definition of interpolation functions. For instance, we have

$$N_1 = N_{x1}N_{y1} = (1-3s^2+2s^3)(1-3t^2+2t^3) \tag{14.9a}$$

for

$$\begin{aligned} s=0,\ t=0, &\quad N_1 = 1,\\ s=1,\ t=0, &\quad N_1 = 0,\\ s=1,\ t=1, &\quad N_1 = 0,\\ s=0,\ t=1, &\quad N_1 = 0,\\ s=\tfrac{1}{2},\ t=\tfrac{1}{2}, &\quad N_1 = \tfrac{1}{4},\\ s=\tfrac{1}{2},\ t=0, &\quad N_1 = \tfrac{1}{2}, \end{aligned}$$

and so on. Now $N_2 = N_{x2}\,N_{y1} = as(s-1)^2(1-3t^2+2t^3)$. The first derivative of N_2 with respect to x is given by[4]

$$\frac{\partial N_2}{\partial x} = \frac{a}{a}(3s^2-4s+1)(1-3t^2+2t^3). \tag{14.9b}$$

Then for

	N_2	$\partial N_2/\partial x$ (rad)
$s=0,\ t=0,$	0.0	1.0
$s=1,\ t=0,$	0.0	0.0
$s=1,\ t=1,$	0.0	0.0
$s=0,\ t=1,$	0.0	0.0

	N_2	$\partial N_2/\partial x$ (rad)
$s=\frac{1}{4}, t=0,$	$9a/64$	$\frac{3}{16}$
$s=\frac{1}{2}, t=0,$	$8a/64$	$-\frac{1}{4}$
$s=\frac{3}{4}, t=0,$	$3a/64$	$-\frac{5}{16}$
$s=0, t=\frac{1}{2},$	0.0	$\frac{1}{2}$
$s=\frac{1}{2}, t=\frac{1}{2},$	$4a/64$	$-\frac{1}{8}.$

Plots of N_1, N_2, and $\partial N_2/\partial x$ are shown in Figures 14.4(a), (b), and (c). In the case of N_2, the slope corresponding to the degree of freedom θ_{x1} at node 1 is unity. It can thus be shown that all other functions in Equation 14.8a satisfy the definition of interpolation functions.

The strain (gradient)-displacement relation for plate bending is given by[6]

$$\{\varepsilon\} = -z\begin{Bmatrix} \dfrac{\partial^2 w}{\partial x^2} \\ \dfrac{\partial^2 w}{\partial y^2} \\ \dfrac{\partial^2 w}{\partial x \partial y} \end{Bmatrix} = \begin{Bmatrix} w_{xx} \\ w_{yy} \\ 2w_{xy} \end{Bmatrix} = -z[\mathbf{B}_b]\{\mathbf{q}_b\}. \tag{14.10}$$

The strain-displacement transformation matrix is obtained by finding the required derivatives of w from Equation 14.8.

The constitutive law is expressed through the relation between the moments and second derivatives or curvatures:

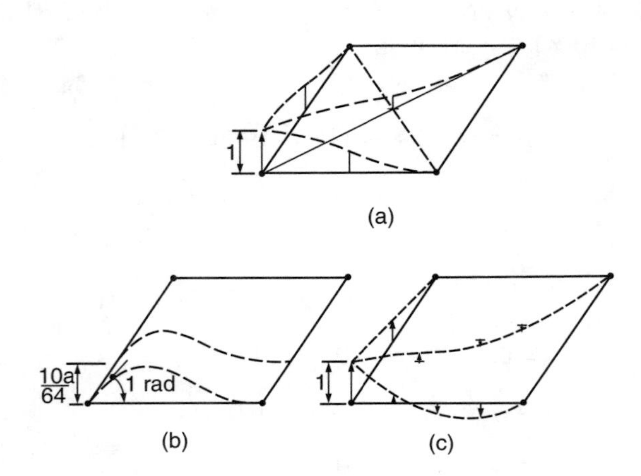

FIGURE 14.4
Plots of typical interpolation functions for plate bending: (a) N_1, (b) N_2, and (c) $\partial N_2/\partial x$.

$$\{\sigma\} = \begin{Bmatrix} M_{xx} \\ M_{yy} \\ M_{xy} \end{Bmatrix} = \frac{Eh^3}{12(1-\nu^2)} \begin{bmatrix} 1 & v & 0 \\ v & 1 & 0 \\ 0 & 0 & \dfrac{1-\nu}{2} \end{bmatrix} \begin{Bmatrix} w_{xx} \\ w_{yy} \\ w_{xy} \end{Bmatrix} \tag{14.11}$$

$$= [\mathbf{C}]\{\varepsilon\}.$$

Here h is the plate thickness.

The potential energy for the plate bending problem is expressed as

$$\Pi_p = \frac{h}{2} \iint_A \{\varepsilon\}^T [\mathbf{C}][\varepsilon] dxdy - h \iint_A \{\overline{\mathbf{X}}\}^T \{w\} dxdy - \int_{S_1} \{\overline{\mathbf{T}}\}^T \{w\} dS. \tag{14.12}$$

The components of $\{\varepsilon\}$ can be found by taking appropriate partial derivatives of w (Equation 14.8) and by using the transformations

$$\frac{\partial}{\partial x} = \frac{1}{a}\frac{\partial}{\partial s} \quad \text{and} \quad \frac{\partial}{\partial y} = \frac{1}{b}\frac{\partial}{\partial t}. \tag{14.13}$$

Substitution of $\{\varepsilon\} = [\mathbf{B}_b]\{\mathbf{q}_b\}$ and $w = [\mathbf{N}_b]\{\mathbf{q}_b\}$in Π_p and taking variation of Π_p with respect to the components of $\{\mathbf{q}_b\}$ lead to its stationary (minimum) value as

$$\partial \Pi_p = 0 \Rightarrow \frac{\partial \Pi_p}{\partial \{\mathbf{q}_b\}} = 0, \tag{14.14a}$$

which leads to the element equations as

$$[\mathbf{k}_b]\{\mathbf{q}_b\} = \{\mathbf{Q}_b\}, \tag{14.14b}$$

where

$$[\mathbf{k}_b] = h \iint [\mathbf{B}_b]^T [\mathbf{C}][\mathbf{B}_b] dxdy$$

and

$$\{\mathbf{Q}_b\} = h \iint [\mathbf{N}]^T \{\overline{\mathbf{X}}\} dxdy$$

$$+ \int_{S_1} [\mathbf{N}]^T \{\overline{\mathbf{T}}\} ds.$$

The coefficients of $[\mathbf{k}_b]$ can be found in closed form by integration. Their values are tabulated in Reference 4. The load vector $\{\mathbf{Q}_b\}$ can also be found in closed form. For instance, for uniform transverse surface load p on the plate,

$$\{\mathbf{Q}_b\}^T = \left[\frac{pab}{4} \quad \frac{pa^2b}{24} \quad \frac{pab^2}{24} \quad \frac{pab}{4} \quad \frac{pa^2b}{24} \quad -\frac{pab^2}{24} \quad \frac{pab}{4} \quad -\frac{pa^2b}{24} \right.$$
$$\left. -\frac{pab^2}{24} \quad \frac{pab}{4} \quad -\frac{pa^2b}{24} \quad \frac{pab^2}{24}\right] \tag{14.15}$$
$$= \frac{pab}{24}[6 \quad a \quad b \quad 6 \quad a \quad -b \quad 6 \quad -a \quad -b \quad 6 \quad -a \quad b].$$

Here we have considered only one of the many possible and available approximations for plate bending. In fact, the function in Equation 14.8 can be improved by adding the additional degree of freedom $\theta_{xy} = \partial^2 w/\partial x \partial y$.[4,5]

Assembly

The assembly of the element equations for beam-columns and plates can be achieved by using the direct stiffness approach, which assures the interelement compatibility of nodal displacements and rotations.

As stated earlier, since there are equal numbers of degrees of freedom for the junctions of plates and beam-columns, the assembly procedure at such junctions gives no difficulty.

The element equations for the beam-column are given in Equation 14.5a, for the membrane behavior of the plate in Equation 13.24, and for the bending behavior of the plate in Equation 14.14b.

The final assemblage equations can be expressed as

$$[\mathbf{K}]\{\mathbf{r}\} = \{\mathbf{R}\}. \tag{14.16}$$

These are modified for the boundary conditions and then solved for the nodal unknowns. Then the strains, stresses, bending moments, and shear forces can be evaluated as secondary quantities.

Representation of Foundation

Figure 14.5 shows an approximate representation for a foundation by using a series of (independent) springs. Structural supports provided by adjacent structures (Figure 14.1) can also be simulated by using this concept.

In the case of foundations, the constants for the springs, k_f, can be evaluated on the basis of laboratory tests on soil samples and/or from field experiments. They can be evaluated also from the concept of subgrade reaction.[7]

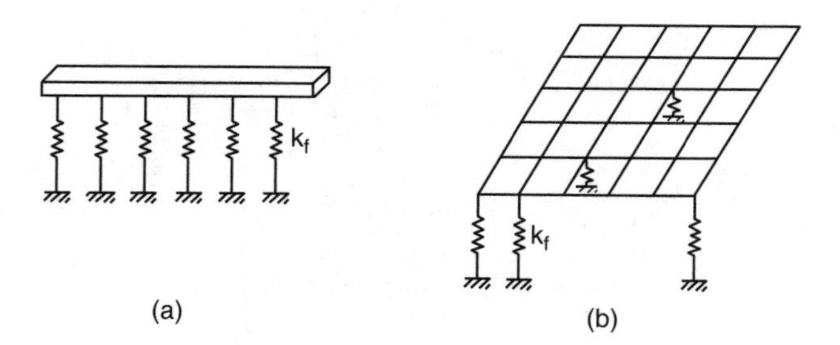

FIGURE 14.5
Idealization of foundation by springs: (a) Beam-column and (b) Plate or slab.

Since the springs are assumed to be independent, their stiffness coefficients can be added directly to the diagonal coefficients of the global matrix [**K**] (Equation 14.16). For instance, if N_{fi}^{x} denotes the stiffness coefficient of spring support at node i in the x direction, then it is added to the diagonal element, $K_{j,j}$, where j denotes the corresponding global degree of freedom. We note that the spring supports can be specified in the direction of translations (x, y, z) and rotations (θ_x, θ_y).

Computer Code

A computer program STFN-FE that allows analysis of building frames and foundations is described in Appendix 3. Some examples solved by using this code are described below.

Example 14.1. Plate with Fixed Edges and Central Load

Figure 14.6 shows a square plate 25.4 cm × 25.4 cm divided into 4 equal elements.[3] The thickness of the plate is 1.27 cm, with $E = 2.1 \times 10^5$ kg/cm^2 and $\nu = 0.3$. The concentrated load $P = 181.2$ kg. According to the closed form approach,[6] the maximum central deflection is given by

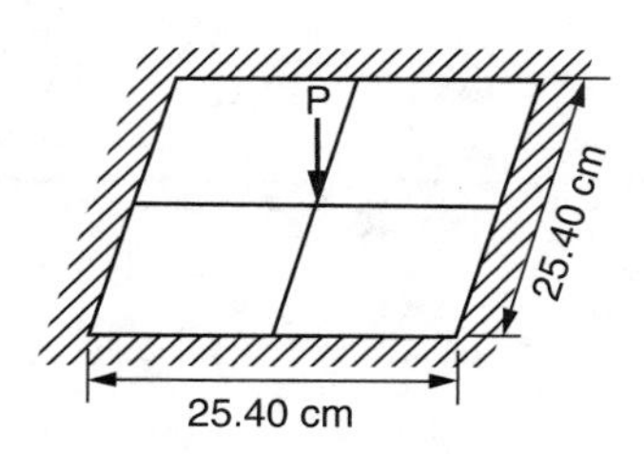

FIGURE 14.6
Square plate with central load.

$$w_{max} = \frac{0.0056 \times P \times (25.4)^2}{D}, \tag{14.17}$$

where

$$D = \frac{Eh^3}{12(1-\nu^2)} = \frac{2.1 \times 10^5 (1.27)^3}{12(1-0.09)} = 3.94 \times 10^4.$$

Therefore

$$w_{max} = \frac{0.0056 \times 181.2 \times 25.4^2}{3.94 \times 10^4} = 0.0166 \text{ cm}.$$

The value of central deflection computed from the finite element analysis is 0.01568 cm.

Example 14.2. Plate (Beam) with Two Fixed Edges, Two Free Edges, and Central Load

Figure 14.7(a) shows a plate fixed at two ends and free at the other two.[3,8] This plate can also be approximately treated as a beam. A concentrated load equal to 0.453 kg is applied at the center point. The value of $E = 7.0 \times 10^5$ kg/cm^2 was assumed.

Figure 14.7(b) shows a finite element mesh for the plate. In Figure 14.8 are shown the computed values of transverse displacements at center section C–C of the plate (Figure 14.7(b)).

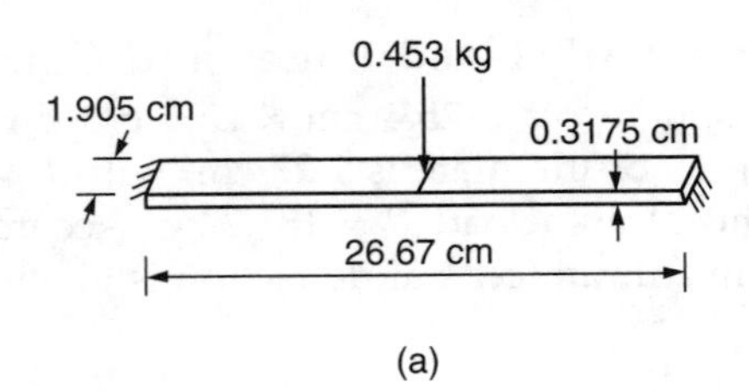

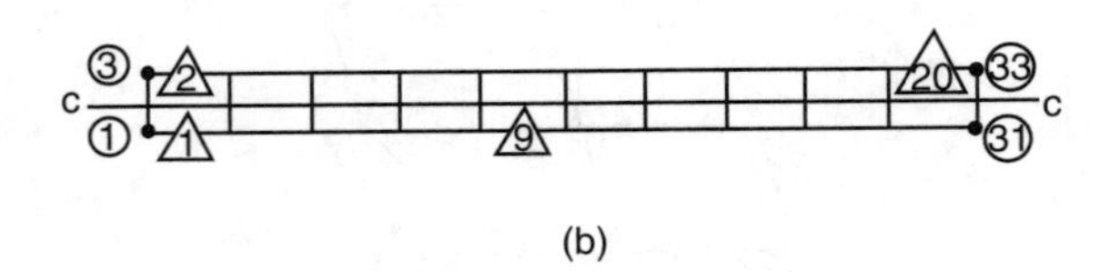

FIGURE 14.7
Analysis for (thin) beam bending:[8] (a) Beam with fixed ends loaded centrally and (b) Finite element mesh.

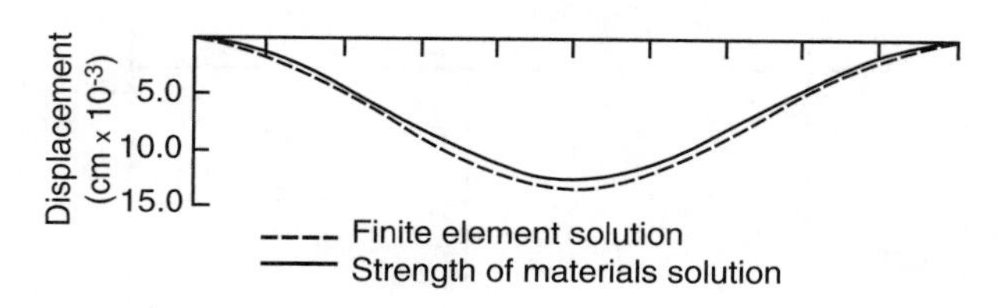

FIGURE 14.8
Comparisons for displacements: computed and closed from solutions.

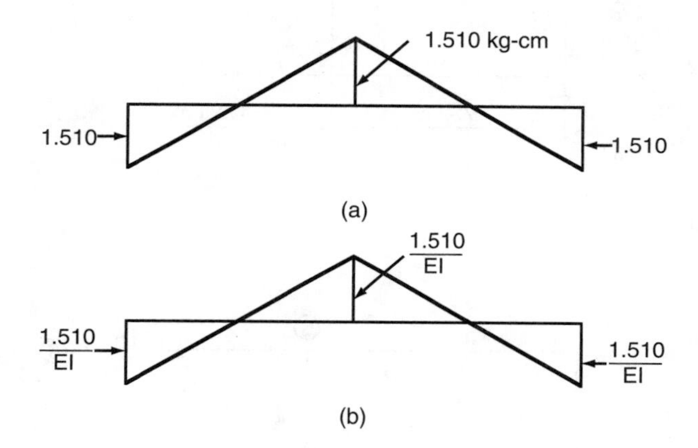

FIGURE 14.9
Procedure for closed form solution for displacements: (a) Bending moment diagram and (b) Conjugate beam.

We can compute the displacement by using the results from strength of materials[9] if the plate is treated approximately as a beam. Figures 14.9(a) and (b) show the distribution of bending moments and conjugate beam, respectively. Displacements computed from this approach are shown in Figure 14.8 in comparison with the finite element solutions. For instance, the maximum deflection at the central section is

$$w_{max} = \frac{PL^3}{192EI} = \frac{0.453 \times 26.67^3}{192 \times 7.0 \times 10^5 \times 0.00508} = 0.0126 \text{ cm}, \tag{14.18}$$

where

$$I = \frac{bh^3}{12} = \frac{1.905 \times 0.3175^3}{12} = 0.00508.$$

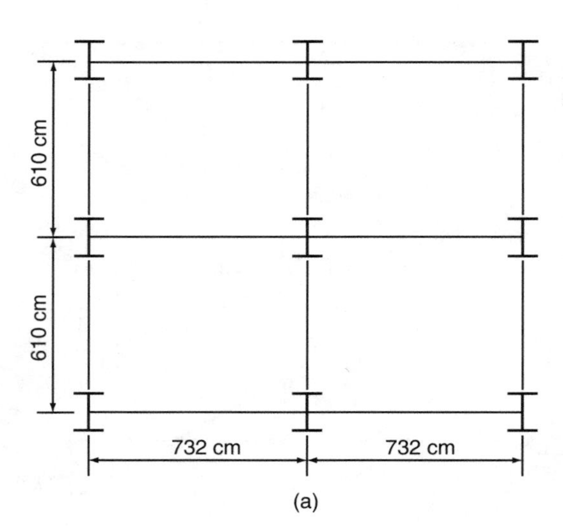

FIGURE 14.10(a)
Details of building frame:[10] Plane layout.

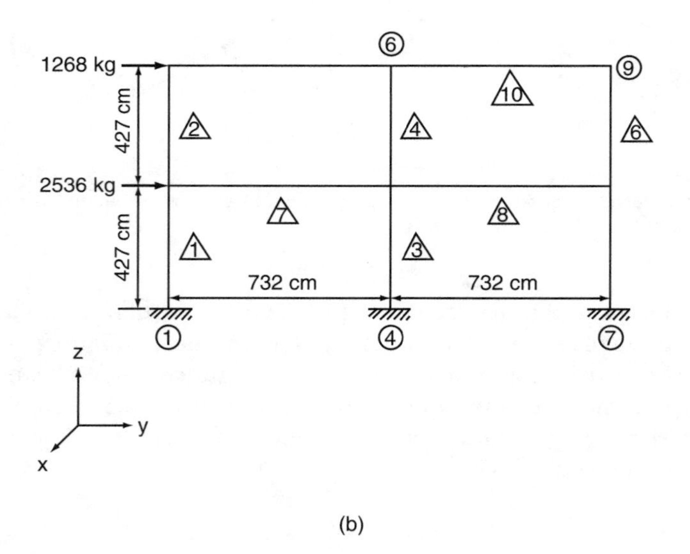

FIGURE 14.10(b)
Single idealized frame.

Example 14.3. Frame with Lateral Wind Loads

A one-dimensional idealization of a part of the building frame (Figure 14.10(a)) is shown in Figure 14.10(b).[10] The loading and other properties are shown in Figure 14.10(b).

Figure 14.11 shows the computed deflections and the deflected shape of the frame. Table 14.1 shows the computed values of bending moments in

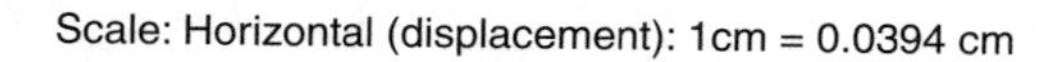

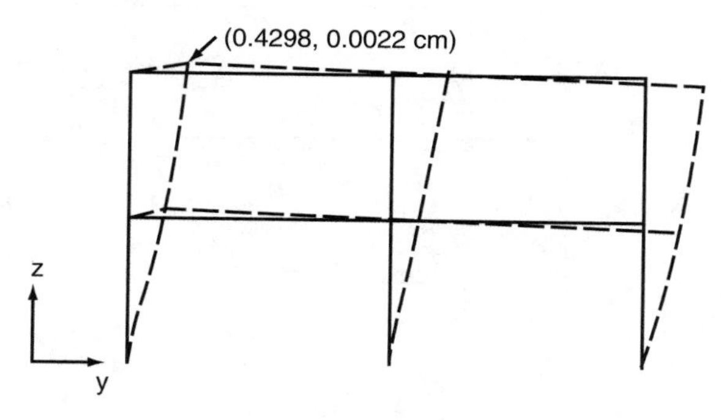

FIGURE 14.11
Deflected shape of frame.

various members in comparison with those from the conventional portal method of frame analysis; the latter is based on the assumption of a point of inflection at the midsection of a member. It can be seen that the values from the two methods differ widely at various locations. The finite element approach shows the redistribution of moments in the frame.

TABLE 14.1
Comparison of Bending Moments M_x (kg-cm) for the Frame (Figure 14.10(b))

Member	End	Portal Method	Finite Element
1	1	203,103	327,224
	2	203,103	191,750
2	1	67,701	21,474
	2	67,701	95.712
3	1	46,205	355,139
	2	406,205	258.680
4	1	135,402	120,860
	2	135,402	17,260
5	1	203,103	312,616
	2	203,103	179,523
6	1	67,701	30,914
	2	67,701	99,495
7	1	270,803	213,223
	2	270,803	190,466
8	1	270,803	189,069
	2	270,803	210,436
9	1	67,7801	95,714
	2	67,701	85,694
10	1	67,701	87,565
	2	67,701	99,495

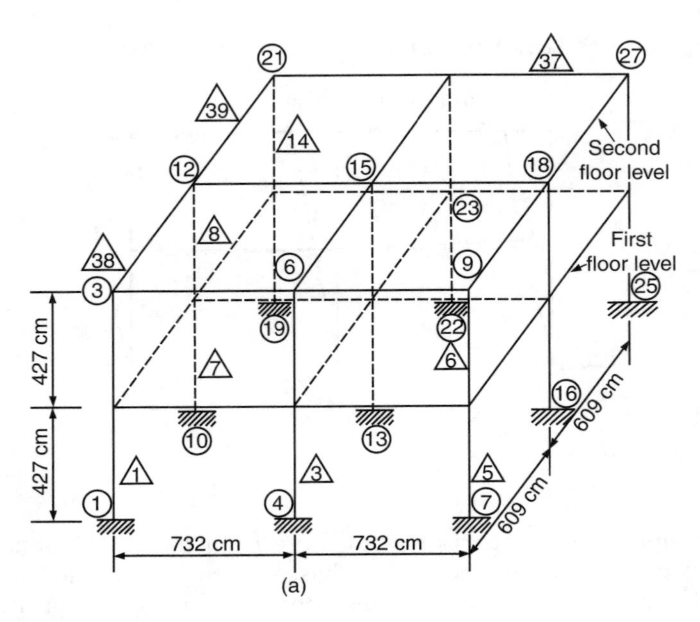

FIGURE 14.12(a)
Analysis of "three-dimensional" building frame:[10] Details of frame.

Example 14.4. Building Frame with Floor Slabs

Figure 14.12(a) shows the layout of a building frame idealized by using a one-dimensional beam-column, two-dimensional plate membrane, and bending elements; Figure 14.12(b) shows the loading at the two floor levels. Two analyses were performed: one without floor plates and the other with floor plates. The properties of the frame material are

$$E = 2.1 \times 10^6 \ \text{kg/cm}^2,$$

$$I_x = 27{,}346 \ \text{cm}^4,$$

$$I_y = 27{,}346 \ \text{cm}^4,$$

$$\text{Area of cross section,} \quad A = 94.84 \ \text{cm}^2,$$

$$\nu = 0.3.$$

Table 14.2 shows a comparison between moments (M_x) at typical nodes for the analyses with and without floor slabs. It can be seen that without floor slabs the middle section of the frame will have higher loads (moments) than those at the edges. Moreover, the displacements at the central section are significantly reduced if floor plates are included. For instance, the y displacement at node 12 with plates is 0.24393 cm and that without plates is 0.40225 cm.

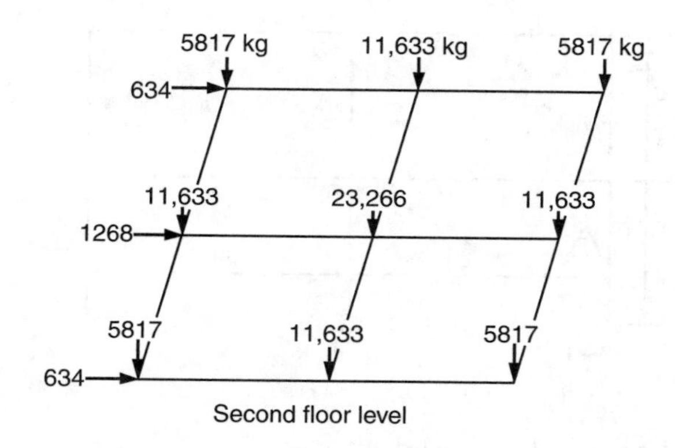

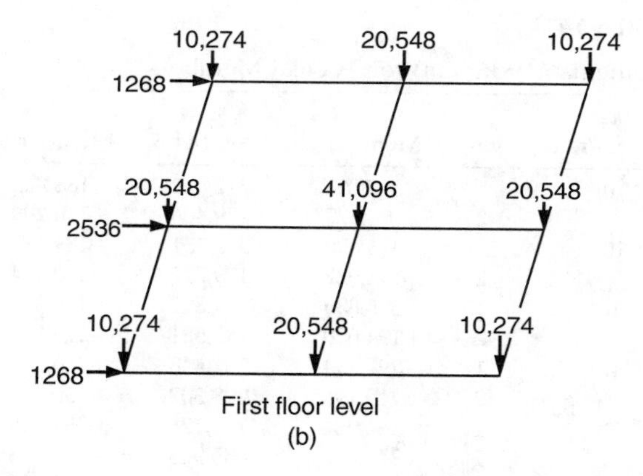

FIGURE 14.12(b)
Loadings at first and second floor levels.

TABLE 14.2
Comparison of Moments (M_x) at Typical Nodes

Node	Plates Not Included (kg-cm)	Plates Included (kg-cm)
3	32,597	31,289
6	93,747	104,875
12	49,077	22,741
15	158,825	116,723
21	32,576	32,386
24	93,712	105,206

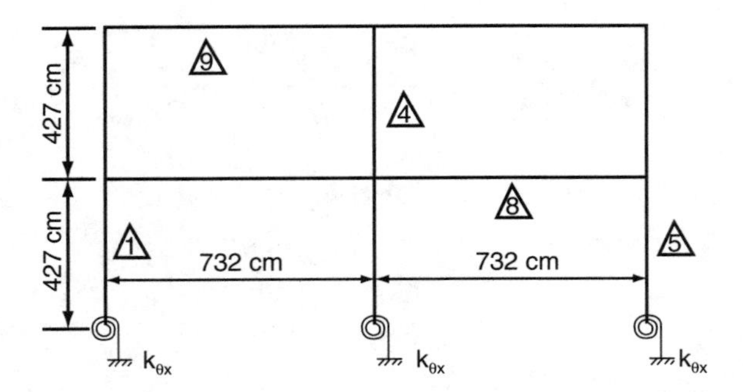

FIGURE 14.13
Frame with rotational restraint represented by spring.

TABLE 14.3
Moments M_x (kg-cm) for Typical Members

$k_{\theta x}$ (kg-cm/rad)	End	Member 1	Member 5	Member 8
10^{15}	1	327,011	312,412	–188,945
	2	191,627	179,406	–210,299
10^{10}	1	326,690	312,151	–189,132
	2	191,927	179,727	–210,524
10^9	1	323,835	309,812	–190,816
	2	194,620	182,581	–212,537
10^8	1	297,794	287,448	–206,185
	2	219,196	208,307	–230,761
10^7	1	164,984	162,128	–284,898
	2	344,492	335,943	–322,438
10^5	1	3,297	3,260	–380,988
	2	497,105	489,437	–433,436
10^3	1	4	3	–382,943
	2	500,217	48,422	–435,691

Example 14.5. Effect of Springs at Supports

To examine the effect of restraint on the behavior of the frame shown in Figure 14.10(b), a rotational spring $k_{\theta x}$ (kg-cm/rad) is introduced at the three supports (Figure 14.13). The magnitudes of $k_{\theta x}$ were varied as shown in Table 14.3. The results indicate that a very low value of $k_{\theta x}$ essentially models a simple support, whereas a very high value produces results close to those with total restraint (see Table 14.1). Near the value of $k_{\theta x}$ between 10^8 and 10^7 the point of inflection can be near the midsection, as assumed in many conventional analyses for frames.

Transformation of Coordinates

As we discussed earlier, it is convenient and useful to adopt a local coordinate system for an element different from the global coordinate system that is used to define the entire body. We note, however, that our final aim is to develop element and assemblage relations in the global or common coordinate system.

Although the local coordinate systems were used for applications in Chapters 3–13, the formulation process involved transformations that yielded the element equations in the global system. For instance, in the case of the one-dimensional line element (Chapters 3–9), the direction of the local system was the same as that of the global system, and the transformations occurred essentially in the derivatives and integrations. The isoparametric formulation involved use of global displacements in the development of element equations. Hence, both these instances did not involve transformation of coordinates.

For certain situations such as inclined beam-columns, nonorthogonal slabs in the building frame, and curved structures (e.g., shells), it is often necessary to use local systems whose directions are different from the global systems. Then it is required to transform the element relations evaluated in the local system to those in the global system. Such transformation is achieved by using a transformation matrix consisting of direction cosines of angles between the local and the global coordinates.

Figure 14.14(a) shows a beam-column element in an (orthogonal) global reference system, x, y, z, with a local (orthogonal) system, x', y', z' attached with the element. For this case, the transformation matrix can be expressed as

$$[\mathbf{t}] = \begin{bmatrix} \ell_x & m_x & n_x \\ \ell_y & m_y & n_y \\ \ell_z & m_z & n_z \end{bmatrix} \tag{14.19a}$$

where l_x, m_x, etc. are the direction cosines of angles between the local and global axes; for instance l_x represents the direction cosine of the angle between x' and x axes and so on.

As a simple illustration, let us consider the beam-column element in the two-dimensional x–y plane, Figure 14.14(b); the local coordinate x' is inclined at an angle of α with the global axes x. Here the transformation matrix $[\mathbf{t}]$ is given by

$$[\mathbf{t}] = \begin{bmatrix} \ell_x & m_x \\ \ell_y & m_y \end{bmatrix} = \begin{bmatrix} \cos\alpha & \sin\alpha \\ -\sin\alpha & \cos\alpha \end{bmatrix} \tag{14.19b}$$

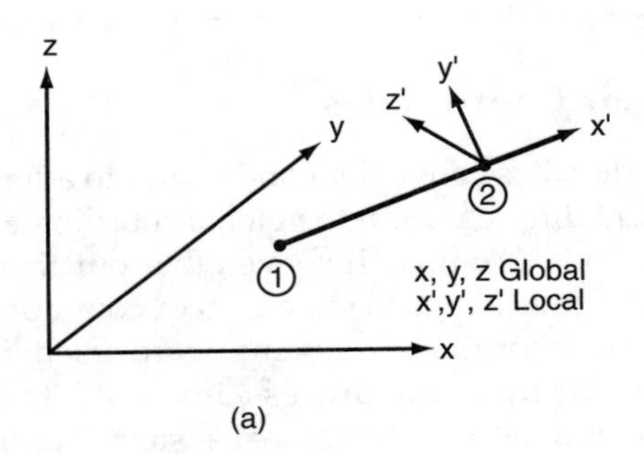

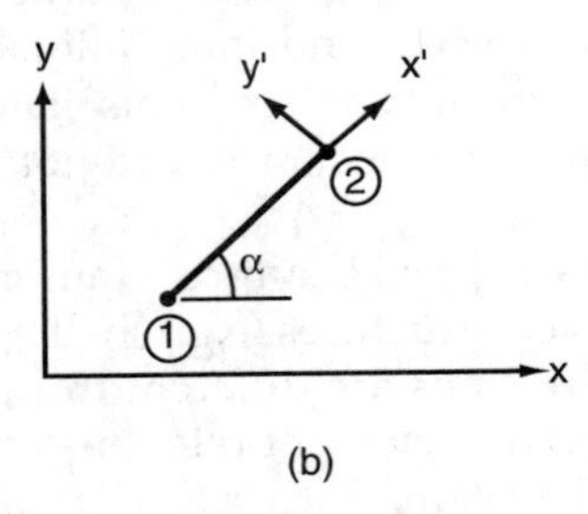

FIGURE 14.14
Transformation of coordinates: (a) Three-dimensional and (b) Two-dimensional.

The transformation matrix [**t**] is orthogonal, that is, its inverse equals its transpose:

$$[\mathbf{t}]^{-1} = [\mathbf{t}]^T \tag{14.19c}$$

We can now write transformations for various quantities. If [$\mathbf{k}_l$] and {$\mathbf{Q}_l$}, respectively, denote element stiffness matrix and load vector evaluated with respect to the local system x', y', z', then

$$[\mathbf{k}_g] = [T]^T [k_\ell][T] \tag{14.20a}$$

and

$$\{\mathbf{Q}_g\} = [T]^T \{Q_\ell\} \tag{14.20b}$$

where [$\mathbf{k}_g$] and {$\mathbf{Q}_g$} are the corresponding matrix and vector referred to the global system, x, y, z.

In the foregoing, the transformation matrix [**T**] is composed of [**t**] in Equation 14.19a; for instance, [T] in Equation 14.20 is given by

$$[T] = \begin{bmatrix} [t] & & & 0 \\ & [t] & & \\ & & [t] & \\ 0 & & & [t] \end{bmatrix}. \tag{14.20c}$$

The transformation between displacements can be written as

$$\{u_l\} = [T]\{u_g\} \tag{14.21}$$

where $[\mathbf{u}_l]$ and $[\mathbf{u}_g]$ are the vectors of displacements at a point for local and global systems, respectively.

The element equations at the global level are now expressed as

$$[k_g]\{q_g\} = \{Q_g\} \tag{14.22}$$

where $\{\mathbf{q}_g\}$ is the vector of element nodal unknowns obtained by using Equation 14.21.

Once the global relations are obtained, they can be assembled by using the direct stiffness assembly procedure by fulfilling the interelement compatibility of the unknowns.

The foregoing concept can be used and extended for other one-dimensional, and two- and three-dimensional elements. The process is usually straightforward.

Problems

14.1. Derive the load vector {**Q**} in Equation 14.6 for uniform body force $\bar{X}$ and uniform traction $\bar{T}$ causing bending. Hint: See Chapter 7.

14.2. Evaluate the matrix $[\mathbf{B}_b]$ in Equation 14.10. Hint: Find second derivatives of w in Equation 14.8a as indicated in Equation 14.10.

14.3. Derive the part of the stiffness matrix for the beam-column element corresponding to the twist about the z axis.

Solution:

$$[\mathbf{k}_s] = GJ \begin{bmatrix} 1 & -1 \\ -1 & 1 \end{bmatrix},$$

where G is the shear modulus and J is the polar moment of inertia. The corresponding degrees of freedom are θ_{z1} and θ_{z2} at nodes 1 and 2, respectively.

References

1. Martin, H. C. 1966. *Introduction to Matrix Methods of Structural Analysis,* McGraw-Hill, New York.
2. Tezcan, S. S. April 1966. "Computer Analysis of Plane and Space Structures," *J. Struct. Div. ASCE,* Vol. 92, No. ST2.
3. Desai, C. S. and Patil, U. K. 1976–1977. "Finite Element Analysis of Building Frames and Foundations," report, Department of Civil Engineering, Virginia Polytechnic Institute and State University, Blacksburg.
4. Bogner, F. K., Fox, R. L., and Schmidt, L. A. 1965. "The Generation of Interelement-Compatible Stiffness and Mass Matrices by the Use of Interpolation Formulas," in Proc. Second Conf. on Matrix Methods in Struct. Mech., Wright Patterson Air Force Base, OH.
5. Desai, C. S. and Abel, J. P. 1972. *Introduction to the Finite Element Method,* Van Nostrand Reinhold, New York.
6. Timoshenko, S. and Kieger, S. W. 1959. *Theory of Plates and Shells,* McGraw-Hill, New York.
7. Terzaghi, K. and Peck, R. B. 1967. *Soil Mechanics in Engineering Practice,* John Wiley & Sons, New York.
8. Durant, D. 1977. "Analysis of a Plate Bending Problem Using Code STFN-FE," course project report. Department of Civil Engineering, Virginia Polytechnic Institute and State University, Blacksburg.
9. Timoshenko, S. 1930. *Strength of Materials,* Van Nostrand Reinhold, New York.
10. Malek-Karam, A. 1976. "Design and Analysis of a Two-Story Building," M.E. project report, Department of Civil Engineering, Virginia Polytechnic Institute and State University, Blacksburg.

APPENDIX 1

Various Numerical Procedures: Solution of Beam Bending Problem

Introduction

To illustrate the use of energy procedures and of the methods of weighted residuals, in this appendix we shall solve an example of beam bending by using these procedures. Moreover, we also consider the Ritz and finite difference methods; the former is based on the minimization concept and is often considered a forerunner of the finite element method. The coverage of different methods herein is intended only as an introduction to various procedures and to give the reader an idea of the available schemes; detailed study of these procedures is beyond the scope of this elementary treatment.

Before the problem of beam bending is considered, we shall give further details of the methods of weighted residuals (MWRs) introduced in Chapter 2. A good introduction and applications of MWR are given by Crandall.[1]

As stated in Equation 2.12, the trial or approximation function in the MWR is expressed as

$$u = \sum_{i=1}^{n} \alpha_i \varphi_i. \tag{2.12}$$

The undetermined parameters α_i are chosen such that the residual $R(x)$ over the domain D vanishes. This is usually done in an average sense by weighting $R(x)$ (Figure 2.7) with respect to weighting functions $W_i(x)$. Thus,

$$\int_D R(x) W_i(x) dx = 0, \quad i = 1, 2, \ldots, n. \tag{A1.1}$$

For a one-dimensional problem, the domain D is simply the linear extent of the body.

Various Residual Procedures

There are a number of ways to choose W_i and depending on the choice of W_i, we obtain different procedures.

In the case of the collocation method,

$$W_i = \delta(x - x_i). \tag{A1.2a}$$

Then

$$R(x_i)\delta(x - x_i) = 0, \quad i = 1, 2, \ldots, n, \tag{A1.2b}$$

where

$$\delta = \begin{cases} 1, & x = x_i, \\ 0, & x \neq x_i \end{cases}$$

is the Dirac delta function. This means that the residual is equated to zero at a selected number of points in the domain. For instance, as shown in Figure A1.1(a), the residual is equated to zero at $n = 5$ points.

The total domain can be divided into a number of subdomains (Figure A1.1(b)), and the residual is integrated and equated to zero over each subdomain. This yields the subdomain method. Here the weighting functions are

$$W_t = \begin{cases} 1, & x_i \leq x \leq x_{i+1}, \\ 0, & x < x_i \text{ and } x > x_{i+1}. \end{cases} \tag{A1.3a}$$

Then

$$\int_{x_4}^{x_{t+1}} R(x)dx = 0, \quad i = 1, 2, \ldots, n-1. \tag{A1.3b}$$

In the case of the least-squares method (Figure A1.1(c)), the weighting functions are chosen to be

$$W_i = \frac{\partial R}{\partial \alpha_i}, \tag{A1.4}$$

and this leads to minimization of the integrated square residual as

$$\int_D R^2(x)dx = 0. \tag{A1.5}$$

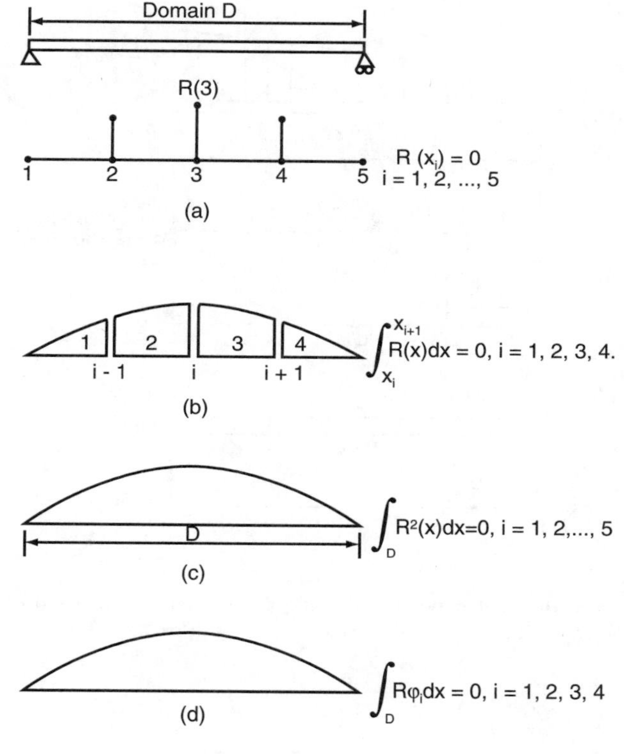

FIGURE A1.1
Methods of weighted residuals: (a) Collocation, (b) Subdomain, (c) Least squares, and (d) Galerkin.

In Galerkin's method (Figure A1.1(d)), the weighting functions are chosen as the coordinate functions from Equation 2.12, and hence

$$\int_D R(x)\varphi_i(x)\,dx = 0, \quad i = 1, 2, \ldots, n. \tag{A1.6}$$

This expression implies that the functions φ_i are made orthogonal to the residual $R(x)$. In finite element applications, the approximation or trial functions are commonly expressed in terms of shape, interpolation, or basis functions N_i; then the N_i are usually chosen as the weighting functions.

Beam Bending by Various Procedures

The beam and details are shown in Figure A1.2. Assuming flexural rigidity to be uniform, the governing differential equation is given by Equation 7.1b, and the residual is

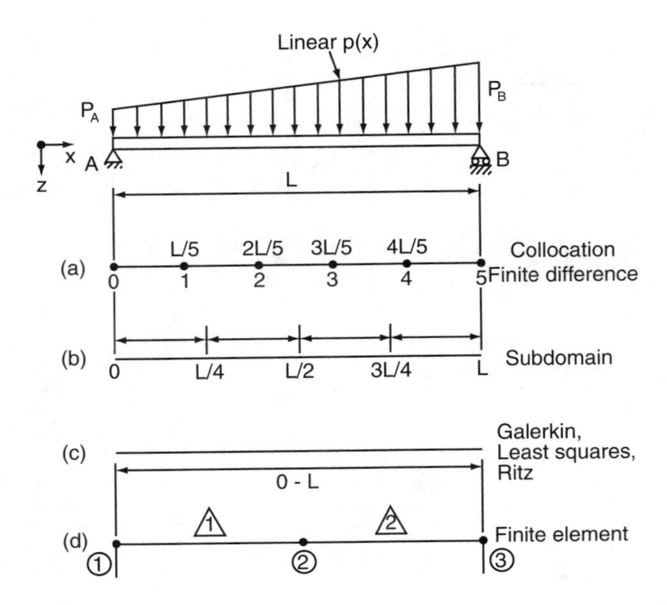

FIGURE A1.2
Beam bending with different methods: (a) Collocation, finite difference, (b) Subdomain, (c) Galerkin, least-squares, Ritz, and (d) Finite element.

$$R(x) = F\frac{d^4w}{dx^4} - p(x), \tag{A1.7}$$

where $F = EI$ is the flexural rigidity, w is the assumed transverse displacement, $p(x)$ is the forcing function, and x is the coordinate. The boundary conditions associated with Equation 7.1b can be expressed as

$$w(x=0) = w(x=L) = 0, \tag{A1.8a}$$

$$\frac{d^2w}{dx^2}(x=0) = \frac{d^2w}{dx^2}(x=L) = 0. \tag{A1.8b}$$

The first set (Equation A1.8a) represents the essential or forced boundary conditions and the second set (Equation A1.8b) represents the natural boundary conditions.

We now choose the following trial or approximation function for the unknown w*:

$$\begin{aligned} w &= \alpha_1 \sin\frac{\pi x}{L} + \alpha_2 \sin\frac{2\pi x}{L} + \alpha_3 \sin\frac{3\pi x}{L} + \alpha_4 \sin\frac{4\pi x}{L} \\ &= \sum_{i=1}^{4} \alpha_i \varphi_i(x), \end{aligned} \tag{A1.9}$$

where the α_i are the undetermined parameters and the φ_i are the known functions. Here we have chosen w^* in terms of the trigonometric functions, whereas in Chapter 7 the approximation function was chosen in terms of interpolation functions N_i.

Note that the function w in Equation A1.9 satisfies the boundary conditions in Equation A1.8 at the two ends of the beam. In Chapter 7, when we used Galerkin's method for each element, then only the geometric boundary conditions (Equation A1.8a) at the ends were used to modify the assemblage equations; as explained in Chapter 3, the natural boundary conditions (Equation A1.8b) are satisfied automatically in an integrated sense.

To express $R(x)$ in terms of w and its derivative, we differentiate the expression in Equation A1.9 four times as

$$\begin{aligned}\frac{d^4w}{dx^4} &= \lambda^4\alpha_1 \sin\frac{\pi x}{L} + 16\lambda^4\alpha_2 \sin\frac{2\pi x}{L} + 81\lambda^4\alpha_3 \sin\frac{3\pi x}{L}\\ &\quad + 256\lambda^4\alpha_4 \sin\frac{4\pi x}{L},\end{aligned} \tag{A1.10}$$

where $\lambda = \pi/L$.

Now we shall consider the solution of the beam bending problem by using a number of different procedures. For this illustration, the following properties are assumed:

$$E = 10 \times 10^6 \text{ psi},$$

$$L = 10 \text{ in.},$$

$$A = 1 \text{ in.} \times 1 \text{ in.} = 1 \text{ in.}^2,$$

$$P_A = 500 \text{ and } P_B = 1000 \text{ lb/in.}$$

Collocation

As shown in Figure A1.2, we chose 4 points at $x_i = L/5$, $2L/5$, $3L/5$, and $4L/5$, $i = 1, 2, 3, 4$, from the end A. Note that the residual is identically zero at the supports. Then we have

$$R(x = x_i) = 0, \quad i = 1, 2, 3, 4. \tag{A1.11a}$$

For instance, for $x_i = L/5$, we have

$$\begin{aligned}&\lambda^4\alpha_1 \sin\frac{\pi}{L}\frac{L}{5} + 16\lambda^4\alpha_2 \sin\frac{2\pi}{L}\frac{L}{5} + 81\lambda^4\alpha_3 \sin\frac{3\pi}{L}\frac{L}{5}\\ &\quad + 256\lambda^4\alpha_4 \sin\frac{4\pi}{L}\frac{L}{5} - \frac{P_B - P_A}{L}\frac{L}{5} - P_A = 0\end{aligned} \tag{A1.11b}$$

and so on. The resulting equations are

$$\begin{aligned}
4.777\times10^3\alpha_1+1.23\times10^5\alpha_2+6.25\times10^4\alpha_3+1.22\times10^6\alpha_4&=600,\\
7.72\times10^3\alpha_1+7.63\times10^4\alpha_2-3.86\times10^5\alpha_3-1.98\times10^6\alpha_4&=700,\\
7.72\times10^3\alpha_1-7.63\times10^4\alpha_2-3.86\times10^5\alpha_3+1.98\times10^6\alpha_4&=800,\\
4.77\times10^3\alpha_1-1.23\times10^6\alpha_2+6.25\times10^5\alpha_3-1.22\times10^6\alpha_4&=900.
\end{aligned}\tag{A1.12}$$

Solution of these equations gives values of α_i as

$$\begin{aligned}
\alpha_1&=0.11374399, &\alpha_2&=-0.00105974,\\
\alpha_3&=0.00033150, &\alpha_4&=00.00001564.
\end{aligned}\tag{A1.13}$$

Hence, the approximate solution according to the collocation method is

$$\begin{aligned}
w=0.11374\sin\frac{\pi x}{L}-0.00106\sin\frac{2\pi x}{L}+0.0003315\sin\frac{3\pi x}{L}\\
-0.00001564\sin\frac{4\pi x}{L}.
\end{aligned}\tag{A1.14}$$

Subdomain Method

Here, we have (Figure A1.2(b))

$$\begin{aligned}
\int_0^{L/4}R(x)dx&=0,\\
\int_{L/4}^{2L/4}R(x)dx&=0,\\
\int_{2L/4}^{3L/4}R(x)dx&=0,\\
\int_{3L/4}^{L}R(x)dx&=0.
\end{aligned}\tag{A1.15}$$

After integrations, the resulting four equations are

$$\begin{aligned}
7.57\times10^3\alpha_1+2.07\times10^5\alpha_2+1.19\times10^6\alpha_3+3.31\times10^6\alpha_4&=1410,\\
1.83\times10^4\alpha_1+2.07\times10^5\alpha_2-4.93\times10^5\alpha_3-3.31\times10^6\alpha_4&=1720,
\end{aligned}\tag{A1.16}$$

$$1.83 \times 10^4 \alpha_1 - 2.07 \times 10^5 \alpha_2 - 4.93 \times 10^5 \alpha_3 + 3.31 \times 10^6 \alpha_4 = 2030,$$

$$7.57 \times 10^3 \alpha_1 - 2.07 \times 10^5 \alpha_2 + 1.19 \times 10^6 \alpha_3 - 3.31 \times 10^6 \alpha_4 = 2340.$$

Solution of these equations leads to the following approximation:

$$\begin{aligned} w = 0.12388 \sin \frac{\pi x}{L} - 0.0015118 \sin \frac{2\pi x}{L} + 0.00078719 \sin \frac{3\pi x}{L} \\ = 0.00004724 \sin \frac{4\pi x}{L}. \end{aligned} \tag{A1.17}$$

Least-Squares Method

In this procedure, the weighting functions are

$$W_i = \frac{\partial}{\partial \alpha_i} [R(x)]. \tag{A1.18}$$

Therefore,

$$\begin{aligned} W_1 = \sin \frac{\pi x}{L}, \quad W_2 = \sin \frac{2\pi x}{L}, \\ W_3 = \sin \frac{3\pi x}{L}, \quad W_4 = \sin \frac{4\pi x}{L}. \end{aligned}$$

According to the least-squares method (Figure A1.2(c)),

$$\int_0^L R(x) \frac{\partial R(x)}{\partial \alpha_i} = 0$$

or

$$\begin{aligned} \int_0^L R(x) \sin \frac{\pi x}{L} dx = 0, \\ \int_0^L R(x) \sin \frac{2\pi x}{L} dx = 0, \\ \int_0^L R(x) \sin \frac{3\pi x}{L} dx = 0, \\ \int_0^L R(x) \sin \frac{4\pi x}{L} dx = 0. \end{aligned} \tag{A1.19}$$

The final four equations and the resulting approximate solution are

$$\begin{aligned}
4.06 \times 10^4 \alpha_1 + 0 \times \alpha_2 + 0 \times \alpha_3 + 0 \times \alpha_4 &= 4774.65 \\
0 \times \alpha_1 + 6.49 \times 10^5 \alpha_2 + 0 \times \alpha_3 + 0 \times \alpha_4 &= -795.77, \\
0 \times \alpha_1 + 0 \times \alpha_2 + 3.29 \times 10^6 \alpha_3 + 0 \times \alpha_4 &= 1591.55, \\
0 \times \alpha_1 + 0 \times \alpha_2 + 0 \times \alpha_3 + 1.04 \times 10^7 \alpha_4 &= -397.89,
\end{aligned} \tag{A1.20}$$

$$\begin{aligned}
w = {} & 0.11760 \sin \frac{\pi x}{L} - 0.001226 \sin \frac{2\pi x}{L} + 0.0004837 \sin \frac{3\pi x}{L} \\
& - 0.00003830 \sin \frac{4\pi x}{L}.
\end{aligned} \tag{A1.21}$$

Galerkin's Method

It is incidental that the weighting functions W_i in the least-squares method are the same as the functions φ_i used for the Galerkin method. Hence, in this specific case, both the Galerkin and least-squares methods yield the same solutions.

Ritz Method

In the Ritz method[1,2] the potential energy in the body (beam) is expressed in terms of the trial functions, and the resulting expression is minimized with respect to α_i. This leads to a set of simultaneous equations in α_i. For instance,

$$\Pi_p = \frac{1}{2} \int_0^L F \left(\frac{d^2 w}{dx^2} \right)^2 dx - \int_0^L pw dx \tag{A1.22}$$

with

$$p(x) = p_A + \frac{x}{L} \left(p_B - p_A \right). \tag{A1.23}$$

Minimization of Π_p with respect to α_i gives

$$\begin{aligned}
\frac{\partial \Pi_p}{\partial \alpha_1} &= 0, \\
\frac{\partial \Pi_p}{\partial \alpha_2} &= 0,
\end{aligned} \tag{A1.24}$$

$$\frac{\partial \Pi_p}{\partial \alpha_3} = 0,$$

$$\frac{\partial \Pi_p}{\partial \alpha_4} = 0.$$

For this specific problem, the equations are the same as in the Galerkin and least-squares methods, and the approximate solution is the same as in Equation A1.21.

Comment

It may be noted that in the Ritz procedure the potential energy is minimized for the entire beam; in other words, the limit of the integral is from 0 to L. The concept is thus similar to the finite element method (Chapters 3–5), except that in the case of the finite element method, the minimization of Π_p is achieved for the domain composed of a patchwork of elements.

Finite Element Method

The solution by the finite element method can be achieved in a manner identical to that covered in Chapter 7, with the subdivision consisting of two elements (Figure A1.2(d)). We can use Equation 7.15 for generating the two-element equations and then performing the assembly with boundary conditions, and the solutions of the resulting assemblage equations are[3]

$$\frac{F}{125}\begin{bmatrix} 12 & 30 & -12 & 30 & 0 & 0 \\ 30 & 100 & -30 & 50 & 0 & 0 \\ -12 & -30 & 24 & 0 & -12 & 30 \\ 30 & 50 & 0 & 200 & -30 & 50 \\ 0 & 0 & -12 & -30 & 12 & -30 \\ 0 & 0 & 30 & 50 & -30 & 100 \end{bmatrix} \begin{Bmatrix} w_1 \\ \theta_1 \\ w_2 \\ \theta_2 \\ w_3 \\ \theta_3 \end{Bmatrix} = \begin{Bmatrix} 1437.5 \\ 1250.0 \\ 3750.0 \\ 417.0 \\ 2312.5 \\ -1875.0 \end{Bmatrix}. \tag{A1.25}$$

Solution after introduction of $w_1 = w_3 = 0$ gives

$$\begin{aligned} &w_1 = 0.000000, \quad w_2 = 0.1171875, \quad w_3 = 0.000000, \\ &\theta_1 = 0.036665, \quad \theta_2 = 0.0007295, \quad \theta_3 = -0.0383335. \end{aligned} \tag{A1.26}$$

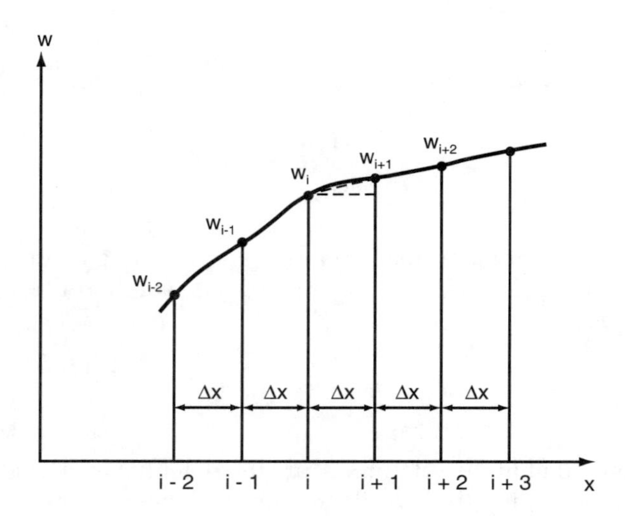

FIGURE A1.3
Finite difference approximation.

Finite Difference Method

Before the era of the finite element method, the finite difference method was the commonly used technique for problems in engineering and mathematical physics. Details of this method are beyond the scope of this book. However, we shall solve the beam problem using this method mainly to complete the discussion of commonly used numerical methods. For study of this method, the reader can refer to various textbooks.[1]

The finite difference method is based on the concept of replacing the continuous derivative in the governing differential equation by approximate finite differences. For instance, various derivatives are approximated as (see Figure A1.3)

$$\begin{aligned}
&\textit{First derivative:} && \frac{dw}{dx} \simeq \frac{w_{l+1} - w_i}{\Delta x} \simeq \frac{w_i - w_{i-1}}{\Delta x} \simeq \frac{w_{i+1} - w_{i-1}}{2\Delta x} \\
&\textit{Second derivative:} && \frac{d^2w}{dx^2} = \frac{w_{i-1} - 2w_i + w_{i+1}}{\Delta x^2} \\
&\textit{Third derivative:} && \frac{d^3w}{dx^3} = \frac{w_{i-2} + 2w_{i-1} - 2w_{i+1} + w_{i+2}}{2(\Delta x)^3} \\
&\textit{Fourth derivative:} && \frac{d^4w}{dx^4} = \frac{w_{i-2} - 4w_{i-1} + 6w_i - 4w_{i+1} + w_{i+2}}{\Delta x^4}.
\end{aligned} \tag{A1.27}$$

Use of Equation A1.27 to replace the fourth-order derivative in Equation 7.1b leads to

$$F\frac{w_{i-2}-4w_{i-1}+6w_i-4w_{i+1}+w_{i+2}}{\Delta x^4}=p_i. \tag{A1.28}$$

Substitution of i = 0, 1, 2, 3, 4, 5 for 6 points in the beam domain divided into 5 segments (Figure A1.2(a)) gives six simultaneous equations. Introduction of the boundary conditions (Equation A1.8) finally gives four simultaneous equations in w_1, w_2, w_3, and w_4:

$$\begin{aligned} 5w_1-4w_2+w_3+0&=0.0115,\\ -4w_1+6w_2-4w_3+w_4&=0.0134,\\ w_1-4w_2+6w_3-4w_4&=0.0154,\\ 0+w_2-4w_3+5w_4&=0.0173. \end{aligned} \tag{A1.29}$$

Solution:

$$\begin{aligned} w_1&=0.070656, & w_2&=0.114432,\\ w_3&=0.115968, & w_4&=0.073344, \end{aligned} \tag{A1.30}$$

with $w_0 = w_5 = 0$.

Comparisons of the Methods

To compare results from various methods, we first state the closed form solution for the displacement based on the strength of materials theory:[4]

$$w=\frac{p_A x}{24F}\left(L^3-2Lx^2+x^3\right)+\frac{(p_B-p_A)x}{180FL^2}\left(3x^4-10L^2x^2+7L^4\right). \tag{A1.31}$$

Results for displacements at typical locations on the beam by using various methods and the closed form solution (Equation A1.31) are compared in Table A1.1.

The results for the first five methods and the last procedure are obtained by substituting various values of x into Equations A1.14, A1.17, A1.21, and A1.31. Since we did not have a point at x = 5, there is no direct result at that point for the finite difference method. In the case of the finite element

TABLE A1.1

Comparisons for Displacements

Method	Deflection, Location from End *A*				
	2 in.	**4 in.**	**5 in.**	**6 in.**	**8 in.**
Collocation	0.067	0.110	0.113	0.109	0.068
Subdomain	0.072	0.117	0.123	0.118	0.075
Galerkin	0.068	0.111	0.117	0.112	0.071
Least-squares	0.068	0.111	0.117	0.112	0.071
Ritz	0.068	0.111	0.117	0.112	0.071
Finite element	0.067	0.110	0.117	0.116	0.069
Finite difference	0.071	0.114	—	0.116	0.073
Strength of materials	0.068	0.111	0.117	0.112	0.071

method, once the nodal displacements and slopes are obtained, values at other points can be obtained by substitution of (local) coordinates into Equation 7.2. We can also compute moments and shear forces by using the results with second and third derivatives, as we did in Chapter 7.

For examples of solutions of problems similar to the beam bending and other problems solved by using different procedures presented herein, the reader can refer to various publications such as Crandall.[1]

For the foregoing beam bending problem, the results for displacements from various procedures are close to each other and to the result from the closed form solution. These comparisons are presented only for the sake of introducing the reader to some of the available numerical procedures. The merits of the finite element method become evident as we solve problems with greater complexities in factors such as material and geometric properties and loading characteristics.

References

1. Crandall, S. H. 1956. *Engineering Analysis,* McGraw-Hill, New York.
2. Abel, J. F. Private communication.
3. Desai, C. S. and Abel, J. P. 1972. *Introduction to the Finite Element Method,* Van Nostrand Reinhold, New York.
4. Timoshenko, S. 1956. *Strength of Materials,* Van Nostrand Reinhold, New York.

APPENDIX 2

Solution of Simultaneous Equations

Introduction

Most problems solved by using numerical methods result in a set of algebraic simultaneous equations of the form (Equations 2.21 and 2.22)

$$[\bar{\mathbf{K}}]\{\bar{\mathbf{r}}\} = \{\bar{\mathbf{R}}\}. \tag{A2.1}$$

These equations can be linear or nonlinear; in this text we have essentially dealt with linear problems.

In general, Equation A2.1 is often expressed in matrix notation as

$$[\mathbf{A}]\{\mathbf{x}\} = \{\mathbf{b}\} \tag{A2.2a}$$

or simply as

$$Ax = b, \tag{A2.2b}$$

where $[\mathbf{A}]$ is the matrix of (known) coefficients such as K_{ij} (Equation 2.22), $\{\mathbf{x}\}$ is the vector of unknowns such as $\{\mathbf{r}\}$, and $\{\mathbf{b}\}$ is the vector of (known) forcing parameters such as $\{\mathbf{R}\}$. In expanded form Equation A2.2 can be written as

$$\begin{aligned}
a_{11}x_1 + a_{12}x_2 + \cdots + a_{1n}x_n &= b_1, \\
a_{21}x_1 + a_{22}x_2 + \cdots + a_{2n}x_n &= b_2, \\
\cdots\cdots\cdots \\
a_{n1}x_1 + a_{n2}x_2 + \cdots + a_{nn}x_n &= b_n,
\end{aligned} \tag{A2.3}$$

where n is the number of unknowns and denotes the total number of equations.

Methods of Solution

The two common methods for solution of Equation A2.3 are the direct and iterative procedures. Gaussian elimination and a number of its modifications are examples of direct methods, and Jacobi, Gauss–Seidel, successive overrelaxation (SOR), and symmetric successive overrelaxation (SSOR) are examples of the iterative techniques.[1-3] We shall briefly illustrate some of these techniques.

Gaussian Elimination

This is perhaps the simplest and a common method for solving linear equations. It is based on the idea of creating a sequence of equivalent systems of equations by using a number of steps of elimination, and then solutions for the unknowns x are obtained by a process of back substitution. The equivalent systems have the same solutions, but each successive sequence is simpler than the previous one. Let us consider the system of equations in Equation A2.3 and denote the initial sequence by superscript (1) as

$$\begin{aligned} &a_{11}^{(1)}x_1 + a_{12}^{(1)}x_2 + \cdots + a_{1n}^{(1)}x_n = b_1^{(1)}, \\ &a_{21}^{(1)}x_1 + a_{22}^{(1)}x_2 + \cdots + a_{2n}^{(1)}x_n = b_2^{(1)}, \\ &\cdots\cdots\cdots \\ &a_{n1}^{(1)}x_1 + a_{n2}^{(1)}x_2 + \cdots + a_{nn}^{(1)}x_n = b_n^{(1)}, \end{aligned} \tag{A2.4a}$$

or

$$A^{(1)}x = b^{(1)}. \tag{A2.4b}$$

In the first step of elimination, we use the first equation in Equation A2.4a and multiply it by appropriate multipliers so as to annihilate the first terms of all the subsequent equations. For instance, if the first equation is multiplied by $\lambda_{21} = -(a_{21}^{(1)}/a_{11}^{(1)})$ and then added to the second equation, we have

$$\overbrace{\left(a_{21}^{(1)} + \lambda_{21}a_{11}^{(1)}\right)}^{0}x_1 + \left(a_{22}^{(1)} + \lambda_{21}a_{21}^{(1)}\right)x_2 + \cdots + \left(a_{1n}^{(1)} + \lambda_{21}a_{1n}^{(1)}\right)x_n = \left(b_1^{(1)} + \lambda_{21}b_1^{(1)}\right) \tag{A2.5}$$

or

$$0 + a_{22}^{(2)}x_2 + \cdots + a_{2n}^{(2)}x_n = b_2^{(2)},$$

where $a_{22}^{(1)} = a_{22}^{(1)} + \lambda_{21} a_{12}^{(1)}$ and so on and the superscript (2) denotes the second sequence. Similarly, we can multiply the first equation by

$$\lambda_{31} = -\frac{a_{31}^{(1)}}{a_{11}^{(1)}}, \ \lambda_{41} = -\frac{a_{41}^{(1)}}{a_{11}^{(1)}}, \ \ldots, \ \lambda_{n1} = -\frac{a_{n1}^{(1)}}{a_{11}^{(1)}}$$

and add to the second, third, etc. equations to obtain a modified sequence at the end of the first step of elimination as

$$\begin{aligned} a_{11}^{(1)} x_1 + a_{12}^{(1)} x_2 + \cdots + a_{1n}^{(1)} x_n &= b_1^{(1)}, \\ 0 + a_{22}^{(2)} x_2 + \cdots + a_{2n}^{(2)} x_n &= b_2^{(2)}, \\ 0 + a_{32}^{(2)} x_2 + \cdots + a_{3n}^{(2)} x_n &= b_3^{(2)}, \\ \cdots\cdots\cdots \\ 0 + a_{n2}^{(2)} x_2 + \cdots + a_{nn}^{(2)} x_n &= b_n^{(2)}. \end{aligned} \tag{A2.6a}$$

In the next step, we multiply the second equation in Equation A2.6a by

$$\lambda_{32} = -\frac{a_{32}^{(2)}}{a_{22}^{(2)}}, \ \lambda_{42} = -\frac{a_{42}^{(2)}}{a_{22}^{(2)}}, \ \ldots, \ \lambda_{4n} = -\frac{a_{n2}^{(2)}}{a_{22}^{(2)}}$$

and add the results of the third, fourth, etc. equations, which leads to

$$\begin{aligned} a_{11}^{(1)} x_1^{(1)} + a_{12}^{(1)} x_2 + a_{13}^{(1)} x_3 + \cdots + a_{1n}^{(1)} x_n &= b_1^{(1)}, \\ 0 + a_{22}^{(2)} x_2 + a_{23}^{(2)} x_3 + \cdots + a_{2n}^{(2)} x_n &= b_2^{(2)}, \\ 0 + 0 \quad + a_{33}^{(3)} x_3 + \cdots + a_{3n}^{(3)} x_n &= b_3^{(3)}, \\ 0 + 0 \quad + a_{43}^{(3)} x_3 + \cdots + a_{4n}^{(3)} x_n &= b_4^{(3)}, \\ \cdots\cdots\cdots \\ 0 + 0 \quad + a_{4n}^{(3)} x_3 + \cdots + a_{nn}^{(3)} x_n &= b_n^{(3)}. \end{aligned} \tag{A2.6b}$$

Finally, at the end of $n - 1$ elimination steps, we shall have

$$
\begin{aligned}
a_{11}^{(1)}x_1^{(1)} + a_{12}^{(1)}x_2 + a_{13}^{(1)}x_3 + a_{14}^{(1)}x_4 + \cdots + a_{1n}^{(1)}x_n &= b_1^{(1)}, \\
0 + a_{22}^{(2)}x_2 + a_{23}^{(2)}x_3 + a_{24}^{(2)}x_4 + \cdots + a_{2n}^{(2)}x_n &= b_2^{(2)}, \\
0 + 0 + a_{33}^{(3)}x_3 + a_{34}^{(3)}x_4 + \cdots + a_{3n}^{(3)}x_n &= b_3^{(3)}, \\
0 + 0 + 0 + a_{44}^{(3)}x_4 + \cdots + a_{4n}^{(4)}x_n &= b_4^{(4)}, \\
\vdots \qquad & \quad \vdots \\
a_{n-1,n-1}^{(n-1)}x_{n-1} + a_{n-1,n}^{(n-1)}x_n &= b_{n-1}^{(n-1)}, \\
a_{n,n}^{(n)}x_n &= b_n^{(n)}.
\end{aligned}
\tag{A2.6c}
$$

We note that Equations A2.6a, A2.6b, and A2.6c are equivalent to the original Equation A2.4 in the sense that they all have the same solution. Equation A2.6c is a triangular system of equations, and the solution for the unknown x_n can be obtained directly from the last equation.

Back Substitution

The first step here is the solution for x_n from the last equation of Equation A2.6c as

$$x_n = \frac{b_n^{(n)}}{a_{n,n}^{(n)}}. \tag{A2.7a}$$

Now the solution for x_{n-1} can be found from the $(n - 1)$th equation as

$$x_{n-1} = \frac{b_{n-1}^{(n-1)}}{a_{n-1,n-1}^{(n-1)}} - \frac{a_{n-1,n}^{(n-1)}}{a_{n-1,n-1}^{(n-1)}} x_n. \tag{A2.7b}$$

Since x_n is known from Equation A2.7a, x_{n-1} can be found easily. The process can be repeated until the solution for x_1 is obtained.

The foregoing is the basic Gaussian elimination technique. A number of modifications and alternatives such as Crout, Jordan, Aitken, and Gauss–Doolittle can be used depending on the characteristics of the system equations.[1-3]

Banded and Symmetric Systems

In finite element applications, very often we encounter systems of equations that are banded and symmetric. In banded matrices, nonzero coefficients occur only on the main and adjacent diagonals, and other locations have zero coefficients. Moreover, the system is very often symmetric; that is, $a_{ij} = a_{ji}$. For instance, the following represents a banded (tridiagonal) symmetric system of equations:

$$\begin{aligned}
a_{11}x_1 + a_{12}x_2 + 0 + 0 + 0 &= b_1, \\
a_{12}x_1 + a_{22}x_2 + a_{23}x_3 + 0 + 0 &= b_2, \\
0 + a_{23}x_2 + a_{33}x_3 + a_{34}x_4 + 0 &= b_3, \\
0 + 0 + a_{34}x_3 + a_{44}x_4 + a_{45}x_5 &= b_4, \\
0 + 0 + 0 + a_{35}x_4 + a_{55}x_5 &= b_5.
\end{aligned} \tag{A2.8}$$

Bandedness and symmetry of the equations allow significant simplifications in the foregoing general elimination procedure. It is necessary to store only the nonzero elements in the computer and only the coefficients on the main diagonal and the upper or lower diagonals.[4]

Solution Procedure

Almost all systems of equations resulting from the finite element analysis are relatively large and require the use of the computer. In fact, a major portion of computational effort in a finite element solution is spent in the solution of these equations. It is not difficult to program the elimination and the subsequent iterative procedures on the computer.

For sake of introduction, we shall now illustrate the use of the foregoing elimination process.

Example A2.1. Solution by Gaussian Elimination

Consider the system of three equations, Equation 3.38. In Chapter 3, we solved these equations by a form of Gaussian elimination. Here we follow the foregoing general procedure:

$$\begin{aligned}
100^{(1)}x_1 - 100^{(1)}x_2 + 0^{(1)}x_3 &= 7.5^{(1)}, \\
-100^{(1)}x_1 + 200^{(1)}x_2 - 100^{(1)}x_3 &= 15.0^{(1)}, \\
0^{(1)}x_1 - 100^{(1)}x_2 + 200^{(1)}x_3 &= 15.0^{(1)}.
\end{aligned} \tag{A2.9}$$

Details:

$$\lambda_{21} = \frac{+100^{(1)}}{100^{(1)}} = 1.$$

Therefore we have the second equation as

$$0 + (200 - 100)^{(2)} x_2 - 100^{(1)} x_3 = 15.0^{(1)} + 7.5^{(1)}$$

or

$$0 + 100^{(2)} x_2 - 100^{(2)} x_3 = 22.5^{(2)}.$$

Since the third equation already has zero in the first term, the modified equations are

$$100^{(1)} x_1 - 100^{(1)} x_2 + 0 = 7.5^{(1)},$$

$$0 + 100^{(2)} x_2 - 100^{(2)} x_3 = 22.5^{(2)},$$

$$0 - 100^{(2)} x_2 + 200^{(1)} x_3 = 15.0^{(2)}.$$

Now $\lambda_{32} = +100/100 = 1$. Therefore, the third equation becomes

$$0 + 0 + (200 - 100)^{(3)} x_3 = 15.0^{(2)} + 22.5^{(2)}$$

or

$$0 + 0 + 100^{(3)} x_3 = 37.5^{(3)}.$$

Hence, the final modified equations are

$$100^{(1)} x_1 - 100^{(1)} x_2 + 0 = 7.5^{(1)},$$

$$0 + 100^{(2)} x_2 - 100^{(2)} x_3 = 22.5^{(2)},$$

$$0 + 0 + 100^{(3)} x_3 = 37.5^{(3)}.$$

The back substitution gives

$$\begin{aligned} x_3 &= \frac{37.5}{100} \\ x_2 &= \frac{22.5}{100} + \frac{100}{100} \times \frac{37.5}{100} \\ &= \frac{60}{100} \\ x_1 &= \frac{7.5}{100} + \frac{100}{100} \times \frac{60}{100} \\ &= \frac{67.5}{100}. \end{aligned} \tag{A2.10}$$

Iterative Procedures

In the iterative procedure, an estimate is made for the unknowns x and is successively corrected in a series of iterations or trials. The iterative procedure is continued until convergence, which is often defined by selecting a small acceptable number by which the final solution differs from the solution in the previous iteration. The simplest iterative procedure is the Jacobi scheme.[1-3] To illustrate this procedure, consider the set of equations (A2.3) and express it as

$$\begin{aligned}
&a_{11}x_1^{(0)} + a_{12}x_2^{(0)} + a_{13}x_3^{(0)} + \cdots + a_{1n}x_n^{(0)} = b_1,\\
&a_{21}x_1^{(0)} + a_{22}x_2^{(0)} + a_{23}x_3^{(0)} + \cdots + a_{2n}x_n^{(0)} = b_2,\\
&a_{31}x_1^{(0)} + a_{32}x_2^{(0)} + a_{33}x_3^{(0)} + \cdots + a_{3n}x_n^{(0)} = b_3,\\
&\cdots\cdots\cdots\\
&a_{n1}x_1^{(0)} + a_{n2}x_2^{(0)} + a_{n3}x_3^{(0)} + \cdots + a_{nn}x_n^{(0)} = b_n,
\end{aligned} \tag{A2.11a}$$

where the superscript (0) denotes initial estimates. In the first iteration, we compute the values of x as

$$\begin{aligned}
x_1^{(1)} &= \qquad\qquad\quad -\frac{a_{12}}{a_{11}}x_2^{(0)} - \frac{a_{13}}{a_{11}}x_3^{(0)} - \cdots - \frac{a_{1n}}{a_{11}}x_n^{(0)} + \frac{b_1}{a_{11}}\\
x_2^{(1)} &= -\frac{a_{21}}{a_{22}}x_1^{(0)} \qquad\qquad - \frac{a_{32}}{a_{22}}x_3^{(0)} - \cdots - \frac{a_{2n}}{a_{22}}x_n^{(0)} + \frac{b_2}{a_{22}}\\
x_3^{(1)} &= -\frac{a_{31}}{a_{33}}x_1^{(0)} - \frac{a_{32}}{a_{33}}x_2^{(0)} \qquad\qquad - \cdots - \frac{a_{3n}}{a_{33}}x_n^{(0)} + \frac{b_3}{a_{33}}\\
&\cdots\cdots\cdots\\
x_n^{(1)} &= -\frac{a_{n1}}{a_{nn}}x_1^{(0)} - \frac{a_{n2}}{a_{nn}}x_2^{(0)} - \frac{a_{n3}}{a_{nn}}x_3^{(0)} - \cdots \qquad\qquad + \frac{b_3}{a_{33}}.
\end{aligned} \tag{A2.11b}$$

Now we compare $x_i^{(1)}$ with $x_i^{(0)}$ by finding the difference

$$\left| x_i^{(1)} - x_i^{(0)} \right| \le \varepsilon, \quad i = 1, 2, \ldots, n, \tag{A2.12a}$$

where ε is a small number. If this condition is satisfied, we accept $x_i^{(1)}$ as the approximate solution; otherwise, we proceed to the next iteration. In general at the mth iteration we have

$$
\begin{aligned}
x_1^{(m)} &= \qquad\qquad -\frac{a_{12}}{a_{11}}x_2^{(m-1)} - \frac{a_{13}}{a_{11}}x_3^{(m-1)} - \cdots - \frac{a_{1n}}{a_{11}}x_n^{(m-1)} + \frac{b_1}{a_{11}} \\
x_2^{(m)} &= -\frac{a_{21}}{a_{22}}x_1^{(m-1)} \qquad\qquad - \frac{a_{23}}{a_{22}}x_3^{(m-1)} - \cdots - \frac{a_{2n}}{a_{22}}x_n^{(m-1)} + \frac{b_2}{a_{22}} \\
x_3^{(m)} &= -\frac{a_{31}}{a_{33}}x_1^{(m-1)} - \frac{a_{32}}{a_{33}}x_2^{(m-1)} \qquad\qquad - \cdots - \frac{a_{3n}}{a_{33}}x_n^{(m-1)} + \frac{b_3}{a_{33}} \\
&\cdots\cdots\cdots \\
x_n^{(m)} &= -\frac{a_{n1}}{a_{nn}}x_1^{(m-1)} - \frac{a_{n2}}{a_{nn}}x_2^{(m-1)} - \frac{a_{n3}}{a_{nn}}x_3^{(m-1)} - \cdots \qquad\qquad + \frac{b_n}{a_{nn}}.
\end{aligned}
\tag{A2.11c}
$$

and the convergence condition is

$$
\left| x_i^{(m)} - x_i^{(m-1)} \right| \le \varepsilon, \quad i = 1, 2, \ldots, n. \tag{A2.12b}
$$

The foregoing procedure can be slow to converge; that is, it may require a large number of iterations before an acceptable solution is obtained. The rate of convergence can be improved by using the Gauss–Siedel procedure in which the solution for an unknown during an iteration is used in the computation of subsequent unknowns:

$$
\begin{aligned}
x_1^{(m)} &= \qquad\qquad -\frac{a_{12}}{a_{11}}x_2^{(m-1)} - \frac{a_{13}}{a_{11}}x_3^{(m-1)} - \cdots - \frac{a_{1n}}{a_{11}}x_n^{(m-1)} + \frac{b_1}{a_{11}}, \\
x_2^{(m)} &= -\frac{a_{21}}{a_{22}}x_1^{(m)} \qquad\qquad - \frac{a_{23}}{a_{22}}x_3^{(m-1)} - \cdots - \frac{a_{2n}}{a_{22}}x_n^{(m-1)} + \frac{b_2}{a_{22}}, \\
x_3^{(m)} &= -\frac{a_{31}}{a_{33}}x_1^{(m)} - \frac{a_{32}}{a_{33}}x_2^{(m)} \qquad\qquad - \cdots - \frac{a_{3}}{a_{22}}x_n^{(m-1)} + \frac{b_3}{a_{33}}, \\
&\cdots\cdots\cdots \\
x_n^{(m)} &= -\frac{a_{n1}}{a_{nn}}x_2^{(m)} - \frac{a_{n2}}{a_{nn}}x_2^{(m)} - \frac{a_{n3}}{a_{nn}}x_3^{(m)} - \cdots \qquad\qquad + \frac{b_n}{a_{nn}}.
\end{aligned}
\tag{A2.13}
$$

Example A2.2. Solution by Gauss-Siedel Procedure
Consider the equations in Equation A2.9. We guess a solution set as

$$
x_1^{(0)} = 0.50, \quad x_2^{(0)} = 0.50, \quad x_3^{(0)} = 0.50.
$$

Then for the first iteration we have

$$\begin{aligned}
x_1^{(1)} &= \frac{100}{100}(0.5) + \frac{7.5}{100} = 0.575,\\
x_2^{(1)} &= \frac{100}{200}(0.575) + \frac{100}{200}(0.5) + \frac{15}{200} = 0.6125,\\
x_3^{(1)} &= \frac{100}{200}(0.6125) + \frac{15}{200} = 0.38125.
\end{aligned} \tag{A2.14a}$$

If we choose $\varepsilon = 0.005$, then the convergence check is

$$\left.\begin{aligned}
\left|x_1^{(1)} - x_1^{(0)}\right| &= 0.075 > 0.005\\
\left|x_2^{(1)} - x_2^{(0)}\right| &= 0.1125 > 0.005\\
\left|x_3^{(1)} - x_3^{(0)}\right| &= 0.04375 > 0.005
\end{aligned}\right\} \text{not acceptable.} \tag{A2.15a}$$

Second iteration:

$$\begin{aligned}
x_1^{(2)} &= \frac{100}{100}(0.6125) + \frac{7.5}{100} = 0.6875,\\
x_2^{(2)} &= \frac{100}{200}(0.6875) + \frac{100}{200}(0.38125) + \frac{15}{200} = 0.609375,\\
x_3^{(2)} &= \frac{100}{200}(0.609375) + \frac{15}{200} = 0.3796875.
\end{aligned} \tag{A2.14b}$$

Convergence check:

$$\left.\begin{aligned}
\left|x_1^{(3)} - x_1^{(2)}\right| &= 0.003125 < 0.005\\
\left|x_2^{(3)} - x_2^{(2)}\right| &= 0.002344 < 0.005\\
\left|x_3^{(3)} - x_3^{(2)}\right| &= 0.001172 < 0.005
\end{aligned}\right\} \text{acceptable.} \tag{A2.15b}$$

Third iteration:

$$\begin{aligned}
x_1^{(3)} &= \frac{100}{100}(0.609375) + \frac{7.5}{100} = 0.684375,\\
x_2^{(3)} &= \frac{100}{200}(0.684375) + \frac{100}{200}(0.3796875) + \frac{15}{200} = 0.6070312,\\
x_3^{(3)} &= \frac{100}{200}(0.6070312) + \frac{15}{200} = 0.3785156.
\end{aligned} \tag{A2.14c}$$

Convergence check:

$$\left.\begin{aligned} \left|x_1^{(2)} - x_1^{(1)}\right| &= 0.1125 \quad > 0.005 \\ \left|x_2^{(2)} - x_2^{(1)}\right| &= 0.003125 \ < 0.005 \\ \left|x_3^{(2)} - x_3^{(1)}\right| &= 0.0015625 < 0.005 \end{aligned}\right\} \text{ not acceptable.} \tag{A2.15c}$$

Hence the final solution at the end of three iterations is

$$x_1 = 0.6844,$$
$$x_2 = 0.6070,$$
$$x_3 = 0.3785.$$

We can improve on this solution by setting a more severe condition on convergence, say $\varepsilon = 0.001$. Then additional iterations are needed for an acceptable solution.

For large systems of equations generated in the finite element applications, it becomes necessary to use improved iterative schemes. One such improved scheme is successive overrelaxation (SOR).[2] To show the motivation of this method, we first write the Jacobi and Gauss–Siedel schemes in matrix notation as follows:[2]

Jacobi:

$$x^{(m)} = (L + U)x^{(m-1)} + \beta. \tag{A2.16a}$$

Gauss-Siedel:

$$x^{(m)} = Lx^{(m)} + Ux^{(m-1)} + \beta. \tag{A2.16b}$$

L and U are the lower and upper triangular parts of the matrix $[A]$ and β is the constant term at a given stage of iteration. For instance, in Example A2.2, at $m = 1$,

$$L = \begin{bmatrix} 0 & & \\ \frac{100}{200} & 0 & \\ 0 & \frac{100}{200} & 0 \end{bmatrix},$$

$$U = \begin{bmatrix} 0 & \frac{100}{100} & 0 \\ & 0 & \frac{100}{200} \\ & & 0 \end{bmatrix},$$

$$\beta = \frac{1}{100}\begin{Bmatrix} 7.5 \\ 15 \\ 15 \end{Bmatrix}.$$

We can now write

$$x^{(m)} = x^{(m-1)} + \omega\left[\left(Lx^{(m)} + Ux^{(m-1)} + \beta\right) - x^{(m-1)}\right], \qquad \text{(A2.16c)}$$

where ω is the overrelaxation factor and the second term on the right-hand side denotes the correction term. Rearrangement of terms in Equation A2.16c gives

$$x^{(m)} = (I - \omega L)^{-1}\left[(1-\omega)I + \omega U\right]x^{(m-1)} + \omega(I - \omega L)^{-1}\beta, \qquad \text{(A2.17)}$$

which is a statement of SOR.[2] SOR accelerates the solution procedure significantly. For details the reader may consult References 1 and 2.

Comment

In the foregoing, we have presented only a very elementary introduction to some of the solution procedures. There are also available a number of other schemes and subschemes. Moreover, there are a number of aspects related to the numerical characteristics of the set of equations that can influence the accuracy and reliability of a procedure. For instance, the magnitudes of the diagonal elements a_{ii} which become pivots in the elimination procedure can influence the computational characteristics. The initial guess or estimate of the unknowns and the value of ω become important in the iterative procedures. Detailed descriptions of the available procedures and the characteristics of the equations are beyond the scope of this book. The reader can refer to many publications[1-4] for further study.

References

1. Fox, L. 1965. *An Introduction to Numerical Linear Algebra,* Oxford University Press, New York.
2. Young, D. M. and Gregory, R. T. 1972. *A Survey of Numerical Mathematics,* Vol. 2, Addison-Wesley, Reading, MA.
3. Bathe, K. J. and Wilson, E. L. 1976. *Numerical Methods in Finite Element Analysis,* Prentice-Hall, Englewood Cliffs, NJ.
4. Desai, C. S. and Abel, J. F. 1972. *Introduction to the Finite Element Method,* Van Nostrand Reinhold, New York.

APPENDIX 3

Computer Codes

Introduction

Descriptions of a number of computer codes are included in this appendix. Most of the codes described herein have been prepared such that the beginner can understand and use them with relative ease. They are relevant to many topics covered in this book, and hence can be used by the teacher and the student for solving specific problems. A brief description of the advanced codes is given below.

Three of the codes, DFT/C-1DFE, PLANE-2D, and FIELD-2D, are available on the web* (www.crcpress.com). Other codes including advanced 2-D and 3-D codes for nonlinear static and dynamic problems with unified disturbed state concept (DSC) constitutive models[1] can be made available to the reader for individual teaching, research, and application needs. Details of these codes can be obtained at http://www.dscfe.com and/or by writing to Dr. C. S. Desai.

Descriptions of some of the codes are given below; details of advanced (other) codes can be obtained at http://www.dscfe.com.

Advanced Codes

The advanced codes include two- and three-dimensional finite element programs for static, repetitive, and dynamic (coupled) analysis for solid, civil, geotechnical, and mechanical engineering, pavement engineering, and electronic packaging problems. They are based on the unified disturbed state concept (DSC) constitutive model[1] for materials and interfaces/joints, and provide hierarchical options for factors such as elastic, plastic, and creep behavior, microcracking and fracture, softening or degradation, cyclic fatigue, instability and liquefaction, and simulation of construction

*The user of these codes agrees to the conditions and cautions provided on the web; in fact, this concurrence constitutes the signed agreement by the user for utilizing the codes. If there are any questions, the user must contact Dr. C. S. Desai before using the codes.

sequences. The stress analysis codes provide options for a number of material models such as linear elastic, nonlinear elastic, elasto-plastic (classical: von Mises, Drucker–Prager, and Mohr Coulomb, and hardening: critical state and hierarchical single surface — HiSS), creep (elastoviscoplastic overlay), and the unified DSC for microcracking fracture and softening.

Chapter 5

- *CONS-1DFE: One-Dimensional Consolidation with Nonlinear Properties*

  This is a modification of DFT/C-1DFE described in Chapter 6 for time dependent settlement analysis of foundations with nonlinear material properties.

Chapter 7

- *BMCOL-1DFE: Analysis of Axially and Laterally Loaded Beam-Columns such as Piles and Retaining Walls; Linear and Nonlinear Analysis*

  This code can solve problems such as beams, beam-columns (Chapter 7), beams on deformable foundations, axially and laterally loaded piles, and retaining walls idealized as one-dimensional. Linear and nonlinear stress–strain behavior for deformable supports such as soil foundations can be included.

  The output is in the form of nodal displacements and rotations, and bending moments.

Chapter 8

- *MAST-1 DFE: One-Dimensional Diffusion-Convection*

  This code is based on the formulation described in Chapter 8.

Other Codes on Diffusion-Convection

This is a two-dimensional code for solution of situations such as salt water intrusion and other concentration problems.

Chapter 9

- *WAVE-1 DFE: One-Dimensional Wave Propagation*

  This code is based on the formulation described in Chapter 9. Wave propagation for problems idealized as one-dimensional such as in bars and soil media can be solved.

Chapters 11 and 12

- *FIELD-2D: Analysis of Two-Dimensional Steady State Field Problems: Torsion, Potential Flow, Seepage, Heat Flow available on the web (www.crcpress.com).*

Problem	Option Code NTYPE	Relevant Problems	Output Quantities
Torsion (Chapter 11)	1	Torsion of bars of regular or irregular cross sections	Nodal stress functions, shear stresses, twisting moment
Potential flow—velocity potential	2	Flow through pipes and open channels	Nodal velocity potentials, velocities, quantity of flow
Potential flow—stream function	3	Flow through pipes and open channels	Nodal stream functions, quantity of flow
Seepage (flow through porous media)	4	Steady state confined seepage through pipes, foundations of dams and sheet piles, and earth banks and wells	Nodal heads, velocities, quantity of flow
Heat flow	5	Steady state heat flow in bars, plates, slabs, and other plane bodies.	Nodal temperatures, quantity of heat

This is a common code for solution of a number of steady state field problems governed by similar differential equations (Chapters 11 and 12). The following table gives the details of each problem and its corresponding option code.

The formulation is based on a four-node isoparametric quadrilateral element with linear constitutive laws; for instance, for seepage, Darcy's law is assumed to be valid.

Other Codes on Flow Problems

1. Codes for two-dimensional transient free (phreatic) surface seepage through plane bodies such as earth banks, dams, and wells.
2. Code for three-dimensional steady and transient free surface flow through arbitrary three-dimensional porous bodies such as earth banks, dams, and junctions between various structures, e.g., abutment and dam.
3. Code for two-dimensional time dependent heat flow in bodies such as plates and bars.

Chapter 13

- *PLANE-2D: Two-Dimensional Plane Stress-Deformation: Linear Analysis available on the web (www.crcpress.com).*

  This code is capable of analysis for problems idealized as plane stress, plane strain, and axisymmetric (Chapter 13). It can handle linear elastic analysis of engineering problems such as (approximate) bending analysis of beams and inplane or membrane analysis of flat plates or slabs, shear walls, earth and concrete dams and slopes, foundations, underground pipes and tunnels, and retaining walls.

Surface and nodal point loads as well as self weight load can be applied. A version of the code with a graphic option that can plot zones of equal stress and displacement intensities is also available.

The finite element formulation is based on a four-node isoparametric quadrilateral element (see Chapter 13) and on linear elastic stress–strain law. The output is in the form of nodal displacements and element stresses.

Chapter 14

- *STFN-FE: Analysis of Frame Structures and Foundations: Linear Analysis*

  In a general building frame type of problem the beams and columns are approximated as one-dimensional beam-column elements, the slabs as plates subjected to inplane and bending (Chapter 14), and the foundation is replaced by equivalent spring elements. The spring elements can also be used to simulate supports provided by members such as adjoining structures.

  Surface and point loads can be applied in all the three coordinate directions.

  The output is in the form of nodal displacements and rotations, element stresses, and bending moment and shear forces.

  A modified version of this code includes nonlinear characteristics for foundation media.

Reference

1. Desai, C. S. 2001. *Mechanics of Materials and Interfaces: The Disturbed State Concept*, CRC Press, Boca Raton, FL.

Index

A

B

C

D

E

F

G

H

I

J

L

P

Q

R

S

U

V

W

Y

Z